Progress in Botany/Fortschritte der Botanik 43

43

Progress in Botany

Morphology · Physiology · Genetics
Taxonomy · Geobotany

Fortschritte der Botanik

Morphologie · Physiologie · Genetik
Systematik · Geobotanik

Editors/Herausgeber

Heinz Ellenberg, Göttingen
Karl Esser, Bochum
Klaus Kubitzki, Hamburg
Eberhard Schnepf, Heidelberg
Hubert Ziegler, München

Springer-Verlag
Berlin Heidelberg New York 1981

With 25 Figures

ISBN-13:978-3-642-68299-5 e-ISBN-13:978-3-642-68297-1
DOI: 10.1007/978-3-642-68297-1

Softcover reprint of the hardcover 1st edition 1981

Contents

A. MORPHOLOGY

I. Cytology

List of Editors

Section A: Professor Dr. E. SCHNEPF, Lehrstuhl für Zellenlehre der Universität Heidelberg, Berliner Str. 15, D 6900 Heidelberg

Section B: Professor Dr. H. ZIEGLER, Institut für Botanik und Mikrobiologie, Technische Universität München, Arcisstr. 21, D 8000 München 2

Section C: Professor Dr. K. ESSER, Lehrstuhl für Allgemeine Botanik, Ruhr-Universität, Postfach 10 2148, D 4630 Bochum 1

Sections D and FII: Professor Dr. K. KUBITZKI, Institut für Allgemeine Botanik und Botanischer Garten, Universität Hamburg, Postfach 302 722, D 2000 Hamburg 36

Sections E and FI: Professor Dr. Dr. h.c. Dr. h.c. H. ELLENBERG, Lehrstuhl für Geobotanik, Systematisch-Geobotanisches Institut, Untere Karspüle 2, D 3400 Göttingen

A. Morphology

I. Cytology

a) General and Molecular Cytology

By ANTHONY W. ROBARDS

The report this year considers plastids and mitochondria of higher plants.

1. Ontogeny and Development of Plastids

The subject of the formation, replication, and development of plastids is one that has generated many publications without, as yet, being fully understood. The major reference work in this field is by KIRK and TILNEY-BASSETT (1978) and the whole subject of plastids is (despite the title *Chloroplasts*) well reviewed in REINERT (1980). Three recent reviews have provided up-to-date information more specifically on replication and development: POSSINGHAM (1980) has dealt with the subject of plastid replication and development in the life cycle of higher plants, while THOMSON and WHATLEY (1980) have considered the development of nongreen plastids. SCHNEPF (1980) has also provided a useful discussion of the problems of plastid classification. WHATLEY (1977, 1978) has proposed a pathway of plastid development and inter-relationships with seven stages: eoplasts; amyloplasts; amoeboid plastids; pregranal plastids; chloroplasts; senescing chloroplasts, and chromoplasts. This developmental pathway may have several variations but, in its basic form, envisages the eoplast developing through amyloplasts and amoeboid plastids to full chloroplasts if the appropriate conditions prevail. In higher plants there is an apparent economy in the utilization of resources in that fully differentiated chloroplasts are only formed at sites, and under conditions, where they can be utilized. The early, nongreen plastids (eoplasts, amyloplasts and amoeboid plastids) are all regarded as "proplastids" by conventional terminology and may transform readily from one type to another. These would all have formed part of the population of leucoplasts (nonpigmented plastids) as defined in earlier classifications. Changes in the "normal" pattern of development can provide a wide variety of plastid structural types, often reflecting changes in biochemical processes within the organelle. This, together with the possibility for plastid dedifferentiation – as well as increasing evidence for plastid redifferentiation (e.g., IKEDA, 1979; MESQUITA, 1974-76a,b) – means that the total range of plastid form and function seen in higher plant cells is extremely large.

In common with many other cellular organelles, all too little is known and understood about plastid replication. It is generally accepted that plastids arise from existing plastids and are passed from parent cell to daughter cell during division (POSSINGHAM, 1980). Plastids may be found associated together around the nucleus prior to division. Many electron micrographs have been published showing plastids apparently in a state of division, either by constriction or (less well-documented) by "budding". Such evidence is less than satisfactory and much more work is required in this area, particularly using the best methods of modern light microscopy.

Even though little is known about structural aspects of plastid replication, this still dwarfs the amount of information about the behavior of DNA during division and development. Although the last 10 years have seen an explosion in the amount of information about plastid DNA (ptDNA) (HERRMANN and POSSINGHAM, 1980), how the plastid DNA segregates and replicates is almoust totally unknown. Most plastids have a circular strand of DNA about 40 µm in circumference with a molecular weight of 10^8; it is a common feature that the gene sequences are highly reiterated. During fertilization there is usually, but not universally (e.g, *Pelargonium* and *Oenothera*), maternal inheritance of plastid characters but little is known about any processes of plastid fusion in higher plants. While the structural description of plastids in general, and chloroplasts in particular, has been relatively well documented, elucidation of the developmental pathways between different plastid types, and the understanding of plastid replication remain at a low level. Papers such as those by LEESE and LEECH (1977) have considered biochemical changes during development while TEVINI (1977), for example, has been more specifically concerned with changes in the lipid composition. Some authors (e.g., VERBELEN and DE GREEF, 1979) have described chloroplast development in one particular species *(Phaseolus vulgaris)* but more work is to be encouraged using appropriate techniques in the observation of carefully chosen plant systems.

2. The Structure of Plastids

In view of what has been stated above, it will be seen that the precise inter-relationships among and between different plastid types to a large extent remains a matter of debate (SCHNEPF, 1980; THOMSON and WHATLEY, 1980). For that reason it is probably most useful if the different types are dealt with one-by-one.

a) Proplastids

"Proplastid" is the word that has been used to describe almost any undifferentiated nongreen plastid, about 1 - 1.5 µm in diameter, found present in meristematic cells. There may be 10 - 20 per cell in dicotyledonous stem apex cells and about 20 - 40 per cell in roots. Structurally, they are usually small, double membrane-bound bodies containing a few disordered membranes and a relatively homogenous, dense matrix with polyribosomes. In the terminology of WHATLEY (1978) proplastids would embrace eoplasts, amoeboid plastids, and amyloplasts.

b) Eoplasts

WHATLEY (1978) uses the term "eoplast" to describe small, approximately spherical plastids with relatively dense stroma and normally found in meristems or formed during seasonal dormancy.

c) Amyloplasts

Amyloplasts are starch-containing plastids with lack chlorophyll. They are widely distributed and may be storage organelles or found in cells concerned with gravitational perception where they are thought to act as statoliths. They develop directly from eoplasts or may dedifferentiate from chloroplasts.

d) Amoeboid Plastids

Among the speculations made by WHATLEY (1978) and THOMSON and WHATLEY (1980), the identification of "amoeboid" plastids as specific stages in organellar development is of particular interest. The earliest stages of plastids during development are typefied by small, spherical bodies. These subsequently change into organelles that can be seen by light microscopy (and are also suggested by electron microscopy) to undergo constant amoeboid changes in shape. Subsequently the shape is again more fixed: either in the discoid pattern of a chloroplast or, perhaps, as the spherical body of a chromoplast. Amoeboid plastids are, therefore, transitory in nature. It is thought that these changes in shape may be brought about by: changes in the sol-gel state of the stroma; a change in the nature of the envelope; and a change in the surface-volume ratio. The amoeboid stage occurs both in the light as well as in the dark and this, together with the knowledge that plastids require cytoplasmic components for their further development (for example, Fraction I protein is known to be formed in part by chloroplast ribosomes and in part by cytoplasmic ribosomes), supports the possibility that amoeboid plastids may represent a stage where they can especially well "feed" on cytoplasmic materials.

e) Pregranal Plastids

Between the amoeboid plastid and a mature chloroplast may come one of a variety of "pregranal plastids" (THOMSON and WHATLEY, 1980). As their name implies, these organelles are no longer amoeboid – they have a relatively fixed shape – but they do not possess the full granal elaboration of a chloroplast. Among the different varieties of pregranal plastid, the etioplast is the form that has received the greatest attention.

f) Etioplasts

It is well known and established that chloroplasts of most higher plants do not attain full structural elaboration and photosynthetic capacity except in the presence of light. If the normal pathway of plastid development would lead to the production of a chloroplast then, in the absence of light, an etioplast develops.

The development of etioplasts has been discussed in the general reviews cited above as well as by GALEY et al. (1980) who demonstrated that, as etioplasts develop to form thylakoids, so the proportion of galactolipids and sulpholipids increases. The structural changes in developing etioplasts are generally closely associated with biochemical changes. The etioplasts contain chlorophyll precursors as protochlorophyll (the esterified form) and protochlorophyllide (nonesterified) which are converted to chlorophyll and chlorophyllide as the development into chloroplasts takes place (SUNDQVIST et al., 1980). However, there are insufficient precursors in the etioplasts for the full chlorophyll complement and substantial de novo synthesis of chlorophyll occurs soon after the conversion of an etioplast into a chloroplast. Thus LEWANDOWSKA and OQUIST (1980) found that exposure to light for 2h transformed prolamellar bodies into grana before substantial new chlorophyll synthesis commenced. Some of the effects of light on etioplast development have been studied by KOHN and KLEIN (1976) using isolated maize plastids.

The detailed structure of the complex assembly known as the prolamellar body has now been well documented (GUNNING and STEER, 1975; ISRAELACHVILLI and WOLFE, 1980). The tubular-membranous units are arranged in lattice-like forms, most commonly with tetrahedrally branched tubular components. The tetrahedrally branched types give lattices similar to those of zinc blende or wurtzite (mineral forms of zinc sulphide) but five-sided rings can also occur. SIMPSON (1978a) has reported the structure of etioplasts as seen in freeze-fracture preparations: his results are consistent with the models proposed from thin-section studies. It is generally thought that the internal membrane structure of plastids arises by elaboration of the inner plastidal membrane. KESSELMEIER and RUPPEL (1979) and KESSELMEIER (1980) have studied etioplast development in a range of plants including (KESSELMEIER, 1980) *Nicotiana tabacum* suspension cultures. They concluded that steroidal saponins are essential building blocks of prolamellar bodies which decline again as greening occurs (WELLBURN and HAMPP, 1979). HENRY (1979), in common with other authors, has reported that, for *Pisum sativum*, prolamellar body formation occurred by successive differentiation of stroma tubules, tubular strands, lattice formation, and organization of the hexagonal core. The stroma contains many ribosomes and when, eventually, thylakoid formation takes place, it does so by rapid proliferation of membranes from the core of the prolamellar body, spreading out into the plastid stroma.

g) Mature Chloroplasts

More has probably been written about the structure of chloroplasts than about all other plastids in toto. This is hardly surprising in view of the fundamental importance of the photosynthetic apparatus to mankind. Among the vast literature, useful recent reviews will be found in BARBER (1976). KIRK and TILNEY-BASSETT (1978), LEECH (1976) and REINERT (1980). The mature chloroplasts of higher plants are usually discoid in shape, are surrounded by a double membrane (envelope) and contain the *thylakoid membranes* which are regularly stacked as *grana* within the matrix *(stroma)* of the organelle.

The subject of the *plastid envelope* has been comprehensively reviewed by DOUCE and JOYARD (1979). Of the pair of surrounding membranes, each about 7.5 nm thick and separated by a gap of 10 - 20 nm, the outer is far more permeable (up to 10,000 m.w.) than the inner (impermeable to sucrose). Normal C_3 plastids have a relatively smooth inner membrane whereas that in C_4 bundle-sheath plastids can be thrown into folds forming the *peripheral reticulum*. With about 58% dry weight of the envelope as lipid, a protein-lipid ratio of 1:1.2 results (compared with about 1:0.35 and 1:0.8 for the inner and outer mitochondrial membranes respectively). Thus chloroplast envelope membranes are relatively low in protein compared with mitochondrial membranes or compared with thylakoid membranes (1:0.5). Qualitatively, the polar lipids of envelope and thylakoid membranes are identical, although the proportions of mono- and disalactosyldiacylglycerol, phosphatidylcholine and phosphatidylglycerol vary. The envelope fatty acids are more saturated than those in the thylakoid membranes. PRIESTLEY and WOOLHOUSE (1980) reported, in confirmation of other findings, that the main envelope constituents were 16:0 and 18:3 acyl fatty acids. The envelope membranes do not contain chlorophyll but do possess some carotenoids; they also appear to have the biosynthetic machinery to produce their own galactolipids while one or two of their polypeptides are synthesized on chloroplast ribosomes.

The granal stacks comprise a series of thylakoids arranged one above the other in the form of disc-shaped cisternae (flattened sacs). Some of the thylakoid membranes extend from one granal stack to the next: these interconnecting membranes have been referred to as *stroma lamellae* or *intergranal frets*. The membrane pair between two adjacent thylakoids within a granal stack is termed the *partition*. Among the many publications reporting on the structure of granal membranes as seen by conventional, transmission electron microscopy of thin sections, the recent work of OLESEN (1978), who used fixatives incorporating tannic acid, is of particular interest in that particles in membranes are demonstrated by this technique and OLESEN was able to distinguish populations of particles with different diameters. This work correlates with the freeze-fracture studies described below, where attempts are made to correlate specific particles with specific functions. The precise three-dimensional arrangement of the granal and intergranal membranes has been an active area of recent work. JOPE et al. (1980) performed a computer analysis of a "spiral string-of-grana model" and, using a granal diameter:thickness ratio of 5:1, investigated the similarities of a computer-generated model to electron micrographs of actual chloroplasts. They concluded that grana overlap each other, may have starch grains intercalated between them, have varying thickness and occasionally irregular shape, and may occur in long, interconnected series. MUSTARDY and JANOSSY (1979) also supported such a "helical" arrangement using scanning electron microscopy. It is, however, questionable whether this arrangement occurs generally.

A great deal of attention has been focused on correlating the ultrastructure, biochemistry, and function of the various granal components. Freeze-etching has demonstrated the complexity of the granal membranes which have a number of different populations of intercalated particles. An important goal is to identify the functional nature of the morphological features revealed by freeze-fracture studies. Among the many workers in this field, the recent papers by STAEHELIN (1976), MACHOLD, SIMPSON and HOYER-HANSEN (1977), SIMPSON (1978b), SIMPSON et al. (1978b), STEINBACK et al. (1979), MILLER and CUSHMAN (1979), CARTER and STAEHELIN (1980), HILLER and GOODCHILD (1980), MULLET et al. (1980) and SIMPSON (1980) are particularly germane. The work from the Carlsberg Laboratory, Copenhagen, (including that of SIMPSON and colleagues) has been well reviewed by WETTSTEIN (1981). Of the 43 or more polypeptides from photosynthetic membranes that can be resolved by gel electrophoresis, at least 30 are implicated in photosynthetic function, and 15 can now be assigned a specific role in photosynthesis. By using mutant varieties (for example, of barley, *Hordeum*) which are deficient for some activities (photosystem I) while having others normal (photosystem II), it is possible to make a correlation between presence or absence of particles on freeze-fractured membranes and the presence or absence of a specific photosynthetic function. In normal (wild-type) thylakoid membranes, four fracture faces can be distinguished on the basis of particle size and density. The EF face has large particles which are more abundant in stacked (EFs) than unstacked (EFu) regions. Similarly, the PF face, while having smaller particles, has more of them on stacked (PFs) than unstacked (PFu) faces. The mutant lacking photosystem I activity has only 1120 particles μm^{-2} on the PFs face whereas there are 6260 particles μm^{-2} on the same face in the wild type. It is therefore concluded that the light-harvesting chlorophyll a/b protein is a major component of many of the particles on the PFs face. This conclusion suggests that the light-harvesting chlorophyll a/b protein should be positioned asymmetrically toward the stroma side of the membrane. The large particles on the EFs face can also be exposed by etching down onto the ESs surface: they can then be shown to be tetrameric and, sometimes, to form arrays. Using a similar approach, MILLER and CUSHMAN (1979) showed that photosystem II deficient membranes in a variegated mutant of tobacco *(Nicotiana)* lack the characteristic large particle associated with the EFs face and also lack a tetrameric particle on the ESs surface, so suggesting, again, that these particles represent an amalgam of the photosystem II reaction complex. Experiments such as these are gradually leading to the establishment of morphological equivalence of particles, seen within and on thylakoid membranes, with specific photosynthetic activities.

The enzyme ribulose-1,5-bisphosphate carboxylase/oxygenase (fraction I protein) participates in both the photosynthetic carbon reduction cycle and also in the photorespiratory carbon oxidation cycle. This major enzyme can be dissociated into two subunits: a large subunit (m.w. 55,000), which is maternally inherited, and a small subunit (m.w. 12,000 - 15,000) which is inherited in a Mendelian fashion. The gene (which has now been mapped in full) for the large subunit has been located on chloroplast DNA in maize *(Zea)* and the protein is synthesized on chloroplast ribosomes; the small subunit is synthesized on cytoplasmic ribosomes in precursor form. Thus, as in other situations within the plastid, there is a close co-operation between nuclear and plastidal control of protein synthesis – even to the level that two components of a single enzyme are coded and constructed separately. Of the 30 known thylakoid polypeptides, nine are thought to be coded for by chloroplast DNA and synthesized on chloroplast ribosomes while seven are believed to be coded by nuclear DNA and synthesized on cytoplasmic ribosomes prior to transport into the chloroplast. Even the chloroplast ribosome itself is also assembled by the co-operative activity of both nuclear and plastidal genes (WETTSTEIN, 1981). In addition to the papers cited above, further information on the expression of the chloroplast genome can be found in KUNG (1977); further, some authors are now attempting to study the effects of different experimental or environmental conditions on the biochemical constitution of the thylakoids |e.g., OQUIST et al. (1978) who investigated the effects of season and low temperature on polypeptides of *Pinus sylvestris* thylakoids|. Within the stroma may be found *plastoglobuli* – the lipid-storing bodies – which are usually positioned between intergranal fret membranes (SIMPSON and LEE, 1976; SPREY, 1978). DNA is present as a 40 µm diameter, single, circular double-strand (BEDBROOK and KOLODNER, 1979; HERRMANN and POSSINGHAM, 1980). Many copies of this chromosome are present with, in theory, sufficient bases to code for several hundred polypeptides of about 20,000 molecular weight. However, the genes are extensively reiterated. RNA is part of the chloroplast 70S ribosomes (FILLIPOVICH et al., 1975; FILLIPOVICH et al., 1976; WOLLGIEHN and PARTHIER, 1980). Starch is also often present but tends, in functional chloroplasts, to reflect the transient physiological state of the plastid rather than its long-term nature as in amyloplasts proper.

In C_4 plants there are two different chloroplast types: those found in the mesophyll and those found in the bundle-sheath cells. The former are similar to those found in C_3 plants with stacked grana, intergranal frets etc. The latter have a much simpler lamellar arrangement with little, if any, stacking into grana and, often, differentiation of the inner plastid envelope membrane into a peripheral reticulum. MILLER et al. (1977) used the freeze-etching technique to compare mesophyll and bundle-sheath chloroplasts. They found that the bundle-sheath thylakoids contained virtually the same number of small (PF) particles as mesophyll thylakoids but only 40% as many of the large (EF) tetrameric particles. This, they suggested, reflects the relative deficiency of bundle-sheath chloroplasts in photosystem II activity.

h) Senescing Chloroplasts

According to THOMSON and WHATLEY (1980), senescing chloroplasts are those which are differentiating to a stage beyond normal photosynthetic capability with full thylakoid elaboration (see also WHATLey, 1977; 1978). This may involve conversion to a chromoplast or to another form of nonphotosynthesizing organelle. WOLINSKA (1976) is among those who have specifically studied chloroplasts during senescence.

i) Chromoplasts

Chromoplasts have been given considerable attention over the years and are generally thought to develop mainly, but not exclusively, from degenerating chloroplasts. They lack chlorophyll but accumulate carotenoids and provide the coloration to many flowers, aging leaves, roots, and fruits. Their function is unclear but may represent a co-evolutionary development as animal attractants. Chromoplasts are seen with many different, irregular shapes and may be amoeboid. According to WHATLEY (1978), they can develop from most other plastid types. They may differentiate to eoplasts (which they are considered to resemble closely, only differing in their possession of carotenoids), to amyloplasts or, possibly, not only from but also to chloroplasts. Chromoplasts of four main types can be distinguished: globular (carotenoids which may either be new products or breakdown materials in plastoglobuli); tubular (with internal tubular/membranous structures); membranous (internal disorganized membrane structures, possibly derived from the inner membrane of the envelope); and crystalline (with crystalline accumulations of pigments such as lycopene or β-carotene). It is possible to monitor both degradation of some components simultaneously with synthesis of others as chromoplasts differentiate. It has been reported that gibberellic acid will induce the redifferentiation of chloroplasts from chromoplasts. Among the recent papers dealing specifically with chromoplast structure and interconversions are: LJUBESIC and RADIC (1979); SIMPSON et al. (1977, 1978a); WUTTKE (1976); SITTE (1977) and SCHNEPF (1980). FALKE (1976) and WINKENBACH et al. (1976) reported on the structure and development of chromoplasts in *Tropaeolum majus* as well as isolation and characterization of the lipoprotein elements. They found that chromoplasts of unfolding petals contained tubular elements of about 17 nm diameter. It was shown by isolation and density gradient centrifugation that the fibrils (tubules) contained orange pigments. On analysis, the filaments and isolated "elongated bodies" were shown to contain 15% - 33% protein, large amounts of carotenoids and their esters, glyco- and phospholipids and minor amounts of tocopherols. LIEDVOGEL et al. (1976) and LIEDVOGEL and KLEINIG (1976) studied the flower chromoplasts of daffodil (*Narcissus pseudonarcissus*). They found that the chromoplasts contained up to 20 closely appressed membranes which they considered might be homologous with envelope (rather than thylakoid) membranes of chloroplasts. Up to 90% of the chromoplast's volume was occupied by such membranes which consisted of approximately 74% lipid and 26% protein; they also contained almost all the chromoplast carotenoids although these accounted for only 6% of the lipids. Digalactosyl diglyceride was the most abundant lipid. On the basis of the lipid and protein patterns – which differ from thylakoid patterns – the authors again concluded that the inner chromoplast membranes were most likely to have their origin at the inner plastid membrane.

j) Elaioplasts

Elaioplasts are lipid-storing plastids other than those that are clearly chromoplasts. They are particularly found in the epidermal cells of some monocotyledons (Liliaceae, Orchidaceae) and in stems of cacti.

k) Proteoplasts

Proteoplasts (or "proteinoplasts" as they are also often termed) are protein storage organelles. The protein can often be accumulated as fibrils or crystals within membrane-lined vesicles within the plastid. It has been suggested (e.g., HURKMANN and KENNEDY, 1976) that the stored protein may serve for subsequent new membrane synthesis. The similarity between membranes and tubules in some proteoplasts and those in some etioplasts prior to prolamellar body formation has been noted with the suggestion that such plastids could be intermediate stages in etioplast/prolamellar body development. This is possible but there is little evidence to support such an idea.

l) Sieve Element Plastids

The plastids found in the sieve elements of higher plants have a characteristic structure which varies from species to species. It has been suggested that they can develop at any time from any type of plastid. Thylakoids degenerate, ribosomes become lost, storage materials can accumulate (such as protein fibrils and crystals) and the organelle becomes spherical or irregularly shaped. The close relationship between the specific appearance of such plastids and the particular plant within which they exist has led some authors (notably BEHNKE) to use them as a systematic aid (e.g., BEHNKE and DAHLGREN, 1976).

3. Mitochondria

Whereas plastids in general, and chloroplasts in particular, have received considerable attention in numerous publications, the subject of mitochondrial structure and function *specifically in higher plant cells* has been relatively poorly documented. Among other reasons, this probably arises from the difficulties of isolating mitochondria or carrying out experimental studies with them in many plant tissues. It is thus significant that some reviews of mitochondria in plant cells necessarily draw their evidence more from animals than from plants (e.g., PALMER, 1976). Other reviews are provided by ÖPIK (1974), GUNNING and STEER (1975) and MUNN (1974).

The double membrane around the mitochondrion is, in many respects, similar to that around the plastid. Each membrane is approximately 5 - 7 nm thick and the two are separated by a space of variable width but which is typically seen in micrographs of sectioned material to be in the range 10 - 20 nm across. Some freeze-fracture views indicate that this perimitochondrial space may be narrower in vivo than as seen in fixed and embedded tissue. The outer membrane is rather freely permeable to molecules up to m.w. 10,000 whereas the inner membrane, as in plastids, is much less permeable, only allowing molecules up to about 150 m.w. to pass through. The outer membrane has a composition of approximately 55% protein and 45% lipid, whereas the inner membrane is composed of about 75% protein and only 25% lipid. Freeze-fracture studies show it to have particles on the outer (P) face in the range 700 - 1500 μm^{-2} and on the inner (E) face of about 2800 - 3200 μm^{-2}. This compares with frequencies of about 2000 - 3000 μm^{-2} on the outer (E) half of the inner membrane and up to 6000 μm^{-2} on the inner (P) half (MUNN, 1974). These are generalized data and do not necessarily apply to any particular mitochondrion in plants. However, they do support the now well-established belief that the frequency of particles seen on membrane fracture faces is an indication of the functional activity of that membrane.

The matrix or stroma of mitochondria is composed of numerous protein molecules, including 70S ribosomes. Calcium phosphate granules are also frequent inclusions. Mitochondrial DNA is also present in a form similar to that found in plastids: that is, a circular strand of double-helix DNA about 30 µm in circumference (plant mitochondrial DNA appears to be longer than most animal equivalents). In much the same way that many of the chloroplast proteins are coded for by either the nuclear genome or the plastid genome or by the co-operative action of both, so is the case for mitochondria. Unfortunately, however, these aspects have been little studied in higher plants (for a detailed review on mitochondrial DNA see Progr. Bot. 42, 214).

Among the handful of recent papers on the structure of plant mitochondria, most have concerned themselves with observations of mitochondria in relation either to a specific situation or to an experimental treatment. Thus ÖPIK (1975a,b) reported the reactions of mitochondria with diaminobenzidine in an effort to locate cytochrome oxidase activity. DUCKETT and TOTH (1977) commented on a giant mitochondrion in a periclinal chimaera of *Ficus elastica* and found that what appeared to be mitochondrial aggregations were, in fact, two individuals, tightly intertwined. BAJRACHARYA et al. (1976) reported that the development of mitochondria from promitochondria is regulated by phytochrome in the cotyledons of *Sinapis* seedlings and showed that dark-grown mitochondria had arrays of parallel, plate-like cristae while "typical" plant mitochondria with irregularly orientated sacculi formed in the light. POMEROY (1977) showed a somewhat similar relationship between mitochondrial structure and temperature: mitochondria of winter wheat *(Triticum)* grown at 24°C had better-developed cristae than those grown at 2°C. GUERN and GENEVES (1979) described reversible alterations to mitochondria of *Helianthus tuberosus* cultured cells following a short exposure to a temperature of 53°C.

Again, as for plastids, mitochondrial replication is poorly studied and understood. Mitochondria divide by constriction so that numbers are generally maintained as cells divide. Promitochondrial initials have sometimes been suggested, but there is little evidence to support their position in a developmental pathway. AVETISOVA and CHEL'TSOVA (1979) used wheat to study mitochondrial changes during ontogenesis of the shoot apical meristem while RAKHMANKULOV and AZIZHODZHAEV (1977) reported changes in relation to hybridization of cotton and LUCK and JORDAN (1980) provided a detailed description of changes in both plastid and mitochondrial structure during microsporogensis in *Hyacinthoides non scripta*. In this latter report, mitochondria were found to change from large organelles with small cristae to smaller, elongate types with well-developed cristae. They passed through a period of disorganization followed by reorganization at the time of meiosis, supporting the idea of cytoplasmic reorganization at the time of meoisis associated with changes to the haploid nuclear state. WARMKE and LEE (1977) reported mitochondrial degeneration in the male sterile anthers of *Zea*.

In conclusion, too little is known about the structure and behavior of mitochondria in higher plants. Although the direct observation of structure is simple and well-reported, the detailed structure of membranes and, particularly, the association of specific functions with identifiable components of the organelle remains almost exclusively by extrapolation from animal or yeast mitochondria. There remains much to be done to rectify this substantial gap in botanical knowledge.

References

AVETISOVA, L.V., CHEL'TSOVA, L.P.: Ontogeny 10, 491-501 (1979).

BAJRACHARYA, D., FALK, H., SCHOPFER, P.: Planta (Berl.) 131, 253-261 (1976) - BARBER, J. (ed.): The Intact Chloroplast. 476 pp. Amsterdam: Elsevier 1976. - BEDBROOK, J.R., KOLODNER, R.: A. Rev. Pl. Physiol. 30, 593-620 (1979). - BEHNKE, H.D., DAHLGREN, R.: Bot. Not. 129, 287-296 (1976).

CARTER, D.P., STAEHELIN, L.A.: Arch. Biochem. Biophys. 200, 374-386 (1980).

DOUCE, R., JOYARD, J.: Structure and function of the plastid envelope, 1-116. In: Advances in Botanical Research, Vol. 7, ed. H.W. WOOLHOUSE. London: Academic Press 1979. - DUCKETT, J.G., TOTH, R.: Ann. Bot. (Lond.) 41, 903-912 (1977).

FALK, H.: Planta (Berl.) 128, 15-22 (1976). - FILIPPOVICH, I.I., NOZDRINA, V.N., OPARIN, A.I.: Dokl. Akad. Nauk. SSRR Ser. Biol. 225, 223-226 (1975). -FILIPPOVICH, I.I., NOZDRINA, V.N., KUPCHINENKO, V.V., OPARIN, A.I.: Biokhimya 41, 708-717 (1976).

GALEY, J., FRANCKE, B., BAHL, J.: Planta 142, 433-439 (1980). - GUERN, N., GENEVES, L.: C.R. Hebd. Acad. Sci. Ser. D. Sci. Nat. 282, 965-968 (1979).- GUNNING, B.E.S., STEER, M.W.: Ultrastructure and the Biology of Plant Cells. 312 pp. London: Arnold 1975.

HENRY, E.W.: Cytologia (Tokyo) 44, 727-738 (1979). - HERRMANN, R.G., POSSINGHAM, J.V.: Plastid DNA - The plastome, 45-96. In: Chloroplasts, ed. J. REINERT (Results and Problems in Cell Differentiation, Vol.10). Berlin-Heidelberg-New York: Springer 1980. - HILLER, R.G., GOODCHILD, D.J.: Thylakoid membrane and pigment organisation. In: Plant Biochemistry - an Advanced Treatise. Vol. VIII: Photosynthesis, eds. N.K. BOARDMAN, M.D. HATCH. London: Academic Press 1980. - HURKMAN, W.J., KENNEDY, G.S.: Protoplasma 89, 171-184 (1976).

IKEDA, T.: Bot. Mag. Tokyo 92, 23-30 (1979). - ISRAELACHVILI, J.N., WOLFE, J.: Protoplasma 102, 315-322 (1980).

JOPE, C.A., ATCHISON, B.A., PRINGLE, R.C.: Bot. Gaz. 141, 37-47 (1980).

KESSELMEIER, J.: Protoplasma 104, 295-306 (1980). - KESSELMEIER, J., RUPPEL, H.G.: Z. Pflanzenphysiol. 93, 171-184 (1979). - KIRK, J.T.O., TILNEY-BASSETT, R.A.E.: The Plastids: Their Chemistry, Structure, Growth and Inheritance. 960 pp. Amsterdam: Elsevier/North Holland 1978. - KOHN, S., Klein, S.: Planta (Berl.) 132, 169-175 (1976). - KUNG, S.D.: Annu. Rev. Plant. Physiol. 28, 401-437 (1977).

LEECH, R.M.L.: The photosynthetic apparatus of higher plants. 125-143. In: Plant Structure, Function and Adaptation, ed. M.A. HALL. London: Macmillan 1976. - LEESE, B.M., LEECH, R.M.: Biochem. Soc. Trans. 5, 1266-1269 (1977). - LEWANDOWSKA, M., OQUIST, G.: Physiol. Plant. 48, 39-46 (1980). -LIEDVOGEL, B., KLEINIG, H.: Planta (Berl.) 129, 19-21 (1976). - LIEDVOGEL, B., SITTE, P., FALK, H.: Cytobiologie 12, 155-174 (1976). -LJUBESIC, N., RADIC, M.: Acta Bot. Croat. 38, 29-34 (1979). - LUCK, B.T., JORDAN, E.G.: Ann. Bot. 45, 511-514 (1980).

MACHOLD, O., SIMPSON, D.J., HOYER-HANSEN, G.: Carlsberg Res. Commun. 42, 499-516 (1977). - MESQUITA, J.F.: Rev. Biol. (Lisb.) 10, 127-150

(1974-76a); - Rev. Biol. (Lisb.) 10, 151-160 (1974-76b). - MILLER, K.R., CUSHMAN, R.A.: Biochim. Biophys. Acta 546, 481-497 (1979). - MILLER, K.R., MILLER, G.J., MCINTYRE, K.R.: Biochim. Biophys. Acta 459, 145-156 (1977). - MULLET, J., BURKE, J.J., ARNTZEN, C.J.: Pl. Physiol. (Bethesda) 65, 814-822 (1980). - MUNN, E.A.: The Structure of Mitochondria, 465 pp. London: Academic Press 1974. - MUSTARDY, L.A., JANOSSY, A.G.S.: Plant Sci. Lett. 16, 281-284 (1979).

ÖPIK, H.: J. Cell Sci. 17, 43-56 (1975a); - Planta (Berl.) 122, 269-271 (1975b); - Mitochondria, 52-83. In: Dynamic Aspects of Plant Ultrastructure, ed. A.W. ROBARDS. London: McGraw-Hill 1974. - OLESEN, P.: Biochem. Physiol. Pflanzen 172, 319-342 (1978). - OQUIST, G., MARTIN, B., MARTENSSON, O., CHRISTERSSON, L., MALMBERG, G.: Physiol. Plant. 44, 300-306 (1978).

PALMER, J.M.: Structures associated with catabolism, 91-124. In: Plant Structure, Function and Adaptation, ed. M.A. HALL. London: Macmillan 1976. - POMEROY, M.K.: Plant Physiol. (Bethesda) 59, 250-255 (1977). - POSSINGHAM, J.V.: Annu. Rev. Plant. Physiol. 31, 113-129 (1980). - PRIESTLEY, D.A., WOOLHOUSE, H.W.: J. Exp. Bot. 31, 437-447 (1980).

RAKHMANKULOV, S., AZIZHODZHAEV, A.: Uzb. Biol. Zh. 5, 74-76 (1977). - REINERT, J. (ed.): Chloroplasts. 240 pp. Berlin-Heidelberg-New York: Springer 1980.

SCHNEPF, E.: Types of plastid: their development and interconversions, 1-27. In: Chloroplasts, ed. J. REINERT (Results and Problems in Cell Differentiation, Vol.10). Berlin-Heidelberg-New York: Springer 1980. - SIMPSON, D.J.: Carlsberg Res. Commun. 43, 145-179 (1978a); - 43, 365-389 (1978b); - 45, 201-210 (1980). - SIMPSON, D.J., LEE, T.H.: Cytobios. 15, 139-147 (1976). - SIMPSON, D.J., BAQAR, M.R., LEE, T.H.: Ann. Bot. (Lond.) 41, 1101-1108 (1977); - Aust. J. Bot. 26, 783-792 (1978a). - SIMPSON, D.J., MOLLER, B.L., HOYER-HANSEN, G.: Freeze-fracture structure and polypeptide composition of thylakoids of wild-type and mutant barley plastids. In: Chloroplast Development, eds. G. AKOYUNOGLOU, J.H. ARGYROUOI-AKOYUNOGLOU. Amsterdam: North Holland 1978b. - SITTE, P.: Biologie in unserer Zeit 65-74 (1977). - SPREY, B.: Z. Pflanzenphysiol. 83, 159-180 (1978). -STAEHELIN, L.A.: J. Cell Biol. 71, 136-158 (1976). - STEINBACK, K.E., BURKE, J.J., ARNTZEN, C.J.: Arch. Biochem. Biophys. 195, 546-557 (1979). - SUNDQVIST, C., RYBERG, H.: Physiol. Plant. 47, 124-128 (1979). - SUNDQVIST, C., BJORN, L.O., VIRGIN, H.I.: Factors in chloroplast differentiation, 201-224. In: Chloroplasts, ed. J. REINERT (Results and Problems in Cell Differentiation, Vol.10). Berlin-Heidelberg-New York: Springer 1980.

TEVINI, M.: Light, function and lipids during plastid development, 121-145. In: Lipids and Lipid Polymers in Higher Plants, eds. M. TEVINI, H.K. LICHTENTHALER. Berlin-Heidelberg-New York: Springer 1977. - THOMSON, W.W., WHATLEY, J.M.: Annu. Rev. Plant. Physiol. 31, 375-394 (1980).

VERBELEN, J.P., DE GREEF, J.A.: Biol. Jahrb. 42, 123-129 (1979).

WARMKE, H.E., LEE, S-L.J.: J. Hered. 68, 213-222 (1977). - WELLBURN, A.R., HAMPP, R.: Biochim. Biophys. Acta 542, 380-397 (1979). - WETTSTEIN, W. von: Chloroplast and nucleus: concerted interplay between genomes of different cell organelles, 250-272. In: International Cell Biology 1980-1981, ed. H.G. SCHWEIGER. Berlin-Heidelberg-New York: Springer 1981. - WHATLEY, J.M.: New Phytol. 78, 407-420 (1977); - New Phytol. 80, 489-502 (1978). - WINKENBACH, F., FALK, H., LIEDVOGEL, B., SITTE, P.: Planta (Berl.) 128, 23-28 (1976). - WOLINSKA, D.: Acta Soc.

Bot. Pol. 45, 341-352 (1976). - WOLLGIEHN, R., PARTHIER, B.: RNA and protein synthesis in plastid differentiation, 97-145. In: Chloroplasts, ed. J. REINERT (Results and Problems in Cell Differentiation, Vol.10). Berlin-Heidelberg-New York: Springer 1980. - WUTTKE, H-G.: Z. Naturforsch. Sect. C. Biosci. 31, 456-460 (1976).

Dr. ANTHONY W. ROBARDS
Department of Biology
University of York
GB Y01 5DD Heslington, York

b) Special Cytology: Differentiated Cells and Cell Development in Higher Plants

By EBERHARD SCHNEPF

The great area of electron microscopy in which nearly every study on a new cell type provided much new information on cell ultrastructure has left many more detailed problems concerning the correlation between structure and function and between structure and molecular composition as well as questions of morphogenesis, development, and differentiation to be resolved. In this attempt, the interplay between cell components and cells proves a more and more important subject of cytological research. The present report includes some examples of differentiated cells, selected mainly with respect to their function, and of cell development in higher plants.

1. Secretory Cells

A detailed and comprehensive, well-illustrated review on secretory tissues in plants is given by FAHN (1979a).

a) Mucilage-Secreting Cells

For cells whose main secretory products are polysaccharides rather than protein, ROBINSON (1980) and KRISTEN (1980a,b) denied the endomembrane concept and the common interpretation of dicotyosomal structure, i.e., the existence of transition vesicles between the ER and the dictyosomes and the polar organization of the dictyosome with "forming" and "maturing" face. The published serial sections through dictyosomes of suspension-cultured sycamore cells and maize root cap cells (ROBINSON, 1980), however, indicate polar structural differences within the dictyosome and do not prove the lack of transition vesicles unambiguously. They are clearly demonstrated by KRISTEN (1980a) in the ligula cells of *Isoetes lacustris*, which probably secrete a protein-containing mucilage.

The mucilage papillae of *Marchantia* contain not only a well-developed Golgi apparatus with hypersecretory dictyosomes, but also an extended rough ER which is believed to provide the dictyosomes with raw material. Its development keeps pace with dictyosomal proliferation. Polysaccharides are detectable within the secretory Golgi vesicles, not within the ER (GALATIS and APOSTOLAKOS, 1977). In contrast, the stigmatic exudate of *Aptenia cordifolia*, containing mono- and disaccharides (mainly glucose and fructose) and polysaccharides with a high portion of uronic acids as well as proteins, obviously is secreted by exocytosis of vesicles that are directly derived from rough ER cisternae (KRISTEN et al., 1979); the Golgi apparatus is developed only poorly in the papilla.

The ovary glands of *Aptenia cordifolia* discharge polysaccharides, mono- and oligosaccharides and glycoproteins in the beginning of the secretory process, obviously via the Golgi apparatus. The composition of the exudate is changed later when the gland cells disintegrate, releasing the content of "storage vesicles" (vacuoles) and of degradation products of the cell wall into the ovary (KRISTEN and LIEBEZEIT, 1980; KRISTEN et al., 1980). KRISTEN (1980b) points out that mass secretion of proteins in higher plants, if it occurs, is done mainly in the holocrine way. The protein-polysaccharide mucilage in resin ducts of *Mango* fruits is released in a similar way (JOEL and FAHN, 1980a,b,c).

b) Nectaries

In a review, FAHN (1979b) discusses the ultrastructure of nectaries in relation to the mechanism of nectar secretion. He comes to the conclusion that the prenectar reaches the secretory cell mainly via the symplast and that the nectar is eliminated exocytotically by vesicles derived from the ER alone or from the ER and the Golgi apparatus. These conclusions are supported by observations on the fine structure of banana nectaries in different developmental stages (FAHN and BENOUAICHE, 1979). However, it cannot be excluded that the process of nectar secretion is not uniform in the various types of nectaries which differ considerably also in anatomy. Many nectary cells have apical wall ingrowths which characterize them as transfer cells. A special kind of transfer cell is realized in the nectaries of *Asclepias curassavica* where the apical plasmalemma is highly infolded and cell wall material is lacking or occurs only sparsely between the infoldings (SCHNEPF and CHRIST, 1980).

c) Salt Glands and Hydathodes

Salt glands generally have cutinized or suberinized wall areas with the function of a Casparian strip, other wall areas – transfusion areas – are not incrusted with lipophilic substances. The salt glands of *Frankenia* (CAMPBELL and THOMSON, 1976) have a ring around each transfusion area, the interfascial apparatus; it contains conspicuous lamellae which seem to consist of waxes (OLESEN, 1979).

In the passive hydathodes of young barley plants, the guttated fluid passes from vessels directly into adjacent intercellular spaces and, thereafter, through modified stomata. The occurrence of plasma-rich mesophyll and, especially, of xylem parenchyma cells along the terminal vessels and the intercellular spaces allows a modification of the guttation stream along the route (DIEFFENBACH et al., 1980a), an assumption which is supported by comparisons with exudates of stumps when leaves were cut off (DIEFFENBACH et al., 1980b).

d) Glands of Carnivorous Plants

A review on carnivorous plants is given by HESLOP-HARRISON (1978).

In continuing their studies on the glands of the trap of *Utricularia monanthos* (FINERAN and LEE, 1974), FINERAN and LEE (1980) and FINERAN (1980) describe the structure of the external glands and their ontogeny. They are found also on other parts of the plant and have wall protuberances. It is, as yet, not clear whether they absorb solutes from the external medium or have a secretory function. With lanthanum and uranyl salts the apoplastic pathway for transport in different *Utricularia* glands was visualized by FINERAN and GILBERTSON, 1980).

The structure of the glands of the newly detected carnivorous plant, *Triphyophyllum* (Dioncophyllaceae) is similar to that of *Drosophyllum* glands (MARBURGER, 1979; GREEN et al., 1979).

e) Glands with Terpenoid, Flavonoid and Phenolic Secretions

For most glands it is not known exactly where oils, resins, flavonoids and other lipophilic secretory substances are synthesized and in which way they leave the protoplast. At least it was proved by DELL and McCOMB (1978) that the glandular hairs of *Newcastelia* and not the surrounding tissue are the site of terpene resin synthesis.

After electron microscopical studies, BENAYOUN and FAHN (1979) suggest for the resin-secreting cells of *Pinus halepensis* roots that the ER collects the resin fractions produced by the plastids, mitochondria, and nuclear envelope and releases them via exocytosis (for similar results in *Mango* resin ducts see JOEL and FAHN, 1980a,b,c). It has been well known for some time that in many glands the ER ensheathes the plastids, usually leucoplasts. A detailed three-dimensional analysis of the spatial relationship between these two structures is made by ROHR et al. (1980) after serial sections.

The resinous coat on the surface of the leaves and stems of the creosote bush, *Larrea*, consisting mainly of phenols, is likewise derived from glandular trichomes (THOMSON et al. 1979). They extrude the resin mainly after senescence when they collapse after they have become isolated from symplastic and apoplastic continuity with other cells by the development of a suberin layer internal to the primary wall of the trichome.

Lytic processes seem to be involved also in the liberation of the secretory products of *Chrysanthemum* trichomes (VERMEER and PETERSON, 1979) and in the secretory cavities of *Ruta graveolens* callus (PETERSON et al., 1978).

From the results of cytological methods, as well as from chemical and biochemical analyses of isolated organelles of flavonoid glands in *Populus* buds, CHARRIÈRE-LADREIX (1979) deduced that (1) different flavonoid pools occur in plastids, in the ER, and in the plastid-ER association; (2) the flavonoids cross different cytomembranes during their intracellular transport; (3) the flavonoids are modified during membrane passage; and (4) the intracellular accumulation sites are temporary ones.

The products from glands that differ in structure (HEINRICH, 1977: *Monarda* glandular hairs) or in localization (HEINRICH et al., 1980: *Poncirus*) may differ in composition. The apex of the style of Apocynaceae bears a glandular epithelium that produces an adhesive which fixes pollen grains to the mouth parts of flower-visiting insects. It consists of terpenoids and of a polysaccharide. The latter seems to be produced by the Golgi apparatus. The glue has a foamy fine structure (SCHICK, 1980). Similarly, the translator of *Asclepias* and *Gomphocarpus* (Asclepiadaceae) which attaches two pollinia of adjacent anthers to insects searching nectar, is an emulsion of a lipophilic and a hydrophilic, presumably polysaccharidic material which has become rigid (by drying?) after secretion. It is produced by a glandular epithelium (SCHNEPF et al., 1979). Presumably, the lipophilic component is synthesized by the ER and secreted in an eccrine way, whereas the hydrophilic material is formed by the Golgi apparatus and extruded exocytotically. The morphogenesis of the secretion product, the translator, which has a complex shape, is the result of a highly coordinated activity of the cells of the secretory epithelium.

2. Idioblasts

a) Oil Idioblasts

The development of the oil bodies in the oil body cells of liverworts was studied repeatedly. Although their structure and formation differ in detail in different taxa, they have in common that the oil is included within a special vacuole (LEHMANN and JASTER, 1981) together with some hydrophilic material, mainly polysaccharides (GALATIS et al., 1978a,b,c). The oil is believed to be synthesized within the vacuole or at its membrane, the polysaccharides are produced by the Golgi apparatus. The young developing oil body of *Marchantia* that arises through the fusion of Golgi vesicles is associated with many microtubules (GALATIS et al., 1978a).

In contrast, the oil droplet in the oil idioblasts of *Laurus nobilis* (and presumably also of other seed plants) is secreted to the outside of the protoplast. It accumulates at the "cupule" and forms a droplet there. The cupule is a special wall differentiation consisting mainly of cellulose; the wall of the oil cell contains lamellated suberin layers and an inner cellulosic layer in which patches of suberin are included (MARON and FAHN, 1979).

b) Oxalate Crystal Idioblasts

It is well established now that calcium oxalate crystals are formed within a non-cytoplasmic compartment, often within a vacuole (or a corresponding "protein body", BUTTROSE and LOTT, 1978) though the developing crystal within the vacuole can initially be ensheathed by a thin layer of cytoplasm as in the bark and in the secondary xylem of *Acacia senegal* (WATTENDORF, 1978). The cell walls of the crystalliferous cells consist in part of a "filling cork"-like, non-carbohydratous material. In *Capsicum* anthers, the druse crystals form within the vacuoles in association with membrane complexes, paracrystalline bodies that are considered to be a nucleation site for druse crystal formation (HORNER and WAGNER, 1980). The differentiation of leaf crystal idioblasts in *Canavalia* can be influenced by gibberellic acid, presumably by modifying calcium availability (ZINDLER-FRANK, 1980).

3. Epidermis

a) Guard Cell Structure

The question whether mature guard cells are connected with their neighbouring cells via plasmodesmata seems to be answered now. As other authors before, WILMER and SEXTON (1979) demonstrated that the symplasmic connections are lost during the differentiation of the guard cells (cf. Progr. Bot. 42, 1). Only incomplete fragments, often branched, are left. Also in the fern, *Ophioglossum*, the walls of mature guard cells lack plasmodesmata (PETERSON and HABLETON, 1978).

In *Eucalyptus* leaves, the cuticle lying over the guard cells breaks during stomatal development. The remnants form the outer part of the stomatal pore (CARR and CARR, 1978). In some species, wall thickenings press the cuticle outwards and occlude the anterior chamber, leaving only a small passage (CARR and CARR, 1980).

Inclusions in the guard cell wall of different plants may in part be cytoplasmic remnants and in part consist of cutin (CORK and NELMES, 1979). Silica is deposited in specific cell wall layers in the stomatal apparatus of sugar cane as shown by X-ray analysis (SAKAI and THOM, 1979).

The tonoplast of *Allium* guard cells displays a green intrinsic fluorescence when irradiated with broad-band blue light, indicating the occurrence of a flavin or flavoprotein, perhaps related to the blue light-enhanced stomatal opening observed in onion (ZEIGER and HEPLER, 1979).

b) Guard Cell Development

Analysing the development of the stomatal pattern in *Vinca* leaves, SACHS (1979) observed that the presence of developing and mature stomata may influence only the orientation of divisions in neighbouring cells and not the initiation and maturation of stomata. Obviously, continuous short-range interactions between the future guard cells and the adjoining cells, rather than interactions between future stomata or a fixed programme of development, are essential for the formation of the pattern of functional stomata in the mature leaf.

A detailed analysis of stomatal development in *Vigna sinensis* , based on light and electron microscopy, is given by GALATIS and MITRAKOS (1979). In *Vigna*, guard cells are the result of a series of differential divisions which, in part, show a premitotic polarization of the cell, an eccentric position of the nucleus and differences in size and organization of the daughter cells. These asymmetrical differential divisions occur in the ordinary protodermal cells. Differential divisions in which asymmetry is less conspicuous are observed in meristemoids. In every case a pre-prophase band of microtubules indicates exactly the sites where the expanding cell plate will join the mother cell wall. This was observed also by BUSBY and GUNNING (1980) in sugar cane which had been reported to have a pre-prophase band aligned vertically to the plane of division in guard cell development (SINGH, 1977; SINGH et al., 1977).

When young *Vigna* leaves are treated with colchicine, the development of stomatal meristemoids and of guard cell mother cells into a pair of guard cells is inhibited. Instead "persistent stomatal meristemoids" and "persistent guard cell mother cells" result which, nevertheless, have guard cell-like chloroplasts, guard cell-like wall thickenings, and which, like guard cells, accumulate K^+-ions in light (GALATIS, 1977).

During the morphogenesis of guard cells, microtubules obviously play an important role. They line regions of local wall thickening and run parallel to the newly deposited microfibrils in the cell wall (GALATIS, 1980; GALATIS and MITRAKOS, 1980).

c) Cuticle

A review on cutinized and suberinized cell walls is given by WATTENDORF (1980).

Investigating the effect of temperature on wet cuticles of *Citrus aurantium* and *Hedera helix* with calorimetry, densitometry, spin-label electron-spin-resonance-spectroscopy, photo-bleaching, and microscopical methods, ECKL and GRULER (1980) stated a transition at 16.3°C that is believed to be related to a reorientation of parts of the polymer matrix and several transitions at about 40° - 50°C, presumably mainly in connection with changes in the fluidity of the soluble lipids and their orientation. Based on the well-founded assumption that a part of the cuticle has a lamellar structure (for a recent, detailed study see WATTENDORF and HOLLOWAY, 1980: *Agave americana*), it is suggested that at low temperatures the soluble lipids are oriented mainly parallel to the cuticle surface and at high temperatures parallel to the normal. The layers of soluble lipids provide the main resistance to water penetration. At high temperatures the lamellae have hydrophilic pores. The changes assumed to occur in the polymer matrix are more difficult to understand. Probably, molecules melt at high temperatures.

With electron microscopy of thin sections HOCH (1979, see Progr. Bot. 42, 1) could identify polysaccharide microfibrils (or strands) in the cuticle of *Malus pumila* leaves that originate at the outer part of the cellin wall and occasionally extend to the cuticle surface (in contrast to the *Agave* cuticle, WATTENDORF and HOLLOWAY, 1980). Uranyl and silver ions seem to penetrate into the cuticle along the polysaccharide strands, whereas phenyl mercuric acetate deposits are not found associated with them. The cytochemistry of the developing cuticle of grape fruits is described by CONSIDINE and KNOX (1979).

4. Aleurone Cells, Endosperm

PYLIOTIS et al. (1979) localized electron microscopically acid phosphatase activity in the ER of barley aleurone cells after treatment with gibberellic acid. It is lacking in non-treated cells. Also in these studies it remains unclear how the enzyme is discharged into the periplasmic space, because neither a vesiculation of the ER nor a direct continuity between ER cisternae and the plasmalemma could be recognized with certainty.

These observations confirm the results of earlier studies on aleurone cells in which α-amylase was detected within the ER with biochemical methods (e.g., LOCY and KENDE, 1978) or by immunohistochemistry (JONES and CHEN, 1976). They explain also the changes in aleurone cells during gibberellic acid-induced enzyme secretion which mainly consist of an increase in the amount of the ER (JONES, 1980, biochemical experiments using the marker enzyme cytochrome c-reductase) but imply also structural variations of the ER (OBATA, 1979, formation of stacks of rough ER).

Studying the development of protein inclusions in wheat endosperm, PARKER (1980) observed myelin-like whorls of membranes and, in one cultivar and only transiently, membrane lattices of branched tubules with a basic cubic-repeat unit of 44.8 nm in close contact with the protein mass.

5. Motor Cells for Leaf Movement; Gravity Sensitive Cells

Looking for relationships between motor cell ultrastructure and leaf movement in *Mimosa* and other Leguminosae, and because of the ion fluxes during movement (CAMPBELL et al., 1979) it was observed by CAMPBELL and THOMSON (1977) that the inner cells of the tertiary pulvinus of

Mimosa have a large vacuole, whereas the outer ones have many small vacuoles. In *Albizzia* , the motor cells have a central vacuole when expanded but become multivacuolate during the time the cell volume decreases (CAMPBELL and GARBER, 1980).

In the pulvini of *Samanea* the outer cells contain chloroplasts with large starch grains at the beginning of the dark periods but are devoid of starch at the beginning of the light period (MORSE and SATTER, 1979). The rhythm in starch content might be related to ion movements that are the basis for leaf movement. In the primary pulvinus of *Mimosa*, the amyloplasts in the starch sheath cell sediment by gravity and seem to play a role in graviperception (FLEURAT-LESSARD, 1981).

PERBAL and RIVIÈRE (1980) describe the development of the statenchyma cells in *Asparagus officinalis* epicotyls and come to the conclusion that the ability of perceiving gravity may be linked not only to the development of sedimenting, starch-rich plastids and the polar distribution of the ER, as described repeatedly (Progr. Bot. 51, 153), but also to other cytological features of the cell.

The role of the ER in graviperception is indicated especially clearly in the studies of OLSEN and IVERSEN (1980) on statocytes of pea root caps. Normally reacting plants show the usual arrangement of sedimenting amyloplasts and a complex of ER at the distal cell pole, whereas in an ageotropic mutant the ER is distributed more evenly through the cell. CASPERS and SIEVERS (1979) likewise showed by centrifugation experiments that the graviresponse of *Lepidium* roots depends on the specific, polar arrangement of the ER in the statocytes (see also HENSEL and SIEVERS, 1979). A certain exception seems to be represented by cells of the gravity-sensitive leaf sheath base of the grass, *Echinochloa*. They also contain readily sedimenting starch statoliths but lack the distal ER complex on which the leucoplasts come to lie in normal geotropic position. In these cells they come into contact with the plasmalemma or peripheral microtubules (PARKER, 1979a). The cells of the lower side of horizontally held stalks of *Echinochloa*, including the collenchymatous bundle cap fibres, can elongate up to threefold in response to geostimulation (PARKER, 1979b).

6. Animal Food Tissue, Galls

a) Cells of Ant and Beatle Food Tissue

Continuing his studies on cells on which ants feed (Beltian bodies of *Acacia*, RICKSON, 1975, Müllerian bodies of *Cecropia* , RICKSON, 1976a,b), RICKSON (1980) investigated the fine structure of the ant food bodies of *Macaranga* (Euphorbiaceae). They consist of cells rich in lipids and starch. In contrast, the cells of the *Calycanthus* flowers which are eaten by trapped beatles store proteins in the vacuole (in part in the form of fibrils and of tubules), as well as in the form of long tubules free in the cytoplasm (RICKSON, 1979a). It should be added in this connection that ants may also feed a plant (RICKSON, 1979b, see also HUXLEY, 1980). Breakdown products of radioactive-labelled insects, introduced into stem cavities of *Hydnophytum formicarium*, are taken up by the plant though the cavities are covered with a periderm layer. In nature, captured insects are imported by ants.

b) Galls

The modified root tip cells of *Ficus carica* that are induced by the attack of the ectoparasitic nematode, *Xiphinema index*, are characterized (like nutritive cells of other zoocecidia, REY, 1979a) by a large, lobed nucleus and a dense cytoplasm with many ribosomes, plastids and mitochondria. Often cytokinesis is incomplete so that bi- or multinucleate cells with wall fragments attached to the mother cell wall are formed. At the periphery of the galled root tip and around necrotic cells transfer cells develop (WYSS et al., 1980). Wall ingrowths are also common in nutritive cells of insect galls (REY, 1979b).

7. Seasonal Variations in Cell Structure

An increasing amount of publications deals with the problem of how seasonally varying cell activities, especially dormancy and adaptation to frost, are reflected by variations in cell ultrastructure. Several Finnish authors report on changes in the starch content, the amount of ribosomes, vacuolar content and in the structure of the nucleus and the nucleoles of conifers (SOIKKELI, 1978, 1980; KUPILA-AHVENNIEMI et al., 1978; KUPILA-AHVENNIEMI and HOHTOLA, 1979, 1980). In general, these investigations do not include exact quantitative data, as they are given, e.g., by MAUSETH (1980) for the shoot apical meristem of *Echinocereus engelmannii* at germination.

In freeze-fracture studies, CRAGG and WILLISON (1980) and WILLISON and CRAGG (1980) observed unusually large nuclear pores in quiescent buds of *Tilia* (diameter up to 175 nm).

8. Cell Development

a) General Considerations, Models

A general discussion on problems of morphogenesis of higher organisms is given by MEINHARDT (1979). He presents a model in which gradients control the spatial pattern of cell differentiation. Positive feedback mechanisms and activator-inhibitor gradients play the most important role in his theory (see also the general considerations of MÜLLER, 1979). Polar fluxes of auxin, gradients, and positive feedback mechanisms are likewise suggested by SACHS (1981) to be the cause for the differentiation of conductive cells in stems. In the control system for initiating and maintaining polarity of WILSON (1980, see also THORNLY et al., 1980) also diffusible morphogens that mutually control their rate of synthesis are taken to explain polar development.

These ideas are supported to a certain extent by SPIEGEL and COX (1980). Pattern formation implies a regular spacing. Observing fruit body formation in haploid and diploid strains of the cellular slime mould *Polysphondylium pallidum*, the cell sizes of which differ considerably, SPIEGEL and COX (1980) could show that the spacing is not achieved by a cell-counting mechanism. Instead, it appears to depend on distance. To determine distances, gradients seem to be used widely. The above-mentioned publications are not primarily focused on *cell* development. However, as GUNNING (1980) points out, cell morphogenesis and tissue and organ development are closely interconnected: order begets order. Continuing his investigations on cell development and on the distribution of microtubules in *Azolla* roots, using conventional and high voltage electron microscopy, he observed microtubule nucleating sites concentrated along the edges of the cells. They are conspicuous only when the interphase microtubule system is generated in early post-cytokinesis and when the pre-prophase band is formed. GUNNING (1980) points out that cell morphogenesis is based on a hierarchy of order-generating systems in which the lattice of cell edges and nucleation sites is but one participant, helping to produce ordered arrays of microtubules that in turn have roles in cell morphogenesis and, moreover, in tissue and organ development.

A detailed analysis of the change of cell shape during development, based on a comparison of shapes of ideal cells with those of real cells and considering the conditions for cell division and expansion, is given by KORN (1980).

b) Spore Germination

A review on the cytology, physiology and biochemistry of the germination of fern spores is given by RAGHAVAN (1980). QUATRANO (1978) reports on the development of cell polarity, using mainly the zygote of *Fugus* and *Pelvetia* and other cells of lower plants as examples. WEISENSEEL (1979, 1980) reviews the induction of polarity and the polar growth with special emphasis on electrical phenomena (see also JAFFE, 1981). DeMAGGIO et al. (1980) observed that during the germination of spores of the fern *Onoclea* there is no temporal separation of heterotrophic and autotrophic reactions. Chlorophyll content and activities of the glyoxylate and the glycolate cycle increase concurrently.

It has been well known for a long time that environmental factors such as gravity or light determine the site where the spore of a moss or a fern develops its first outgrowth and hence orient the polar axis. Steady electrical fields and forced calcium entry caused by the ionophore A 23187 likewise can be used to control the point at which the rhizoids emerge from *Funaria* spores (CHEN and JAFFE, 1979; see also ROBINSON and CONE, 1980, for germination of the zygotes of the brown alga, *Pelvetia*). Obviously, the plasmalemma is the receptor for various information which reaches the cell from the environment. As a result of perception, local and initially labile differences in the plasmalemma are caused. These local differences lead to a transcellular electrical current which, by a positive feedback, increases the differences within the membrane and, in addition, polarizes the cytoplasm. An initially labile pattern is stabilized and becomes a fixed spatial pattern in connection with structural modifications. Spatial differentiation, hence, seems to develop centripetally, it begins in the cell periphery and only later includes the nucleus (WEISENSEEL, 1980).

In further development, the nucleus interacts with these processes and, obviously also acquires a position within a system of feedback mechanisms. In germinating spores of the fern, *Onoclea sensibilis*, the nucleus moves from its initial position in the centre of the spore to the proximal side and then along the raphe to that end of the spore where the first – asymmetric – division occurs (VOGELMANN and MILLER, 1980). The smaller cell develops directly into a rhizoid, the larger one and its derivatives give rise to the prothallus (MILLER and BASSEL, 1980; for other ferns see RAGHAVAN and HUCKABY, 1980). Under the influence of caffeine, cross-wall formation is greatly disturbed while mitosis is largely unaffected. In consequence, multinucleate, single cells arise which grow but do not form rhizoids, though occasionally a rhizoid-like outgrowth develops. It does not contain a nucleus because that daughter nucleus which initially comes to lie within the outgrowth after mitosis moves through the incomplete cross-wall back to the centre of the spore. In consequence, the outgrowth does not develop furhter: for rhizoid formation to occur, a nucleus must be confined in that region (MILLER and BASSEL, 1980). These observations and experiments show that the establishment of an internal order, implying the exact positioning of the nuclei, is the precondition for regular cell development.

c) Tip Growth

In the caulonema of the moss, *Funaria*, the site of side branch formation and hence the site of tip growth initiation and the orientation of a new polar axis is determined by the position of the nucleus of the mother cell which, in turn, is controlled by unknown internal positioning factors (distance from the distal cross-wall) and by gravity. Generally it is the third cell of a filament which initiates a side branch; as shown by experiments with the inhibitor, cytochalasin B, it is the age of the cell rather than its distance from the tip which induces side branch formation. The bulging out of the cell induces the vectorial movement of the nucleus into the outgrowth. The migration depends on an intact microtubule system. The elongation of the outgrowth ceases when the nucleus cannot reach its normal position, e.g., when the microtubules are affected or when, after displacement of the nucleus by centrifugation or by treatment with D_2O, the remigration is interrupted by mitosis. In these cases, a new outgrowth in an abnormal position and occasionally even with reversed polar axis is formed (SCHMIEDEL and SCHNEPF, 1979a,b).

A similar interdependence between the site of growth, the orientation of the polar axis, and the position of the nucleus is observed in the tip cell. Under normal conditions, the nucleus maintains a constant distance from the tip, and remigrates into this position soon after experimental displacement. When, because of the onset of mitosis or because of the destruction of the microtubule system, the nucleus is inhibited to remigrate, growth and polar axis may be reoriented (SCHMIEDEL and SCHNEPF, 1980): Another example for developmental control by feedback mechanisms (SCHNEPF, 1981). Under normal conditions, the plasmalemma controls the position of the nucleus, as shown in centrifugation experiments; when the position of the nucleus cannot be adjusted, the nucleus, on the contrary, controls the plasmalemma and initiates a new growth center. Growth abnormalities after colchicine treatment may likewise be caused by affecting these interrelationships. The interdependence of nuclear position and tip growth is not observed in cells in which a strict coordination of growth and spatial arrangement is not required for regular development, i.e., in cases, where daughter cells of uniform size are not cut off (SCHNEPF and HEINZAMNN, 1980).

New and more detailed examples for the highly polarized organization of the cytoplasm in tip growing cells, implying differential distribution of organelles, are given by SEAGULL and HEATH (1979, 1980a), who describe root hair fine structure with special emphasis on the distribution of microtubules and microfilaments, and by SCHMIEDEL and SCHNEPF (1980) and SCHNEPF (1981) for *Funaria* caulonema tip cells in which qualitative differences between organelles in different position (higher starch content in the chloroplasts and more cisternae per dictyosome in the tip region) also are observed.

The conspicuous apical accumulation of vesicles, mainly of Golgi vesicles which contain cell wall precursors, seems to be related to the growth rate (SCHMIEDEL and SCHNEPF, 1979a), it is not observed in slowly growing cells and disappears when tip growth is inhibited (REISS and HERTH, 1979a; SCHMIEDEL and SCHNEPF, 1980). In the vectorial transport of vesicles electrophoretic processes could be involved (QUATRANO, 1978), besides contractile mechanisms: A detailed analysis of the microfilament system in root hairs and of their involvement in cytoplasmic streaming is given by SEAGULL and HEATH (1980b).

Cells with tip growth are traversed by natural H^+ currents (WEISENSEEL et al., 1979) and show a polar gradient in the distribution of calcium (BOSCH et al., 1980: proton microprobe anaylsis; REISS and HERTH, 1978, 1979b: visualization by fluorescence with chlorotetracycline).

The calcium ionophore A 23187 stops tip growth, but not cytoplasmic streaming in pollen tubes of *Lilium longiflorum* (HERTH, 1978) and affects localized wall secretion in the tip region (REISS and HERTH, 1979b). Experiments on pollen tubes with the broad range ionophore X-537A indicate that oriented exocytosis depends extensively on a specific distribution of calcium ions, whereas vesicle transport and formation also depend on other ions (REISS and HERTH, 1980). The polar distribution of calcium ions seems to play an important role in organizing morphological gradients (CHEN and JAFFE, 1979; REISS and HERTH, 1979a,b).

The different external factors which determine the direction of tip growth are not treated in this report. It should, however, be added that strong magnetic fields are able to orient growth of pollen tubes (SPERBER et al., 1981).

d) The Plane of Division

The relationship between the position of the pre-prophase band of microtubules and the subsequent orientation of the cell plate is investigated by BUSBY and GUNNING (1980). They show that in the basal meristematic cells of the uniseriate hairs of *Salvinia* and in the intercalarly dividing cells of the uniseriate hairs of *Tradescantia* stamen filaments pre-prophase bands, predicting the site of cytokinesis, occur. They are lacking, however, in the tip growing moss caulonema cells (SCHMIEDEL et al., 1981) though they occur in moss leaflets.

In the developing protonema of *Onoclea* the orientation of the plane of cell division results in a new wall that has the least area. From plasmolysis studies, MILLER (1980) concludes that the primary factor which determines the plane of division is the local pattern of stress. COOKE and PAOLILLO (1980) similarly assume that the control of the plane of division in fern gametophytes depends on the thermodynamics of the surface area.

e) Senescence, Abscission, Formation of Intercellular Spaces

The opening and closing of the flowers of *Ipomoea tricolor* are given by differential growth as well as by asymmetric turgor changes in a specific group of inner epidermal cells at the ribs. In contrast to normal epidermis cells they display a series of structural modifications (PHILLIPS and KENDE, 1980) that seem to affect their turgor status: They enlarge, modify their shape, reduce their wall thickness, vacuolize extensively while showing autophagic activity and changing from a ribosome-rich cytoplasm with stacked ER and many dictyosomes to one with only a few organelles. At the onset of senescence that begins in the bud and leads to cell death in about 3 days, the tonoplast becomes intensely stained in certain areas.

Cell death as a functional event in the development of intercellular spaces in *Avicennia* leaves is described by CHALAIN and BERJAK (1979), the formation of "air blisters" in *Pilea* by VAUGHN et al. (1980) and SEXTON (1979) studied the cell separation in the abscission zone of *Impatiens* leaves with the light microscope. He came to the conclusion that there is little requirement for cell-to-cell contact in the temporal and spatial integration of cell wall breakdown. By treatment with commercial cellulase, KAWASE (1979) could achieve the development of aerenchyma-like intercellular spaces in sunflower stems resembling those which are formed by anaerobic stimulation of ethylene production which increases internal cellulase activity.

In contrast to the flower and leaf abscission of most plants, that of the explosive fruit of the dwarf mistletoe, *Arceuthobium*, shows no degradation of cell walls of the abscission-zone parenchyma. This zone consists of a layer of thin-walled cells arranged in a flat ring around the base of the fruit (TOTH and KUIJT, 1978). When the fruit explodes, the walls break at random, presumably without the participation of enzymatically induced cell wall dissolution.

References

BENAYOUN, J., FAHN, A.: Ann. Bot. 43, 179-181 (1979). - BOSCH, F., EL GORESY, A., HERTH, W., MARTIN, B., NOBILING, R., POVH, B., REISS, H.D., TRAXEL, K.: Nuclear Sci. Applications 1, 1-39 (1980). - BUSBY, C.H., GUNNING, B.: Eur. J. Cell Biol. 21, 214-223 (1980). - BUTTROSE, M.S.; LOTT, J.N.A.: Can. J. Bot. 56, 2083-2091 (1978).

CAMPBELL, N.A., GARBER, R.C.: Planta 148, 251-255 (1980). - CAMPBELL, N.A., THOMSON, W.W.: Ann. Bot. 40, 681-686 (1976); - Ann. Bot. 41, 1361-1362 (1977). - CAMPBELL, N.A., STIKA, K.M., MORRISON, G.H.: Science 204, 185-187 (1979). - CARR, D.J., CARR, S.G.M.: Protoplasma 96, 127-148 (1978); - Protoplasma 104, 239-251 (1980). - CASPERS, L., SIEVERS, A.: Eur. J. Cell Biol. 20, 119 (1979). - CHALAIN, T.M.B. DE, BERJAK, P.: New Phytol. 83, 147-155 (1979). - CHARRIÈRE-LADREIX, Y.: Thèse, Grenoble 1979. - CHEN, T.-H., JAFFE, L.F.: Planta 144, 401-406 (1979). - CONSIDINE, J.A., KNOX, R.B.: Protoplasma 99, 347-365 (1979).- COOKE, T.J., PAOLILLO, D.J., Jr.: Am. J. Bot. 67, 1320-1233 (1980). - CORK, R.J., NELMES, B.J.: J. Cell Sci. 38, 83-95 (1979). - CRAGG, F.J., WILLISON, J.H.M.: Can. J. Bot. 58, 1804-1813 (1980).

DELL, B., MCCOMBE, A.J.: J. Exp. Bot. 29, 89-95 (1978). - DEMAGGIO, A.E., GREENE, C., STETLER, D.: Plant Physiol. 66, 922-924 (1980). - DIEFFENBACH, H., KRAMER, D., LÜTTGE, U.: Ann. Bot. 45, 397-401 (1980a). - DIEFFENBACH, H., LÜTTGE, U., Pitman, M.G.: Ann. Bot. 45, 703-712 (1980b).

ECKL, K., GRUDER, H.: Planta 150, 102-113 (1980).

FAHN, A.: Secretory Tissues in Plants. 302 pp. London-New York-San Francisco: Academic Press 1979a; - Am. J. Bot. 66, 977-985 (1979b). - FAHN, A., BENOUAICHE, P.: Ann. Bot. 44, 85-93 (1979). - FINERAN, B.A.: Protoplasma 105, 9-25 (1980). - FINERAN, B.A., GILBERTSON, J.M.: Eur. J. Cell Biol. 23, 66-72 (1980). - FINERAN, B.A., LEE, M.S.L.: J. Ultrastruct. Res. 48, 162-166 (1974); - Protoplasma 103, 17-34 (1980). - FLEURAT-LESSARD, P.: Protoplasma 105, 177-184 (1981).

GALATIS, B.: Planta 136, 103-114 (1977); - J. Cell Sci. 45, 211-244 (1980).- GALATIS, B., APOSTOLAKOS, P.: Can. J. Bot. 55, 772-795 (1977). -

GALATIS, B., MITRAKOS, K.: J. Cell Sci. 37, 11-37 (1979); - Am. J. Bot. 67, 1243-1261 (1980). - GALATIS, B., APOSTOLAKOS, P., KATSAROS, C.: Can. J. Bot. 56, 2252-2267 (1978a). - GALATIS, B., KATSAROS, C., APOSTOLAKOS, P.: Can. J. Bot. 56, 2268-2285 (1978b). - GALATIS, B., APOSTOLAKOS, P., KATSAROS, C.: Protoplasma 97, 13-29 (1978c). - GREEN, S., GREEN, T.L., HESLOP-HARRISON, Y.: Bot. J. Linn. Soc. 78, 99-116 (1979). - GUNNING, B.E.S.: Eur. J. Cell Biol. 23, 53-65 (1980).

HEINRICH, G.: Biochem. Physiol. Pflanz. 171, 17-24 (1977). - HEINRICH, G., SCHULTZE, W., WEGENER, R.: Protoplasma 103, 115-129 (1980). - HENSEL, W., SIEVERS, A.: Eur. J. Cell Biol. 20, 121 (1979). - HERTH, W.: Protoplasma 96, 275-282 (1978). - HESLOP-HARRISON, Y.: Sci. Am. 238(2), 104-115 (1978). - HOCH, H.C.: Planta 147, 186-195 (1979). - HORNER, H.T., Jr., WAGNER, B.L.: Am. J. Bot. 67, 1347-1360 (1980). - HUXLEY, C.: Biol. Rev. Cambr. Phil. Soc. 55, 321-340 (1980).

JAFFE, L.F.: The role of ion currents in establishing developmental gradients, 507-511. In: International Cell Biology 1980-1981, ed. H.G. SCHWEIGER. Berlin-Heidelberg-New York: Springer 1981. - JOEL, D.M., FAHN, A.: Ann. Bot. 46, 225-233 (1980a); - ibid. 46, 779-783 (1980b); - ibid. 46, 785-790 (1980c). - JONES, R.L.: Planta 150, 70-81 (1980). - JONES, R.L., CHEN, R.-F.: J. Cell Sci. 20, 183-198 (1976).

KAWASE, M.: Am. J. Bot. 66, 183-190 (1979). - KORN, R.W.: Ann. Bot. 46, 649-666 (1980). - KRISTEN, U.: Eur. J. Cell Biol. 23, 16-21 (1980a); - Ber. Dtsch. Bot. Ges. 93, 587-594 (1980b). - KRISTEN, U., LIEBEZEIT, G.: Z. Pflanzenphysiol. 99, 241-250 (1980). - KRISTEN, U., BIEDERMANN, M., LIEBEZEIT, G., DAWSON, R., BÖHM, L.: Eur. J. Cell Biol. 19, 281-287 (1979). - KRISTEN, U., BIEDERMANN, M., LIEBEZEIT, G.: Z. Pflanzenphysiol. 96, 239-249 (1980). - KUPILA-AHVENNIEMI, S., HOHTOLA, A.: Protoplasma 100, 289-301 (1979); - Protoplasma 105, 107-119 (1980). - KUPILA-AHVENNIEMI, S., PIHAKASKI, S., PIHAKASKI, K.: Planta 144, 19-29 (1978).

LEHMANN, H., JASTER, B.: Protoplasma 106, 109-119 (1981). - LOCY, R., KENDE, H.: Planta 143, 89-99 (1978).

MARBURGER, J.E.: Am. J. Bot. 66, 404-411 (1979). - MARON, R., FAHN, A.: Bot. J. Linn. Soc. 78, 31-40 (1979). - MAUSETH, J.D.: Am. J. Bot. 67, 173-181 (1980). - MEINHARDT, H.: Biol. in u. Zeit 9, 33-39 (1979). - MILLER, J.H.: Am. J. Bot. 67, 534-542 (1980). - MILLER, J.H., BASSEL, A.R.: Physiol. Plant. 50, 213-220 (1980). - MORSE, M.J., SATTER, R.L.: Physiol. Plant. 46, 338-346 (1979). - MÜLLER, W.A.: Biol. in u. Zeit 9, 135-140 (1979).

OBATA, T.: Ann. Bot. 44, 333-337 (1979). - OLESEN, P.: Protoplasma 99, 1-9 (1979). - OLSEN, G.M., IVERSEN, T.-H.: Physiol. Plant. 50, 275-284 (1980).

PARKER, M.L.: Planta 145, 471-477 (1979a); - Can. J. Bot. 57, 2399-2407 (1979b); - Ann. Bot. 46, 29-36 (1980). - PERBAL, G., RIVIÈRE, S.: Biol. Cellulaire 39, 91-98 (1980). - PETERSON, R.L., HAMBLETON, S.: Can. J. Bot. 56, 2836-2852 (1978). - PETERSON, R.L., SCOTT, M.G., ELLIS, B.E.: Can. J. Bot. 56, 2717-2729 (1978). - PHILLIPS, H.L., Jr., KENDE, H.: Protoplasma 102, 199-215 (1980). - PYLIOTIS, N.A., ASHFORD, A.E., WHITECROSS, M.I., JACOBSEN, J.V.: Planta 147, 134-140 (1979).

QUATRANO, R.S.: Annu. Rev. Plant Physiol. 29, 487-510 (1978).

RAGHAVAN, V.: Int. Rev. Cytol. 62, 69-118 (1980). - RAGHAVAN, V., HUCKABY, C.S.: Am. J. Bot. 67, 653-663 (1980). - REISS, H.-D., HERTH,

W.: Protoplasma 97, 373-377 (1978); - Planta 145, 225-232 (1979a); ibid. 146, 615-621 (1979b); - ibid. 147, 295-301 (1980). - REY, L.: Bull. Soc. Bot. Fr. 126, 179-190 (1979a); - Beitr. Biol. Pfl. 54, 315-327 (1979b). - RICKSON, F.R.: Am. J. Bot. 62, 913-922 (1975); - ibid. 63, 1266-1271 (1976a); - ibid. 63, 1272-1279 (1976b); - ibid. 66, 80-86 (1979a); - ibid 66, 87-90 (1979b); - ibid. 67, 285-292 (1980). - ROBINSON, D.G.: Eur. J. Cell Biol. 23, 22-36 (1980). - ROBINSON, K.R., CONE, R.: Science 207, 77-78 (1980). - ROHR, R.;DEXHEIMER, J., KIEFFER, M.: Can. J. Bot. 58, 1859-1871 (1980).

SACHS, T.: Ann. Bot. 43, 693-700 (1979); - Polarity changes and tissue organization in plants, 489-496. In: International Cell Biology 1980-1981, ed. H.G. SCHWEIGER. Berlin-Heidelberg-New York: Springer 1981. - SAKAI, W.S., THOM, M.: Ann. Bot. 44, 245-248 (1979). - SCHICK, B.: Flora 170, 394-432 (1980). - SCHMIEDEL, G., SCHNEPF, E.: Protoplasma 100, 367-383 (1979a); - Protoplasma 101, 47-59 (1979b); - Planta 147, 405-413 (1980). - SCHMIEDEL; G., REISS, H.-D., SCHNEPF, E.: Protoplasma (in press) (1981). - SCHNEPF, E.: Polarity and gradients in tip growing plant cells, 485-488. In: International Cell Biology 1980-1981, ed. H.G. SCHWEIGER. Berlin-Heidelberg-New York 1981. - SCHNEPF, E., CHRIST, P.: Protoplasma 105, 135-148 (1980). - SCHNEPF, E., HEINZMANN, J.: Biochem. Physiol. Pflanz. 175, 67-76 (1980). - SCHNEPF, E., WITZIG, F., SCHILL, R.: Über Bildung und Feinstruktur des Translators der Pollinarien von *Asclepias curassivica* und *Gomphocarpus fruticosus* (Asclepiadaceae). Tropische und subtropische Pflanzenwelt, Vol. 25, Akad. Wiss. Lit. Mainz, Math.-Naturwiss. Klasse. 1979. - SEAGULL, R.W., HEATH, I. B.: Eur. J. Cell Biol. 20, 184-188 (1979); - Protoplasma 103, 205-229 (1980a). - Protoplasma 103, 231-240 (1980b). - SEXTON, R.: Protoplasma 99, 53-66 (1979). - SINGH, A.P.: Can. J. Bot. 55, 2801-2809 (1977). - SINGH, A.P., SHAW, M., HOLLINS, G.: Cytologia 42, 611-620 (1977). - SOIKKELI, S.: Can. J. Bot. 56, 1932-1940 (1978); - Protoplasma 103, 241-252 (1980). - SPERBER, D., DRANSFELD, K., MARET, G., WEISENSEEL; M.H.: Naturwissenschaften 68, 40-41 (1981). - SPIEGEL, F.W., COX, E.C.: Nature 286, 806-807 (1980)

THOMSON, W.W., PLATT-ALOIA, K., KOLLER, D.: Bot. Gaz. 140, 249-260 (1979). - THORNLEY, J.H.M., WILSON, J.W.; COLLEY, E.: Ann. Bot. 46, 713-717 (1980). - TOTH, R., KUIJT, J.: Bot. Gaz. 139, 158-164 (1978).

VAUGHN, K.C., DOWNS, B.D., WILSON, K.G.: Ann. Bot. 46, 221-224 (1980). - VERMEER, J., PETERSON, R.L.: Can. J. Bot. 57, 714-729 (1979). - VOGELMANN, T.C., MILLER, J.H.: Am. J. Bot. 67, 648-652 (1980).

WATTENDORFF, J.: Protoplasma 95, 193-206 (1978); - Biol. in u. Zeit 10, 81-90 (1980). - WATTENDORFF, J., HOLLOWAY, P.J.: Ann. Bot. 46, 13-28 (1980). - WEISENSEEL, M.H.: Induction of polarity, 485-505. In: Encyclopaedia of Plant Physiology, N.S., Vol. 7, eds. W. HAUPT, M.E. FEINLEIB. Berlin-Heidelberg-New York: Springer 1979; - Biol. in u. Zeit 10, 39-44 (1980). - WEISENSEEL, M.H., DORN, A., JAFFE, L.F.: Plant Physiol. 64, 512-518 (1979). - WILLISON, J.H.M.: Plant Sci. Lett. 18, 121-126 (1980). - WILLISON, J.H.M., CRAGG, F.J.: Can. J. Bot. 58, 1814-1819 (1980). - WILLMER, C.M., SEXTON, R.: Protoplasma 100, 113-124 (1979). - WILSON, J.W.: Ann. Bot. 46, 701-711 (1980). - WYSS, U., LEHMANN, H., JANK-LADWIG, R.: J. Cell Sci. 41, 193-208 (1980)

ZEIGER, E., HEPLER, P.K.: J. Cell Sci. 37, 1-10 (1979). - ZINDLER-FRANK, E.: Z. Pflanzenphysiol. 98, 43-52 (1980).

Professor Dr. EBERHARD SCHNEPF
Zellenlehre, Universität
Im Neuenheimer Feld 230
D 6900 Heidelberg

c) Oberflächenskulpturen bei höheren Pflanzen

Von WILHELM BARTHLOTT und RAINER SCHILL

1. Allgemeines und Methodik

Dieser Bericht berücksichtigt vorwiegend zwischen 1977 und 1980 erschienene raster-elektronenmikroskopische Arbeiten zur Oberflächenskulptur von Epidermen (Sproß, Blatt, Blüte, Samen) und Pollen von Tracheophyten. Mit berücksichtigt sind einige ultrastrukturelle Arbeiten (TEM) sowie einige Publikationen, die sich mit der Chemie epicuticularer Sekrete (Wachse u.ä.) beschäftigen. Arbeiten, bei denen das REM primär zur Klärung morphologisch-entwicklungsgeschichtlicher Fragen eingesetzt wird (stellvertretend seien ENDRESS, 1980, sowie LEINS und ERBAR, 1980, genannt) fallen in das Kapitel "Allgemeine Morphologie".

Abgesehen von Vereinfachungen der Bedienung wurden am Untersuchungsinstrument (REM) in den vergangenen Jahren keine wesentlichen Neuerungen eingeführt; auch eine Steigerung des Auflösungsvermögens wurde nicht erreicht. Als allgemeine Einführungen in die Methodik sind im Berichtzeitraum Neuauflagen von OHNSORGE und HOLM (1978) und REIMER und PFEFFERKORN (1978) sowie eine übersichtliche "Introduction to Biological Scanning Electron Microscopy" von HAYAT (1978) erschienen.

Bei der Critical-Point-Trocknung werden die Freon/Frigen-Medien verdrängt durch den wesentlich schonenderen, billigen, schnelleren und umweltfreundlichen Austausch im CO_2-Verfahren. Eine besonders elegante Methode der Dehydrierung wurde dabei von GERSTERBERGER und LEINS (1978) beschrieben: Ausgenutzt wird die Eigenschaft von Formaldehyd-dimethyl-acetal (Dimethoxymethan) mit Wasser eine Reaktion unter der Bildung von Methanol und Formaldehyd einzugehen. Eine Methode zum Aufkleben feiner Objekte (Pollen, Samen) mit oxydierenden Ölen (z.B. sog. "Anlegeöl") beschreiben BARTHLOTT und ZIEGLER (1980): Es werden damit spiegelartig-glatte und dennoch fest haftende Klebeschichten erzielt.

Die Anfangszeit der mikromorphologischen Forschung, die oft durch die Publikation ästhetischer "dreidimensionaler" Bilder altbekannter Strukturen ohne qualifizierte Interpretation gekennzeichnet war, dürfte als beendet angesehen werden. Die REM hat sich in Kombination mit TEM und histochemischen Methoden als unentbehrliche Methode zum Studium struktureller und taxonomischer Fragen bewährt.

2. Epidermale Oberflächen

Die Arbeiten zur Oberflächenskulptur von Epidermen sind im folgenden unter mikromorphologisch-strukturellen, unter systematisch-taxonomischen und unter funktionell-ökologischen Aspekten zusammengefasst. An zusammenfassenden Darstellungen sei auf die entsprechenden Kapitel in der Neuauflage von METCALFE und CHALK (1980) sowie auf BARTHLOTT (1981) verwiesen.

a) Mikromorphologische Aspekte

Zellmuster, also die taxon-spezifische Anordnung von Epidermiszellen und die Verteilung idioblaster Elemente (Trichome, Stomata etc.) können oftmals zur Charakterisierung bis zur Artebene dienen. Studien solcher "Dermogramme" wurden an Poaceae von VIGNAL (1979) und von BARTHLOTT und MARTENS (1979) publiziert, CLARK und JERNSTEDT (1978) beschreiben ein besonders charakteristisches Muster für die Samenschalen von *Eschscholzia* (Papaveraceae).

Eine Übersicht und revidierte Klassifikation von Trichomen geben THEOBOLD et al. (1980). Diese wichtige Arbeit bringt zumindest ein anwendbares deskriptives System (basierend vorwiegend auf Dikotylen), obwohl sich gerade das äußerst vielfältige fließende Formen-Kontinuum von Trichomen jedem starren Klassifikations-Versuch widersetzt. Als wichtige Neuerscheinung zur Struktur von Drüsenhaaren und sekretorischen Geweben ist das Buch von FAHN (1979) zu nennen. Anzufügen bleiben einige neuere Arbeiten über Drüsenhaare bei *Pinguicula* (HESLOP-HARRISON und HESLOP-HARRISON, 1981), *Solanum* (LYSHEDE, 1980) und über die Riesenzellen der Epidermis von *Mesembryanthemum* (KRAMER, 1979). Vorwiegend taxonomische Aspekte der Trichome werden an Asteraceae-Anthemideae (NAPP-ZINN und EBLE, 1980), *Quercus* (HARDIN, 1979a,b) und den Familien der Dipsacales (WEBERLING, 1977, 1978) und Ranales (ALEYKUTTY und INAMDRAR, 1980) dargestellt. Über helicale Strukturen an Trichomen einiger Dikotylen berichten LERSTEN und CURTIS (1977). DEHGAN (1980) bildet bei seinen cuticulartaxonomischen Untersuchungen an *Jatropha* (Euphorbiaceae) Trichome ab, die im REM-Bild einzellig aussehen, in Wirklichkeit aber multicellular-uniseriat aufgebaut sind: ein weiteres Beispiel, daß REM-Untersuchungen durch andere Methoden abgesichert werden müssen.

Eine Übersicht und Revision der Stoma-Typen von Dicotylen bietet WILKINSON (1980), auf die als zusammenfassende Arbeit verwiesen werden kann. Bemerkenswerte komplizierte Cutinleisten als Verschlußmechanismus der Stomata von *Eucalyptus* beschreiben CARR und CARR (1979), als Nektardrüsen funktionierende geringfügig modifizierte Stomata untersuchen TEUBER et al. (1980) an *Medicago* und SCHAFFNER (1979) und *Cuscuta*. Auf das Vorkommen von Spaltöffnungsapparaten an Samenschalen *(Eschscholzia)* machen JERNSTEDT und CLARK (1979) aufmerksam.

Skulpturen der Cuticula im engeren Sinne (cuticulare Faltungsmuster) benutzt BAAGØE (1977) als taxonomisches Merkmal bei der Beschreibung der Ligula von Asteraceen. Die Entwicklungsgeschichte cuticularer Faltungsmuster wird von BARTHLOTT (1980) an Blütentrichomen von *Antirrhinum* untersucht. Es ergeben sich keinerlei Hinweise für einen Zusammenhang mit einer Über-Produktion von Cutin oder Wandspannungsdifferenzen als Ursache für die Entstehung solcher Skulpturen.

Einen kurzen allgemeinverständlichen Überblick der Feinstruktur und Stratifikation von Cutinschichten gibt WATTENDORF (1980), einige neuere vorwiegend transmissions-elektronenmikroskopische Arbeiten seien angefügt: ECKL und GRULER, 1980; ESPELIE et al., 1980; HOCH, 1979; SCHÖNHERR, 1976; SCHÖNHERR und ECKL, 1979. Die Rückseite der isolierten Cuticula zeigt im REM-Bild oftmals charakteristische Eigenstrukturen von großem systematischen Interesse und erlaubt meist eine Unterscheidung zwischen Angio- und Gymnospermen (BARTHLOTT, 1981); in einer umfangreicheren Arbeit werden von MIRANDA und CHAPHEKAR (1980) die Rückseiten isolierter Cuticulae von 82 Arten Pinaceae beschrieben. Die zahlreichen Arbeiten zum Feinbau der Cuticula lassen die Tendenz erkennen, daß pflanzliche Cuticulae offensichtlich wesentlich vielfältiger aufgebaut sind als früher angenommen und sich ihre

Stratifikation nicht auf ein einfaches einheitliches Schema reduzieren lässt.

Gut untersucht sind epidermale Vorkommen (intra- und subcuticular, intrazellulär) von Silikaten und Oxalaten, die sich besonders gut durch energiedispersive Röntgenanalysen darstellen lassen. SAKAI und THOM (1979), HAYWARD und PARRY (1980), LANNING et al. (1980) beschäftigen sich dabei meist mit den Silikaten der Poaceen, DENGLER und LIN (1980) mit der Kieselsäure in den Epidermen von *Selaginella*. FRANCESCHI und HORNER (1980) geben einen Überblick der Verbreitung von Calciumoxalat in Pflanzen.

Helicale Wandverdickungen an Epidermis-Außenwänden beschreiben LERSTEN und CURTIS (1977) an Trichomen einiger Rosaceen, Betulaceen und Rubiaceen, BARTHLOTT und ZIEGLER (1980, 1981) an den Samenschalen von Orchidaceen. Weitere Hinweise für das Vorkommen helicaler Strukturen an Samenschalen finden sich in den Arbeiten von GRUBERT (1980) und KAPIL et al. (1980).

Epicuticulare Wachse und ähnliche Sekrete wurden seit der grundlegenden Arbeit von JEFFREE et al. (1975) in zahlreichen Publikationen behandelt (vgl. auch WILKINSON, 1980). In einer umfangreicheren zusammenfassenden Arbeit werden mikromorphologische, chemische, taxonomische und einige ökologische Aspekte epicuticularer Wachse von BARTHLOTT und WOLLENWEBER (1981) dargestellt (dort auch Literatur-Übersicht).

Wachse können subcuticular, intracuticular und besonders auffällig als epicuticulare Beläge an Epidermen auftreten. Epicuticular-Wachse sind als dünne filmartige Schichten offensichtlich weit verbreitet, im REM-Bild dann aber kaum oder nicht nachweisbar. Häufig kristallisieren sie jedoch auf der Cuticula als Stäbchen, Fäden oder in anderen Formen aus und zeigen dann eine außerordentlich hohe mikromorphologische Diversität (z.B. Stäbchen-Kristalloide mit zusätzlicher mikrokristalliner Oberflächenskulptur). Chemisch handelt es sich um recht heterogene Substanzgruppen; cyclische Verbindungen (pentacyclische Triterpene, Flavonoide) scheinen epicuticular weiter verbreitet zu sein als bisher angenommen. Häufiger sind auch "wachsartige" Epicuticularskulpturen, die in organischen Lösungsmitteln gänzlich unlöslich sind und über deren chemische Natur nichts bekannt ist. Epicuticulare Wachse können wertvolle taxonomisch-systematische Merkmale sein, sie sind aber bis zu einem gewissen Grade abhängig von Umwelteinflüssen (vgl. z.B. RAO und RAJAREDDY, 1980).

b) Taxonomische Aspekte

Epidermale Merkmale können ausgezeichnete taxonomische Kriterien liefern und werden allgemein in zunehmendem Maße berücksichtigt (vgl. Neuauflage von METCALFE und CHALK, 1980, sowie HARDIN, 1980). Überraschenderweise sind sie relativ unabhängig von den Umweltbedingungen, unter denen eine Pflanze wächst. Evidenzen für eine strenge genetische Kontrolle über epidermale Merkmale haben CUTLER (1979) und CUTLER und BRANDHAM (1977) gefunden. Eine zusammenfassende Darstellung der taxonomisch-systematischen Signifikanz und Anwendbarkeit epidermaler Merkmale wurde an anderer Stelle ausführlicher gegeben (BARTHLOTT, 1981). Im folgenden sei eine Liste wichtiger neuer Arbeiten unter systematischen Aspekten gegeben (die Angiospermen-Familien sind rein alphabetisch geordnet), die eine schnelle Orientierung ergmöglicht:

Bryophyta: PROCTOR, 1979a,b; SCHÖNHERR und ZIEGLER, 1975; SHARP et al., 1979; BARTHLOTT und SCHULTZE-MOTEL, 1981.

Pteridophyta: DENGLER und LIN, 1980; SMITH, 1980; BARTHLOTT und WOLLENWEBER, 1981.

Gymnospermae: MIRANDA und CHAPHEKAR, 1980; CHABOT und CHABOT, 1977; LANGHAMMER und SCHULZE, 1977; FRANICH et al., 1977; PAGE, 1980.

Angiospermae: *Acanthaceae:* NAPP-ZINN und HEINS, 1979. *Asteraceae:* BAAGØE, 1977; COUTURIER und GODEAU, 1980; KARRFALT und KREITNER, 1980; NAPP-ZINN und EBLE, 1980; SEMPLE et al., 1980. *Bignoniaceae:* HESSE und MORAWETZ, 1980. *Bombacaceae:* O'DOWD, 1980. *Brassicaceae:* TITZ und SCHNATTINGER, 1980; HOLLOWAY et al., 1977a. *Bromeliaceae:* EHLER, 1977; NAPP-ZINN et al., 1978. *Cactaceae:* BARTHLOTT und VOIT, 1979. *Cannabidaceae:* TURNER et al., 1980. *Campanulaceae:* GESLOT, 1980. *Caryophyllaceae:* MELZHEIMER, 1980. *Celastraceae:* DEN HARTOG und BAAS, 1978. *Combretaceae:* STACE, 1980. *Cucurbitaceae:* JEFFREY, 1980. *Cuscutaceae:* HUTCHISON und ASHTON, 1979. *Crassulaceae:* T'HART und BERENDSEN, 1980; MANHEIM und MULROY, 1978; ELLER und WILLI, 1977. *Dichapetalaceae:* BOESEWINKEL und BOUMAN, 1980. *Dioncophyllaceae:* MARBURGER, 1979. *Erythroxylaceae:* BOESEWINKEL und GEENEN, 1980. *Euphorbiaceae:* DEHGAN, 1980. *Fagaceae:* HARDIN, 1979a,b. *Gentianaceae:* BOUMAN und SCHIER, 1979. *Goodeniaceae:* CAROLIN, 1980. *Hydatellaceae:* HAMANN et al., 1979. *Lamiaceae:* RUDALL, 1980. *Lentibulariaceae:* HESLOP-HARRISON und HESLOP-HARRISON, 1981. *Liliaceae:* CUTLER, 1977, 1979; CUTLER et al., 1980. *Linaceae:* BOESENWINKEL, 1980. *Melastomataceae:* JACQUES-FELIX und LEUENBERGER, 1980. *Meliaceae:* FABOYA et al., 1980. *Menispermaceae:* WILKINSON, 1978. *Mesembryanthemaceae:* HARTMANN, 1979; KRAMER, 1979. *Myrtaceae:* CARR und CARR, 1979; JOHNSON, 1980. *Musaceae:* NAPP-ZINN und FRANZ, 1980. *Nymphaeaceae:* ALEYKUTTY und INAMDAR, 1980; SCHNEIDER und FORD, 1978; COLLISON, 1980. *Onagraceae:* SEAVY et al., 1977; BOWMAN, 1980; DENFORD, 1980. *Orchidaceae:* ARDITTI et al., 1979, 1980; BARTHLOTT und ZIEGLER, 1980, 1981; HEALEY et al., 1980; SWANSON et al., 1980. *Papilionaceae:* HOLLOWAY et al., 1977b. *Papaveraceae:* CLARK und JERNSTEDT, 1978; JERNSTEDT und CLARK, 1979. *Polygalaceae:* VERKERKE und BOUMAN, 1980. *Pyrolaceae:* OLSON, 1980. *Poaceae:* BARTHLOTT und MARTENS, 1979; VIGNAL, 1979; SCHOLZ, 1979; SAKAI und THOM, 1979; BIANCHI et al., 1980; BAUM et al., 1980; TULLOCH et al., 1980; LANNING et al., 1980; HAYWARD und PARRY, 1980. *Ranunculaceae:* ALEYKUTTY et al., 1980. *Rosaceae:* LERSTEN und CURTIS, 1977. *Rutaceae:* BOESENWINKEL, 1978; ESPELIE et al., 1980. *Salicaceae:* BINNS und BLUNDEN, 1980; CURTIS und LERSTEN, 1980. *Solanaceae:* LYSHEDE, 1980. *Scrophulariaceae:* SUTTON, 1980; BARTHLOTT, 1980. *Turneraceae:* KLOOS und BOUMAN, 1980. *Zosteraceae:* BARNABAS, 1977.

c) Funktionell-ökologische Aspekte

Die Epidermis bzw. Cuticula/Epicuticula bildet die topographische und funktionelle Grenzschicht zwischen lebender Pflanze und ihrer Umwelt, über die jede Interaktion laufen muß. Die außerordentlich hohe strukturelle Diversität läßt nach der Funktion der Oberflächenskulpturen fragen. Ihre Funktionen können dabei so divers wie die Skulpturen selbst sein, wie JOHNSON (1975) am Beispiel der Trichome diskutiert hat. Wichtige Aspekte sind offensichtlich die Permeabilität der Oberfläche, Kontaminations-hemmende Eigenschaften, Temperaturkontrolle sowie inhibitorische bzw. toxische Eigenschaften der Oberflächenchemie auf Mikroorganismen. Es zeichnen sich hierbei besonders interessante Resultate ab (Diskussion einiger zusammenfassender Daten bei BARTHLOTT und WOLLENWEBER, 1981, und BARTHLOTT, 1981). Es ist zu erwarten, daß funktionelle Aspekte verstärkt untersucht werden und sich dabei wesentliche neue Erkenntnisse ergeben.

Oberflächen-Temperaturen hängen von einer großen Anzahl verschiedener Eigenschaften des betroffenen Organs ab und können taxonspezifisch am gleichen Standort äußerst verschieden sein (SMITH, 1978). Einen wesentlichen Einfluß auf Absorption/Reflexion der Globalstrahlung haben epicuticulare Sekrete, Trichome und andere Skulpturen, wie ELLER (1979) in einer ausführlichen Arbeit zusammenfassend darstellt (vgl. auch SCHULZE et al., 1980). Unter Insolation liegen Evidenzen für einen Zusammenhang zwischen Oberflächen-Skulptur ("Rauhigkeit") und Turbulenzbildung der umgebenden kühleren Luft und damit einer erhöhten Energie-Abgabe vor (GRACE, 1977, 1978; REICOSKY und HANOVER, 1976; BARTHLOTT, 1981; BARTHLOTT und WOLLENWEBER, 1981). Die Cactacee *Copiapoa* der chilenischen Küstenwüste besitzt nach MOONEY et al. (1977) einen Transpirationskühlungsmechanismus durch cuticular/epicuticular absorbiertes Wasser.

Die Permeabilität von Oberflächen scheint von einer Reihe verschiedenster Faktoren abzuhängen und ist nicht nur eine Funktion der Dicke der Cuticula oder der Präsenz/Absenz von Oberflächenwachsen (neuere Daten bei SCHÖNHERR, 1976; SCHÖNHERR et al., 1979; HASS und SCHÖNHERR, 1979; HOCH, 1979; ESPELIE et al., 1980; O'TOOLE et al., 1979; Übersicht über Oberflächenwachse und Permeabilität bei Pflanzen und Tieren bei HADLEY, 1980). LYSHEDE (1978) beschreibt Mikrokanäle in Epidermis-Außenwänden, die mit der Funktion der Wasseraufnahme gesehen werden. In einer bemerkenswerten Arbeit geht PROCTOR (1979) auf die Funktion von Moos-Oberflächen ein, die häufig infolge des mangelhaft ausgebildeten inneren Leitsystems externe Oberflächen-Leitsysteme für Wasser ausbilden. Ein solches strukturell sehr eigentümliches Oberflächen-Leitsystem wird von BARTHLOTT und SCHULTZE-MOTEL (1981) für die Hedwigiaceae beschrieben. Mit Fragen der schweren Benetzbarkeit mit Wasser beschäftigen sich SCHÖNHERR und ZIEGLER (1975) an den Atemporen von Lebermoosen. Auf den bisher nicht genügend berücksichtigten bzw. nur unzureichend bekannten Zusammenhang zwischen schwerer Benetzbarkeit und verringerter Kontaminationsfähigkeit von Oberflächen gehen RENTSCHLER (1973, 1977) und BARTHLOTT und WOLLENWEBER (1981) ein; weitere Fragen der Kontamination werden von ELLWOOD et al. (1979), SHARMA et al. (1980) und BITTON und MARSHALL (1980) behandelt. Einigen Oberflächen-Substanzen scheint eine inhibitorische bzw. toxische Wirkung auf pathogene Mikroorganismen zuzukommen (CAMPBELL et al., 1980; HARBORNE, 1980; WEINHOLD und HANCOCK, 1980).

Bei speziellen Anpassungen ist der Zusammenhang zwischen Struktur und Funktion gelegentlich relativ einfach. So etwa bei den "Fallstrukturen" mancher Insektivoren (ADAMS und SMITH, 1977) oder spezialisierten Trichomen (z.B. die "pearl bodies" bei *Ochroma*, die von Ameisen eingesammelt werden: O'DOWD, 1980). STAR (1980) glaubt, daß den epicutularen Flavonoid-Sekreten der Blätter von *Pityrogramma* eine allelopathische Wirkung zuzuschreiben sei. KRULIK (1980) diskutiert im Zusammenhang mit den optischen Eigenschaften der "Fensterblätter" bestimmter Sukkulenten die Funktion ihrer Oberflächenskulpturen; BREHM und KRELL (1975) untersuchen den Zusammenhang zwischen Flavonoid-Gehalt der Epidermiszellen von Blütenblättern und der UV-Reflexion.

Als Überleitung zum folgenden Abschnitt über Pollen seien einige neuere Arbeiten angefügt, die sich mit der mikromorphologischen Interaktion zwischen Narben-Oberfläche und Pollen beschäftigen: HESLOP-HARRISON, 1979; HESLOP-HARRISON und SHIVANNA, 1977; SHIVANNA et al., 1978; STEAD et al., 1979; SCHMID, 1980; TILTON und HORNER, 1980.

3. Pollen

Die Bemühungen einer Vereinheitlichung und Präzisierung der Sporoderm-Terminologie gehen weiter. Beispiele dafür sind PRAGLOWSKI und RAJ (1979) (Oncus, Verruca, Annulus, Pseudoannulus, prolat, oblat und dem Begriffspaar Skulptur und Struktur) wie auch STRAKA (1975), FAEGRI (1978, Polarachse), STRAKA (1980, Ektexine). NILSSON und MULLER (1978) berichten über empfohlene und nicht empfohlene Termini im Rahmen der ICP Arbeitsgruppe.

Erfreulich für den zu berichtenden Zeitraum ist, daß die kombinierte Untersuchung mit Lichtmikroskop (LM), Transmissions-Elektronenmikroskop (TEM) und Raster-Elektronenmikroskop (REM) auf dem Gebiet der Pollenmorphologie beinahe zur Regel geworden ist. Nur so ist eine optimale Darstellung der mannigfaltigen Strukturen und Skulpturen in diesen Größenordnungen möglich. Entsprechend dazu sind dann auch die Arbeiten, die sich mit Präparationsmethoden befassen. DAMBLON (1975) betont nochmals die Überlegenheit des "Sputtering", LEUENBERGER und SCHILL (1975) weisen auf mögliche Bedampfungsartefakte hin. Eine Kritische-Punkt-Trocknung für alle Pollenuntersuchungen, wie von LYNCH und WEBSTER (1975) empfohlen, muß aufgrund unserer langjährigen Erfahrung allerdings abgelehnt werden. ROWLEY und JARAI-KOMLODI (1976) präparieren ein und dasselbe Pollenkorn für alle drei Mikroskoparten, PACINI et al. (1980) gleichzeitig für das TEM und REM. BRISSON und PETERSON (1975) sowie FERGUSON (1978a) schildern die Möglichkeiten von in Methacrylat bzw. Epoxyharz eingebettetem Material auch für die raster-elektronenmikroskopische Untersuchung.

Im Bereich der Exine-Entwicklung und der Aufklärung von Exinefeinstrukturen gibt es eine Fülle neuer Arbeiten. Bei einigen Zingiberales, Bsp. *Heliconia*, scheint die Exine, wie KRESS et al. (1978) zeigen, auf kleinste Fortsätze und eine nur 0,08 µm dicke Schicht reduziert zu sein, ein exineloser Pollen?

TAKAHASHI (1980) demonstriert die Bildung reticulater Skulpturen an *Hemerocallis*-Pollen. SKVARLA und NOWICKE (1976, 1979) untersuchen die Exine bei *Nigella* und einigen Centrospermen-Familien, BEHNKE (1977) speziell bei den Gattungen *Gisekia*, *Limeum* und *Hectorella*. Des weiteren SKVARLA et al. (1976, Onagraceae), STAINIER und HORVATH (1978a,b, *Phaseolus*). KÖHLER und LANGE (1979) unterscheiden aufgrund von Exinemerkmalen Wild- und Kultur-Getreide, ROGERS (1979, 1980) findet einen Pollendimorphismus bei distylen *Linum*-Arten, HANSEN (1980) bei *Lophophytum*.

HESSE (1978a,b, 1979a,b,c, 1980a-d) zeigt die Entwicklungsgeschichte und Ultrastruktur des Pollenkitts von über einem Dutzend verschiedener Familien der Mono- und Dikotylen: Er schildert dabei die Rolle des Tapetums und die Wanderung der kleinen Tröpfchen in die Exinehohlräume. Die Entstehung und Auswirkungen der unterschiedlichen Pollenklebrigkeit und Fragen der Anheftung des Pollens an blütenbesuchende Insekten mittels Pollenkitt ergänzen das Spektrum. Pollenkörner ohne Pollenkitt finden sich bei *Gnetum* (HESSE, 1980a). Eine Übersicht der Chemie des Sporopollenins geben BROOKS und SHAW (1978), MAY et al. (1975) speziell die Analyse von solchem bei zwei Vertretern der Gattung *Ambrosia*. Aus Sporopollenin sind nach SKVARLA et al. (1978) die an der Oberfläche der Exine haftenden Viscin-Fäden (threads) bei den Onagraceae. In einer sehr interessanten und ausführlichen, ausgezeichnet illustrierten Studie, werden dabei 17 Gattungen mit 148 Arten der Familie untersucht. Mit Ausnahme weniger Vertreter der Circeae sind die Fäden für die Familie typisch. Sie sind ca. 2000 µm lang und 0,04 µm dick und lassen sich in drei morphologische Typen unterscheiden: glatte, aus glatten Untereinheiten zusammengesetzte und segmentierte. Neben der taxonomischen Verwertbarkeit wird auch die Entwicklung diskutiert. Die glatten sollen dabei die ursprünglichen, die anderen Anpassungen an einen wirksameren Transport der Bestäuber darstellen. Eine kurze Übersicht über die Viscin-Fäden der Ericaceae rundet die Arbeit ab. VIJAYARAGHAVAN und SHUKLA (1980) finden bei der Orchideengattung *Zeuxine* ebenfalls solche Fäden.

Von den zahlreichen, meist reich illustrierten, palynologischen Veröffentlichungen mit systematisch-taxonomischem Hintergrund werden zunächst Beispiele einzelner Arten und Gattungen, dann die monographischen Bearbeitungen ganzer Familien vorgestellt.

Aus den Apocynaceae bearbeitet COUSIN (1979) *Vinca*, CAMPO et al. (1979) *Tabernaemontana* s.l.. Araliaceae: *Schefflera* (TSENG und SHOUP, 1978). Asteraceae: *Urospermum* (LACK und LEUENBERGER, 1979); die westindischen *Veronica*-Arten (KEELEY und JONES, 1977). Bombacaceae: *Patinoa* (NILSSON und ROBYNS, 1975). Buxaceae: die kubanischen Vertreter von *Buxus* (KÖHLER, 1979). Caesalpiniaceae (GRAHAM et al., 1980). Campanulaceae: BADRE et al. (1975) über die Gattung *Berenice*. Chenopodiaceae: *Suaeda* (BASSETTE und CROMPTON, 1978). Euphorbiaceae: *Phyllanthus* (BOR, 1979). Fabaceae: *Cranocarpus* (FERGUSON und SKVARDA, 1979). Haloragaceae: *Myriophyllum* (AIKEN, 1978; MATHEWES, 1978). Hypericaceae: *Hypericum* (CLARKE, 1975). Linaceae: *Cinum* (XAVIER et al., 1980). Menispermaceae: Coscinieae (FERGUSON, 1978); Moraceae (NIEZGODA und NOWACZYK, 1976: *Acanthinophyllum*, *Clarisia*, *Soroceae* und *Trophis*; PUNT, 1978a: *Scyphosyce* und *Dorstenia*). Myoporaceae: *Ranopisoa* (LOBREAU-CALLEN, 1978). Papaveraceae: Hinweis auf die große Variabilität hinsichtlich Anzahl und Position der Aperturen innerhalb der Gattung *Argemone* von LAYKA (1976). Polemoniaceae: *Collomia* (CHUANG et al., 1978). Proteaceae: *Bleasdalea* (HAAS, 1975). Rosaceae: *Crataegus* und *Mespilus* (BYATT, 1976) bzw. *Potentilla* (EIKELAND, 1979). Solanaceae: *Solanum* (ANDERSON und GENSEL, 1976). Urticaceae: *Hesperoenide* und *Urtica* (WOODLAND et al., 1976).

Bei den Familienbearbeitungen müssen zunächst die im Rahmen der "World Pollen and Spore Flora" erschienenen genannt werden: Juglandaceae (STONE and BROOME, 1975); Schisandraceae (PRAGLOWSKI, 1976); Cornaceae (FERGUSON, 1977); Winteraceae (PRAGLOWSKI, 1979); Entolomataceae (PEGLER und YOUNG, 1978). Auch weitere Lieferungen der NW-europäischen Pollenflora sind erschienen (vgl. Progr. Botany 40, 59).

Außerdem seien erwähnt: Aceraceae (BIESBOER, 1975); Apiaceae (ROLAND-HEYDACKER und CERCEAU-LARRIVAL, 1978); Amarathaceae (ZANDONELLA und LECOCQ, 1977); Annonaceae (THOMAS und LUGARDON, 1976); Asclepiadaceae (SCHILL und JÄKEL, 1978, umfassend; DANDICKO-ZAFIMAHOVA, 1978, *Calotropis*); Bignoniaceae (BUURMAN, 1977); Campanulaceae (DUNBAR, 1975a,b); Caryophyllaceae (IWARDSSON, 1977); Cistaceae (RIVAS, 1979); Cneoraceae (LOBREAU-CALLEN et al., 1981); Connaraceae (DICKINSON, 1979); Flacourtiaceae (KEATING, 1975); Orchidaceae (WILLIAMS und BROOME, 1976; NEWTON und WILLIAMS, 1978; ACKERMAN und WILLIAMS, 1980; CASPERS und CASPERS, 1976; JONSSON, 1979; BALOGH, 1979: neben neuen Pollendaten, z.B. Aperturen, immer deren Verwertbarkeit für die Systematik der Gruppe); Polemoniaceae (TAYLOR und LEVIN, 1975); Urticaceae (HAMILTON, 1976); Valerianaceae (PATEL und SKVARLA, 1979).

Zu einigen neueren Arbeiten mit phylogenetischem Hintergrund, werden mit Hilfe der Pollenmorphologie Evolutionstendenzen herausgearbeitet.

Verwandschaftliche Zusammenhänge zwischen höheren Kategorien, Familien, Ordnungen, studieren NOWICKE und SKVARLA (1979). In einer sehr ausführlichen Untersuchung (Pollen von 650 Arten aus 25 Familien) der Ordnungen Centrospermae, Plumbaginales, Polygonales, Primulales, Ranunculales und Dilleniales wird auf den engeren oder weiteren Zusammenhang einzelner Vertreter innerhalb des Komplexes hingewiesen: z.B. Ausschluß der Batidaceae, Gyrostemonaceae und Theligonaceae aus den Centrospermae; Verneinung von engen Verwandschaftsbeziehungen der Plumbaginaceae, Polygonaceae oder Primulaceae zu den Centrospermae bzw. untereinander. Dasselbe gilt auch für die Familien der Ranunculales; Herausstellung des besonderen Status von *Berberis* und *Mahonia*. NAIR postuliert drei voneinander unabhängige Entwicklungsäste der Angiospermen, den der Monocotylen und zwei der Dicotylen (Magnoliales/Ranales). Innerhalb der Loganiaceae weist PUNT (1978b) auf bestimmte Entwicklungstrends hin; CLARKE (1975) studiert die Verwandschaftsverhältnisse bei den Valerianaceae, ENDRESS und HONEGGER (1980) bei den Austrobaileyaceae.

Literatur

ACKERMAN, J.D., WILLIAMS, N.H.: Grana 19, 7-18 (1980). - ADAMS, R.M.; SMITH, G.W.: Am. J. Bot. 64, 265-272 (1977). - AIKEN, S.G.: Can. J. Bot. 56, 976-982 (1978). - ALEYKUTTY, K.M., INAMDAR, J.A.: Fedd. Rep. 91, 95-108 (1980). - ANDERSON, G.J., GENSEL, P.G.: Pollen Spores 18, 533-552 (1976). - ARDITTI, J., MICHAUD, J.D., HEALEY, P.L.: Am. J. Bot. 66, 1128-1137 (1979); - Am. J. Bot. 67, 347-360 (1980).

BAAGØE, J.: Bot. Tidsskr. 71, 193-224 (1977), 72, 125-147 (1978). - BADRE, F., CADET, T., CUSSET, G., HIDEUX, M.: Adansonia, ser. 2, 15, 139-146 (1975). - BALOGH, P.: Orquidea (Mex.) 7, 241-260 (1979). - BARNABAS, A.D.: Z. Pflanzenphysiol. 85, 417-427. - BASSETTE, I.J., CROMPTON, C.W.: Can. J. Bot. 56, 581-591 (1978). - BAUM, B.R., TULLOCH, A.P., BAILEY, L.G.: Can. J. Bot. 58, 2467-2480 (1980). - BARTHLOTT, W.: Ber. Dtsch. Bot. Ges. 93, 379-390 (1980); - Nord. J. Bot. 1, 345-355 (1981). - BARTHLOTT, W., MARTENS, B.: Trop. Subtrop. Pflanzenw. 30, 1-113 (1979). - BARTHLOTT, W., SCHULTZE-MOTEL, W.: Willdenowia 11, 3-11 (1981). - BARTHLOTT, W., VOIT, G.: Plant Syst. Evol. 132, 205-229 (1979). - BARTHLOTT, W., WOLLENWEBER, E.: Trop. Subtrop. Pflanzenw. 32, 1-67 (1981). - BARTHLOTT, W., ZIEGLER, B.: Ber. Dtsch. Bot. Ges. 93, 391-403 (1980); - Ber. Dtsch. Bot. Ges. 94, 270-282 (1981). - BEHNKE, H.-D.: Plant Syst. Evol. 128, 227-235 (1977). - BIANCHI, G., LUPOTTO, E., BORGHI, B., CORBELLINI, M.: Planta 148, 328-331 (1980). - BINNS, W.W., BLUNDEN, G.: Bot. J. Linn. Soc. 81, 205-214 (1980). - BIESBOER, D.D.: Grana 15, 19-27 (1975). - BITTON, G., MARSHALL, K.C. (Eds.): Adsorption of Microorganisms to Surfaces. New York: Wiley 1980. - BLECKMANN, C.A., HULL, H.M., HOSHAR, R.W.: Bot. Gaz. 141, 1-8 (1980). - BOESEWINKEL, F.D.: Acta Bot. Neerl. 27, 341-354 (1978); - Acta Bot. Neerl. 29, 17-32 (1980). - BOESEWINKEL, F.D., BOUMAN, F.: Acta Bot. Neerl. 29, 103-115 (1980). - BOESEWINKEL, F.D., GEENEN, J.: Acta Bot. Neerl. 29, 231-241 (1980). - BOR, J.: Rev. Palaeobot. Palynol. 27, 149-172 (1979). - BOUMAN, F., SCHIER, S.: Acta Bot. Neerl. 28, 467-478 (1979). - BOWMAN, R.N.: Am. J. Bot. 67, 671-686 (1980). - BREHM, B.G., KRELL, D.: Science 190, 1221-1223 (1975). - BRISSON, J.D., PETERSON, R.L.: Microscop. Soc. Can. 2, 64-65 (1975). - BROOKS, J., SHAW, G.: Grana 17, 91-97 (1978). - BUURMAN, J.: Pollen Spores 19, 447-519 (1977). - BYATT, J.I.: Pollen Spores 18, 335-349 (1976).

CAMPBELL, C.L., HUANG, J.-S., PAYNE, G.A.: Defense at the perimeter: the outer walls and the gates, 103-120. In: Plant Disease, Vol. 5, ed. J.G. HORSFALL, E.B. COWLING. London: Academic Press 1980. - CAMPO, M., NILSSON, S., LEEUWENBERG, A.J.M.: Grana 18, 5-14 (1979). - CAROLIN, R.C.: Aust. J. Bot. 28, 123-137 (1980). - CASPERS, N., CASPERS, L.: Pollen Spores 18, 203-215 (1976). - CARR, S.G.M., CARR, D.J.: Ann. Bot. 44, 239-243 (1979). - CHABOT, J.F., CHABOT, B.F.: Can. J. Bot. 55, 1064-1075 (1977). - CHUANG, T.J., HSIEH, W.C., WILKEN, D.H.: Am. J. Bot. 65, 450-458 (1978). CLARK, C., JERNSTEDT, J.A.: Syst. Bot. 3, 386-402 (1978). CLARKE, G.C.S.: Grana 15, 117-125 (1975). - CLARKE, G.: Grana 17, 61-75 (1978). - COLLINSON, M.E.: Ann. Bot. 46, 603-632 (1980). - COUSIN, M.T.: Grana 18, 115-128 (1979). - COUTURIER, A., GODEAU, M.: Rev. Gen. Bot. 87, 297-335 (1980). - CUTLER, D.F.: Trop. Subtrop. Pflanzenw. 28, 1-32 (1979). - CUTLER, D.F., BRANDHAM, P.E.: Kew Bull. 32, 23-42 (1977). - CUTLER, D.F., BRANDHAM, P.E., CARTER, S., HARRIS, S.J.: Bot. J. Linn. Soc. 80, 293-317 (1980). - CURTIS, J.D., LERSTEN, N.R.: Am. J. Bot. 67, 1289-1296 (1980).

DAMBLON, F.: Grana 15, 137-144 (1975). - DAN DICKO-ZAFIMAHOVA, L.: Adansonia, ser. 2, 17, 455-463 (1978). - DEHGAN, B.: Bot. J. Linn. Soc. 80, 257-278 (1980). - DENFORD, K.E.: Experientia 36, 299-300

(1980). - DENGLER, N.G., LIN, E.Y.-G.: Can. J. Bot. 58, 2459-2466 (1980). DEN HARTOG, R.M., BAAS, P.: Acta Bot. Neerl. 27, 355-388 (1978). - DICKISON, W.C.: Pollen Spores 21, 31-79 (1979). - O'DOWD, D.J.: Am. J. Bot. 67, 543-549 (1980). - DUNBAR, A.: Bot. Notiser 128, 73-101 (1975a); - Bot. Notiser 128, 102-118 (1975b).

ECKL, K., GRULER, H.: Planta 150, 102-113 (1980). - EHLER, N.: Trop. Subtrop. Pflanzenw. 20, 1-40 (1977). - EIKELAND, F.E.: Grana 18, 189-192 (1979). - ELLER, B.M.: Flora 168, 146-192 (1979). - ELLER, B.M., WILLI, P.: Flora 166, 461-474 (1977). ELLWOOD, D.C., MELLING, J., RUTTER, P.R.: Adhesion of Microorganisms to Surfaces. London: Academic Press (1979). - ENDRESS, P.K.: Plant Syst. Evol. 134, 79-120 (1980). - ENDRESS, P.K., HONEGGER; R.: Grana 19, 177-183 (1980). - ESPELIE, K.E., DAVIS, R.W., KOLATTUKUDY, P.E.: Planta 149, 498-511 (1980).

FABOYA, O.-O.P., OKOGUN, J.J., GODDARD, D.R.: Phytochemistry 19, 1226-1228 (1980). - FAEGRI, K.: Grana 17, 15-16 (1978). - FAHN, A.: Secretory Tissues in Plants. London: Academic Press 1979. - FERGUSON, I.K.: World Pollen Spores Flora, Vol. 6. Stockholm: Almqüist & Wikrell (1977); - Ann. Mines Belgiques, 2e horaison, 33-37 (1978a); - Kew Bull. 32, 339-346 (1978b). - FERGUSON, I.K., SKVARLA, J.J.: Grana 18, 15-20 (1979). - FRANCESCHI, V.R., HORNER, H.T., Jr.: Bot. Review 46, 361-427 (1980). - FRANICH, R.A., WELLS, L.G., BARNETT, J.R.: Ann. Bot. 41, 621-626 (1977).

GERSTERBERGER, P., LEINS, P.: Ber. Dtsch. Bot. Ges. 91, 381-387 (1978). - GESLOT, A.: Adansonia, ser. 2, 19, 307-318 (1980). - GRACE, J.: Plant Response to Wind. London: Academic Press 1977; - Plant, Cell, Environment 1, 35-38 (1978). - GRAHAM, A., BARKER, G., DA SILVA, M.F.: Grana 19, 79-84 (1980). - GRUBERT, M.: Plant Syst. Evol. 135, 137-149 (1980).

HAAS, E.J.: Pollen Spores 17, 213-222 (1975). - HAAS, K., SCHÖNHERR, J.: Planta 146, 399-403 (1979). - HADLEY, N.F.: Am. Scientist 68, 546-553 (1980). - HAMANN, U., KAPLAN, K., RÜBSAMEN, T.: Bot. Jahrb. Syst. 100, 555-569 (1979). - HAMILTON, A.C.: Pollen Spores 18, 28-66 (1976). - HANSEN, B.: Grana 19, 189-191 (1980). - HARBORNE, J.B.: Plant phenolics, 329-402. In: Secondary Plant Products, eds. E.A. BELL, B.V. CHARLWOOD (Encyclopedia of Plant Physiology, Vol. 8). Berlin-Heidelberg-New York: Springer 1980. - HARDIN, J.W.: Am. J. Bot. 66, 576-585 (1979a); - Bull. Torrey Bot. Club 106, 313-325 (1979b); - Proceed. Dendrology East. Deciduous Forest Biome (Virginia Polytechn. Inst. State Univ., Blacksburg), 17-39 (1980). - t`Hart, H., BERENDSEN, W.: Plant Syst. Evol. 135, 107-117 (1980). - HARTMANN, H.: Trop. Subtrop. Pflanzenwelt 28, 33-55 (1979). - HAYAT, M.A.: Introduction to Biological Scanning Electron Microscopy. Baltimore: University Park Press 1978. - HAYWARD, D.M., PARRY, D.W.: Ann. Bot. 46, 541-548 (1980). - HEALEY, P.L., MICHAUD, J.D., ARDITTI, J.: Am. J. Bot. 67, 508-518 (1980). - HESLOP-HARRISON, J.: Ann. Bot. 44, Supl. 1, 1-47 (1979). - HESLOP-HARRISON, Y., HESLOP-HARRISON, J.: Ann. Bot. 47, 293-319 (1981). - HESLOP-HARRISON, Y., SHIVANNA, K.R.: Ann. Bot. 41, 1233-1258 (1977). - HESSE, M.: Plant Syst. Evol. 129, 13-30 (1978a); Plant Syst. Evol. 130, 13-42 (1978b); - Plant Syst. Evol. 131, 277-289 (1979a); - Plant Syst. Evol. 132, 107-139 (1979b); - Pollen Spores 21, 399-413 (1979c); - Plant Syst. Evol. 134, 229-267 (1980a); - Plant Syst. Evol. 135, 253-263 (1980b); - Plant Syst. Evol. 133, 135-148 (1980c); Plant Syst. Evol. 136, 41-46 (1980d). - HESSE, M., MORAWETZ, W.: Plant Syst. Evol. 135, 1-10 (1980). - HOCH, H.C.: Planta 147, 186-195 (1979). - HOLLOWAY, P.J., BROWN, G.A., BAKER, F.A., MACEY, M.J.K.: Chem. Phys. Lipids 19, 114-127 (1977a). - HOLLOWAY, P.J., HUNT, G.M., BAKER, E.A., MACEY, M.J.K.: Chem. Phys. Lipids 20, 141-155 (1977b). - HORVAT, F., STAINIER, F.: Pollen Spores 21, 17-30 (1979). - HUTCHISON, J.M., ASHTON, F.M. Am. J. Bot. 66, 40-46 (1979).

IWARDSSON, M.: Grana 16, 15-22 (1977).

JACQUES-FELIX, H., LEUENBERGER, B.E.: Willdenowia 10, 27-31 (1980). - JEFFREE, C.E., BAKER, E.A., HOLLOWAY, P.J.: New Phytologist 75, 539-549 (1975). - JEFFREY, C.: Bot. J. Linn. Soc. 81, 233-247 (1980). - JERNSTEDT, J.A., CLARK, C.: Am. J. Bot. 66, 586-590 (1979). - JOHNSON, C.T.: Austr. J. Bot. 28, 77-104 (1980). - JOHNSON, H.B.: Bot. Rev. 41, 233-258 (1975). - JONSSON, L.: Bot. Notiser 132, 511-519 (1979).

KAPIL, R.N., BOR, J., BOUMAN, F.: Bot. Jahrb. Syst. 101, 555-573 (1980). - KARRFALT, E.E., KREITNER, G.L.: Can. J. Bot. 58, 61-67 (1980). - KEATING, R.C.: Grana 15, 29-49 (1975). - KEELEY, S.C., JONES, S.B.: Am. J. Bot. 64, 576-584 (1977). - KLOOS, A., BOUMAN, F.: Beitr. Biol. Pflanzen 55, 49-66 (1980). - KÖHLER, E.: Wiss. Z. Friedrich Schiller-Univ. Jena, Math. Naturwiss. Reihe 4, 683-689 (1979). - KÖHLER, E., LANGE, E.: Grana 18, 133-140 (1979). - KRAMER, D.: Flora 168, 193-204 (1979). - KRESS, W.J., STONE, D.E., SELLERS, S.C.: Am. J. Bot. 65, 1064-1076 (1978). - KRULIK, G.A.: Can. J. Bot. 58, 1591-1600 (1980).

LACK, H.W., LEUENBERGER, B.: Pollen Spores 21, 415-425 (1979). - LANGHAMMER, L., SCHULZE, G.: Willdenowia 8, 41-48 (1977). - LANNING, F.C., HOPKINS, T.L., LOERA, J.C.: Ann. Bot. 45, 549-554 (1980). - LAYKA, S.: Pollen Spores 18, 351-375 (1976). - LEINS, P., ERBAR, C.: Beitr. Biol. Pflanzen 55, 11-22 (1980). - LERSTEN, N.R., CURTIS, J.D.: Can. J. Bot. 55, 128-132 (1977). - LEUENBERGER, B., SCHILL, R.: Pollen Spores 17, 611-615 (1975). - LOBREAU-CALLEN, D.: Adansonia, ser. 2, 18, 237-241 (1978). - LOBREAU-CALLEN, D., NILSSON, S., ALBERS, F., STRAKA, H.: Grana 17, 125-139 (1978). - LYNCH, S.P., WEBSTER, G.L.: Grana 15, 127-136 (1975). - LYSHEDE, O.B.: New Phytol. 80, 421-426 (1978a); - Am. J. Bot. 46, 519-526 (1980b).

MANHEIM, B.S., MULROY, T.W.: Phytochemistry 17, 1799-1800 (1978). - MARBURGER, J.E.: Am. J. Bot. 66, 404-411 (1979). - MATHEWES, R.W.: Can. J. Bot. 56, 1372-1380 (1978). - MAY, G., SOUTHWORTH, D., DICKINSON, D.: Grana 15, 149-153 (1975). - MELZHEIMER, V.: Bot. Jahrb. Syst. 101, 153-190 (1980). - METCALFE, C.R., CHALK, L.: Anatomy of the Dicotyledons, Vol. I, 2nd ed. Oxford: Oxford Univ. Press 1980. - MIRANDA, V., CHAPHEKAR, M.: Bot. J. Linn. Soc. 81, 61-78 (1980). - MOONEY, H.A., WEISSER, P.J., GULMON, S.L.: Flora 166, 117-124 (1977).

NAIR, P.K.K.: Grana 18, 141-144 (1979). - NAPP-ZINN, K., EBLE, M.: Plant Syst. Evol. 136, 169-207 (1980). - NAPP-ZINN, K., FRANZ, A.: Flora 170, 565-569 (1980). - NAPP-ZINN, K., HEINS, W.: Trop. Subtrop. Pflanzenwelt 26, 1-53 (1979). - NAPP-ZINN, K., SCHMIDT, R., GENSCHER, H.: Trop. Subtrop. Pflanzenwelt 24, 1-87 (1978). - NEWTON, G.D., WILLIAMS, N.H.: Selbyana 2, 169-182 (1978). - NIEZGODA, C.J., NOWACZYK, J.: Pollen Spores 18, 513-522 (1976). - NILSSON, S., MULLER, J.: Grana 17, 55-58 (1978). - NILSSON, S., ROBYNS, A.: Pollen Spores 17, 59-66 (1975). - NOWICKE, J.W., SKVARLA, J.J.: Ann. Missouri Bot. Gard. 66, 633-700 (1979).

OLSON, A.R.: Am. J. Bot. 67, 968-974 (1980). - OHNSORGE, J., HOLM, R.: Rasterlektronenmikroskopie. - Eine Einführung für Mediziner und Biologen. Stuttgart: Thieme 1978.

PACINI, E., CIAMPOLINI, F., CRESTI, M.: Grana 19, 193-195 (1980). - PAGE, C.N.: Plant Syst. Evol. 135, 71-79 (1980). - PATEL, V.C., SKVARLA, J.J.: Pollen Spores 21, 81-103 (1979). - PEGLER, D.N., YOUNG, T.W.K.: World Pollen Spore Flora, Vol. 7. Stockholm: Almquist & Wiksell 1978. - PRAGLOWSKI, J.: World Pollen Spore Flora, Vol. 5. Stockholm: Almquist & Wiksell 1976; - World Pollen Spore Flora, Vol. 8. Stockholm:

Almquist & Wiksell 1979. - PRAGLOWSKI, J., RAJ, B.: Grana 18, 109-113 (1979). - PROCTOR, M.C.F.: Structure and eco-physiological adaptations in Bryophytes. In: Bryophyte Systematics, eds. G.C.S. CLARKE, J.G. DUCKETT. London: Academic Press 1979a; - J. Bryol. 10, 531-538 (1979b). - PUNT, W.: Grana 17, 77-79 (1978a); - Rev. Palaeobot. Palynol. 26, 313-335 (1978b).

RAO, J.V.S., RAJAREDDY, K.: Indian J. Exp. Biol. 18, 495-500 (1980). - REICOSKY, D.A., HANOVER, J.W.: Am. J. Bot. 63, 449-456 (1976). - REIMER, L., PFEFFERKORN, G.: Raster-Elektronenmikroskopie. Berlin-Heidelberg-New York: Springer 1978. - RENTSCHLER, J.: Proceed. 3rd Int. Clean Air Congr., A 139-A 142 (1973); - Proceed. 4th Int. Clean Air Congr., 99-102 (1977). - RIVAS, C.S.: Grana 18, 91-98 (1979). - ROGERS, C.M.: Plant Syst. Evol. 131, 127-132 (1979); - Grana 19, 19-20 (1980). - ROLAND-HEYDACKER, F., CERCEAU-LARRIVAL, M.-T.: Grana 17, 81-89 (1978). - ROWLEY, J.R., JARAIKOMLODI, M.: Acta Bot. Acad. Sci. Hung. 22, 449-461 (1976). - RUDALL, P.: Bot. J. Linn. Soc. 80, 319-340 (1980).

SAKAI, W.S., THOM, M.: Ann. Bot. 44, 245-248 (1979). - SCHAFFNER, G.: Ber. Dtsch. Bot. Ges. 92, 721-729 (1979). - SCHILL, R., JÄKEL, U.: Trop. Subtrop. Pflanzenwelt 22, 1-122 (1978). - SCHMID, R.: Bot. Soc. Am. Misc. Ser. 158, 100 (1980). - SCHNEIDER, E.L., FORD, E.G.: Bull. Torrey Bot. Club 105, 192-200 (1978). - SCHOLZ, H.: Willdenowia 8, 511-515 (1979). - SCHÖNHERR, J.: Planta 131, 159-164 (1976). - SCHÖNHERR, J., ZIEGLER, H.: Planta 124, 51-60 (1975). - SCHÖNHERR, J., ECKL, K., GRULER, H.: Planta 147, 21-26 (1979). - SCHULZE, E.-D., ELLER, B.M., THOMAS, D.A., WILLERT, D.J.v., BRINCKMANN, E.: Oecologia 44, 258-262 (1980). - SEAVEY, S.R., MAGILL, R.E., RAVEN, P.H.: Ann. Missouri Bot. Gard. 64, 18-47 (1977). - SEMPLE, J.C., BLOK, V.C.: Can. J. Bot. 58, 147-163 (1980). - SHARMA, G.K., CHANDLER, C., SALEMI, L.: Ann. Bot. 45, 77-80 (1980). - SHARP, A.J., GARDNER, R., GIESY, M.: Phytomorphology 28, 329-331 (1979). - SHIVANNA, K.R., HESLOP-HARRISON, Y., HESLOP-HARRISON, J.: Acta Bot. Neerl. 27, 107-109 (1978). - SKVARLA, J.J., NOWICKE, J.W.: Plant Syst. Evol. 126, 55-78 (1976); - Am. J. Bot. 66, 162-165 (1979). - SKVARLA, J.J., RAVEN, P.H., PRAGLOWSKI, J.: Linn. Soc. Symp. Ser. 1, 447-479 (1976). - SKVARLA, J.J., RAVEN, P.H., CHISSOE, W.F., SHAPP, M.: Pollen Spores 20, 5-143 (1978). - SMITH, P.M.: Bull. Torrey Bot. Club 107, 134-145 (1980). - SMITH, W.K.: Science 201, 614-616 (1978). - STACE, C.A.: Bot. J. Linn. Soc. 81, 185-203 (1980). - STAINIER, F., HORVAT, F.: Pollen Spores 20, 196-214 (1978a); - Pollen Spores 20, 341-349 (1978b). - STAR, A.E.: Bull. Torrey Bot. Club 107, 146-153 (1980). - STEAD, A.D., ROBERTS, J.N., DICKINSON, H.G.: Planta 146, 211-216 (1979). - STRAKA, H.: Soc. Bot. Fr. Coll. Palynolog. 19-44 (1975); - Grana 19, 157-158 (1980). - SUTTON, D.A.: Bot. J. Linn. Soc. 81, 169-184 (1980). - SWANSON, E.S., CUNNINGHAM, W.P., HOLMAN, R.T.: Am. J. Bot. 67, 784-789 (1980).

TAKAHASAI, M.: Grana 19, 3-5 (1980). - TAYLOR, T.N., LEVIN, D.A.: Grana 15, 91-112 (1975). - TEUBER, L.R., ALBERTSEN, M.C., BARNES, D.K., HEICHEL, G.H.: Am. J. Bot. 67, 433-439 (1980). - THEOBALD, W.L., KRAKULIK, J.L., ROLLINS, R.C.: Trichome description and classifications. In: Anatomy of the Dicotyledons, Vol. I, 2nd ed., eds. C.R. METCALFE, L. CHALK. Oxford: Oxford Univ. Press 1980. - THOMAS, A., LUGARDON, B.: Adansonia, ser. 2, 15, 543-572 (1976). - TILTON, V.R., HORNER, H.T.jr.: Am. J. Bot. 67, 1113-1131 (1980). - TITZ, W., SCHNATTINGER, R.: Plant Syst. Evol. 134, 269-286 (1980). - O'TOOLE, J.C., CRUZ, R.T., SEIBER, J.N.: Physiol. Plant. 47, 239-244 (1979). - TULLOCH, A.P., BAUM, B.R., HOFMAN, L.L.: Can. J. Bot. 58, 2602-2615 (1980). - TURNER, J.C., HEMPHILL, J.K., MAHLBERG, P.G.: Am. J. Bot. 67, 1397-1406 (1980). - TSENG, C.C., SHOUP, J.R.: Am. J. Bot. 65, 384-394 (1978).

VERKERKE, W., BOUMAN, F.: Bot. Gaz. 141, 277-282 (1980). - VIGNAL, CH.: Adansonia, ser. 2, 19, 39-70 (1979). - VIJAYRAGHAVAN, M.R., SHUKLA, A.K.: Grana 19, 173-175 (1980).

WATTENDORF, J.: Biol. in unserer Zeit 10, 81-90 (1980). - WEBERLING, F.: Beitr. Biol. Pflanz. 53, 61-89 (1977); - Beitr. Biol. Pflanz. 54, 29-32 (1978). - WEINHOLD, A.R., HANCOCK, J.G.: Defense at the perimeter: extruded chemicals, 121-138. In: Plant Disease, Vol. 5, eds. J.G. HORSFALL, E.B. COWLING. London: Academic Press 1980. - WILLIAMS, N.H., BROOME, C.R.: Am. Orchid Soc. Bull. 8/76, 699-707 (1976). - WILKINSON, H.P.: Kew Bull. 32, 347-360 (1978); - The plant surface, 97-165. In: Anatomy of the Dicotyledons, Vol. I, 2nd ed., eds. C.R. METCALFE, L. CHALK. Oxford: Oxford Univ. Press 1980. - WOODLAND, D.W., BASSET, I.J., CROMPTON, C.W.: Can. J. Bot. 54, 374-383 (1976).

XAVIER, K.S., MILDNER, R.A., ROGERS, C.M.: Grana 19, 183-188 (1980).

ZANDONELLA, P., LECOCQ, M.: Pollen Spores 19, 119-141 (1977).

Priv.-Doz. Dr. WILHELM BARTHLOTT
Professor Dr. RAINER SCHILL
Institut für Systematische Botanik
und Pflanzengeographie
der Universität Heidelberg
Im Neuenheimer Feld 328
D 6900 Heidelberg

II. Reproductive Structures of the Flowering Plants

By PETER K. ENDRESS

1. Introduction

This report deals with literature mainly published in 1980, as a previous survey of inflorescences and flowers has been presented by HAGEMANN in Progr. Bot. 42 (1980). The reproductive structures of the angiosperms are considered, with the exception of embryological and palynological features and special adaptations to pollination ecology.

2. Inflorescences

The classic work of TROLL on inflorescences is indispensable as a reference in any inflorescence study. If evolutionary or systematic viewpoints are in the foreground, however, other aspects also need to be considered, e.g., pollination and dispersal biology, and respective adaptive potentials of the inflorescences. An important paper by BRIGGS and JOHNSON (1979) treats the inflorescences of the large and diverse family Myrtaceae. As in other woody groups, the behavior of the flowering shoots is more complex than in herbaceous plants and their description faces more difficulties. The authors introduce some new, descriptive terms which allow the characterization of the myrtaceous inflorescences in a precise way and prove to be helpful for evolutionary considerations. Important terms include: "anthotelic" (of a shoot ending in a flower), "blastotelic" (ending in a bud or sprout); blastotelic being further subdivided into "auxotelic" (growing at the end) and "anauxotelic" (not growing at the end). Very useful is the coining of the terms "pherophyll" – finally – (for "Tragblatt") and "metaxyphyll" (for "Zwischenblatt"). In contrast to so many papers on inflorescence evolution, it is appealing that BRIGGS and JOHNSON do not place a richly branched panicle or thyrse as the evolutionary starting point, but rather a basically flexible condition which may lead, in many separate trends, to diverse stabilized patterns of elaboration: reduction or enrichment in various ways. This paper will undoubtedly stimulate other studies on woody plants.

Another group with problematical inflorescences are the Solanales which are complicated by metatopies and accessory shoots. Several authors have devoted studies to this group. In a detailed ontogenetical study HUBER (1980) compares the "front" flowers (as elements of the "pair-flowered" cymes) of the Gesneriaceae and Scrophulariaceae with the cymose partial inflorescences (pre- and subflorescences sensu TROLL) of the Solanaceae. He further discusses the possibilities of a transition of monopodia into sympodia and vice versa. In *Schizoboea* (Gesneriaceae-Didymocarpeae) WEBER (1980) found the inflorescences to be polythelic anthocladial shoot systems as in some genera of the Klugieae, the florescences being two "pair-flowered" cymes forming an open thyrse. The often discussed inflorescence structure of *Calceolaria* (Scrophulariaceae) is compared by ANDERSON and MOLAU (1980) through several groups of the genus with different growth forms, and respective evolutionary trends are discussed.

For the Betulaceae, another woody family, JÄGER (1980) shows some new aspects of evolutionary trends. The most conspicuous is the tendency toward monopodial or sympodial proliferation of the synflorescence. Monopodial proliferation is also shown for other anemophilous trees (in contrast to their entomophilous relatives), e.g., *Acer*, *Fraxinus*. Another problem arising in many woody groups is the delimination of a "synflorescence". JÄGER points out that it is useful, in some instances, to consider such a unit as consisting of more than an annual shoot. This problem of delimination occurs also in *Choysia ternata* (Rutaceae) (GUÉDÈS, 1980), and in *Helianthemum oelandicum* (WIDÉN, 1980).

In a theoretical article, SELL (1980) associates the morphological phenomenon of open and terminated inflorescences with the ontogenetic process of basipetal and acropetal effloration and with physiological interactions between various organs. According to him, basipetal flowering arises from simpler interactions and limits the amount of flowering; it precedes (evolutionarily) acropetal flowering.

Numerous articles, some of them taxonomic revisions, deal with the inflorescence structure and evolutionary trends within larger or smaller groups: *Sabia* (VAN DE WATER, 1980), *Begonia* sect. *Squamibegonia* (DE WILDE and ARENDS, 1980), *Sogerianthe* (Loranthaceae) (KUIJT, 1980), *Cornus sericea* (BARABÉ and VIETH, 1980a,b), *Jatropha* (Euphorbiaceae) (DEHGAN and WEBSTER, 1979), *Carissa grandiflora* (Apocynaceae) (COHEN and ARZEE, 1980), *Duchesnea* (Rosaceae) (OGURA and KUMURA, 1980), *Chlorogalum pomeridianum* (Liliaceae) (JERNSTEDT, 1980).

FISHER and MOGEA (1980) show that the inflorescence in *Salaca* (Palmae) is enclosed in a chamber within the leaf base and is exserted through a slit on the abaxial surface of the leaf base. VON BOTHMER et al. (1980) depict the surface of the basal part of spikelets of *Hordeum* sect. *Anisolepsis* species by SEM micrographs. NAPP-ZINN and FRANZ (1980) show some anatomical pecularities of the petaloid bracts of *Heliconia revoluta*.

3. Flowers

a) Systematic and Evolutionary Aspects

In his new outline of angiosperm classification, TAKHTAJAN (1980) includes a survey on evolutionary trends of the reproductive organs. The question "How did the archaic flower of the angiosperm look"? is answered by SCHWEITZER (1980) in a less critical way than in his original paper on the spectacular *Irania* (Iraniales or Czekanovskiales, Gymnospermae). Ovules and pollen of *Irania* are not known, and the reconstruction of the "flower" is based on several extrapolations. A comparison of the male structures of *Irania* with the stamen groups of Dilleniaceae is not appropriate, since the sporangia are not angiosperm-like. To the relatively well-preserved inflorescences of Eocene angiosperms recently described by CREPET and collaborators, CREPET and DAGHLIAN (1980) add a further one of Castaneoid affinities. Apparently, at this time, some of the recent families with highly specialized flowers were already present. Review articles of early flowers with aspects of the pollination biology are presented by CREPET (1979) and DILCHER (1979). BAKER (1980) gives a lucid survey on the development of anthecology in a most important symposium volume, which will be reviewed elsewhere.

On the basis of a comprehensive account of floral and other characters, SCHMID (1980) proposes to treat *Psiloxylon* as a new subfamily in Myrtaceae and to elevate the Chamaelaucieae to subfamilial status. PHILIPSON and STONE (1980) give a brief description of the floral vascular anatomy of the monotypic new family Aralidiaceae. These and other features show it to be best placed between Araliaceae and Cornaceae. Its floral and fruiting structures point to the inclusion of the Austrobaileyaceae into the Magnoliales (near Annonaceae) rather than Laurales (ENDRESS, 1980).

b) Flower Ontogeny and Phyllotaxis of Floral Organs

Analyses of the apex during the transition phase from the vegetative to the reproductive state were carried out in *Sinapsis alba* by PIERARD et al. (1980) at the biochemical level (changes in the protein content) and in *Xanthium strumarium* by HAVELANGE (1980) at the cytological level (changes in organelle number and differentiation). THORNLEY and COCKSHULL (1980) construct a mathematical model for the morphogenetic change during the transition of a vegetative apex into the reproductive state. That the apical meristem controls the initiation of floral organs in tobacco is pointed out by McHUGHEN (1980).

Careful morphological analyses of the initiation of the floral organs were given by HUBER (1980) for many Solanaceae, by TUCKER (1980) for *Peperomia* species (Piperaceae) and by LEINS and ERBAR (1980) for *Monodora crispata* (Annonaceae). The SEM technique, which has been used for studies in floral ontogeny for the past 10 years, has become an increasingly important tool in the study of the initiation sequence of floral organs. It has also led to new insight on the phyllotaxis of floral organs. The detailed studies of *Silene coeli-rosa* by LINDON (1978a,b, 1979) have revealed that even the organs of these flowers, known as "strictly" whorled, arise in a spiral sequence. HUBER (1980) made similar observations in some Solanaceae. It seems that the shorter the plastochrones, the more the position of a primordium is influenced by the size and position of its immediate neighboring young organs, and less by the immediately precedent primordia. This leads, then, to deviations from the regular spiral, namely either to alternate whorls or – in polymerous flowers – to more irregular organ positions (cf. *Monodora*, LEINS and ERBAR). Other groups of the Magnoliidae, apparently with longer plastochrones, show a regular spiral phyllotaxis with approximate divergence angles of 137.5°: e.g., Austrobaileyaceae, *Hortonia* of Monimiaceae (ENDRESS, 1980). An acropetal sequence with slight modifications was seen in some Bignoniaceae by JAIN and SINGH (1980).

While studying the organogenesis of flowers of some Labiatae, SHARMA and SINGH (1980) observed that no procambium is differentiated for primordia of the floral appendages which become obscured in mature flowers. This shows, once more, that the doctrine of "vascular conservatism" (which postulates that the external form of an organ is less conservative than its vasculature) is not tenable.

RITTERBUSCH (1980a,b,c,d) shows methods to tackle the development of form exemplified by complicated flowers *(Calceolaria)*. He uses "Bildscharen" und trajectories and their transformations and normalizations. It is expected that they will be a valuable tool not only for a visualizing representation of developmental sequences, but also for the detection of growth patterns in complex biological systems.

c) Perianth

The opening process and anatomical changes of the perianth are described by JERNSTEDT (1980) for *Chlorogalum pomeridianum* (Liliaceae) and by PHILLIPS and KENDE (1980) for *Ipomoea tricolor* (Convolvulaceae). In *Ipomoea*, a group of inner epidermal cells on the corolla ribs undergo dynamic structural changes two days prior to flowering, during flowering as well as flower fading. They possibly initiate the opening and closing of the flower. In cleistogamous flowers of *Lamium amplexicaule* cell expansion and cell division are arrested in the upper corolla compared with chasmogamous flowers (LORD, 1980). BARTHLOTT (1980) analyzed the development of the complex cuticular fold pattern in the unicellular clavate hairs of the petals of *Antirrhinum majus* (Scrophulariaceae). On the basis of the vascular anatomy, GREY-WILSON (1980) concludes that in the flower of *Impatiens* the upper sepal pair (in the resupinate flower) is absent and that the four lateral petals are united into two pairs.

d) Androecium

Green (1980) gives a revised terminology for the spore-containing parts of anthers: "microsporangium" (a discrete sporogenous region of the anther), "loculus" (a chamber releasing pollen through a single stomium), "locellus" (a compartment of a loculus resulting from secondary septation of the microsporangium). This terminology may, however, present difficulties when dealing with anthers with "septate microsporangia", as there is no sharp boundary between "primary" and "secondary" septation. Functional aspects are considered by HUFFORD (1980), who depicts the staminal vascular architecture in five species with basifixed and dorsifixed anthers. The results support an earlier hypothesis of SCHMID, that the orientation of the phloem and xylem is adapted for efficiency to the particular form of the stamen. HORNER and WAGNER (1980) show in the presumptive stomium region in each anther theca of *Capsicum annuum* formation of calcium oxalate druse crystals at precise times during anther development; their formation might aid in degradation of the cell walls between the locules and contribute to the pollen release mechanism.

With old and new examples of spontaneous and experimentally induced terata DUPUY and GUÉDÈS (1980) exemplify some morphological problems. One such problem is the diplophyllous nature of the stamens. However, as the range of the teratological morphs – in the selected samples – is far beyond the expected ones, the value of this method may be questioned. HUBER (1980) found that the stamens of several Solanaceae do not show diplophyllous traits in their ontogeny.

Several studies are again devoted to stamen initiation patterns in polyandrous androecia. In *Monodora crispata* (Annonaceae) the ca. 200 stamen primordia arise rapidly on a massive, dome-like apex in a centripetal succession (without an orderly spiral phyllotaxis, cf. Chap. 3.2, LEINS and ERBAR (1980). VISHENSKAIA (1980a,b) studied two theaceous species: In *Thea sinensis* the stamens arise on a circular ridge in a centrifugal direction, while in *Stuartia pseudocamellia*, the centrifugal inception takes place on five separate primordia. Different androecium initiation patterns occur in the Papaveroideae (NEMIROVICH-DACHENKO, 1980). In the multistaminate Papavereae the stamens arise on a massive, ringlike primordium; this polyandrous condition is interpreted as evolutionarily secondary.

In several large taxa, the androecium morphology was studied from comparative and evolutionary viewpoints. WOON and KENG (1979) give an account on the Dipterocarpaceae and point to evolutionary trends including number of stamens, elongation of connective tips, elaboration of appendages, and reduction in number of sporangia. DE MENEZES (1980) considers the androecium in the Velloziaceae. That flowers with 18 anthers (six fascicles with three anthers each), the anthers dehiscing by four longitudinal slits should be more primitive than six-staminate flowers, the anthers dehiscing by two lateral slits, is, however, not convicing. LASSEIGNE (1979) studies the structure of the heteromorphic anthers in numerous species of *Cassia*, DULBERGER and ORNDUFF (1980) in *Cyanella* (Tecophilaceae). REDDY et al. (1980) interpret the pair of fertile stamens to be anterolateral in *Elytraria crenata*, but posterolateral in *Nelsonia campestris* (both Nelsonieae, Acanthaceae).

e) Gynoecium

The gynoecium ontogeny has been analyzed carefully in various groups. TUCKER (1980) interprets the gynoecium of *Peperomia* as unicarpellate. It arises as a circular rim giving rise to a tube with an adaxial and an abaxial lobe. The ovule develops on the ground of the tube. Species differ greatly as to whether the abaxial or the adaxial lobe predominates in late stages of carpel development. In the Solanaceae, HUBER (1980) gives implicit evidence that in early developmental stages there is an unexpected interspecific (and sometimes even intraspecific) variation as to the inception timing of the synascidiate region. This result may question the heavy weight which is usually given to the morphological meaning of the initial ontogenetic stages. HUBER further analyzes the gynoecium ontogeny of *Nolana paradoxa*: five primary carpel primordia subdivide and give rise to numerous secondary carpels. This process may be compared, in some sense, with the secondary subdivision of androecial primordia in complex polyandrous androecia. In *Monodora crispata* (Annonaceae) LEINS and ERBAR (1980) show that the "paracarpous" gynoecium behaves as a single carpel in early developmental stages. They interpret it as a unicarpellate structure. However, it could also be looked upon as a carpel with secondary subdivision, in some way similar to *Nolana*. Whether the much-disputed gynoecium of *Takhtajania* (Winteraceae) also falls into this category, cannot be answered as yet due to lack of appropriate material of this almost or completely extinct Madagascan relic (VINK, 1978; TUCKER and SAMPSON, 1979; LEROY, 1980).

There is unending discussion on the structure of the vexing papaveraceous gynoecium. Dimery and tetramery are alternatingly propagated. The contribution of GONNERMANN (1980) pleads for tetramery with all four carpels being fertile. The teratological and anatomical arguments, as well as the evolutionary scheme, however, are not fully convincing. TERABAYASHI (1979) compares the gynoecial vasculature of *Epimedium* and *Vancouveria* with other Berberidaceae. Differences are explained on the basis of diverging fruit differentiation.

In the perianthless female flowers of *Myrica pennsylvanica* wax-secreting papillae surround the ovary. MACDONALD (1980) supposes, based on an ontogenetical study, that they arise on reduced axes of the next higher order than the flower, and that they are homologous to inflorescence bracts. TOBE (1980a,b) finds the carpels and stamens of *Clematis* species to have one vascular trace each and to be unilacunar, in contrast to the three sepal traces. A short review on the structure of the cactus gynoecium is given by BOKE (1980).

Aspects on the differentiation of the stigma and pollen tube transmitting tissue are further actively investigated, mainly in connection with incompatibility problems. However, as they are likewise important for the understanding of the gynoecium constructions, they cannot be left out here. A remarkable study by SCHICK (1980) deals with the functional organization of the apocynaceous flower, especially with the differentiation of the gynoecium apex. It is also a contribution toward the evolutionary understanding of the extremely elaborated pollination apparatus of the Asclepiadaceae. The adhesive substance produced by the gynoecial apex in *Allamanda cathartica* (Apocynaceae) and the translator in Asclepiadaceae are secreted in a similar way and have a similar chemical composition. Basic papers on the stigma in grasses by J. HESLOP-HARRISON (1980) and J. and Y. HESLOP-HARRISON (1980) demonstrate a discontinuous cuticle and absence of a tonoplast, the pollen tube-transmitting medium being constituted of acidic pectic polysaccharides and a protein component, seemingly an internal secretory product of the basal parts of the stigma branch cells. Another important series of publications deals with the stigma of submarine Hydrocharitaceae by PETTITT (1980) and PETTITT et al. (1981). It is a "dry" type stigma. However, in contrast to terrestrial plants, the stigma secretions do not disperse in water, but retain their structural integrity. Noteworthy is the observation by GHOSH and SHIVANNA (1980) that in the heterostylous *Linum grandiflorum*, the pin morph is of the "dry" type, but the thrum morph resembles the "wet" type.

Further observations on stigmatic differentiation are provided for *Tradescantia pallida* (HERD and BEADLE, 1980) (dry type), *Nelumbo pentapetala* (SCHNEIDER and BUCHANAN, 1980), *Anthurium* (CROAT, 1980) (both wet type and with possible additional nectary function), *Anthyllis* species (COUDERC, 1980), *Alchemilla vulgaris* and *Sibbaldia procumbens* (SAID, 1980) (all wet type). The factors controlling the sticking of pollen to the stigma (surface tension, wind force, electrostatical force, electrodynamical force, gravity, inertial force, pollen wall sculpture) are studied in "dry" and "wet" stigmas by WOITTIEZ and WILLEMSE (1979).

KRISTEN et al. (1980) and KRISTEN and LIEBEZEIT (1980) analyze the sequence of secretion patterns and secretion products of the pollen tube transmitting medium (below the stigma) around anthesis in *Apentia cordifolia* (Aizoaceae); HERRERO and DICKINSON (1980) in *Petunia hybrida*; TILTON and HORNER (1980) in *Ornithogalum caudatum*. TILTON (1980) further found in *Ornithogalum caudatum* the micropyle filled with an exudate secreted by the nucellar cap and inner integument, the exostomium being sealed by a secreted membrane. He hypothesizes that the exudate and membrane probably function as surface for localized deposition of synergid-synthesized chemotropic agents.

Different pathways of the pollen tube in spinach are shown by WILMS (1980). In *Mangifera indica* an outgrowth of the ovary wall towards the ovule, formed after pollination, mediates pollen tube transmission (JOEL and EISENSTEIN, 1980).

4. Fruits and Seeds

Out of the large number of studies on seeds and fruits those are selected where the structure is analyzed in regard to function, evolution, or systematics. Some important papers deal with arils. An evolutionary interpretation of the aril (elaiosome) in Papilionoideae based on careful structural and functional analysis of *Kennedia* and *Hardenbergia* species by BERG (1979) shows the rim, formed as a regular byproduct around the "counter-palisade" tissue (which functions in the closure of the detachment scar in the seed coat) as a preadaptation to the parallel evolution of many arillate groups in the Papilionoideae. DELL (1980) deals with the ontogeny of the same region in *Albizzia lophantha* (Mimosoideae). A critical survey of different types of seed appendages is given by KAPIL et al. (1980); the authors characterize the five forms: arils, wings, hairs, opercula, and jaculators. Aril ontogeny is shown in *Polygala* species (VERKERKE and BOUMAN, 1980), *Passiflora suberosa* and *Turnera ulmifolia* (KLOOS and BOUMAN, 1980). In two *Calathea* species (Marantaceae) the arils were found to be differentiated as elaiosomes (HORVITZ and BEATTIE, 1980); this is the first well-documented case of myrmecochory for wet tropical rain forest floor herbs.

GOTTSBERGER et al. (1980) in a study of the reproductive biology of *Drimys brasiliensis* (Winteraceae) with colored, indehiscent, ornithochorous fruitlets consider this type of fruitlets as the most primitive type of diaspores in the angiosperms. The fruits of some *Tropaeolum* species exhibit a marked color contrast between the black berries and the persistent red spur, most probably an adaptation toward ornithochory (WEBERLING and MÜLLER, 1980). SCHNETTER et al. (1980) analyze the hygrochastic "legumes" of *Haematoxylon brasiletto* (Caesalpinioideae), which dehisce with two longitudinal lateral slits. In *Platystemon californicus* (Papaveraceae) the heteromericarpous fruits show a differential germination behavior of the seeds in the central chamber of the fruit and those enclosed in the carpel segments (HANNAN, 1980). GRUBERT (1980) studies the epidermal structure of mucilage-producing seeds and fruits before and after contact with water. The seed testa of *Chiloschista lunifera* (Orchidaceae) is provided with extensible helical wall thickenings of two different types which constitute an unusual hygroscopic attachment mechanism of the seeds (BARTHLOTT and ZIEGLER, 1980). LUCAS (1979) gives, besides an anatomical description of hooked spikelet emergences in *Pseudochinolaena* (Gramineae), a survey of grass spikelets acting as burrs.

Evidence from seed characters and other features further supports the placement of *Pakaraimaea dipterocarpaceae* in the Dipterocarpaceae (MAGUIRE and ASHTON, 1980). Pericarp and testa structure of Hydatellaceae support the transfer of *Trithuria filamentosa* into the genus *Hydatella* (HAMANN et al., 1979). PIMENOV and KIRILLINA (1980) discuss the limits of the genus *Ferula* (Umbelliferae) on the basis of pericarp anatomy, CHERNIK (1980) the systematics of Ulmaceae. The study of the Thymelaeaceae by TAN (1980) points to relationships with Malvales and Euphorbiaceae rather than with Elaeagnaceae or Proteaceae. COLLINS (1980) provides a review of recent and Tertiary nymphaeaceous seeds.

In several groups the development from ovule to seed has been studied: *Linum usitatissimum* (BOESEWINKEL, 1980), *Dichapetalum mombuttense* (BOESEWINKEL and BOUMAN, 1980), *Erythroxylum coca* (BOESEWINKEL and GEENEN, 1980), *Polyalthia longifolia* (MOHANA RAO, 1979), *Cassytha filiformis* (MOHANA ROA, 1980), *Austrobaileya scandens* (ENDRESS, 1980).

Critical studies on fruit or seed surfaces and/or anatomy within a genus or a group of genera: several orchid genera (ARDITTI et al., 1980; HEALEY et al., 1980), *Zieria* (POWELL and ARMSTRONG, 1980), *Jaracanda* and other Bignoniaceae (HESSE and MORAWETZ, 1980), *Sedum* (t'HART and BERENDSEN, 1980), *Cypripedium* (BUTZIN, 1980), *Brassica* sect. *Brassica* (STORK et al., 1980), *Arabis* (TITZ, 1980; TITZ and SCHNATTINGER, 1980), *Manglietia* (TIEP, 1980), Aegean *Ercdium* species (DAHLGREN, 1980).

5. Historical Surveys

PETROVA (1980) reviews reproductive anatomical work on the angiosperms carried out in the USSR during the last ten years, VAN HEEL (1979) morphological studies at the Rijksherbarium, Leiden.

References

ANDERSSON, L., MOLAU, U.: Bot. Not. 133, 21-32 (1980). - ARDITTI, J., MICHAUD, J.D., HEALEY, P.L.: Am. J. Bot. 67, 347-360 (1980).

BAKER, H.G.: New Zeal. J. Bot. 17, 431-440 (1980). - BARABÉ, D., VIETH, J.: Bull. Soc. Bot. Fr. 127, Lettres Bot., 47-52 (1980a); - Can. J. Bot. 58, 918-935 (1980b). - BARTHLOTT, W.: Ber. Dtsch. Bot. Ges. 93, 379-390 (1980). - BARTHLOTT, W., ZIEGLER, B.: Ber. Dtsch. Bot. Ges. 93, 391-403 (1980). - BERG, R.Y.: Norw. J. Bot. 26, 229-254 (1979). - BOESEWINKEL, F.D.: Acta Bot. Neerl. 29, 17-32 (1980). - BOESEWINKEL, F.D., BOUMAN, F.: Acta Bot. Neerl. 29, 103-115 (1980). - BOESEWINKEL, F.D., GEENEN, J.: Acta Bot. Neerl. 29, 231-241 (1980). - BOKE, N.H.: Bioscience 30, 605-610 (1980). - BOTHMER, R. von, JACOBSEN, N., NICORA, E.: Bot. Not. 133, 539-554 (1980). - BRIGGS, B.G., JOHNSON, L.A.S.: Proc. Linn. Soc. N.S. W. 102, 157-256 (1979). - BUTZIN, F.: Willdenowia 10, 51-66 (1980).

CHERNIK, V.V.: Bot. Zh. (Leningrad) 65, 521-531 (1980). - COHEN, L., ARZEE, T.: Bot. Gaz. Crawfordsville 141, 258-263 (1980). - COLLINSON, M.E.: Ann. Bot. 46, 603-632 (1980). - COUDERC, H.: Bull. Soc. Bot. Fr. 127, Lettres Bot., 139-149 (1980). - CREPET, W.L.: Rev. Palaeobot. Palynol. 27, 213-238 (1979). - CREPET, W.L., DAGHLIAN, CH.P.: Am. J. Bot. 67, 739-757 (1980). - CROAT, TH.B.: Am. J. Bot. 67, 888-904 (1980).

DAHLGREN, G.: Bot. Not. 133, 491-514 (1980). - DEHGAN, B., WEBSTER, G.L.: Univ. Calif. Publ. Bot. 74, 1-73 (1979). - DELL, B.: Am. J. Bot. 67, 556-563 (1980). - DILCHER, D.L.: Rev. Palaeobot. Palynol. 27, 291-328 (1979). - DULBERGER, R., ORNDUFF, R.: New Phytol. 86, 45-56 (1980). - DUPUY, P., GUÉDÈS, M.: Bull. Mus. Nat. Hist. Nat. Bot. Sér. 4, 2, Sect. B, 2, 83-144 (1980).

ENDRESS, P.K.: Bot. Jahrb. Syst. 101, 393-433 (1980).

FISHER, J.B., MOGEA, J.P.: Bot. J. Linn. Soc. 81, 47-59 (1980).

GHOSH, S., SHIVANNA, K.R.: Planta 149, 257-261 (1980). - GONNERMANN, C.: Fedde Rep. 91, 593-613 (1980). - GOTTSBERGER, G., SILBERBAUER-GOTTSBERGER, I., EHRENDORFER, F.: Plant Syst. Evol. 135, 11-39 (1980). - GREEN, J.W.: New Phytol. 84, 401-406 (1980). - GREY-WILSON, C.: Kew Bull. 35, 221-227 (1980). - GRUBERT, M.: Plant Syst. Evol. 135, 137-149 (1980). - GUÉDÈS, M.: Bot. J. Linn. Soc. 80, 243-255 (1980).

HAMANN, U., KAPLAN, K., RÜBSAMEN, T.: Bot. Jahrb. Syst. 100, 555-563 (1979). - HANNAN, G.L.: Madrono 27, 163-170 (1980). - 't HART, H., BERENDSEN, W.: Plant Syst. Evol. 135, 107-117 (1980). - HAVELANGE, A.: Am. J. Bot. 67, 1171-1178 (1980). - HEALEY, P.L., MICHAUD, J.D., ARDITTI, J.: Am. J. Bot. 67, 508-518 (1980). - HEEL, W.A. VAN: Blumea 25, 107-113 (1979). - HERD, Y.R., BEADLE, D.J.: Ann. Bot. 45, 251-256 (1980). - HERRERO, M., DICKINSON, H.G.: Planta 148, 138-145 (1980). - HESLOP-HARRISON, J.: N. Z. J. Bot. 17, 537-546 (1980). - HESLOP-HARRISON, J., HESLOP-HARRISON, Y.: Acta Bot. Neerl. 29, 261-276 (1980). - HESSE, M., MORAWETZ, W.: Plant Syst. Evol. 135, 1-10 (1980). - HORNER, H.T., WAGNER, B.L.: Am. J. Bot. 67, 1347-1360 (1980). - HORVITZ, C.C., BEATTIE, A.J.: Am. J. Bot. 67, 321-326 (1980). - HUBER, K.: Dissertationes Botanicae, Vol. 55, 252 S. Vaduz: Cramer 1980. - HUFFORD, L.D.: Proc. Iowa Acad. Sci. 87, 96-102 (1980).

JÄGER, E.J.: Flora 170, 91-113 (1980). - JAIN, D.K., SINGH, V.: J. Ind. Bot. Soc. 59, 10-16 (1980). - JERNSTEDT, J.A.: Am. J. Bot. 67, 824-

832 (1980). - JOEL, D.M., EISENSTEIN, D.: Acta Bot. Neerl. 29, 203-206 (1980).

KAPIL, R.N., BOR, J., BOUMAN, F.: Bot. Jahrb. Syst. 101, 555-573 (1980). - KLOOS, A., BOUMAN, F.: Beitr. Biol. Pflanz. 55, 49-66 (1980). - KRISTEN, U., LIEBEZEIT, G.: Z. Pflanzenphysiol. 99, 241-250 (1980). - KRISTEN, U., BIEDERMANN, M., LIEBEZEIT, G.: Z. Pflanzenphysiol. 96, 239-249 (1980). - KUIJT, J.: Blumea 26, 403-410 (1980).

LASSEIGNE, A.: Iselya 1, 141-160 (1979). - LEINS, P., ERBAR, C.: Beitr. Biol. Pflanz. 55, 11-22 (1980). - LEROY, J.-F.: Adansonia, Sér. 2, 20, 9-20 (1980). - LORD, E.M.: Am. J. Bot. 67, 1430-1441 (1980). - LUCAS, M.A.: Iselya 1, 115-139 (1979). - LYNDON, R.F.: Ann. Bot. 42, 1343-1348 (1978a); - Ann. Bot. 42, 1349-1360 (1978b); - Ann. Bot. 43, 539-551 (1979).

MACDONALD, A.D.: Can. J. Bot. 58, 2001-2006 (1980). - MAGUIRE, B., ASHTON, P.S.: Taxon 29, 225-231 (1980). - McHUGHEN, A.: Ann. Bot. 46, 633-635 (1980). - MENEZES, N.L. DE: Linn. Soc. Symp. Ser. 8, 117-138 (1980). - MOHANA RAO, P.R.: Bot. Jahrb. Syst. 101, 83-89 (1979); - Israel J. Bot. 28, 44-50 (1980).

NAPP-ZINN, K., FRANZ, A.: Flora 170, 565-569 (1980). - NEMIROVICH-DANCHENKO, E.N.: Bot. Zh. Leningrad 65, 1088-1100 (1980).

OGURA, H., KIMURA, CH.: J. Jap. Bot. 55, 270-279 (1980).

PETROVA, L.R.: Bot. Zh. Leningrad 65, 1681-1695 (1980). - PETTITT, J.M.: Ann. Bot. 45, 257-271 (1980). - PETTITT, J.M., DUCKER, S., KNOX, B.: Sci. Am. 244,3, 92-101 (1981). - PHILIPSON, W.R., STONE, R.C.: Taxon 29, 391-416 (1980). - PHILLIPS, H.L., Jr., KENDE, H.: Protoplasma 102, 199-215 (1980). - PIERARD, D., JAQMARD, A., BERNIER, G., SALMON, J.: Planta 150, 397-405 (1980). - PIMENOV, M.G., KIRILLINA, N.A.: Bot. Zh. Leningrad 65, 1756-1766 (1980). - POWELL, J.M., ARMSTRONG, J.A.: Telopea 2, 85-112 (1980).

REDDY, M.I., RADHAKRISHNAIAH, M., NARAYANA, L.L., GHOSH, R.B.: Bot. Not. 133, 311-317 (1980). - RITTERBUSCH, A.: Flora 169, 299-308 (1980a); - Flora 169, 309-315 (1980b); - Flora 169, 405-423 (1980c); - Flora 169, 498-509 (1980d).

SAID, C.: Bull. Soc. Bot. Fr. 127, Lettres Bot., 227-236 (1980). - SCHICK, B.: Flora 170, 394-432 (1980). - SCHMID, R.: Taxon 29, 559-595 (1980). - SCHNEIDER, E., BUCHANAN, J.D.: Am. J. Bot. 67, 182-193 (1980). - SCHNETTER, R., HILGER, H.H., RICHTER, U.: Bot. Jahrb. Syst. 101, 135-142 (1979). - SCHWEITZER, H.-J.: Spektrum Wiss. 1980, 12, 22-33 (1980). - SELL, Y.: Flora 169, 282-294 (1980). - SHARMA, M., SINGH, V.: Ann. Bot. 46, 119-123 (1980). - STORK, A.L., SNOGERUP, S., WÜEST, J.: Candollea 35, 421-450 (1980).

TAKHTAJAN, A.: Bot. Rev. 46, 225-359 (1980). - TAN, K.: Notes R. Bot. Gard. Edinburgh 38, 149-164 (1980). - TERABAYASHI, S.: Acta Phytotax. Geobot. 30, 153-168 (1979). - THORNLEY, J.H.M., COCKSHULL, K.E.: Ann. Bot. 46, 333-341 (1980). - TIÊP, N.V.: Fedde Rep. 91, 497-576 (1980). - TILTON, V.R.: Can. J. Bot. 58, 1872-1884 (1980). - TILTON, V.R., HORNER, H.T., Jr.: Am. J. Bot. 67, 1113-1131 (1980). - TITZ, W.: Ber. Dtsch. Bot. Ges. 93, 405-415 (1980). - TITZ, W., SCHNATTINGER, R.: Plant Syst. Evol. 134, 269-286 (1980). - TOBE, H.: Bot. Mag. (Tokyo) 93, 39-54 (1980a); - Bot. Mag. (Tokyo) 93, 125-133 (1980b). - TUCKER, S.C.: Am. J. Bot. 67, 686-702 (1980). - TUCKER, S.C., SAMPSON, F.B.: Science 203, 920-921 (1979).

VERKERKE, W., BOUMAN, F.: Bot. Gaz. Crawfordsville 141, 277-282 (1980). - VINK, W.: Blumea 24, 521-525 (1978). -VISHENSKAIA, T.D. Bot. Zh. Leningrad 65, 39-50 (1980a); - Bot. Zh. Leningrad, 65, 948-957 (1980b).

WATER, TH.P.M. VAN DE: Blumea 26, 1-64 (1980). - WEBER, A.: Plant Syst. Evol. 134, 183-192 (1980). - WEBERLING, F., MÜLLER, L.: Flora 169, 295-298 (1980). - WIDÉN, B.: Bot. Not. 133, 99-115 (1980). - WILDE, J.J.F.E. DE, ARENDS, J.C.: Misc. Pap. Landbouwhogeschool Wageningen 19, 377-421 (1980). - WILMS, H.J.: Acta Bot. Neerl. 29, 243-260 (1980). WOITTIEZ, R.D., WILLEMSE, M.T.M.: Phytomorphology 29, 57-63 (1979). - WOON, C., KENG, H.: Gard. Bull. Singapore 32, 1-55 (1979).

Professor Dr. PETER K. ENDRESS
Institut für Systematische Botanik
der Universität
Zollikerstrasse 107
CH 8008 Zürich

B. Physiology

I. Photosynthesis

The Photosynthetic Reaction Center

By J. AMESZ

1. Introduction

A photosynthetic reaction center may be defined as the structure in which the primary photochemical reaction and associated reactions take place. In previous reviews (AMESZ, 1977, 1979) we discussed some of the characteristics of reaction centers and of the membrane in which the reaction centers are contained; in this section we shall give a brief resume of some of the most important of these properties in order to provide a framework for the more detailed discussions to be given below.

As a rule, the photosynthetic pigments (chlorophylls, bacteriochlorophylls, and carotenoids) are contained in the photosynthetic membrane. For algae and higher plants this membrane is usually called the thylakoid membrane; in photosynthetic bacteria it is the cytoplasmic membrane (in the case of purple bacteria also called the chromatophore membrane) that contains the reaction center and associated photosynthetic machinery. The so-called light-harvesting pigments are contained in pigment-protein complexes (AMESZ, 1979; THORNBER et al., 1979); light energy absorbed by these pigments is transferred to the reaction center. Even when photosynthesis is proceeding at maximum efficiency, a small proportion of the excitation energy of the pigments is lost as heat or fluorescence of chlorophyll or bacteriochlorophyll. These losses will increase when the reaction center is unable to function efficiently, because secondary electron transport is too slow to effect a sufficiently rapid turnover of the primary reactants. This may be the case in subcellular preparations, or, with intact systems, at high light intensity or low temperature or in the presence of inhibitors of secondary electron transport.

Fluorescence measurements have traditionally provided an important tool to study the mechanism of photosynthesis (see, e.g., review by LAVOREL and ETIENNE, 1977). Recently photoacoustic spectroscopy was introduced as a potentially useful method to obtain information about heat-dissipating processes in photosynthesis (MALKIN and CAHEN, 1979; INOUE et al., 1979; LASSER-ROSS et al., 1980). In this method, modulated radiation is absorbed in a given sample, and the modulated heat generated by the radiationless transitions is sensed by a microphone as acoustic energy in the gas surrounding the sample. The method can be used to measure absorption spectra of pigments contained in virtually opaque samples, while the depth of measurement can be adjusted by proper choice of the frequency of modulation. Another feature of the method is that it only measures the fraction of absorbed radiation that is converted into heat; energy stored or dissipated in other ways (as, e.g., by fluorescence or photochemistry) is not recorded. Fast and slow processes can be distinguished by varying the modulation frequency.

Three (or four) different photosynthetic reaction centers may be distinguished (see AMESZ, 1977; PARSON, 1978; SAUER, 1979; HOFF, 1981). These are the reaction centers of photosystems 1 and 2 of oxygen-evolving organisms, and the bacterial reaction centers, which may be divided in two types; the reaction center of purple and that of green bacteria, respectively. There is ample evidence that in all cases the primary electron donor is a chlorophyll a or bacteriochlorophyll dimer, which upon excitation transfers an electron to an acceptor molecule and is converted to the corresponding cation radical. This primary electron donor is named after the wavelength of maximum bleaching in the red or near-infrared region. Thus, the primary donor for photosystem 1 is called P-700; for system 2 it is called P-680. The primary donor of purple bacteria, which is a bacteriochlorophyll a dimer, is called P-870, except for a few species that contain bacteriochlorophyll b. The most conspicuous property that distinguishes bacteriochlorophyll b from bacteriochlorophyll a is that it absorbs at longer wavelengths; the primary electron donor in these species is called P-985 or P-960 (HOLT and CLAYTON, 1965; HOLTEN et al., 1978; DAVIS et al., 1979). In green photosynthetic bacteria the primary electron donor is again a bacteriochlorophyll a dimer. The mechanism of photosynthesis in these bacteria has been most extensively studied in species belonging to the genera *Chlorobium* and *Prosthecochloris* (Chlorobiaceae) (OLSON, 1980). The primary electron donor is called P-840 in these species. In the green gliding bacteria (Chloroflexaceae) the primary electron donor bleaches maximally near 860 nm (PIERSON and CASTENHOLZ, 1974) and might be more properly called P-860

The primary reaction of photosynthesis consists of the transfer of an electron from the primary electron donoer to an acceptor. The so-called primary electron acceptor is called Q (system 2), X or A_2 (system 1) or X or Q (photosynthetic bacteria). The designation "primary acceptor" is mainly used for historical reasons; there is various evidence that there is at least one earlier, so-called intermediary acceptor (I) involved in the photochemical electron transfer reaction. The existence and properties of this acceptor are most extensively documented for purple bacteria, where I is bacteriopheophytin a, as was discussed in an earlier review in this series (AMESZ, 1977) and more extensively in recent reviews by HOFF (1981), SAUER (1979), and PARSON (1978). Recent experiments by Russian and American groups indicate that a special bacteriochlorophyll a molecule may function as an even earlier acceptor in purple bacteria (AKHMANOV et al., 1980; HOLTEN et al., 1980).

In the present review we shall deal with some of the properties of the reaction center of photosystem 2, and of the associated electron transport. In addition, a brief review will be given of delayed fluorescence (luminescence) in photosynthesis. Delayed fluorescence is a phenomenon associated with all four types of reaction centers mentioned above. However, it has been most extensively studied in relation to photosystem 2 and most of the discussion will be centered to system 2 luminescence. Finally, some recent results will be discussed concerning photosynthesis and primary electron transport in green photosynthetic bacteria.

2. Photosystem 2 Electron Transport

The primary electron acceptor of photosystem 2, Q, is a plastoquinone molecule, which upon reduction is converted into the plastosemiquinone anion radical (Q^-), as indicated by absorption changes in the near-ultraviolet that occur upon illumination (VAN GORKOM, 1974; VAN GORKOM et al., 1975). From Q the electron is transferred to a secondary acceptor, called R (VELTHUYS and AMESZ, 1974) or B (BOUGES-BOCQUET, 1973), which is also a plastoquinone molecule (PULLES et al., 1976; MATHIS and HAVEMAN, 1977).

Inhibitors of system 2 like DCMU [3-(3',4'-dichlorophenyl)-1,1-dimethyl urea] and atrazine appear to act near R. DCMU was shown to lower the redox potential of the R/R^- couple (VELTHUYS and AMESZ, 1974). More recent evidence with mutants that are resistant to atrazine indicates that this inhibitor binds to the same protein to which R is bound (PFISTER and ARNTZEN, 1979) and that the mutation affects the redox properties of R (BOWES et al., 1980). The binding sites for the inhibitors and for R appear to be close to the binding site involved in the well-known (GOVINDJEE and VAN RENSEN, 1978; see also SAROJINI and GOVINDJEE, 1981) stimulation of electron transport by CO_2: chloroplasts prepared from atrazine-resistant mutants show a diminished CO_2 stimulation (KHANNA et al., 1981). The same is true for chloroplasts treated with trypsin (KHANNA et al., 1981), which enzyme preferentially attacks the R-binding protein (RENGER, 1976; BÖGER and KUNERT, 1979; VAN RENSEN and KRAMER, 1979).

There is evidence that the acceptor R is associated with the so-called α-centers of system 2 (see previous review, AMESZ, 1979, and THIELEN and VAN GORKOM, 1981); the less abundant β-centers appear to lack this quinone (VAN GORKOM et al., 1981, and unpublished experiments). The oxidation-reduction potential of Q is higher in these centers than in the α-centers (MELIS, 1978; HORTON and CROZE, 1979; MELIS and SCHREIBER, 1979; HORTON and BAKER, 1980). The implications for electron transport to system 1 are still unclear.

The reduction of Q is accompanied by band shifts of a pigment or pigments absorbing near 420, 545, and 685 nm (BUTLER, 1973; VAN GORKOM, 1974; VAN GORKOM et al., 1975), which pigment is possibly pheophytin a. Experiments, mainly by KLIMOV and coworkers, have shown that pheophytin a also functions as intermediary electron acceptor, like bacteriopheophytin in purple bacteria. Illumination of preparations enriched in the reaction center of photosystem 2 at an oxidation-reduction potential where Q was reduced in the dark already produced absorbance changes near 420-450, 545, and 685 nm that indicated reduction of pheophytin a (KLIMOV et al., 1977). Under the same conditions an ESR signal was observed with a g-value of 2.0033, that could be attributed to the pheophytin a anion radical (KLIMOV et al., 1980). The reduction was accompanied by a decrease in the yield of chlorophyll a fluorescence (KLIMOV et al., 1977). These conclusions were more recently confirmed and extended by flash spectroscopy (SHUVALOV et al., 1980), the results of which indicate that pheophytin is reduced in less than 1 ns. In the presence of Q^- the signal decayed in about 4 ns, presumably by a back reaction of reduced pheophytin with P-680$^+$.

The electron donors on the oxidizing side of photosystem 2 and intermediates of oxygen evolution have not been identified yet, and most of our knowledge about the electron transport from water to system 2 is based on indirect and kinetic experiments (see review by VELTHUYS, 1980). About 10 years ago JOLIOT et al. (1971) and KOK et al. (1970; FORBUSH et al., 1971; see also review by RADMER and CHENIAE, 1977) showed by measurement of the amount of oxygen released in a series of short, saturating light flashes that each reaction center of system 2 is able to accumulate four positive charges. The corresponding oxidation states of the "oxygen-evolving enzyme" are designated S_0, S_1, S_2, S_3, and S_4. The first two of these are stable in the dark, S_2 and S_3 revert slowly to S_1, and S_4 is converted to S_0 in about 1 ms with concomitant production of oxygen. Experiments of, amongst others, CHENIAE and coworkers (CHENIAE and MARTIN, 1967, 1970, 1978; RADMER and CHENIAE, 1977) and of WYDRZINSKY et al. (1976, 1978) have indicated that manganese is implicated in oxygen evolution (see review by SAUER, 1980), but direct demonstration of changes of the oxidation state of manganese in the S-state transitions is lacking so far.

A promising new development in the study of the mechanism of oxygen evolution comes from the work of SPECTOR and WINGET (1980), who reported the isolation of a manganese-containing protein from spinach chloroplasts. The protein was obtained by cholate extraction, followed by purification on a Sephadex column. It has a molecular weight of 65,000 and atomic absorption analysis indicated the presence of two Mn atoms per protein molecule. The manganese could be removed by treatment with Tris [tris(hydroxymethyl)amino ethane], which treatment is known to abolish oxygen evolution and to release manganese in chloroplasts (YAMASHITA and BUTLER, 1968; CHENIAE and MARTIN, 1978). The extracted grana material, when incorporated into phospholipid vesicles (called "depleted photosomes") showed high rates of photosystem 1 and photosystem 2 electron transport, but no oxygen evolution. The latter activity was restored upon reconstitution of the depleted photosomes with the Mn protein. The reconstituted vesicles ("photosomes") showed the normal pattern of O_2 evolution in flashes, suggesting that the S-state pattern worked normally and also chlorophyll fluorescence transients and the electrochromic absorbance change at 515 nm (see review AMESZ, 1977) were restored (GOVINDJEE et al., 1980). As far as we know these results, though impressive, have not yet been confirmed by others. NAKATANI et al. (1980), using a similar isolation method, obtained a protein with partially restored oxygen evolution and which had about the same molecular weight as the protein isolated by SPECTOR and WINGET. However, the protein was reported to contain one Fe per molecule, but only little Mn.

Kinetic measurements of P-680 and ESR studies indicate that there are at least two intermediates between P-680 and the oxygen-evolving enzyme. The primary electron donor to P-680 is usually called Z. Specific information about its identity is lacking so far, but kinetic experiments indicate that it reacts very rapidly with P-680. Fluorescence measurements, as well as direct observation of the concentration of the $P\text{-}680^+$ radical by means of absorbance measurements in the near-infrared, indicate that $P\text{-}680^+$ decays in 40 to about 400 ns, depending on the number of flashes given (VAN BEST and MATHIS, 1978; SONNEVELD et al., 1979). Another electron donor to system 2 was first observed by ESR in chloroplasts treated with Tris and high pH. In these "Tris-treated" chloroplasts, BABCOCK and SAUER (1975) observed a light-induced ESR signal with $g = 2.00$ that could be ascribed to oxidation of an electron donor to system 2. The electron donor was rapidly rereduced by added reductants such as *p*-phenylenediamine and diphenylcarbazide (BABCOCK and SAUER, 1975; YERKES and BABCOCK, 1980) as indicated by the decay of the ESR signal. In untreated chloroplasts a similar signal was observed. After a flash given in states S_2 or S_3 it decays in about 0.5 ms (BABCOCK et al., 1976). At first it was thought that the signal ("Signal II_{vf}") was due to oxidized Z, but subsequent experiments showed that its generation by a short flash of light takes about 20 µs (BLANKENSHIP et al., 1977a). Since the decay of $P\text{-}680^+$ is much faster, one thus has to assume that Z and the substance responsible for Signal II_{vf} are different compounds.

3. Delayed Fluorescence of Chlorophyll

a) Emission Spectra

Delayed fluorescence of chlorophyll was accidentally discovered 30 years ago by STREHLER and ARNOLD (1951), and has since been studied extensively in many laboratories. For more detailed discussions of many aspects of this phenomenon we may refer to the reviews by LAVOREL (1975), MALKIN (1977a,b), and AMESZ and VAN GORKOM (1978). Here we shall confine ourselves to some general aspects and to recent literature on the subject.

Delayed fluorescence ("delayed light" or "luminescence") by chlorophyll or bacteriochlorophyll in photosynthetic material denotes fluorescence emission at a time scale that is distinctly longer than that of "normal" or "prompt" fluorescence. Emission spectra have shown that the emission comes from the lowest excited singlet state (AZZI, 1966; CARITHERS and PARSON, 1973), but whereas the lifetime of the excited state of chlorophyll or bacteriochlorophyll is a few ns in vitro (CERMAK and KAPLANOVA, 1980), and even less in photosynthetic material (TREDWELL et al., 1978), time constants of up to several minutes have been discerned in the decay of delayed fluorescence. This shows that the population of the excited state does not occur by direct absorption of light, but by slower and more complicated pathways. STREHLER and ARNOLD (1951) already advanced the hypothesis that delayed fluorescence is brought about by a reversal of photosynthetic electron transfer reactions. The "recombination hypothesis", which explains delayed fluorescence by a reversal of the primary charge separation, is now widely accepted. Convincing evidence for the hypothesis has been obtained from experiments by amongst others CARITHERS and PARSON (1975), SHUVALOV (1976), and DE GROOTH and VAN GORKOM (1981; DE GROOTH, 1980). CARITHERS and PARSON showed that the decay of the oxidized primary electron donor P-895$^+$ and the reduced quinone acceptor (X^-) of the photosynthetic bacterium *Rhodopseudomonas viridis* after a flash of light was quantitatively related to light emission. Analogous results were reported by SHUVALOV with chloroplast preparations enriched in photosystem 1. The experiments of DE GROOTH and VAN GORKOM will be discussed below. It has also been shown (FLEISCHMAN, 1969; ETIENNE and LAVOREL, 1975; FLEISCHMAN, 1978; LAVOREL, 1980) that in the absence of illumination, fluorescence can be induced by "chemical" oxidation of the primary electron donor and reduction of the primary acceptor. On the other hand, treatments or reactants that diminish the concentrations of these substances suppress or decrease delayed fluorescence (BENNOUN, 1970). Finally, it should be noted that the recently discovered effects of a magnetic field also provide strong evidence for the recombination hypothesis (RADEMAKER et al., 1979; VOZNYEK et al., 1978; SONNEVELD et al., 1980a,b, 1981).

In algae and higher plants most of the delayed fluorescence comes from photosystem 2. At room temperature, the same is true for prompt fluorescence, and therefore it is not surprising that in some cases the emission spectra of prompt and delayed fluorescence are very similar, as was, for example, observed for green algae by AZZI (1966) and, more recently, for fast-decaying components in luminescence by SONNEVELD et al. (1980a). However, in many instances the emission spectra are different, and reflect the different mechanisms involved in both processes. This applies, for example, to phycobilin-containing algae, like red and blue-green algae (AZZI, 1966; MURATA, 1977; SONNEVELD et al., 1980a) where the emission spectrum of prompt fluorescence shows a strong contribution by phycocyanin which is absent or weak in the delayed fluorescence emission. In the latter case the excitation energy is generated in the reaction center; it may spread into antenna chlorophyll a, but so-called "uphill" transfer of excitation energy from chlorophyll to phycocyanin is energetically unfavorable. Similar reasoning applies to the emission spectra of spinach chloroplasts at liquid nitrogen temperature (SONNEVELD et al., 1980b, 1981). At low temperature there is a strong emission from system 1 near 735 nm in the spectrum of prompt fluorescence (see AMESZ, 1979) which is relatively low in that of delayed fluorescence. In fact, in these and similar cases (see also CLAYTON, 1965) the emission spectra of delayed fluorescence resemble those of the "variable fluorescence", i.e., the increment of prompt fluorescence observed upon closing the reaction center traps.

b) Decay Kinetics and Temperature Dependence

Several exponential components have been observed in the decay of delayed fluorescence. Even if we restrict ourselves to green algae (of which *Chlorella* has been the most widely used species in these studies) and chloroplasts of higher plants, the literature on this subject is too large to allow a comprehensive discussion here. Therefore, we confine ourselves to components faster than about 1 ms, which are probably related to electron transfer reactions occurring in or near the reaction center. At least two different decay components may be discerned in the microsecond region, with halftimes of 150-200 μs and 15-35 μs, respectively (ZANKEL, 1971; LAVOREL, 1973; DUYSENS et al., 1975; WONG et al., 1978; MAROTI et al., 1978; SONNEVELD et al., 1980a). An 8-10 μs component was also reported (WONG et al., 1978). These decays may reflect components observed in the re-reduction of $P\text{-}680^+$ formed in the light as indicated by studies of the reaction kinetics of P-680 after a flash or a series of flashes (GLÄSER et al., 1974; CONJEAUD et al., 1979). Most of the $P\text{-}680^+$ after a flash, however, decays in less than a microsecond by re-reduction with a secondary electron donor (Z) (DUYSENS et al., 1975; GLÄSER et al., 1976; VAN BEST and MATHIS, 1978; SONNEVELD et al., 1979), so it seems possible that luminescence in the μs region mainly reflects emission from those reaction centers where the reaction between $P\text{-}680^+$ and secondary electron donor(s) to system 2 is relatively slow or does not occur at all.

Experiments in several laboratories (BARBIERI et al., 1970; JOLIOT et al., 1971; ZANKEL, 1971; DUYSENS et al., 1975) have shown that delayed light emission is stimulated by the higher S-states. For the submillisecond components the highest intensity is after the third and again after the seventh or eighth flash, which suggests that the strongest simulation occurs in state S_4. The mechanism of this stimulation is not quite clear yet, but it is interesting to note that the amplitude of the relatively slowly decaying components of $P\text{-}680^+$ shows a similar flash dependence (GLÄSER et al., 1976).

Very fast components of delayed fluorescence of 600 and 100 ns, respectively, were recently observed under conditions where the primary acceptor of system 2 (Q) was reduced already before the flash (VAN BEST and DUYSENS, 1977; SONNEVELD et al., 1980a,b). The experiments indicate that photochemical electron transfer does also occur when Q is reduced, and that there is presumably at least one earlier electron acceptor which upon photochemical reduction normally transfers its electron to Q. This acceptor may be identical to the "intermediate" acceptor studied by KLIMOV and coworkers by fluorescence and flash spectroscopy, and which is thought to be pheophytin a (see preceding section). Delayed fluorescence then would be caused by a back reaction between $P\text{-}680^+$ and the reduced acceptor.

The 600 ns component was found to be inhibited by hydroxylamine, which blocks the reduction of $P\text{-}680^+$ by the secondary electron donor Z (DEN HAAN et al., 1974), but not the 100 ns component (SONNEVELD et al., 1981). This suggests that the first component involves a reoxidation of P-680 by oxidized Z, and that its halftime reflects the disappearance of Z^+ after the flash (SONNEVELD et al., 1981). The interpretation of the 100 ns luminescence is still uncertain.

The effects of cooling upon the intensity and kinetics of fluorescence, although complicated, may be generally understood by slowing down and inhibition of secondary electron transport. As was observed already by TOLLIN et al. (1958) long-lived components disappear first upon cooling. The intensity in the ms range may first go through a maximum (TOLLIN et al., 1958; ITOH and MURATA, 1974; ITOH, 1980; GOLTSEV et al., 1980) but drops rapidly beyond about 220 K and below about

125 K it usually becomes too weak to measure (AMESZ and DE GROOTH, 1979;, ITOH, 1980). There is a slowing down of the decay in the micro- to millisecond time region, and below about 200 K the most prominent decay component is one of about 300 μs (ITOH and MURATA, 1974; AMESZ and DE GROOTH, 1979) the intensity of which drops rapidly with decreasing temperature.

The 600 ns component is absent at liquid nitrogen temperature (SONNEVELD et al., 1980b, 1981), but the 100 ns luminescence component was found to be much less sensitive to cooling. As a matter of fact, below 150 K its intensity increased strongly with decreasing temperature until it reached a plateau between about 100 and 50 K with an about tenfold increased intensity as compared to room temperature (SONNEVELD et al., 1980b; 1981). At still lower temperature the intensity decreased again. The halftime of the decay was little affected by cooling.

Apparently conflicting results have been reported concerning the effects of multiple flash excitation at low temperature. According to GOLTSEV et al. (1980) the "S-state dependence" in chloroplasts is lost below about 240 K, and luminescence, measured 1 ms after the flash was virtually independent of the number of flashes between 180 and 230 K. However, AMESZ and DE GROOTH (1979) observed a fairly large stimulation with increasing number of flashes for the first five or ten flashes both at 40 ms and at 70 μs at 240 K, and also at lower temperatures. Even at 150 K a significant stimulation was observed.

c) Stimulation by an Electric Field

It was shown already several years ago that the generation of a membrane potential across the thylakoid (BARBER and KRAAN, 1970; KRAAN et al., 1970) or bacterial (FLEISCHMAN, 1971) membrane stimulates delayed fluorescence. The effect can be observed by the addition of a salt of a permeant cation to a suspension of grana or bacterial chromatophores after illumination, which addition causes a severalfold temporary increase in the intensity of delayed fluorescence. In such an experiment the sign of the membrane potential is positive inside, and since the primary electron acceptor is located closer to the outside, the electron donor closer to the inside of the thylakoid or chromatophore, the luminescence stimulation could be explained by an enhanced rate of the reversal of the primary electron transfer reaction. However, in this type of experiment the stimulation is measured a relatively long time after the illumination, and effects on the rates of reversal or secondary electron transport cannot therefore be ruled out. In fact, JURSINIC et al. (1978) obtained evidence that the stimulation is much stronger in the millisecond than in the microsecond time region. Other effects, like a change in surface charge, may also be involved (WONG et al., 1978). From measurements of the stimulation in the presence of DCMU, which compound inhibits secondary electron transport from the acceptor Q, VENEDIKTOV et al. (1980) obtained an estimated distance, perpendicular to the membrane, of 10-12 Å between the secondary donor (Z) and Q. This result, however, is at variance with the kinetics of the changes in pigment absorption that sense the electric field in the membrane (electrochromic absorbance changes) (see AMESZ, 1977). The results of these studies indicate that Q and the donors of system 2 are located close to the lipid surface on opposite sides of the membrane (CONJEAUD et al., 1979), and thus at larger distance from each other.

A more dramatic stimulation, first observed by ARNOLD and AZZI (1971) and studied in more detail by ELLENSON and SAUER (1976), is obtained if electrodes are used to apply an "external" electric field to a suspension of swollen grana. These so-called "blebs" are produced by hypotonic treatment and are large balloon-like structures of up to 25 μm diameter, mainly consisting of a single membrane (DE GROOTH et al., 1980). The advantage of this method is that the field can be turned on and off in a very short time and that for large blebs the field across the membrane may be up to a few thousand times larger than the average field between the electrodes . DE GROOTH and VAN GORKOM (1981; DE GROOTH, 1980) were able to explain the stimulations obtained in this way quantitatively by means of a model involving an increase of the rate of the reversal of the primary electron transfer reaction by a lowering of the activation energy. Measurements of the polarization of delayed fluorescence supported this model (DE GROOTH and VAN GORKOM, 1981; DE GROOTH, 1980; see also FARKAS et al., 1980). The yield of luminescence per electron transferred (the "exciton yield") was found not to be affected by the field. Stimulation by an external field was also obtained with dry films of chloroplasts (ORTOIDZE et al., 1975).

d) Thermoluminescence

Thermoluminescence may be defined as fluorescence emitted at various temperatures during warming of a sample that had been illuminated during or immediately before cooling. This phenomenon, which was first reported by ARNOLD and SHERWOOD (1957) is probably related to delayed fluorescence, and caused by the recombination reaction of products formed in the light and frozen in during cooling.

In green plant material, the most extensively studied emissions are the so-called A, B_1 and B_2 bands, with peak temperatures at about -10°, +25° and +45°C, respectively (ARNOLD and AZZI, 1971; ICHIKAWA et al., 1975; LURIE and BERTSCH, 1974a; SANE et al., 1974) which bands appear to be associated with photosystem 2 and the oxygen-evolving system (INOUE et al., 1976; INOUE, 1976; ICHIKAWA et al., 1975; DEMETER et al., 1979; SANE et al., 1977). Altogether, as many as 8 bands may be distinguished, peaking at about -45°, -25°, -10°, +10°, +17°, +25°, +45° and +70°C (VASS et al., 1980, 1981; TATAKE et al., 1981). Activation energies varying between 0.5 and 1.3 eV have been calculated (LURIE and BERTSCH, 1974b; VASS et al., 1980, 1981; TATAKE et al., 1981). For the A band this value is 0.65 eV; for the B_1 band an activation energy of 1.1 eV was obtained.

INOUE and SHIBATA (1978a, 1979) observed that the intensities of the A and B bands are dependent on the number of flashes given before cooling. The B band (mainly B_1) showed a periodicity of four with flash number, with maxima after 2, 6, and 10 flashes, indicating that this luminescence originates from a recombination reaction occurring in state S_3. The A band needed more extensive illumination in order to become apparent, and appeared to be only indirectly dependent on the S-state. It could be activated by illumination at low temperature only if the chloroplasts were brought in state S_3 before cooling (INOUE, 1981). These effects were used (INOUE and SHIBATA, 1978b) to obtain information about the temperature dependence of the various S-state conversions, by applying flashes at low temperature and comparing the effects to those of flash illumination at 17°C. It was concluded that the conversion of S_3 to S_4 and of S_2 to S_3 is blocked below -20° and -35°C, respectively, but that S_1-S_2 conversion is less temperature-dependent, and occurs even below -65°C (210 K).

4. The Reaction Center of Green Photosynthetic Bacteria

The green photosynthetic bacteria are a somewhat neglected class of organisms in the study of photosynthesis. Although their unusual pigment composition attracted the attention of several workers (see, e.g.,

VAN NIEL, 1931; LARSEN, 1953; and recent reviews by OLSON, 1980, and PIERSON and CASTENHOLZ, 1978), studies of the mechanism of photosynthetic energy conversion have been far less extensive for green than for purple bacteria. The main reasons for this neglect are undoubtedly the larger antennae (mainly contained in the so-called chlorosomes, see previous review, AMESZ, 1979) and corresponding low reaction center to pigment ratio, and (until recently) the lack of a preparation method to obtain photochemically active pigment-protein complexes from these bacteria. For other classes of photosynthetic organisms such pigment-protein complexes have proved valuable material to study primary and associated electron transport.

A recent review of photosynthesis in green bacteria is given by OLSON (1980). Here we shall confine ourselves mainly to studies with isolated membranes and pigment-protein complexes from *Chlorobiaceae*. The absorption spectrum of the primary electron donor and the antenna bacteriochlorophyll of *Chloroflexus aurantiacus* (PIERSON and CASTENHOLZ; 1974; OLSON, 1980; SCHMIDT, 1980) suggests that the *Chloroflexaceae* (green gliding bacteria) may in some respects be intermediate between the *Chlorobiaceae* and the purple bacteria, but otherwise little is known about photosynthetic electron transport in these bacteria.

Some ten years ago FOWLER et al. (1971) and OLSON et al. (1973) obtained a membrane vesicle preparation from the green bacteria *Prosthecochloris aestuarii* and *Chlorobium limicola* which contained only little bacteriochlorophyll c or d (the pigments present in the chlorosome) and about 80 bacteriochlorophyll a molecules per reaction center. Recently, SWARTHOFF and AMESZ (1979), by detergent treatment of such a preparation, isolated a photochemically active pigment-protein complex (PP complex), which had about the same number of bacteriochlorophyll a molecules per reaction center as Complex I. Except for a lower carotenoid content, its pigment composition and absorption spectrum in the visible region of the spectrum was similar to that of Complex I; the complex had a molecular weight of about 600,000. Treatment with a chaotropic agent removed about half of the antenna bacteriochlorophyll a and reduced the molecular weight of the complex to 350,000. The preparation thus obtained was called the RCPP complex.

Figure 1 compares the absorption spectra measured at 4 K of intact cells of *P. aestuarii*, of the PP and of the RCPP complex. The first spectrum is dominated by bacteriochlorophyll c, with maximum absorption at 753 nm at 4 K, but long-wave bands of bacteriochlorophyll a are easily discerned in the region 800-840 nm. The latter bands are better resolved in the spectra of the PP and RCPP complexes, together with bands due to carotenoid (450-550 nm). The bands near 670 nm are due to bacteriopheophytin c and a possibly related, as yet unidentified pigment, as indicated by thin layer chromatography (SWARTHOFF, unpublished experiments).

Illumination of Complex I, the PP or the RCPP complex produces a complicated difference spectrum in the near-infrared and visible region, which is thought to be due to oxidation of the primary electron donor P-840 and concomitant reduction of an acceptor molecule (FOWLER et al., 1971; OLSON et al., 1977; SWARTHOFF and AMESZ, 1979; SWARTHOFF et al., 1981a). Although it is not easy to explain the origin of all of the bands in this spectrum in detail, it is fairly certain that the absorbance changes are mainly due to or related to the oxidation of P-840, which reaction causes a decrease in absorption in the near-infrared region and near 600 nm, due to disappearance of the Q_y and Q_x bands of P-840 upon conversion to the cation radical P-840$^+$. Other bands are probably due to electrostatic (electrochromic) interaction of P-840$^+$ with other pigment molecules, including bacteriochlorophyll a and the

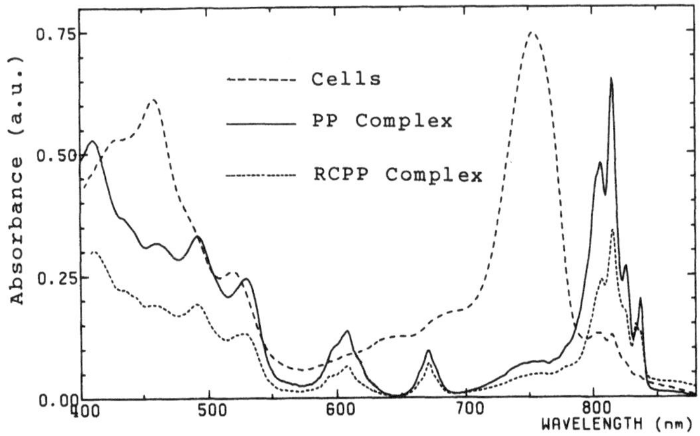

Fig. 1. Absorption spectra, measured at 4 K of intact cells of the green photosynthetic bacterium *Prosthecochloris aestuarii* and of pigment-protein complexes (PP and RCPP complex) prepared from this bacterium. The major pigment of intact cells is bacteriochlorophyll c, which is responsible for the broad, intense absorption band at 753 nm. This pigment is virtually absent in the PP and RCPP complexes; the narrow bands at 800-840 nm in these preparations are due to bacteriochlorophyll a, the same pigment present in purple photosynthetic bacteria

pigment(s) absorbing near 670 nm (see also SWARTHOFF et al., 1981c). In addition, the spectrum contains a contribution due to oxidation of cytochrome c-553, which reaction does not occur at low temperature (SWATHOFF et al., 1981a). Absorbance changes due to reduction of the acceptor are probably small and constitute only a minor contribution to the spectrum (SWARTHOFF et al., 1981b). ESR measurements (OLSON et al., 1977; SWARTHOFF et al., 1981d) indicate that P-840 is a dimer, as is the primary electron donor in purple bacteria.

In the RCPP complex, the efficiency of charge separation was found to be much lower than in the other preparations, as indicated by a much lower rate of accumulation of $P\text{-}840^+$ (SWARTHOFF and AMESZ, 1979). The most plausible explanation for this phenomenon was not that the efficiency of the electron transfer from P-840 to the first electron acceptor is lower, but that the generation of a stable charge separation by subsequent electron transfer to other acceptors is impaired by the isolation procedure. This hypothesis was confirmed by experiments with short flashes and at low temperature, which demonstrated the formation of a bacteriochlorophyll a triplet. The near-infrared absorption difference spectrum indicated that the triplet was that of P-840, and not of antenna bacteriochlorophyll (SWARTHOFF et al., 1981a,b,c). Triplet formation, but to a smaller extent, was also observed in the PP particle. ESR measurements (SWARTHOFF et al., 1981d) showed that the triplet is not generated by intersystem crossing from the singlet excited bacteriochlorophyll in the antenna, but that it is formed in the reaction center by a back reaction between P-840 and an early electron acceptor (the "radical pair mechanism"), in agreement with the optical experiments. It is thus quite understandable that impairment of electron transport at the acceptor side favors triplet formation. Triplet formation by the radical pair mechanism was earlier observed in preparations from purple bacteria (PARSON and MONGER, 1977; HOFF et al., 1977; BLANKENSHIP et al., 1977b).

Optical and ESR measurements have also provided information about the nature of the electron acceptors. JENNINGS and EVANS (1977) observed an ESR signal at g = 1.90 upon illumination of a cell-free preparation of *Chlorobium limicola* at 10 K, that was ascribed to the reduction of an iron-sulfur protein. Similar experiments (KNAFF et al., 1979), however, did not confirm these results. Recent experiments with the PP complex (SWARTHOFF and GAST, unpublished experiments) indicate that there are two iron-sulfur centers that function as electron acceptors, with different oxidation-reduction potentials and different ESR spectra in the reduced form. One of these is probably identical to the one observed by KNAFF and MALKIN (1976) upon chemical reduction of a cell-free preparation of *C. limicola*. Reduction of this iron-sulfur protein was also observed by optical measurements; experiments of SWARTHOFF et al. (1981b) showed that it can be accumulated in the reduced form upon illumination in the presence of a strong reductant at room temperature. These data support the hypothesis that the reaction center of green photosynthetic bacteria resembles more that of photosystem 1 than of purple bacteria, where the electron acceptors are quinone-iron complexes, rather than iron-sulfur proteins. Illumination of the PP complex under conditions where both iron-sulfur centers were reduced produced an ESR signal that may be caused by a reduced porphyrin (bacteriochlorophyll or bacteriopheophytin). Even this compound does not appear to be the first electron acceptor of the green bacterial reaction center, since triplet formation was not impaired by its reduction, indicating that the earliest electron transfer step still does occur when the porphyrin is in the reduced form (SWARTHOFF and GAST, unpublished observations.

Acknowledgment. The author is indebted to Dr. H.J. van Gorkom, Dr. A. Sonneveld and T. Swarthoff for reading parts of the manuscript.

References

AKHMANOV, S.A., BORISOV, A.Y., DANIELUS, R.V., GANODAS, R.A., KOZLOWSKI, V.S., PISKARSKAS, A.S., RAZJIVIN, A.P., SHUVALOV, V.A.: FEBS Lett. 114, 149-152 (1980).- AMESZ, J.: Prog. Bot. 39, 49-61 (1977);- Prog. Bot. 41, 55-70 (1979).- AMESZ, J., VAN GORKOM, H.J.: Annu. Rev. Plant Physiol. 29, 47-66 (1978).- AMESZ, J., DE GROOTH, B.G.: J. Luminescence 18/19, 548-603 (1979).- ARNOLD, W., AZZI, J.: Photochem. Photobiol. 14, 233-240 (1971).- ARNOLD, W., SHERWOOD, H.K.: Proc. Natl. Acad. Sci. USA 43, 105-114 (1957).- AZZI, J.R.: The Delayed Light and Fluorescence Emission Spectra for Green, Blue-Green, Red Algae, and Nonphotosynthetic Mutants. 31 pp. Oak Ridge Natl. Lab. Rep. ORNL-TM-1534, 1966.

BABCOCK, G.T., SAUER, K.: Biochim. Biophys. Acta 396, 43-62 (1975).- BABCOCK, G.T., BLANKENSHIP, R.E., SAUER, K.: FEBS Lett. 61, 286-289 (1976).- BARBER, J., KRAAN, G.P.B.: Biochim. Biophys. Acta 197, 49-59 (1970).- BARBIERI, G., DELOSME, R., JOLIOT, P.: Photochem. Photobiol. 12, 197-206 (1970).- BENNOUN, P.: Biochim. Biophys. Acta 216, 357-363 (1970).- BLANKENSHIP, R.E., McGUIRE, A., SAUER, K.: Biochim. Biophys. Acta 459, 617-619 (1977a).- BLANKENSHIP, R.E., SCHAAFSMA, T.J., PARSON, W.W.: Biochim. Biophys. Acta 461, 297-305 (1977b).- BÖGER, P., KUNERT, K.-J.: Z. Naturforsch. 34c, 1015-1025 (1979).- BOUGES-BOCQUET, B.: Biochim. Biophys. Acta 314, 250-256 (1973).- BOWES, J., CROFTS, A.R., ARNTZEN, C.J.: Arch. Biochem. Biophys. 200, 303-308 (1980).- BUTLER, W.L.: Acc. Chem. Res. 6, 177-184 (1973).

CARITHERS, R.P., PARSON, W.W.: Biochim. Biophys. Acta 387, 194-211 (1975).- CERMAK, K., KAPLANOVA, M.: Czech. J. Physics 30, 713-716 (1980).- CHENIAE, G.M., MARTIN, I.F.: Brookhaven Symp. Biol. 19, 406-417 (1967);- Biochim. Biophys. Acta 197, 219-239 (1970);- Biochim.

Biophys. Acta 502, 321-344 (1978).- CLAYTON, R.K.: J. Gen. Physiol. 48, 633-646 (1965).- CONJEAUD, H. MATHIS, P., PAILLOTIN, G.: Biochim. Biophys. Acta 549, 280-291 (1979).

DAVIS, M.S., FORMAN, A., HANSON, L.K., THORNBER, J.P., FAJER, J.: J. Phys. Chem. 83, 3325-3332 (1979).- DE GROOTH, B.G.: Optical and Electrical Studies on the Organization of the Photosynthetic Membrane. 128 pp. Thesis Univ. Leiden 1980.- DE GROOTH, B.G., VAN GORKOM, H.J.: Biochim. Biophys. Acta 635, 445-456 (1981). - DE GROOTH, B.G., VAN GORKUM, H.J., MEIBURG, R.F.: Biochim. Biophys. Acta 589, 299-314 (1980).- DEMETER, S., HERCZEG, T., DROPPA, M., HORVATH, G.: FEBS Lett. 100, 321-324 (1979).- DEN HAAN, G.A., DUYSENS, L.N.M., EGBERTS, D.J.N.: Biochim. Biophys. Acta 368, 409-421 (1974).- DUYSENS, L.N.M., DEN HAAN, G.A., VAN BEST, J.A.: Rapid reactions of photosystem 2 as studied by the kinetics of the fluorescence and luminescence of chlorophyll a in *Chlorella pyrenoidosa*, 1-12. In: Proc. 3rd Int. Congr. Photosynthesis, Rehovot 1974, ed. M. AVRON. Amsterdam: Elsevier 1975.

ELLENSON, J.L., SAUER, K.: Photochem. Photobiol. 23, 113-123 (1976).- ETIENNE, A.-L., LAVOREL, J.: FEBS Lett. 57, 276-279 (1975).

FARKAS, D.L., KORENSTEIN, R., MALKIN, S.: FEBS Lett. 120, 236-242 (1980).- FLEISCHMAN, D.E.: Chemiluminescence in photosynthetic bacteria, 952-955. In: Progress in Photosynthesis Research, ed. H. METZNER. Tübingen: Laupp 1969;- Photochem. Photobiol. 14, 277-286 (1971);- Delayed fluorescence and chemiluminescence, 513-523. In: The Photosynthetic Bacteria, eds. R.K. CLAYTON, W.R. SISTROM. New York: Plenum Press 1978.- FORBUSH, B., KOK, G., McGLOIN, M.P.: Photochem. Photobiol. 14, 307-321 (1971).- FOWLER, C.F., NUGENT, A., FULLER, R.C.: Proc. Natl. Acad. Sci. USA 68, 2278-2282 (1971).

GLÄSER, M., WOLFF, C., RENGER, G.: Z. Naturforsch. 31c, 712-721 (1976).- GLÄSER, M., WOLFF, C., BUCHWALD, H.E., WITT, H.T.: FEBS Lett. 42, 81-85 (1974).- GOLTSEV, V.N., ORTOIDZE, T.V., SOCOLOV, Z.N., MATORIN, D.N., VENEDIKTOV, P.S.: Plant Sci. Lett. 19, 339-346 (1980).- GOVINDJEE, VAN RENSEN, J.J.S.: Biochim. Biophys. Acta 505, 183-213 (1978).- GOVINDJEE, FORK, D., WYDRZYNSKI, T., SPECTOR, M., WINGET, G.D.: Photobiochem. Photobiophys. 1, 347-351 (1980).

HOFF, A.J.: Photooxidation of the reaction center chlorophylls. In: Light Reaction Paths of Photosynthesis, ed. F.K. FONG. Berlin: Springer 1981, in press.- HOFF, A.J., RADEMAKER, H., VAN GRONDELLE, R., DUYSENS, L.N.M.: Biochim. Biophys. Acta 460, 547-554 (1977).- HOLT, A.S., CLAYTON R.K.: Photochem. Photobiol. 4, 829-831 (1965).- HOLTEN, D., WINDSOR, M.W., PARSON, W.W., THORNBER, J.P.: Biochim. Biophys. Acta 501, 112-126 (1978).- HOLTEN, D., HOGANSON, C., WINDSOR, M.W., SCHENCK, C.C., PARSON, W.W., MIGUS, A., FORK, R.L., SHANK, C.V.: Biochim. Biophys. Acta 592, 461-477 (1980).- HORTON, P., BAKER, N.R.: Biochim. Biophys. Acta 592, 559-564 (1980).- HORTON, P., CROZE, E.: Biochim. Biophys. Acta 545, 188-201 (1979).

ICHIKAWA, T., INOUE, Y., SHIBATA, K.: Biochim. Biophys. Acta 408, 228-239 (1975).- INOUE, Y.: FEBS Lett. 72, 279-282 (1976);- Biochim. Biophys. Acta 634, 309-320 (1981).- INOUE, Y., SHIBATA, K.: Thermoluminescence bands of chloroplasts as characterized by flash excitation, 211-221. In: Photosynthesis '77, Proc. 4th Internat. Congr. Photosynth., ed. D.O. HALL, J. COOMBS, T.W. GOODWIN. London: The Biochemical Society 1978a;- FEBS Lett. 85, 193-197 (1978b);- Trends Biochem. Sci. 4, 182-184 (1979).- INOUE, Y., WATANABE, A., SHIBATA, K.: FEBS Lett. 101, 321-323 (1979).- INOUE, Y., FURUTA, S., OKU, T., SHIBATA, K.: Biochim. Biophys. Acta 449, 357-367 (1976).- ITOH, S.: Plant Cell Physiol. 21, 873-884 (1980).- ITOH, S., MURATA, N.: Biochim. Biophys. Acta 333, 525-534 (1974).

JENNINGS, J.V., EVANS, M.C.W.: FEBS Lett. 75, 33-36 (1977).- JOLIOT, P., JOLIOT, A., BOUGES, B., BARBIERI, G.: Photochem. Photobiol. 14, 287-305 (1971).- JURSINIC, P., GOVINDJEE, WRAIGHT, C.: Photochem. Photobiol. 27, 61-67 (1978).

KHANNA, R., PFISTER, K., KERESZTIS, A., VAN RENSEN, J.J.S., GOVINDJEE: Biochim. Biophys. Acta 634, 105-116 (1981).- KLIMOV, V.V., DOLAN, E., KE, B.: FEBS Lett. 112, 97-100 (1980).- KLIMOV, V.V., KLEVANIK, A.V., SHUVALOV, V.A., KRASNOVSKY, A.A.: FEBS Lett. 82, 183-186 (1977).- KNAFF, D.B., MALKIN, R.: Biochim. Biophys. Acta 430, 244-252 (1976).- KNAFF, D.B., OLSON, J.M., PRINCE, R.C.: FEBS Lett. 98, 285-289 (1979).- KOK, B., FORBUSH, B., McGLOIN, M.: Photochem. Photobiol. 11, 457-475 (1970).- KRAAN, G.P.B., AMESZ, J., VELTHUYS, B.R., STEEMERS, R.G.: Biochim. Biophys. Acta 223, 129-145 (1970).

LARSEN, H.: On the Microbiology and Biochemistry of the Photosynthetic Green Sulfur Bacteria, Kgl Norske Vidensk. Selsk. Skrift No.1, 200 pp. Trondheim: Bruns Bokhandel 1953.- LASSER-ROSS, N., MALKIN, S., CAHEN, D.: Biochim. Biophys. Acta 593, 330-341 (1980).- LAVOREL, J.: Biochim. Biophys. Acta 325, 213-229 (1973);- Luminescence, 223-317. In: Bioenergetics of Photosynthesis, ed. GOVINDJEE. New York: Academic Press 1975;- Biochim. Biophys. Acta 590, 385-399 (1980).- LAVOREL, J., ETIENNE, A.-L.: In vivo chlorophyll fluorescence, 203-268. In: Primary Processes of Photosynthesis, ed. J. BARBER. Amsterdam: Elsevier 1977.- LURIE, S., BERTSCH, W.F.: Biochim. Biophys. Acta 357, 420-428 (1974a);- Biochim. Biophys. Acta 357, 429-438 (1974b).

MALKIN, S.: Delayed luminescence, 349-431. In: Primary Processes of Photosynthesis, ed. J. BARBER. Amsterdam: Elsevier 1977a;- Delayed luminescence, 473-491. In: Photosynthesis I. Photosynthetic Electron Transport and Photophosphorylation, ed. A. TREBST, M. AVRON. Berlin-Heidelberg-New York: Springer 1977b.- MALKIN, S., CAHEN, D.: Photochem. Photobiol. 29, 803-813 (1979).- MAROTI, P., RINGLER, A., LACZKO, G., SZALAI, L.: Acta Phys. Pol. 54, 789-796 (1978).- MATHIS, P., HAVEMAN, J.: Biochim. Biophys. Acta 461, 167-181 (1977).- MELIS, A.: FEBS Lett. 95, 202-206 (1978).- MELIS, A., SCHREIBER, U.: Biochim. Biophys. Acta 547, 47-57 (1979).- MURATA, N.: Uphill energy transfer from chlorophyll a to phycobilins in the blue-green algae *Anabaena variabilis* and *Anacystis nidulans*, 9-13. In: Photosynthetic Organelles, Special Issue of Plant and Cell Physiology No.3, eds. S. MIYACHI, S. KATOH, Y. FUJITA, K. SHIBATA. Tokyo: Jpn. Soc. Plant. Physiol.- Center Acad. Publ. Japan 1977.

NAKATANI, H.Y., BARBER, J., EVANS, M.C.W.: PS II and a new (?) cholate-extracted protein from thylakoids, 409. In: Abstracts 5th Int. Congr. Photosynth., Halkidiki, Greece 1980.

OLSON, J.M.: Biochim. Biophys. Acta 594, 33-51 (1980).- OLSON, J.M., PHILIPSON, K.D. SAUER, K.: Biochim. Biophys. Acta 292, 206-217 (1973).- OLSON, J.M., PRINCE, R.C., BRUNE, D.C.: Brookhaven Symp. Biol. 28, 238-245 (1977).- ORTOIDZE, T.V., BORISEVITCH, G.P., VENEDIKTOV, P.S., KONONENKO, A.A., MARTORIN, D.N., RUBIN, A.B.: Biochem. Physiol. Pflanz. 174, 85-91 (1975).

PARSON, W.W.: Annu. Rev. Biochem. 47, 635-653 (1978).- PARSON, W.W., MONGER, T.G.: Brookhaven Symp. Biol. 28, 195-212 (1977).- PFISTER, K., ARNTZEN, C.J.: Z. Naturforsch. 34c, 996-1009 (1979).- PIERSON, B.K., CASTENHOLZ, R.W.: Arch. Microbiol. 100, 283-305 (1974);- Photosynthetic apparatus and cell apparatus of the green bacteria, 179-197. In: The Photosynthetic Bacteria, eds. R.K. CLAYTON, W.R. SISTROM. New York: Plenum Press 1978.- PULLES, M.P.J., VAN GORKOM, H.J., WILLEMSEN, J.G.: Biochim. Biophys. Acta 449, 536-540 (1976).

RADEMAKER, H., HOFF, A.J., DUYSENS, L.N.M.: Biochim. Biophys. Acta 546, 248-255 (1979).- RADMER, R., CHENIAE, G.: Mechanisms of oxygen evolution, 303-348. In: Primary Processes of Photosynthesis, ed. J. BARBER. Amsterdam: Elsevier 1977.- RENGER, G.: Biochim. Biophys. Acta 440, 287-300 (1976).

SANE, P.V., TATAKE, V.G., DESAI, T.S.: FEBS Lett. 45, 290-294 (1974).- SANE, P., DESAI, T.S., TATAKE, V.G., GOVINDJEE: Photochem. Photobiol. 26, 33-39 (1977).- SAROJINI, G., GOVINDJEE: Biochim. Biophys. Acta 634, 340-343 (1981).- SAUER, K.: Annu. Rev. Phys. Chem. 30, 155-178 (1979);- Acc. Chem. Res. 13, 249-256 (1980).- SCHMIDT, K.: Arch. Microbiol. 124, 21-31 (1980).- SHUVALOV, V.A. Biochim. Biophys. Acta 430, 113-121 (1976).- SHUVALOV, V.A., KLIMOV, V.V., DOLAN, E., PARSON, W.W., KE, B.: FEBS Lett. 118, 279-282 (1980).- SONNEVELD, A., RADEMAKER, H., DUYSENS, L.N.M.: Biochim. Biophys. Acta 548, 536-551 (1979);- FEBS Lett. 113, 323-327 (1980a).- SONNEVELD, A., DUYSENS, L.N.M., MOERDIJK, A.: Proc. Natl. Acad. Sci. USA 77, 5889-5893 (1980b);- Magnetic field-induced increase of sub-microsecond chlorophyll a delayed fluorescence of photosystem II down to 4.2 K. Evidence for an intermediary acceptor between the primary donor P-680 and acceptor Q. In: Proc. 5th Int. Congr. Photosynth., Halkidiki, Greece 1980, eds. G. AKOYUNOGLOU, G. PAPAGEORGIOU, Jerusalem: Int. Science Services 1981, in press.- SPECTOR, M., WINGET, D.G., Proc. Natl. Acad. Sci. USA 77, 957-959 (1980).- STREHLER, B.L., ARNOLD, W.: J. Gen. Physiol. 34, 809-820 (1951).- SWARTHOFF, T., AMESZ, J.: Biochim. Biophys. Acta 548, 427-432 (1979).- SWARTHOFF, T., VAN DER VEEK-HORSLEY, K.M., AMESZ, J.: Biochim. Biophys. Acta 635, 1-12 (1981a);- Photochemistry of pigment-protein complexes from the green bacterium *Prosthecochloris aestuarii*. In: Proc. 5th Int. Congr. Photosynth., Halkidiki, Greece 1980, eds. G. AKOYUNOGLOU, G. PAPAGEORGIOU. Jerusalem: Int. Sciences Services 1981b, in press.- SWARTHOFF, T., AMESZ, J., KRAMER, H.J.M., RIJGERSBERG, C.P.: Israel J. Chem., in press (1981c).- SWARTHOFF, T., GAST, P., HOFF, A.J.: FEBS Lett. 127, 83-86 (1981d).

TATAKE, V.G., DESAI, T.S., GOVDINJEE, SANE, P.V.: Photochem. Photobiol. 33, 243-250 (1981).- THIELEN, A.P.G.M., VAN GORKOM, H.J.: Biochim. Biophys. Acta 635, 111-120 (1981).- THORNBER, J.P., MARKWELL, J.P., REINMAN, S.: Photochem. Photobiol. 29, 1205-1216 (1979).- TOLLIN, G., FUJIMORI, E., CALVIN, M.: Proc. Natl. Acad. Sci. USA 44, 1035-1047 (1958).- TREDWELL, C.J., SYNOWIEC, J.A., SEARLE, G.F.W., PORTER, G., BARBER, J.: Photochem. Photobiol. 28, 1013-1020 (1978).

VAN BEST, J.A., DUYSENS, L.N.M.: Biochim. Biophys. Acta 459, 187-206 (1977).- VAN BEST, J.A., MATHIS, P.: Biochim. Biophys. Acta 503, 178-188 (1978).- VAN GORKOM, H.J.: Biochim. Biophys. Acta 347, 439-442 (1974).- VAN GORKOM, H.J., PULLES, M.P.J., WESSELS, J.S.C.: Biochim. Biophys. Acta 408, 331-339 (1975).- VAN GORKOM, H.J., THIELEN, A.P.G.M., GORREN, A.C.F.: The secondary electron acceptor of photosystem II. In: Function of Quinones in Energy Conserving Systems, ed. B. TRUMPOWER. New York: Academic Press 1981, in press.- VAN NIEL, C.B.: Arch. Microbiol. 3, 1-112 (1931).- VAN RENSEN, J.J.S., KRAMER, H.J.M.: Plant Sci. Lett. 17, 21-27 (1979).- VASS, I., HORVATH, G., HERCZEG, T., DEMETER, S.: FEBS Lett. 116, 293-297 (1980);- Biochim. Biophys. Acta 634, 140-152 (1981).- VELTHUYS, B.R.: Annu. Rev. Plant Physiol. 31, 545-567 (1980). - VELTHUYS, B.R., AMESZ, J.: Biochim. Biophys. Acta 333, 85-94 (1974).- VENEDIKTOV, P.S., GOLTSEV, V.N., SHINKAREV, V.P.: Biochim. Biophys. Acta 593, 125-132 (1980).- VOZNYAK, W.M., JELFIMOV, J.I., PROSKURYAKOV, I.I.: Doklady Akad. Nauk SSSR 242, 1200-1203 (1978).

WONG, D., GOVINDJEE, JURSINIC, P.: Photochem. Photobiol. 28, 963-974 (1978).- WYDRZYNSKI, T.I., MARKS, S.B., SCHMIDT, P.G., GOVINDJEE,

GUTOWSKY, H.S.: Biochemistry 17, 2155-2163 (1978).- WYDRZYNSKI, T., ZUMBULYADIS, N., SCHMIDT, P.G., GUTOWSKI, H.S., GOVINDJEE: Proc. Natl. Acad. Sci. USA 73, 1196-1198 (1976).

YAMASHITA, T., BUTLER, W.L.: Plant Physiol. 43, 1978-1986 (1968).- YERKES, C.T., BABCOCK, G.T.: Biochim. Biophys. Acta 590, 360-372 (1980).

ZANKEL, K.L.: Biochim. Biophys. Acta 245, 373-385 (1971).

Professor Dr. J. AMESZ
Department of Biophysics
Huygens Laboratory of the
State University
P.O. Box 9504
NL 2300 RA Leiden

II. Metabolism of Organic Acids

By MANFRED KLUGE

The present review deals exclusively with the metabolism and biological significance of malate. Because indeed by far the majority of publications on acid metabolism since the last review (Progr. Bot. 40, 1978) concern malate, this restriction appears to be justified.

The view has emerged from the extensive recent literature dealing with malate that this substance fulfills an extraordinary broad spectrum of different tasks in plant life. Under this aspects malate appears to be rather exceptional in the plant kingdom. According to the major topics of the recent literature on malate, our consideration will treat:
a) metabolic aspects of Crassulacean Acid Metabolism (CAM) syndrome, and
b) the role of malate in nitrogen- and amino acid metabolism.

Not even within the topics mentioned above can this review cover the literature, and other important facets of malate metabolism, for example C_4 photosynthesis, or the importance of malate for the regulation of ion balance in plant cells and organs, have to remain beyond the scope of this year's review.

Crassulacean Acid Metabolism (CAM)

1. Current Reviews on CAM

Several papers have been published which summarize results on CAM obtained (1973 - 1978) by a group of German research workers within in the research program *Biochemical Fundaments of Ecological Adaptations in Plants*. In this context, both DITTRICH (1979b) and KLUGE (1979a) treat the flow of carbon in CAM; WINTER and LÜTTGE (1979) consider the phenomenon of CAM induction by stress in *Mesembryanthemum crystallinum*, and v. WILLERT (1979) the metabolic regulation and ecological behaviour of CAM in various Mesembrynathemaceae.

2. Taxonomic Distribution of CAM

The view that CAM has a convergent evolution is supported by the recent finding of CAM performance by *Isoetes storkii* and other *Isoetes* species (KEELEY et al., 1981). This observation is also important from the ecological point of view, because the *Isoetes* are submerged-aquatic plants, whereas to our knowledge CAM is typically an attribute of xerophytes. Also ZIEGLER (pers. comm.) found CAM in hydrophilic plants, namely *Crassula granvikii* and *C. natans*.

SZAREK provided a supplementary list (cf. SZAREK and TING, 1977) of CAM plants which has been described in the literature during 1976 to 1979. According to this list the CAM-containing taxa now include 3 divisions, 26 families, 159 genera, and 612 species. The above list does not yet include some CAM plants (among them 10 species of *Didieraceae*) of the flora of Madagascar described by WINTER (1979). The report by SANTA-KUMARI et al. (1979) of CAM occurrence in a grain legume (chickpea) needs reevaluation. Characteristics of CAM performance were found also in a C_4-dicot, *Portulaca oleracea* (KOCH and KENNEDY, 1980).

3. Carbon Flow in CAM

a) β-Carboxylation as Initial Step of CAM

Numerous papers deal with PEP-carboxylase which catalyzes β-carboxylation of PEP as the initial step of CAM. The PEP-C of the CAM plant, *Bryophyllum fedtschenkoi*, has now been purified to homogeneity, and some properties of this pure enzyme were studied (JONES et al., 1978, 1981; PAYS et al., 1980). According to the findings of these authors, the CAM plant PEP-C is a tetrameric enzyme with a concentration-dependent dissociation to a dimer. The molecular weight of the monomer was 105,000. Immunological studies on PEP-C obtained from *Sedum morganianum* (MÜLLER et al., 1981) support the view of PEP-C being a tetrameric enzyme. From the results of these studies evidence emerged that the PEP-C dimer consists of two differently sized monomers having molecular weights of 105,000 and 115,000.

As is true for crude PEP-C preparations, the highly purified enzyme also is inhibited by L-malate and activated by glucose-6-P (JONES et al., 1978, 1981; PAYS et al., 1980). This is consistent with the idea that both these effectors are involved in controlling the PEP-C during the diurnal CAM cycle. However, the question arises whether the kinetic properties measured with the highly purified enzyme reflect the real in vivo properties. Namely, WINTER (1980a,b) and JONES et al. (1981) showed that within 1 min after extraction the crude PEP-C was 50 times more susceptible to inhibition by malate than aged extracts or the purified enzyme. Hence, better information on the in vivo properties should be obtained from preparations measured at least within the first 30 min after extraction. Indeed, investigations where this postulate was taken into account revealed the interesting result that in CAM plants the properties of PEP-C may change throughout the diurnal CAM cycle. WINTER (1980a) found a shift in the pH-optimum curve of PEP-C. KLUGE et al. (1980, 1981a), WINTER (1980a,b), v. WILLERT et al. (1979) found the sensitivity of PEP-C to malate to be lower during the night than during the day, and further that the affinity of the enzyme for PEP was lower during the day than during the night (KLUGE et al., 1981a; WINTER, 1980b). The mechanism of this diurnal alteration of PEP-C properties remains unknown, however, apart from some promising hypotheses (WINTER, 1980a). Nevertheless, together with the observation that PEP-C of CAM plants can even be activated by L-malate (v. WILLERT et al., 1979; DITTRICH, pers. comm.), and that also inactivation of PEP-C by light was observed in CAM plants (v. WILLERT and v. WILLERT, 1979), the above results suggest that the originally simple hypothesis of PEP-C being mainly regulated during CAM by malate-dependent feedback inhibition has to be modified (cf. KLUGE and TING, 1978).

The question of endogeneous rhythms if PEP-C capacity is a prerequisite for metabolic CAM control remains controversial. There are further reports in favour of this view (PIERRE and QUEIROZ, 1979; BRULFERT et al., 1979), however, other authors (WINTER, 1980a; JONES et al., 1981) failed to find such capacity rhythms. In *Kalanchoe tubiflora* cultivated in the phytotron at Gif-sur-Yvette, KLUGE et al. (1980, 1981) observed significant diurnal alterations of PEP-C, which, however, did not coincide with the alterations of PEP-C-mediated CO_2 fixation and malate synthesis.

It is worth noting that the use of lyophylized material of CAM plants attempting to measure diurnal courses of PEP-C capacities may create artifacts, i.e., diurnal pseudo-alterations of PEP-C capacity (BUCHANAN-BOLLIG et al., 1980) can be obtained.

Some findings seem to support the idea that the PEP-C involved in CAM is distinguishable from others. In this context, BRULFERT et al. (1979) demonstrated in *K. blossfeldiana* the occurrence of a PEP-C with different molecular weights and kinetic properties when CAM was photoperiodically induced. In *Aloe arborescence*, the PEP-C of the CAM-performing green leaf tissue was kinetically and electrophoretically different from the PEP-C of the water tissue which lacks CAM (KLUGE et al., 1979). A comparison of PEP-C properties in *Sedum* species having either obligate or facultative CAM (see OSMOND, 1978) revealed that the PEP-C obtained from the obligate CAM plants had higher K_m for PEP, higher V_{max} and higher sensitivity to malate (KNOPF and KLUGE, 1979). On the other hand, there is no indication that the PEP-carboxylases from either obligate or facultative *Sedum* species are immunologically different (MÜLLER et al., 1980, 1981).

Until now there existed a discrepancy between the low-temperature optimum of dark CO_2 fixation found in most CAM plants (in the range of 15°C) on the one hand, and the high-temperature optimum of the pure PEP-C (about 45°C) on the other hand. BUCHANAN-BOLLIG and KLUGE (1981) found that in presence of proposed in vivo effectors (malate and G-6-P) and the likely cytoplasmic substrate concentrations the optimum of PEP-C shifts to such reduced temperatures that the high rates of dark CO_2 fixation observed in the range of 15°C are not contradicted by the enzyme activity. As shown in *Opuntia polyacantha* (GERWICK and WILLIAMS, 1979), the growth temperature does not shift the temperature optimum and the apparent E_a of PEP-C. Hence, the properties of the enzyme appear to be not responsible for the different pattern of CO_2 exchange performed by the plant under varied temperature regimes.

b) Intracellular Localization of PEP-C and Other CAM Enzymes

There are controversal results on the localization of PEP-C in the CAM-performing cell. SPALDING et al. (1980) found this enzyme to be not bound to cell organelles. This implicates the localization of the enzyme in the cytoplasm. WINTER (pers. comm.) obtained similar results with PEP-C of *Mesembryanthemum*. SCHNARRENBERGER et al. (1980), however, postulated from his studies in *Kalanchoe daigremontiana* that PEP-C is located nearly exclusively in the chloroplasts. It is not clear why the latter authors found the opposite to SPALDING et al. (1980), as identical techniques (i.e., isopycnic sucrose density gradient centrifugation) were used by both research groups. Since PERROT et al. (1981) by an elegant immunolabeling technique, could directly demonstrate that in C_4 plants PEP-C is located mainly in the cytoplasm, it appears to be more likely that SPALDING et al. (1980) rather than SCHNARRENBERGER et al. (1980) are right.

The isolation of metabolically active chloroplasts from CAM plants remains still a problem. Because of the high starch content which remains even after long darkness, the chloroplasts are destroyed when isolated after the conventional procedures (NISHIDA and SANADA, 1977). Better results can be expected from isolation starting from protoplasts (SPALDING and EDWARDS, 1980; DOADS, 1981). The isolation of CAM mesophyll cells which is a precondition for protoplast isolation has been described (SPALDING and EDWARDS, 1978, 1980; RAO et al., 1980; SEENI and GNANAM, 1980).

There is scarcely any doubt that also in CAM plants pyruvate-Pi-dikinase is a chloroplastidic enzyme (SPALDING et al., 1979). NAD-linked malic enzyme is located in the mitochondria, the NADP-linked malic enzyme, however, in the cytoplasm. Enolase and phosphoglycerate mutase are also enzymes of the soluble cytoplasm (SPALDING et al., 1979).

Phosphoenolpyruvate carboxykinase, which in certain CAM plants initiates the decarboxylation of malate (see DITTRICH, 1979b) is assumed to be also a cytoplasma-linked enzyme (KU et al., 1980).

Since they lack malate dehydrogenase, peroxisomes from CAM plants appear to be different from others. On the other hand, they are normally equipped by the enzymes of the glycolate pathway (HERBERT et al., 1978).

Whether the microtubulous structures found by SALEMA and BRANDAO (1978), SALEMA and SANTOS (1980), and SANTOS and SALEMA (1980) in chloroplasts of CAM plants cells represent in fact deposits of malate dehydrogenase proteins, as the above authors propose, remains to be clarified. There are many electron micrographs of CAM plant cells lacking these structures (cf. KLUGE and TING, 1978). Hence, the above chloroplastidic structures cannot be considered as essential requisites for CAM performance.

c) Nocturnal Storage of Malic Acid

Since KENYON et al. (1978) and KENYON and BLACK (1979) were successful in the attempt to isolate and study vacuoles from a CAM plant *(Sedum telephium)*, there is now further direct evidence that the vacuoles are the site of nocturnal malic acid accumulation in CAM. Is is now generally accepted that the transport of malic acid into the vacuole is an active process (see e.g., LÜTTGE et al., 1975), and considerable progress has been achieved in the attempts to understand the mechanism of nocturnal loading the vacuole with malic acid. Studies of electrochemical properties in the malic acid-accumulating cells, together with application of fusicoccin, has led to the conclusion, that protons are actively pumped rather than malate (LÜTTGE and BALL, 1979; LÜTTGE, 1980). This active H^+ transport across the tonoplast is assumed to be coupled with a passive movement of malate anions, the latter possibly being mediated by a not yet known translocator. The stoichiometry of $2H^+$ per 1 $malate^{2-}$ accumulating nocturnally in the vacuoles has been established repeatedly and convincingly (LÜTTGE and BALL, 1979, 1980). Further it could be shown that this stoichiometry was not altered by lipophilic cations. This indicates that the coupling between transport of protons across the tonoplast must be tight (LÜTTGE and BALL, 1980). Detailed electrochemical studies on compartmentation into cell wall, cytoplasm and vacuole of CAM plant cells led to the postulate that the proton pumps which are involved in loading the vacuole may exchange H^+ and K^+ at the plasmalemma and at the tonoplast (RONA et al., 1980).

The first clear assessment of the energy budget linked with the intracellular transport processes in malate storing CAM cells was recently provided by LÜTTGE et al. (1981). The data of these authors are consistent with the idea that the proton transport is mediated by an ATPase operating with a stoichiometry of $2H^+$ per ATP hydrolyzed. Because the highest rates of oxidative phosphorylation are lower than the rates of malic acid accumulation, ATP supply during the dark period may be limited in malic acid-accumulating CAM cells. For this reason it can be predicted that the genesis of the CO_2-acceptor PEP needed in dark fixation must be provided by phosphorolysis of starch (see SCHILLING and DITTRICH, 1979) and not by hydrolysis which would yield free glucose. Also the abnormal efflux of malate out of the vacuole into the cytoplasm under anaerobiosis (NISHIDA and HAYASHI, 1979) could perhaps be explainable under the aspect of lacking energy to maintain transport against the concentration gradient. It is worth mentioning that the above energetical considerations of LÜTTGE et al. (1981) should in the future be included if quantum requirements of CAM plants are to be considered.

In the context of the above strict $2H^+/1\ malate^{2-}$ stoichiometry during malic acid accumulation, which is one of the most typical criteria of CAM, reports should be mentioned where nocturnally accumulating malate is assumed to be at least partially balanced by anorganic cations rather than by protons (PHILLIPS, 1980; BLOOM, 1979a,b). BLOOM's hypothesis anticipates a diurnally changing shuttle of the antagnonistic cations of malate between the epidermis (guard cells) and the mesophyll. In particular at this point BLOOM's experimental approach is not very convincing and needs further evaluation. Also another report of BLOOM (1979b), namely that the induction of CAM in *Mesembryanthemum crystallinum* depend on a specific salt effect, is contradicted by earlier convincing findings that finally the water availability rather than salt requirement is the major factor of CAM induction in *Mesembryanthemaceae* (WINTER and LÜTTGE, 1979).

d) Malic Acid Consumption in CAM

The pathway of malic acid consumption during the light period is still under discussion. The view is now generally accepted that there are two types of CAM plants as far as malic acid breakdown is concerned (cf. DITTRICH, 1979a,b; KLUGE, 1979b; OSMOND, 1978). One of the two types decarboxylates malic acid by "malic enzyme", yielding CO_2 and pyruvate, whereas in the other type malate is assumed to be metabolized by PEP-carboxykinase. The latter pathway would also produce CO_2, however with PEP instead of pyruvate as the residual C_3 skeleton.

Particularly in the "malic enzyme type" of CAM plants the mitochondria appear to be of major importance for the malic acid breakdown. There were several successful attempts to isolate intact mitochondria from various CAM plants in order to study their properties and enzymatic equipment (ARRON et al., 1979; DAY, 1980; SPALDING et al., 1979; v. WILLERT and SCHWÖBEL, 1978). The above authors agree that the mitochondria of CAM plants are capable of oxidizing malate to CO_2 and pyruvate, presumably via NAD-linked malic enzyme which was shown by ARRON et al. (1979) to be a mitochondrial enzyme. However, SPALDING et al. (1979) and ARRON et al. (1979) observed a limited capacity of the isolated mitochondria to oxidize pyruvate. This observation led therefore to the conclusion that malate-derived pyruvate might be exported out of the mitochondria and converted by gluconeogenesis in the cytoplasm or chloroplasts (HOLTUM, 1979). However, it has to be taken into account that isolated and vigorously washed mitochondria are often depleted of thiamine pyrophosphate (i.e., a cofactor of pyruvate dehydrogenase). As a consequence, the rate of pyruvate oxidation shown by such preparations does not necessarily indicate the in vivo capacity of the mitochondria to oxidize pyruvate (DAY, 1980).

Since DAY found the rate of malate decarboxylation in the mitochondria to be considerably influenced by the pH of the cytoplasm which itself is proposed as fluctuating as a consequence of diurnally changing malic acid levels in the cytoplasm (cf. LÜTTGE et al., 1975; KLUGE and TING, 1978; WINTER, 1980a,b), it is a promising idea (DAY, 1980) that the mitochondria are involved in the control of malic acid breakdown. This idea is further supported by the findings of SPALDING et al. (1979) that the oxidation of malate in the mitochondria is inhibited by high concentrations of $NaHCO_3$, most likely via inhibition of NAD "malic enzyme". This effect could perhaps explain why under conditions of inhibited photosynthesis also the malate consumption is retarded (NISHIDA and HAYASHI, 1980; see also NISHIDA, 1977, 1979). This is so, because initially the malate decarboxylation continues, whereas photosynthetic refixation of the produced CO_2 is blocked. As a consequence of this, the cellular CO_2/bicarbonate concentration should increase (see KLUGE et al., 1979) and hence inhibit further oxidation of malate.

It is an interesting finding by v. WILLERT and SCHWÖBEL (1978) that during the induction of CAM in *Mesembryanthemum crystallinum* the mitochondria of this plant increasingly gained the capacity for direct oxidation of malate. However, this malate oxidation did not result in a large gain of ATP. This finding could support the view that the residual C_3 skeleton is used in toto for gluconeogenesis.

HOLTUM (1978) found that at least in the "malic enzyme type" of CAM plants (see DITTRICH, 1979a,b) the activities of the glycolytic enzymes when assayed in the reverse direction were more than sufficient to allow glyconeogenesis via reversed glycolysis at rates required by the observed malic acid consumption. Also the activity of pyruvate-Pi-dikinase which converts in the "malic enzyme type" of CAM malate-derived pyruvate to PEP, appears to be sufficiently high (HOLTUM, 1979).

In some CAM plants (or under special external conditions?) the malate-derived pyruvate seems to undergo full oxidation to CO_2 instead of being recycled as C_3 skeleton. This was concluded by ANDRE et al. from experiments where simultaneous uptake and output of CO_2 and O_2 by CAM performing leaves were measured.

The regulation of malic acid decarboxylation in CAM remains an unsolved problem. Apart from the view that the release of malic acid from the vacuole, hence the malate concentration at the site of decarboxylation is an important control step (KLUGE and TING, 1978; USUDA, 1980), MOREL (1979) and MOREL and VALON (1979) support the idea that, due to an endogenous rhythm, the activity of "malic enzyme" increases at the beginning of the day. The dawn signal, together with the phytochrome system, is supposed to act as timer for this rhythm. It should be noted, however, that the results of USUDA are contradictory to the hypothesis of endogenous rhythms of malic enzyme being important for the control of decarboxylation.

MOREL (1979) and MOREL and VALON (1979) have also stressed the idea that by decarboxylation of malate, hence by pyruvate production in the light, CAM might play an important anaplerotic role in relation to the TCA cycle and amino acid synthesis. This hypothesis suggests further that the rhythm of the TCA cycle would be modified by the rhythmically changing supply of pyruvate. Hence, CAM is supposed to act as the coordinator of the intermediary metabolism, including the metabolism of amino acids.

MOREL et al. (1980) found that during the photoperiodic induction of CAM in *K. blossfeldiana* the level of arginine and putrescine increased and showed a diurnal rhythm. Putrescine is believed by the above authors to maintain a constant pH of the cytoplasm during periods of higher amounts of malic acid being present in this compartment.

e) Studies of Carbon Isotope Composition in Context with CAM Carbon Flow

In *K. blossfeldiana* var. Tom Thumb, DELEENS et al. (1979) have studied alterations of carbon isotope composition in metabolic fractions during photoperiodic induction of CAM. In long days, CAM activity was low and the leaves substantially accumulated soluble sugars which were poor in ^{13}C, hence must derive from the RuDP-C-mediated CO_2 fixation. After CAM induction by short-day treatment, the carbon isotope composition of the leaves suggest that now 100% of the carbon linked with the starch malate sequence has entered the pathway via PEP-C-mediated β-carboxylation.

The earlier interesting finding (DELEENS et al., 1979) that in CAM plants two functionally different glycolytic pathways are operating was reconfirmed (DELEENS et al., 1981). These authors further postulate that at least in facultative CAM plants two starch-malate sequences are operating: one develops regularly during ageing of the leaves, the other is established by the factors responsible for CAM induction. The same authors argue also in favour of the view that the biosynthesis of starch in the younger leaves depends on import of carbon rich in ^{13}C (hence deriving from β-carboxylation) from older leaves. In spite of substantial transport of carbon between the plant organs, the carbon isotope composition between the organs shows differences: leaves are always poorer in ^{13}C than the shoot. The roots are always richest in ^{13}C (LENHART, 1979).

It is worth mentioning some basic considerations of the mechanism of $[^{13}C]$ discrimination in plants, also with respect to CAM: SCHMIDT and WINKLER (1979) discussed the reason for variation in the range of $|^{13}|$ values found in plants. O'LEARY and OSMOND (1980) have shown that the isotope fractionation is not only due to the carboxylation step rather than completed by diffusionaly fractionation during photosynthetic carboxylation. Nevertheless the carboxylation itself can still be considered as to be of major importance. LENHART (1979) has estimated the kinetic $[^{13}C]$-isotopic effect of RuDP-carboxylase as to be in K_{12}/K_{13} with 1.0213. It was independent whether the carboxylase was obtained from a C_3, C_4 or a CAM plant. Also the kinetic isotopic effect of PEP-carboxylase was independent of the origin (LENHART, 1979). In *K. daigremontiana* (a CAM plant) it was $K_{12}/K_{13} = 1.0024$ and corresponded to values of PEP-C from C_3 and C_4 plants.

The Role of Malate and PEP-Carboxylase-Mediated CO_2 Fixation in Nitrogen Metabolism

There is increasing evidence that malate can be rather directly involved in the reduction of nitrate by leaves. In this process, malate can obviously serve as the source for cytosolic NADH. This is to be concluded by the effect that addition of malate and NAD to crude leaf extracts, or the feeding of malate to leaf discs caused reduction of nitrate (NEYRA and HAGEMANN, 1976). WOO and CANVIN (1980c) found that under aerobic conditions and light, malate stimulated the nitrite accumulation in spinach leaf discs. Further, malate could, under anaerobic conditions and darkness, reverse the inhibition of nitrate reduction caused by iodoacetamide and arsenite.

The transfer of reduction equivalents from the mitochondria into the cytoplasm where nitrate reductase is believed to be located is mediated by the OAA/malate shuttle (WOO et al., 1980). This shuttle not only provides the reduced coenzymes in the cytoplasm but also allows oxidation of Krebs-cycle intermediates independently of the operation of electron transport.

The products of nitrate reduction, i.e., nitrite and finally NH_4^+ are repeatedly shown to stimulate PEP-C-mediated β-carboxylation of PEP (HAMMEL et al., 1979; WOO and CANVIN, 1980a,b; PETERSON and EVANS, 1979; CHRISTELLER et al., 1977; ENGEMANN and BOWN, 1981). The causal chain of PEP-C activation by NH_4^+ is not clarified. HAMMEL et al. (1979) assume that excess of ammonia would increase the pH of the cytoplasm. This increased cytoplasmic pH would then stimulate PEP-C activity, resulting in the neutralization of ammonia by the organic acids deriving from β-carboxylation. A similar explanation has been given by ENGEMANN and BOWN (1981) who found an increase of dark CO_2 fixation by NH_4^+ in *Avena* coleoptiles.

There is general agreement that the physiological advantage of activation of dark CO_2 fixation by NH_4^+ consists in the increased availability of carbon skeletons needed for transamination. Specially high demands for this exist in N_2 fixing endosymbionts by legumes. Dark CO_2 fixation and PEP-C appear therefore to be particularly important in nodules of Fabaceae (CHRISTELLER et al., 1977; De VRIES et al., 1980; PETERSON and EVANS, 1979). GADAL (pers. comm.) showed by immunofluorescence labelling of PEP-C that this enzyme is present only in those parts of the nodules where the symbiontic bacterioids have their typcial shape and are active in fixing N_2. In the older parts of the nodules, where the bacterioids are digested, PEP-C is lacking. In N_2 fixation by the symbiontic blue-green algae of the lichen *Peltigera aphtuosa* PEP-C is also involved (RAI et al., 1981). The enzyme is supposed to be located in the fungus and provides carbon skeletons to the algal pàrtner.

MAKOVACA and SINDELAR (1978) observed that in tobacco plants infected with Tobacco Mosaic Virus the RuDP carboxylase activity decreases, PEP-C activity, however, increases drastically. It is conceivable that the increase of PEP-C reflects a high rate of acid synthesis, in this case due to a probably high demand for amino acids needed in production of virus protein.

References

ANDRÉ, M., THOMAS, D.A., v. WILLERT, D.J., GERBAND, A.: Planta 147, 141-144 (1979). - ARRON, G.F., SPALDING, M.H., EDWARDS, G.H.: Plant Physiol. 64, 182-186 (1979).

BEER, S., WETZEL, R.G.: Plant Sci. Lett. 21, 199-207 (1981). - BLOOM, A.: Plant Physiol. 64, 919-923 (1979a); - Plant Physiol. 63, 749-753 (1979b). - BRULFERT, J., IMHOFF, C., FONTAIN, D.: In: Etudes Biologique Végétale. Hommage à Professeur P. Chouard, pp. 443-455. Paris: 1976. - BRULFERT, J., ARRABACA, M.C., GUERRIER, D., QUEIROZ, O.: Planta 146, 129-133 (1979). - BUCHANAN-BOLLIG, I.C., KLUGE, M.: Planta (1981, in press). - BUCHANAN-BOLLIG, I.C., KLUGE, M., LÜTTGE, U.: Zeitschr. Pflanzenphysiol. 97, 457-470 (1980).

CHRISTELLER, J., WA, L.I., SUTTON, W.: Plant Physiol. 60, 47-50 (1977).

DAY, D.: Plant Physiol. 65, 675-679 (1980). - DELEENS, E., GARNIER-DARDART, J., QUEIROZ, O.: Planta 146, 441-449 (1979); - Proc. of the 5th International Congr. Photosynthesis (in press), 1981. - DITTRICH,

P.: Ber. Dtsch. Bot. Ges. 92, 109-117 (1979a); - Enzymes of Crassulacean acid metabolism, 263-270. In: Enzyclopedia of Plant Phyiology, N.S., Vol. 6, Photosynthesis II, eds. M. GIBBS, E. LATZKO. Berlin-Heidelberg-New York: Springer 1979b. - DOADS, J.H.: Zeitschr. Pflanzenphysiol. 96, 117-181 (1981).

ENGEMANN, J., BOWN, A.W.: Plant Sci. Lett. 21, 253-262 (1981).

GERWICK, B., WILLIAMS, G.J.: Photosynthetica 13, 254-259 (1979). - GERWICK, B., WILLIAMS, G.J., SPALDING, M.H., EDWARDS, G.E.: Plant Sci. Lett. 13, 389-396 (1978).

HAMMEL, K., CORNWELL, K., BASSHAM, J.: Plant Cell Physiol. 20, 1523-1529 (1979). - HERBERT, M., BURKHARD, C., SCHNARRENBERGER, G.: Planta 143, 279-284 (1978). - HOLTUM, J.: Communication given to the Meeting of the Austr. Society of Plant Physiol. 1979.

JONES, R., WILKINS, M.B., COGGINS, J.R., FEWSON, C.H., MALCOLM, A.D.B.: Biochem. Journal 175, 391-406 (1978). - JONES, R., BUCHANAN, I.C., WILKINS, M.B., FEWSON, C.H., MALCOLM, A.D.B.: J. Exp. Bot. 32, 427-443 (1981).

KARADGE, B., JOSHI, G.: Indian J. Exp. Biol. 18, 631-634 (1980). - KEELEY, J.K., MORTON, B., BABCOCK, B., CASTILLO, P., FISH, B., JERAULD, E., JOHNSON, B., LANDRE, L.: Oecologia 48, 332-333 (1981). - KENYON, W.H., BLACK, C.C., Jr.: Plant Physiol. 63, pp 37 (1979). - KENYON, W., KRINGSTAD, R., BLACK, C.: FEBS Lett. 94, 281-283 (1978). - KLUGE, M.: Ber. Dtsch. Bot. Ges. 92, 95-107 (1979a); - The flow of carbon in Crassulacean Acid Metabolism (CAM), 113-115. In: Encyclopedia of Plant Physiology, N.S., Vol. 6, Photosynthesis II, eds. M. GIBBS, E. LATZKO. Berlin-Heidelberg-New York: Springer 1979b. - KLUGE, M., TING, I.: Crassulacean Acid Metabolism (Ecological Studies, Bd. 30), 209 pp. Berlin-Heidelberg-New York: Springer 1978. - KLUGE, M., KNAPP, J., KRAMER, D., SCHWERDTNER, I., RITTER, H.: Planta 145, 357-363 (1979). - KLUGE, M., BÖCHER, M., JUNGNICKEL, G.: Z. Pflanzenphysiol. 97, 197-205 (1980). - KLUGE, M., BRULFERT, J., QUEIROZ, O.: Plant Cell Environ. (1981a, in press). - KLUGE, M., BÖHLKE, Ch., QUEIROZ, O.: Planta 152, 87-92 (1981b). - KNOPF, O., KLUGE, M.: Plant Cell Environ. 2, 73-78 (1979). - KOCH, K., KENNEDY, R.: Plant Physiol. 65, 193-197 (1980). - KU, M.S.B., SPALDING, M.H., EDWARDS, G.E.: Plant Sci. Lett. 19, 1-8 (1980).

LENHART, B.: Dissertation FB Chemie, Biologie und Geowissenschaften der TU München, 1979. - LÜTTGE, U.: In: Plant Membrane Transport, ed. J. DAINTY. Amsterdam: Elsevier/North Holland 1980 (in press). - LÜTTGE, U., BALL, E.: J. Membrane Biol. 47, 401-422 (1979); - Plant Cell Environ. 3, 195-200 (1980). - LÜTTGE, U., KLUGE, M., BALL, E.: Plant Physiol. 56, 613-616 (1975). - LÜTTGE, U., SMITH, J.A.C., MARIGO, G., OSMOND, C.B.: FEBS Lett. 126, 81-84 (1981).

MAKOVCOVA, O., SINDELAR, L.: Biol. Plant 20, 135-137 (1978). - MOREL, C.: Physiol. Veg. 17, 697-712 (1979). - MOREL, C., QUEIROZ, O.: Plant Cell Environ. 1, 141-150 (1978). - MOREL, C., VALON, C.: Physiol. Veg. 17, 713-730 (1979). - MOREL, C., VILLANUEVA, V.R., QUEIROZ, O.: Planta 149, 440-444 (1980). - MÜLLER, D., KLUGE, M., GRÖSCHEL-STEWART, U.: Plant Cell Environ. 3, 249-253 (1980); - Plant Cell Environ. (1981, in press).

NEYRA, C.A., HAGEMANN, R.H.: Plant Physiol. 58, 726-730 (1976). - NISHIDA, K.: Plant Cell Physiol. 18, 927-930 (1977); - Plant Cell Physiol. 20, 259-261 (1979). - NISHIDA, K., HAYASHI, Y.: Plant Cell

Physiol. 20, 1209-1215 (1979); - Plant Sci. Lett. 19, 271-276 (1980). - NISHIDA, K., SANADA, Y.: Photosynth. Organells; Spec. Issue of Plant Cell Physiol. 341-346 (1977).

OSMOND, C.B.: Annu. Rev. Plant Physiol. 29, 379-414 (1978). - O'LEARY, H., OSMOND, C.B.: Plant Physiol. 66, 931-934 (1980).

PAYS, A., JONES, R., WILKINS, M.: Biochim. Biophys. Acta 614, 151-162 (1980). - PERROT, C., VIDAL, J., BURLET, A., GADAL, P.: Planta 151, 226-231 (1981). - PETERSON, J., EVANS, H.: Biochim. Biophys. Acta 567, 445-452 (1979). - PHILLIPS, R.: Nature 287, 727-728 (1980). - PIERRE, J., QUEIROZ, O.: Planta 144, 143-151 (1979).

RAI, A.N., ROWELL, P., STEWART, W.D.P.: Planta 151, 256-264 (1981). - RAO, I., RAJENDRUDU, G., DAS, V.: Curr. Sci. 49, 389-390 (1980). - RONA, J.P., PITMAN, M.G., LÜTTGE, U., BALL, E.: J. Membr. Biol. 57, 25-35 (1980).

SALEMA, R., BRANDAO, J.: J. Ultrastructure Res. 62, 132-136 (1978). - SALEMA, R., SANTOS, J.: Electron Microscopy 2, 242-243 (1980). - SANTAKUMARI, M., REEDY, C., REEDY, A., DAS, V.: Naturwissenschaften 66, 54-55 (1979). - SANTOS, D., SALEMA, R.: Electron Microscopy 2, 246-248 (1980). - SCHILLING, N., DITTRICH, P.: Planta 147, 210-215 (1979). - SCHMIDT, H.L., WINKLER, F.J.: Ber. Dtsch. Bot. Ges. 92, 185-191 (1979). - SCHNARRENBERGER, C., GROSS, D., BURKHARD, C., HERBERT, M.: Planta 147, 477-484 (1980). - SEENI, S., GNANAM, A.: Physiol. Plant 49, 465-472 (1980). - SPALDING, M., EDWARDS, G.: Planta 141, 59-63 (1978); - Plant Physiol. 65, 1044-1048 (1980). - SPALDING, M., SCHMITT, M.R., KU, S.B., EDWARDS, G.E.: Plant Physiol. 63, 738-743 (1979). - SPALDING, M., ARRON, G.P., EDWARDS, G.E.: Arch. Biochem. Biophys. 199, 448-456 (1980). - SZAREK, S.R.: Photosynthetica 13, 467-473 (1979). - SZAREK, S.R., TING, I.P.: Photosynthetica 11, 330-342 (1977).

USUDA, H.: Plant Cell Physiol. 21, 211-217 (1980).

VRIES, G.E. de, VELD, I.T., KIJNE, J.W.: Plant Sci. Lett. 20, 115-123 (1980)

WILLERT, D.J. v.: Ber. Dtsch. Bot. Ges. 92, 133-144 (1979). - WILLERT, D.J. v., SCHWÖBEL, H.: 403-410 in: Plant Mitochondria, eds. G. DUCET, L. LANCE. Amsterdam-New York: Elsevier/North Holland, Biomedical Press 1978. - WILLERT, D.J. v., WILLERT, K. v.: Z. Pflanzenphysiol. 95, 43-50 (1979). - WILLERT, D.J. v., BRINCKMANN, E., SCHEITLER, B., THOMAS, D.A., TREICHEL, S.: Planta 147, 31-36 (1979). - WINTER, K.: Oecologia 40, 104-112 (1979); - Plant Physiol. 65, 792-796 (1980a); - 5th Int. Congress of Photosynthesis 1980b. - WINTER, K., LÜTTGE, U.: Balance between C_3 and CAM pathway of photosynthesis, 323-332. In: Water and Plant Life, eds. O.L. LANGE, L. KAPPEN, E.-D. SCHULZE (Ecological Studies, Vol. 19). Berlin-Heidelberg-New York: Springer 1976; - Ber. Dtsch. Bot. Ges. 92, 117-132 (1979). - WOO, K.C., CANVIN, D.T.: Can. J. Bot. 58, 505-510 (1980a); - Can. J. Bot. 58, 517-521 (1980b); - Can. J. Bot. 58, 511-516 (1980c). - WOO, K.C., JOKINEN, M., CANVIN, D.T.: Plant Physiol. 65, 433-436 (1980).

Professor Dr. MANFRED KLUGE
Botanisches Institut der
Technischen Hochschule
Schnittspahnstr. 3-5
D 6100 Darmstadt

III. Metabolism of Inorganic Nitrogen Compounds

By ERICH KESSLER

1. Dinitrogen Reduction

In 1979 and 1980 there appeared again many reviews on various aspects of nitrogen fixation. These include biochemistry (MORTENSON and THORNELEY, 1979; YATES, 1980), bioenergetic problems (HAGEMAN and BURRIS, 1980; LJONES, 1979), genetics and regulation (BRILL, 1980; ELMERICH, 1979), nitrogen fixation and hydrogen metabolism (ROBSON and POSTGATE, 1980; YOCH, 1978), nitrogen fixation and photosynthesis in blue-green algae (STEWART, 1979, 1980), the symbiosis of legumes (PHILLIPS, 1980), the associations of bacteria with grass roots (VAN BERKUM and BOHLOOL, 1980) and of blue-green algae with eukaryotic plants (STEWART, 1978), and the exploitation of nitrogen fixation (POSTGATE, 1980). Books and symposium volumes were published by HARDY et al. (1979), HEWITT and CUTTING (1979), and STEWART and GALLON (1980).

a) Dinitrogen Reduction by Free-Living Organisms

After a period of rapid and impressive success (cf. Progr. Bot. 38, 108; 41, 71), the pace of progress in the elucidation of the biochemical mechanism of dinitrogen reduction seems to have slowed down. The application of sophisticated modern techniques, like EPR and Mössbauer spectroscopy, tends to produce results that are of immediate interest only to the specialist.

Cyclopropene was found to be a substrate for the nitrogenase of *Azotobacter vinelandii* in vivo and in vitro. It is reduced in an ATP-dependent reaction to propene and cyclopropane (McKENNA and HUANG, 1979; McKENNA et al., 1979).

An analysis of the Fe proteins of the nitrogenases from *Azotobacter vinelandii*, *Clostridium pasteurianum*, and *Klebsiella pneumoniae* revealed a high degree of similarity, but also significant differences, of the amino acid sequences (HAUSINGER and HOWARD, 1980). Antibodies against the MoFe protein of the nitrogenase from *Azotobacter vinelandii* show an immunological cross-reaction with nitrate reductases from bacteria, but not from blue-green algae or higher plants. This suggests an evolution from a common ancestor protein of bacterial nitrogenase and nitrate reductase (NICHOLAS et al., 1979).

ROBSON (1979) isolated and characterized an oxygen-stable nitrogenase from *Azotobacter chroococcum*. It is an equimolar complex of MoFe protein, Fe protein, and a protective substance which was found to be an iron-sulfur protein with a molecular weight of 14,000.

The factor necessary for the activation of the Fe protein of nitrogenase (cf. Progr. Bot. 41, 71) from the purple nonsulfur bacteria *Rhodospirillum rubrum* and *Rhodopseudomonas capsulata* requires manganese (YOCH, 1979; YOCH and CANTU, 1980).

The rapid, reversible inactivation by NH_4^+ of nitrogenase first observed in *Rhodopseudomonas* (cf. Progr. Bot. 41, 72; HILLMER and FAHLBUSCH, 1979) was found also in *Azotobacter* (LAANE et al., 1980). Glutamine synthetase is involved in the regulation by ammonia of nitrogenase activity, which is due to the interconversion of an active and an inactive form of the enzyme (HILLMER and FAHLBUSCH, 1979; JONES and MONTY, 1979; YOCH, 1980; YOCH and CANTU, 1980).

A plasmid carrying the genes for nitrogen fixation (*nif* genes) from *Klebsiella pneumoniae* was introduced into *Escherichia coli* and enabled it to grow without combined nitrogen (PÜHLER et al., 1979). A further transfer of the *nif* genes from *E. coli* to *Enterobacter cloacae* was achieved through mating (KLEEBERGER and KLINGMÜLLER, 1980). Similarly, an expression of the *nif* genes from *Klebsiella pneumoniae* was found to occur in *Serratia marcescens* and *Erwinia herbicola*, but not in *Proteus mirabilis* (KRISHNAPILLAI and POSTGATE, 1980).

Cloned *nif* genes from *Klebsiella pneumoniae* show hybridization with the DNA from 19 strains of nitrogen-fixing bacteria (including *Clostridium*, *Azotobacter*, *Rhizobium*, and *Rhodopseudomonas*), but no hybridization with the DNA from 10 bacteria without nitrogenase (RUVKUN and AUSUBEL, 1980).

In *Rhodopseudomonas capsulata* the reduction of dinitrogen is not obligatorily coupled to photosynthesis. It can occur also in the dark, either with anaerobic fermentation of sugars or with aerobic respiration at low oxygen tension (MADIGAN et al., 1979). Some strains of *Erwinia herbicola* were found to fix nitrogen in anaerobiosis, and a few strains of *Enterobacter agglomerans* under aerobic and anaerobic conditions (PAPEN and WERNER, 1979).

The nitrogenase from the blue-green alga (cyanobacterium) *Anabaena cylindrica* was purified and shown to be similar to bacterial nitrogenases (cf. Progr. Bot. 41, 71). Its MoFe protein has a molecular weight of 220,000 and contains 2 Mo, 20 Fe, and 20 atoms sulfide. Cross-reaction occurs with nitrogenase components from *Azotobacter*, *Rhodospirillum*, and *Plectonema* (HALLENBECK et al., 1979).

In *Anabaena cylindrica* ammonia inhibits only the differentiation of heterocysts, but not the synthesis of nitrogenase in previously formed heterocysts (MURRY and BENEMANN, 1979). A strain of *Anabaena* sp., on the other hand, does not produce heterocysts and nitrogenase in the presence of nitrate, but develops active heterocysts in media containing NH_4^+ (BOTTOMLEY et al., 1979).

A hybridization of the DNA of cloned *nif* genes from *Anabaena* and from *Klebsiella pneumoniae* was demonstrated by MAZUR et al. (1980). Mutants without nitrate reductase of *Cylindrospermum fertilissimum* proved to be defective also in nitrogen fixation. This suggests the existence of a common genetic determinant for nitrate reductase and nitrogenase (DIKSHIT and DIKSHIT, 1980).

The nonheterocystous *Plectonema boryanum* can grow and fix nitrogen in the light under anaerobic conditions either with CO_2 (PEARSON and HOWSLEY, 1980) or in the presence of fructose and sulfide (ROGERSON, 1980).

Coral reef sponges were found to harbor symbiotic blue-green algae capable of reducing N_2 (WILKINSON and FAY, 1979). The diatom *Rhopalodia gibba*, which contains blue-green symbiotic inclusions, is able to reduce C_2H_2 in the light and to grow in the absence of nitrogen compounds (FLOENER and BOTHE, 1980).

A unicellular green alga of the *Chlorella* type from a hot spring grows without combined nitrogen. The organism reduces C_2H_2 and $^{15}N_2$ at low oxygen pressure (YAMADA and SAKAGUCHI, 1980). This surprising result may be due to a (so far undetected) prokaryotic symbiont; otherwise it would represent the first example of dinitrogen reduction in a eukaryotic alga. In this connection it is interesting to note that BRADLEY (1980) obtained growth of carrot callus on a nitrogen-deficient medium only in the presence of a *Chlorella* strain that was apparently capable of nitrogen fixation.

b) Symbiotic Dinitrogen Reduction

Nitrogen fixation by free Rhizobia associated with callus cultures or intact nonleguminous plants (cf. Progr. Bot. 38, 110; 41, 73) continues to attract much attention. Nitrogenase activity in *Rhizobium* is increased by plant growth substances like 6-benzyladenine, α-naphthylacetic acid, and gibberellic acid (HESS and FEUEREISEN, 1980). The nitrogen-fixing bacteria show a polar attachment to the root surface of *Petunia* and wheat plants (GÖTZ, 1980; GÖTZ and HESS, 1980).

Cyclic guanosine 3',5'-monophosphate inhibits nitrogen fixation in free-living *Rhizobium japonicum*. The effect is due to an inhibition of the synthesis of nitrogenase proteins and may be involved in the regulation of the expression of the *nif* genes (LIM et al., 1979).

At least some of the *nif* genes of *Rhizobium* are located on indigenous plasmids (NUTI et al., 1979). They can be transferred (cf. Progr. Bot. 38, 111; 41, 74) also to *Agrobacterium tumefaciens* (STANLEY and DUNICAN, 1979).

The stems of the legume *Aeschynomene indica* exhibit nodules which contain *Rhizobium* and are able to reduce nitrogen (YATAZAWA and YOSHIDA, 1979).

Nodulation of *Parasponia andersonii* (Ulmaceae) occurs with 34 (out of 39) slow-growing *Rhizobium* strains from tropical legumes and with *Rh. japonicum* and *Rh. meliloti*. The resulting nitrogenase activities vary widely. Rhizobia isolated from *Parasponia*, on the other hand, failed to nodulate most legumes (TRINICK and GALBRAITH, 1980).

An increasing number of papers report on the isolation of and successful re-infection with the actinomycetal symbionts from nonlegumes (cf. Progr. Bot. 41, 74). *Frankia* from *Alnus* or *Comptonia* is able to nodulate *Alnus*, *Myrica*, and *Comptonia*, i.e., members of the Betulaceae and Myricaceae (BAKER and TORREY, 1980; LALONDE, 1978, 1979). The symbiont from *Elaeagnus* was isolated, too (BAKER et al., 1979). Free-living *Frankia* from *Comptonia peregrina* reduced C_2H_2 under special culture conditions and only when thick-walled vesicles had been formed, which might be the site of nitrogenase activity (TJEPKEMA et al., 1980). Hydrogenase activity was detected in intact and disrupted nodules from *Alnus* and *Myrica* (BENSON et al., 1979, 1980).

Two nodulated Rosaceae were found to reduce C_2H_2 and $^{15}N_2$: *Chamaebatia foliolosa* and *Cowania mexicana* (HEISEY et al., 1980; RIGHETTI and MUNNS, 1980).

The nitrogen-fixing associations and symbioses in the rhizosphere of grasses (cf. Progr. Bot. 41, 74) continue to be in the center of interest. *Azospirillum brasilense* and *A. lipoferum* (= *Spirillum lipoferum*) are able to reduce dinitrogen either in the presence of very low O_2 concentrations or of nitrate, coupled to respiration or to dissimilatory nitrate reduction (SCOTT et al., 1979). Dihydrogen evolved by the nitrogenase of *Azospirillum* can be recycled by a hydrogenase (BERLIER and LESPINAT, 1980; CHAN et al., 1980). A supply to roots of *Zea mays* and *Setaria* of nitrogen fixed by *Azospirillum* was found by COHEN et al. (1980). Microaerophilic N_2-fixing bacteria from the roots of *Spartina alterniflora* from a Canadian salt marsh belong to the genus *Campylobacter* (MC CLUNG and PATRIQUIN, 1980).

2. Nitrate Reduction

a) Assimilatory Nitrate Reduction

Nitrate reduction was reviewed by BEEVERS and HAGEMAN (1980), LOSADA and GUERRERO (1979), VENNESLAND and GUERRERO (1979), and in the volume edited by HEWITT and CUTTING (1979). In addition, a review on the assimilation of sulfur compounds (SCHMIDT, 1979) should be mentioned.

Nitrate reductases from various organisms were purified and characterized (cf. Progr. Bot. 38, 111; 41, 75): Barley (KUO et al., 1980: molecular weight 221,000, with 2 subunits of about 100,000; SMALL and WRAY, 1980: m.w. 203,000, with subunits of 40,000 and 61,000), *Ankistrodesmus braunii* (DE LA ROSA et al., 1980: m.w. 460,000, with 8 subunits of 58,000), *Chlorella vulgaris* (GIRI and RAMADOSS, 1979: m.w. 280,000, with 3 subunits of 90,000, globular structure), *Aspergillus nidulans* (DOWNEY and STEINER, 1979: 4 proteins of m.w. 360,000, 300,000, 240,000, 118,000, with subunits of 49,000, 50,000, 75,000), *Rhodopseudomonas capsulata* (ALEF and KLEMME, 1979: m.w. 185,000, with 2 subunits of 85,000). The Mo cofactor of the nitrate reductase from *Escherichia coli* has a molecular weight of less than 2,000 and is able to complement the deficient enzyme of a mutant of *Neurospora crassa* (AMY and RAJAGOPALAN, 1979). The Mo cofactors of the nitrate reductases from lupin nodule bacteroids (ALIKULOV et al., 1980) and *Chlorella* (JOHNSON et al., 1980) were found to contain an aromatic component and a pteridine, respectively.

An addition of ammonia at low concentration leads to a rapid, reversible inhibition of nitrate reduction in the light by *Anacystis nidulans* and other blue-green algae. This happens only when the ammonia can be assimilated via the glutamine synthetase-glutamate synthase pathway (FLORES et al., 1980).

The nitrite reductase from spinach contains 6 mol Fe and 4 mol acid-labile sulfide per mol siroheme. The EPR spectrum shows that the enzyme has a tetranuclear iron-sulfur center (LANCASTER et al., 1979). A very stable Fe(II)-NO complex of siroheme is formed during the reduction of nitrite to ammonia by nitrite reductase from *Cucurbita pepo* (FRY et al., 1980).

In agreement with earlier results (cf. Progr. Bot. 36, 102), the nitrite reductase of pea leaves and root tips is localized in plastids, whereas the nitrate reductase is not associated with organelles (WALLSGROVE et al., 1979; EMES and FOWLER, 1979). MOORE and BLACK (1979) confirmed for the C_4 plant *Digitaria sanguinalis* the presence in the mesophyll cells of nitrate reductase and nitrite reductase. Isolated bundle-sheath cells, on the other hand, were unable to reduce nitrate and nitrite.

The repression by NH_4^+ of nitrate reductase in *Neurospora crassa* is not due to ammonia itself, but rather to glutamine produced by its assimilation via glutamine synthetase (DUNN-COLEMAN et al., 1979; PREMAKUMAR et al., 1979). The addition of L-methionine-DL-sulfoximine, an inhibitor of glutamine synthetase, therefore leads to an induction in *Neurospora crassa* and *Cyanidium caldarium* of nitrate reductase even in the presence of ammonia (PREMAKUMAR et al., 1980; RIGANO et al., 1979). In extracts of NH_4^+-grown *Chlorella* cells, which do not possess nitrate reductase activity, a protein immunologically related to nitrate reductase was found to be present (FUNKHOUSER and RAMADOSS, 1980; cf. also JOHNSON, 1979). *Neurospora crassa*, on the other hand, does not contain antigenically detectable nitrate reductase protein after growth with ammonia (AMY and GARRETT, 1980).

Synchronized cultures of *Ankistrodesmus* and *Chlorella* show an increase of nitrate reductase activity after illumination and a decline in the dark. Studies with inhibitors indicate that the increase in the light is due to photosynthesis and de novo synthesis of nitrate reductase (FISCHER and SIMONIS, 1979; GRIFFITHS, 1979; cf. also TISCHNER and LORENZEN, 1979). Further evidence for the control by phytochrome of the synthesis and activity of nitrate reductase in various higher plants comes from the work of KNYPL (1980), RAO et al. (1980), SASAKAWA and YAMAMOTO (1979), VIJAYARAGHAVAN et al. (1979), WHITELAM and JOHNSON (1980), and WHITELAM et al. (1979).

b) Dissimilatory Nitrate Reduction

A review on dissimilatory nitrate reduction, with special emphasis on bioenergetic aspects, was published by ZUMFT and CARDENAS (1979).

The dissimilatory nitrate reductase of *Bacillus licheniformis* was extracted from membranes and purified. Its molecular weight is 193,000, with subunits of 150,000 and 57,000. The enzyme contains 0.93 Mo (in a cofactor of low molecular weight), 6.9 atoms non-heme Fe, and 6.7 atoms acid-labile sulfide (VAN'T RIET et al., 1979).

The dissimilatory nitrate reductases of *Bacillus licheniformis* and *Escherichia coli* are localized on the cytoplasmic side of the plasma membrane, whereas the enzyme of *Klebsiella aerogenes* is a transmembrane protein (GRAHAM and BOXER, 1980; WIENTJES et al., 1979, 1980). In a dentrifying strain of *Rhodopseudomonas sphaeroides* the dissimilatory nitrate and nitrite reductases were found to be located in the periplasmic space (SAWADA and SATOH, 1980).

Pseudomonas aeruginosa is able to carry out assimilatory nitrate reduction and denitrification. Its assimilatory and dissimilatory nitrate reductases are encoded by different genes. Some mutants unable to assimilate nitrate are unaffected in their denitrification (SIAS and INGRAHAM, 1979; SIAS et al., 1980).

The dissimilatory nitrite reductases of *Bacillus licheniformis* and *Escherichia Pseudomonas perfectomarinus* were identified as cytochrome cd (LE GALL et al., 1979; ZUMFT et al., 1979). The enzymes from various bacteria appear rather heterogeneous. Some reduce nitrite to NO (LE GALL et al., 1979; WHARTON and WEINTRAUB, 1980; ZUMFT et al., 1979) and others to N_2O (BESSIÈRES and HENRY, 1980; WHARTON and WEINTRAUB, 1980; ZUMFT and VEGA, 1979), with NAD(P)H serving as hydrogen donor (BESSIÈRES and HENRY, 1980; ZUMFT and VEGA, 1979). Respiratory enzymes which reduce nitrite to ammonia, on the other hand, use H_2 and ferredoxin (STEENKAMP and PECK, 1980; YORDY and DELWICHE, 1979). The dissimilatory nitrite reductase of *Paracoccus denitrificans* is located on the periplasmic side of the plasma membrane (ALEFOUNDER and FERGUSON, 1980; MEIJER et al., 1979).

Denitrification of nitrate to N_2O was found in *Rhizobium* (ZABLOTOWICZ and FOCHT, 1979). It can support nitrogen fixation under anaerobic conditions. The dissimlatory reduction of nitrite to ammonia by anaerobic fermentative bacteria, however, provides little useful energy; the nitrite rather serves as an agent for the reoxidation of reduced pyridine nucleotides (COLE and BROWN, 1980).

3. Nitrification

The bioenergetic aspects of nitrification were reviewed by ZUMFT and CARDENAS (1979).

The uptake by *Nitrosomonas* of NH_4^+ is followed by the extrusion of a proton, and Cu is involved in this energy-dependent process. The oxidation of one mol NH_4^+ proceeds with the consumption of 1.5 mol O_2, and the oxidation of one mol NH_2OH with one mol O_2 (BHANDARI and NICHOLAS, 1979). Studies on the oxidation by *Nitrosomonas* of $^{15}NH_4^+$ to hydroxylamine in the presence of $^{18}O_2$ led to the result that the oxygen of NH_2OH is derived from dioxygen rather than from water (DUA et al., 1979). The hydroxylamine oxidoreductase of *Nitrosomonas* is a c-type cytochrome with a molecular weight of 175,000 (YAMANAKA et al., 1979). In addition to nitrite, the enzyme produces some NO. (HNO) and NO seem to act as intermediates in the oxidation of hydroxylamine to nitrite (HOOPER and TERRY, 1979).

References

ALEF, K., KLEMME, J.-H.: Z. Naturforsch. 34c, 33-37 (1979). - ALEFOUNDER, P.R., FERGUSON, S.J.: Biochem. J. 192, 231-240 (1980). - ALIKULOV, Z. A., LVOV, N.P., BURIKHANOV, S.S., KRETOVICH, V.L.: Izv. Akad. Nauk SSSR, Ser. Biol., 712-718 (1980). - AMY, N.K., GARRETT, R.H.: J. Bacteriol. 144, 232-237 (1980). - AMY, N.K., RAJAGOPALAN, K.V.: J. Bacteriol. 140, 114-124 (1979).

BAKER, D., TORREY, J.G.: Can. J. Microbiol. 26, 1066-1071 (1980). - BAKER, D., TORREY, J.G., KIDD, G.H.: Nature 281, 76-78 (1979). - BEEVERS, L., HAGEMAN, R.H.: Nitrate and nitrite reduction, 115-168. In: Biochemistry of Plants, Vol. 5, ed. B.J. MIFLIN. New York: Academic Press 1980. - BENSON, D.R., ARP, D.J., BURRIS, R.H.: Science 205, 688-689 (1979); - J. Bacteriol. 142, 138-144 (1980). - BERKUM, P. VAN, BOHLOOL, B.B.: Microbiol. Rev. 44, 491-517 (1980). - BERLIER, Y.M., LESPINAT, P.A.: Arch. Microbiol. 125, 67-72 (1980). - BESSIÈRES, P., HENRY, Y.: Compt. Rend. Acad. Sci., Sér. D, 290, 1309-1312 (1980). - BHANDARI, B., NICHOLAS, D.J.D.: Arch. Microbiol. 122, 249-255 (1979). - BOTTOMLEY, P.J., GRILLO, J.F., BAALEN, C. VAN, TABITA, F.R.: J. Bacteriol. 140, 938-943 (1979). - BRADLEY, P.M.: Z. Pflanzenphysiol. 100, 65-67 (1980). - BRILL, W.J.: Microbiol. Rev. 44, 449-467 (1980).

CHAN, Y.K., NELSON, L.M., KNOWLES, R.: Can. J. Microbiol. 26, 1126-1131 (1980). - COHEN, E., OKON, Y., KIGEL, J., NUR, I., HENIS, Y.: Plant Physiol. 66, 746-749 (1980). - COLE, J.A., BROWN, C.M.: FEMS Microbiol. Lett. 7, 65-72 (1980).

DE LA ROSA, M.A., DIEZ, J., VEGA, J.M., LOSADA, M.: Eur. J. Biochem. 106, 249-256 (1980). - DIKSHIT, K.L., DIKSHIT, R.P.: Mol. Gen. Genet. 180, 629-633 (1980). - DOWNEY, R.J., STEINER, F.X.: J. Bacteriol. 137,

105-114 (1979). - DUA, R.D., BHANDARI, B., NICHOLAS, D.J.D.: FEBS Lett. 106, 401-404 (1979). - DUNN-COLEMAN, N.S., TOMSETT, A.B., GARRETT, R. H.: J. Bacteriol. 139, 697-700 (1979).

ELMERICH, C.: Physiol. Vég. 17, 883-906 (1979). - EMES, M.J., FOWLER, M.W.: Planta 144, 249-253 (1979).

FISCHER, S., SIMONIS, W.: Z. Pflanzenphysiol. 92, 143-152 (1979). - FLOENER, L., BOTHE, H.: Nitrogen fixation in *Rhopalodia gibba*, a diatom containing blue-greenish inclusions symbiotically, 541-552. In: Endocytobiology. Endosymbiosis and Cell Biology, Vol. I, eds. W. SCHWEMMLER, H.E.A. SCHENK. Berlin: de Gruyter 1980. - FLORES, E., GUERRERO, M.G., LOSADA, M.: Arch. Microbiol. 128, 137-144 (1980). - FRY, I.V., CAMMACK, R., HUCKLESBY, D.P., HEWITT, E.J.: FEBS Lett. 111, 377-380 (1980). - FUNKHOUSER, E.A., RAMADOSS, C.S.: Plant Physiol. 65, 944-948 (1980).

GIRI, L., RAMADOSS, C.S.: J. Biol. Chem. 254, 11703-11712 (1979). - GÖTZ, E.-M.: Z. Pflanzenphysiol. 98, 465-470 (1980). - GÖTZ, E.-M., HESS, D.: Z. Pflanzenphysiol. 98, 453-458 (1980). - GRAHAM, A., BOXER, D.H.: FEBS Lett. 113, 15-20 (1980). - GRIFFITHS, D.J.: New Phytol. 82, 427-437 (1979).

HAGEMAN, R.V., BURRIS, R.H.: Curr. Top. Bioenerg. 10, 279-291 (1980). - HALLENBECK, P.C., KOSTEL, P.J., BENEMANN, J.R.: Eur. J. Biochem. 98, 275-284 (1979). - HARDY, R.W.F., BOTTOMLEY, F., BURNS, R.C. (eds.): A Treatise on Dinitrogen Fixation. Inorganic and Physical Chemistry and Biochemistry. New York: Wiley-Interscience 1979. - HAUSINGER, R.P., HOWARD, J.B.: Proc. Natl. Acad. Sci. USA 77, 3826-3830 (1980). - HEISEY, R.M., DELWICHE, C.C., VIRGINIA, R.A., WRONA, A.F., BRYAN, B.A.: Am. J. Bot. 67, 429-431 (1980). - HESS, D., FEUEREISEN, S.: Biochem. Physiol. Pflanzen 175, 689-698 (1980). - HEWITT, E.J., CUTTING, C.V. (eds.): Nitrogen Assimilation of Plants. London: Academic Press 1979. - HILLMER, P., FAHLBUSCH, K.: Arch. Microbiol. 122, 213-218 (1979). - HOOPER, A.B., TERRY, K.R.: Biochim. Biophys. Acta 571, 12-20 (1979).

JOHNSON, C.B.: Planta 147, 63-68 (1979). - JOHNSON; J.L., HAINLINE, B.E., RAJAGOPALAN, K.V.: J. Biol. Chem. 255, 1783-1786 (1980). - JONES, B.L., MONTY, K.J.: J. Bacteriol. 139, 1007-1013 (1979).

KLEEBERGER, A., KLINGMÜLLER, W.: Mol. Gen. Genet. 180, 621-627 (1980). - KNYPL, J.S.: Plant Sci. Lett. 18, 319-323 (1980). - KRISHNAPILLAI, V., POSTGATE, J.R.: Arch. Microbiol. 127, 115-118 (1980). - KUO, T., KLEINHOFS, A., WARNER, R.L.: Plant Sci. Lett. 17, 371-381 (1980).

LAANE, C., KRONE, W., KONINGS, W., HAAKER, H., VEEGER, C.: Eur. J. Biochem. 103, 39-46 (1980). - LALONDE, M.: Can. J. Bot. 56, 2621-2635 (1978); - Bot. Gaz. 140, S35-S43 (1979). - LANCASTER, J.R., VEGA, J.M., KAMIN, H., ORME-JOHNSON, N.R., ORME-JOHNSON, W.H., KRUEGER, R.J., SIEGEL, L.M.: J. Biol. Chem. 254, 1268-1272 (1979). - LE GALL, J., PAYNE, W.J., MORGAN, T.V., DER VARTANIAN, D.: Biochem. Biophys. Res. Commun. 87, 355-362 (1979). - LIM, S.T., HENNECKE, H., SCOTT, D.B.: J. Bacteriol. 139, 256-263 (1979). - LJONES, T.: FEBS Lett. 98, 1-8, (1979). - LOSADA, M., GUERRERO, M.G.: The photosynthetic reduction of nitrate and its regulation, 365-408. In: Topics in Photosynthesis, Vol. 3, ed. J. BARBER. Amsterdam: Elsevier 1979.

MADIGAN, M.T., WALL, J.D., GEST, H.: Science 204, 1429-1430 (1979). - MAZUR, B.J., RICE, D., HASELKORN, R.: Proc. Natl. Acad. Sci. USA 77, 186-190 (1980). - MC CLUNG, C.R., PATRIQUIN, D.G.: Can. J. Microbiol. 26, 881-886 (1980). - MC KENNA, C.E., HUANG, C.W.: Nature 280, 609-611 (1979). - MC KENNA, C.E., JONES, J.B., ERAN, H., HUANG, C.W.: Nature

280, 611-612 (1979). - MEIJER, E.M., VAN DER ZWAAN, J.W., STOUTHAMER, A.H.: FEMS Microbiol. Lett. 5, 369-372 (1979). - MOORE, R., BLACK, C.C.: Plant Physiol. 64, 309-313 (1979). - MORTENSON, L.E., THORNELEY, R.N.F.: Annu. Rev. Biochem. 48, 387-418 (1979). - MURRY, M.A., BENEMANN, J.R.: Plant Cell Physiol. 20, 1391-1401 (1979).

NICHOLAS, D.J.D., FERRANTE, J.V., CLARKE, G.R.: Anal. Biochem. 95, 24-31 (1979). - NUTI, M.P., LEPIDI, A.A., PRAKASH, R.K., SCHILPEROORT, R.A., CANNON, F.C.: Nature 282, 533-535 (1979).

PAPEN, H., WERNER, D.: Arch. Microbiol. 120, 25-30 (1979). - PEARSON, H.W., HOWSLEY, R.: Nature 288, 263-265 (1980). - PHILLIPS, D.A.: Annu. Rev. Plant Physiol. 31, 29-49 (1980). - POSTGATE, J.R.: Phil. Transact. Roy. Soc. London B 290, 421-425 (1980). - PREMAKUMAR, R., SORGER, G.J., GOODEN, D.: J. Bacteriol. 137, 1119-1126 (1979); - J. Bacteriol. 143, 411-415 (1980). - PÜHLER, A., BURKHARDT, H.J., KLIPP, W.: Mol. Gen. Genet. 176, 17-24 (1979).

RAO, L.V.M., DATTA, N., SOPORY, S.K., GUHA-MUKHERJEE, S.: Physiol. Plant. 50, 208-212 (1980). - RIET, J. VAN'T, WIENTJES, F.B., VAN DOORN, J., PLANTA, R.J.: Biochim. Biophys. Acta 576, 347-360 (1979). - RIGANO, C., DI MARTINO RIGANO, V., VONA, V., FUGGI, A.: Arch. Microbiol. 121, 117-120 (1979). - RIGHETTI, T.L., MUNNS, D.N.: Plant Physiol. 64, 411-412 (1980). - ROBSON, R.L.: Biochem. J. 181, 569-575 (1979). - ROBSON, R.L., POSTGATE, J.R.: Annu. Rev. Microbiol. 34, 183-207 (1980). - ROGERSON, A.C.: Nature 284, 563-564 (1980). - RUVKUN, G.B., AUSUBEL, F.M.: Proc. Natl. Acad. Sci. USA 77, 191-195 (1980).

SASAKAWA, H., YAMAMOTO, Y.: Plant Physiol. 63, 1098-1101 (1979). - SAWADA, E., SATOH, T.: Plant Cell Physiol. 21, 205-210 (1980). - SCHMIDT, A.: Encyclop. Plant Physiol. 6, 481-496 (1979). - SCOTT, D.B., SCOTT, C.A., DÖBEREINER, J.: Arch. Microbiol. 121, 141-145 (1979). - SIAS, S.R., INGRAHAM, J.L.: Arch. Microbiol. 122, 263-270 (1979). - SIAS, S.R., STOUTHAMER, A.H., INGRAHAM, J.L.: J. Gen. Microbiol. 118, 229-234 (1980). - SMALL, I.S., WRAY, J.L.: Phytochem. 19, 387-394 (1980). - STANLEY, J., DUNICAN, L.K.: Mol. Gen. Genet. 174, 211-220 (1979). - STEENKAMP, D.J., PECK, H.D.: Biochem. Biophys. Res. Commun. 94, 41-48 (1980). - STEWART, W.D.P.: Endeavour 2, 170-179 (1978); - Encyclop. Plant Physiol. 6, 457-471 (1979); - Annu. Rev. Microbiol. 34, 497-536 (1980). - STEWART, W.D.P., GALLON, J.R. (eds.): Nitrogen Fixation. London: Academic Press 1980.

TISCHNER, R., LORENZEN, H.: Planta 146, 287-292 (1979). - TJEPKEMA, J.D., ORMEROD, W., TORREY, J.G.: Nature 287, 633-635 (1980). - TRINICK, M.J., GALBRAITH, J.: New Phytol. 86, 17-26 (1980).

VENNESLAND, B., GUERRERO, M.G.: In: Encyclopedia of Plant Physiology, New Series, Vol. 6, eds. M. GIBBS, E. LATZKO, pp. 425-444. Berlin-Heidelberg-New York: Springer 1979. - VIJAYARAGHAVAN, S.J., SOPORY, S.K., GUHA-MUKHERJEE, S.: Plant Cell Physiol. 20, 1251-1261 (1979).

WALLSGROVE, R.M., LEA, P.J., MIFLIN, B.J.: Plant Physiol. 63, 232-236 (1979). - WHARTON, D.C., WEINTRAUB, S.T.: Biochem. Biophys. Res. Commun. 97, 236-242 (1980). - WHITELAM, G.C., JOHNSON, C.B.: New Phytol. 85, 475-482 (1980). - WHITELAM, G.C., JOHNSON, C.B., SMITH, H.: Photochem. Photobiol. 30, 589-594 (1979). - WIENTJES, F.B., KOLK, A.H.J., NANNINGA, N., VAN'T RIET, J.: Eur. J. Biochem. 95, 61-67 (1979). - WIENTJES, F.B., VAN'T RIET, J., NANNINGA, N.: Arch. Microbiol. 127, 39-46 (1980). - WILKINSON, C.R., FAY, P.: Nature 279, 527-529 (1979).

YAMADA, T., SAKAGUCHI, K.: Arch. Microbiol. 124, 161-167 (1980). - YAMANAKA, T., SHINRA, M., TAKAHASHI, K., SHIBASAKA, M.: J. Biochem. 86, 1101-1108 (1979). - YATAZAWA, M., YOSHIDA, S.: Physiol. Plant 45, 293-295 (1979). - YATES, M.G.: Biochemistry of nitrogen fixation, 1-64. In: Biochemistry of Plants, Vol. 5, ed. B.J. MIFLIN. New York: Academic Press 1980. - YOCH, D.C.: Nitrogen fixation and hydrogen metabolism by photosynthetic bacteria, 657-676. In: The Photosynthetic Bacteria, eds. R.K. CLAYTON, W.R. SISTROM. New York: Plenum Press 1978; - J. Bacteriol. 140, 987-995 (1979); - Biochem. J. 187, 273-276 (1980). - YOCH, D.C., CANTU, M.: J. Bacteriol. 142, 899-907 (1980). - YORDY, D.M., DELWICHE, E.A.: J. Bacteriol. 137, 905-911 (1979).

ZABLOTOWICZ, R.M., FOCHT, D.D.: J. Gen. Microbiol. 111, 445-448 (1979). - ZUMFT, W.G., CÁRDENAS, J.: Naturwissenschaften 66, 81-88 (1979). - ZUMFT, W.G., VEGA, J.M.: Biochim. Biophys. Acta 548, 484-499 (1979). - ZUMFT, W.G., SHERR, B.F., PAYNE, W.J.: Biochem. Biophys. Res. Commun. 88, 1230-1236 (1979).

Professor Dr. ERICH KESSLER
Institut für Botanik und
Pharmazeutische Biologie
der Universität
Schloßgarten 4
D 8520 Erlangen

IV. Secondary Plant Substances

Some Tyrosine- and Phenylalanine-Derived Alkaloids

By HORST-ROBERT SCHÜTTE

1. Introduction

Phenylalanine and tyrosine give rise to many secondary plant products. Phenylalanine is preferably converted to cinnamic acid, the parent substance of the large group of nonnitrogenous phenylpropanoids (Progr. Bot. 40, 126 ff.; 41, 93ff.; SCHÜTTE, 1981). Tyrosine and dopa lead preferably to the corresponding alkaloids. The usual pathway to the alkaloids starts with dopamine (I), the decarboxylation product of dopa, which can give the simple phenethylamines like mescaline (II) (PAUL, 1973; SMITH, 1977; BASMADJIAN et al., 1978) or which can react according to a Mannich reaction with carbonyl compounds to give tetrahydroisoquinolines (cf. LIEBISCH, 1969). Most important is the reaction of dopamine derivatives with phenylpyruvic acid derivatives yielding benzyltetrahydroisoquinolines and the different derived compounds. The principles of this pathway were established in the 1960's (cf. SCHÜTTE, 1969; HERBERT, 1979; FODOR, 1980). Here some recent results in this field should be discussed.

2. Reticuline and Other Benzylisoquinolines

It could be found that the Mannich reaction mentioned above proceeds with corresponding α-keto-acids rather than with corresponding aldehydes, as was earlier postulated. Thus the very important reticuline (X), the specific precursor of morphine type alkaloids as well as of aporphines and tetrahydroberberines, is biosynthesized in *Litsea glutinosa* (Lour.) C. B. Rob. (Lauraceae) from dopamine (I) with 3,4-dihydroxyphenylpyruvic acid (VI), yielding the corresponding norlaudanosoline-1-carboxylic acid (VII) which gives rise to reticuline after decarboxylation, reduction, and methylation via 1,2-dehydronorlaudanosoline (VIII) and norlaudanosoline (IX) (Fig. 1) (TEWARI et al., 1975; BHAKUNI et al., 1977b). O-Methylation precedes N-methylation.

In addition to the benzyltetrahydroisoquinolines some aromatic benzylisoquinolines exist in opium poppy, such as papaverine (XII), one of the major benzylisoquinoline alkaloids of *Papaver somniferum* L. For its biosynthesis norlaudanosoline (IX) has to be O-methylated and dehydrogenated. (-)-Tetrahydropapaverine (XI) is the principal immediate precursor of papaverine in *Papaver somniferum* (BROCHMANN-HANSSEN et al., 1975; UPRETY et al., 1975).

Results similar to those for the norlaudanosoline derivatives have been found in *Annona reticulata* L. (Annonaceae) for the biosynthesis of coclaurine (V) (PRAKASH et al., 1979), an established precursor of proaporphine (BARTON et al., 1967b; BHAKUNI et al., 1974), aporphine (BARTON et al., 1967a), and bisbenzylisoquinoline alkaloids (BARTON et al., 1966; BHAKUNI et al., 1978b,d). Dopamine

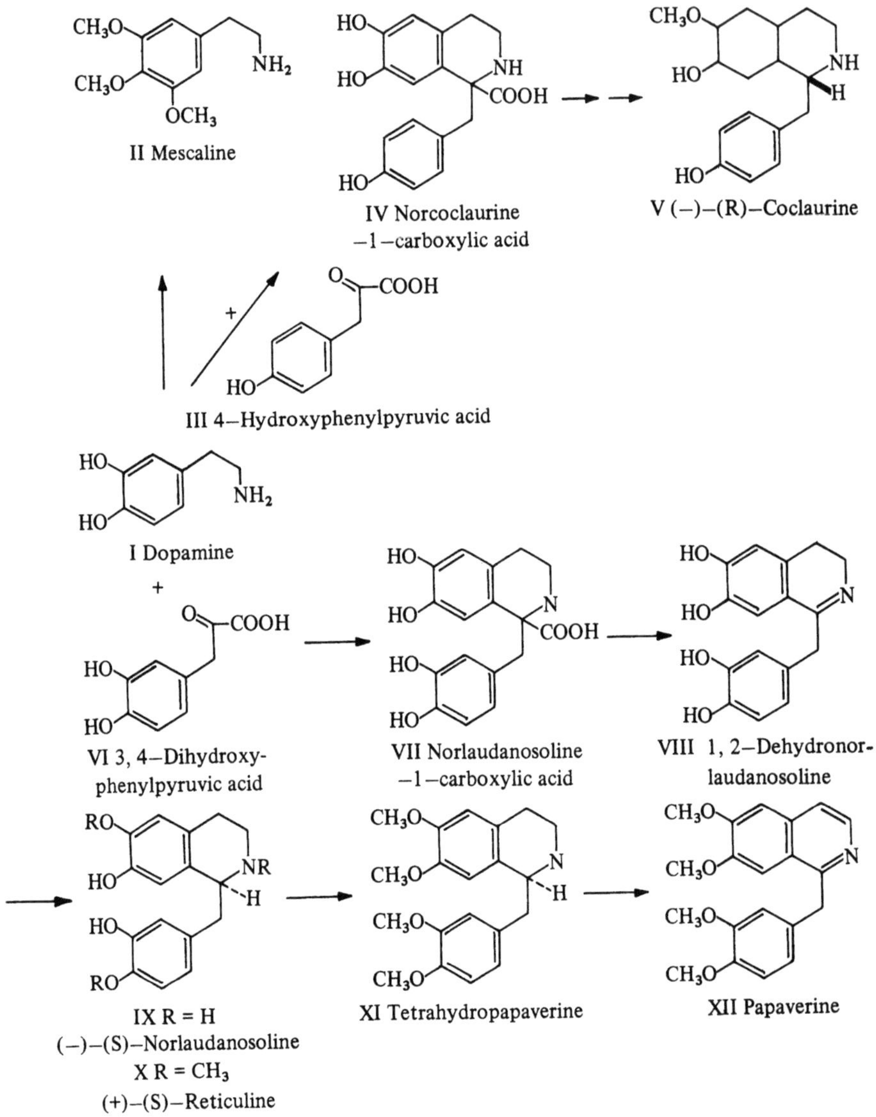

Fig. 1. Biosynthesis of mescaline and some benzylisoquinolines

(I) and 4-hydroxyphenylpyruvic acid (III) derived from tyrosine interact and form norcoclaurine-1-carboxylic acid (IV) which specifically gives rise to coclaurine (V) via 1.2-didehydronorcoclaurine and norcoclaurine (Fig. 1).

Intermolecular and intramolecular oxidative coupling of benzylisoquinolines is a key process which gives rise to the great diversity of skeletal types deriving from this simple structures (Progr. Bot. 34, 165 ff.; BARTON and COHEN, 1957; BATTERSBY, 1967; STAUNTON, 1979).

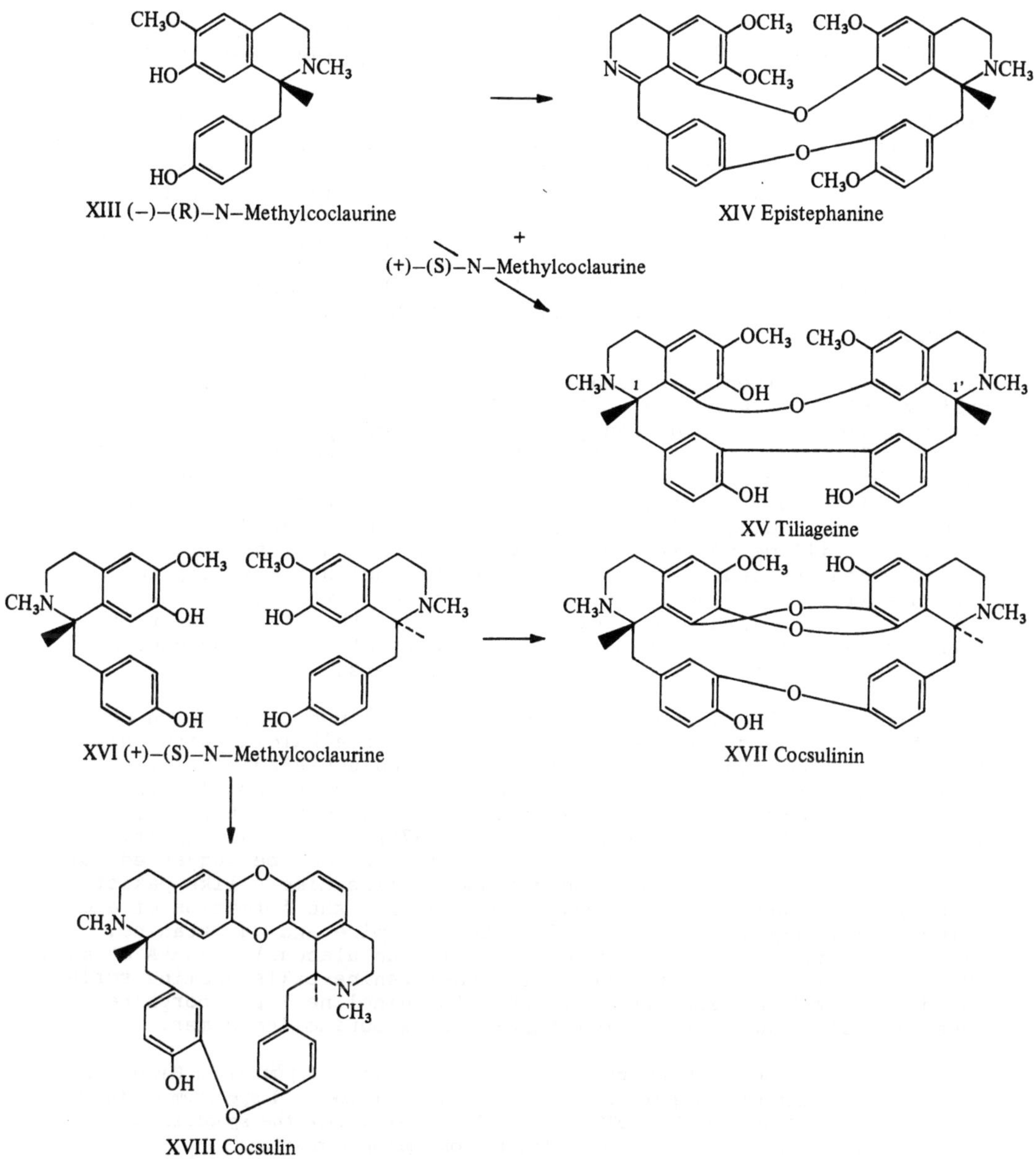

Fig. 2. Biosynthesis of some bisbenzylisoquinolines

3. Bisbenzylisoquinolines

Thus bisbenzylisoquinoline alkaloids can be formed in nature by oxidative coupling of benzylisoquinolines, preferably coclaurine derivatives. In *Stephania japonica* (Menispermaceae) it could be shown that tyrosine, coclaurine, and (-)-(R)-N-methylcoclaurine (XIII) were specifically incorporated into epistephanine (XIV) (BARTON et al., 1966). Similarly

(+)-(S)- and (-)-(R)-N-methylcoclaurines are specifically incorporated into isotetrandrine by *Cocculus laurifolius* DC. (Menispermaceae) (BHAKUNI et al., 1980).

Besides the bisbenzylisoquinolines with two ether bridges such alkaloids with three ether linkages are also known, e.g. cocsulin (XVIII) from *Cocculus laurifolius* and cocsulinin (XVII), an anticancer agent of *Cocculus pendulus* (BHAKUNI and JOSHI, 1975). Both alkaloids are biosynthesized in *C. laurifolius* by oxidative dimerization of two (+)-(S)-N-methylcoclaurine units (XVI) (Fig. 2) (BHAKUNI et al., 1978a,b), which has been isolated from *C. laurifolius*. The O-methyl function from one of the N-methylcoclaurine units is lost and the stereospecificity is maintained in the oxidative dimerization of N-methylcoclaurine into cocsulin.

In *Tiliacora* sp. (Menispermaceae) tiliageine (XV), tiliacorine, and tiliacorinine (TACKIE et al., 1975; SHAMMA et al., 1976) occur, dimeric alkaloids of the bisbenzylisoquinoline group in which the two benzylic "halves" are linked through a direct carbon-to-carbon bond rather than through the more common diarylether bridge. N-Methylcoclaurine is also a specific precursor of these three alkaloids in *Tiliacora racemosa* Colebr. (BHAKUNI and SINGH, 1978a; BHAKUNI et al., 1978d). In tiliageine (XV) and tiliacorine, one part is derived from (+)-(S)- and the other from (-)-(R)-N-methylcoclaurine (XIII) in accordance with the S and R configuration at the asymmetric centers C_1 and $C_{1'}$ in these alkaloids. For tiliacorinine, the diastereomer of tiliacorine, only the S-enantiomer is incorporated in accordance with the S-S configuration at both asymmetric centers.

In addition to the bisbenzylisoquinolines mentioned above which contain two benzylisoquinoline units other dimeric alkaloids with only one benzylisoquinoline unit are known. Thus baluchistanamine from *Berberis baluchistanica* Ahrendt is an isoquinolone-benzylisoquinoline, which seems to be synthesized by oxidation of an oxyacanthine type, also found in this plant (SHAMMA et al., 1974). A similar oxidative phenol coupling leading to bisbenzylisoquinoline may be suggested for the synthesis of proaporphine-benzylisoquinoline dimers like pakistanamine from *Berberis baluchistanica*, as well as for the formation of aporphine-benzylisoquinoline dimers like thalicarpine and pakistanine, mainly occurring in the genus *Thalictrum* (Ranunculaceae) (SHAMMA et al., 1973). The biogenetic sequence in the pakistanine-pakistanamine series could be benzylisoquinoline → bisbenzylisoquinoline → proaporphine-benzylisoquinoline dimer → aporphine-benzylisoquinoline dimer.

The callus tissues derived from the tuber of *Stephanica cepharantha* Hayata can synthesize bisbenzylisoquinolines, but not the full alkaloid spectrum as in the whole plant (AKASU et al., 1976). The callus tissues lack the specific enzymes necessary for O-methylation and methylenedioxy group formation.

4. Aporphine Alkaloids

Aporphines are biosynthesized from phenolic tetrahydrobenzylisoquinolines either by direct oxidative coupling or by an indirect mechanism which utilizes an intermediate dienone. Such a direct oxidative coupling could be carried out either by ortho-para or ortho-ortho reaction yielding compounds of the isoboldine- or bulbocapnine-type (Fig. 3). Isoboldine (XXIII) has been shown to be formed in *Papaver somniferum* by just such a straight forward intramolecular oxidative coupling be-

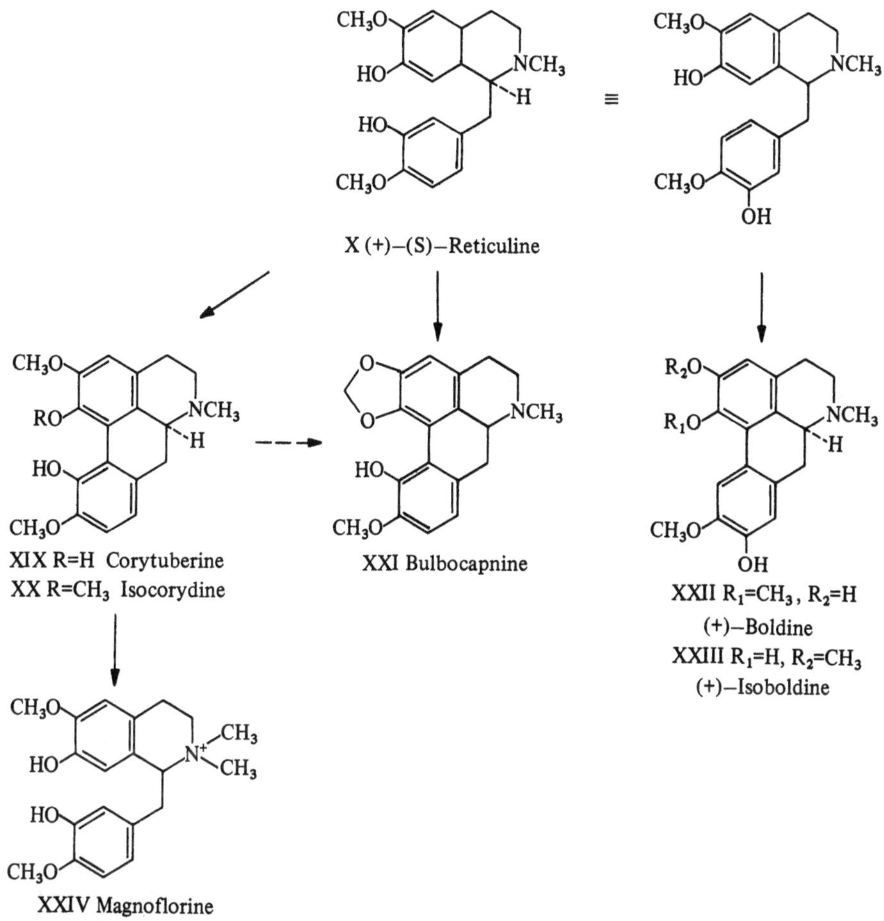

Fig. 3. Biosynthesis of aporphines by direct oxidative coupling

tween the two aryl rings of reticuline (X) BROCHMANN-HANSSEN et al., 1971). The same pathway operates for the synthesis of boldine (XXII), the choleretic principle of *Peumus boldus* Molina (Monimiaceae). (+)-(S)-Reticuline (X) and norreticuline are specific precursors of this alkaloid in *Litsea glutinosa* with (+)-isoboldine (XXIII) as an intermediate, indicating that a change in the methylation pattern takes place in the biosynthesis of boldine, after the oxidative coupling process (Fig. 3) (BHAKUNI et al., 1977c).

These examples are ortho-para couplings. It is suggested that the substitution for ortho-ortho coupling causes considerable steric hindrance to coplanarity, thereby making such a coupling of reticuline very difficult. Nevertheless there is also evidence for direct ortho-ortho oxidative phenol coupling in nature (Fig. 3). Thus in larkspur (*Corydalis cava*, Papaveraceae) reticuline (X) could be specifically incorporated into bulbocapnine (XXI) (BLASCHKE, 1968, 1970; BLASCHKE et al., 1974), probably via corytuberine (XIX) as an intermediate. Also isocorydine (XX) is formed in *Annona squamosa* by a direct ortho-ortho oxidative coupling of (+)-norreticuline and reticuline (X), probably via corytuberine (XIX) from which isocorydine could be formed by methyla-

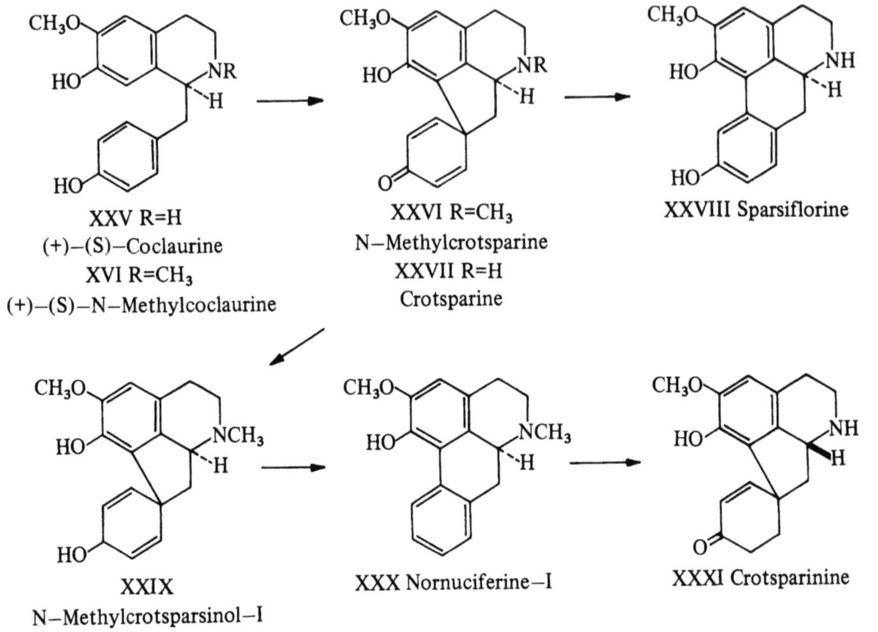

Fig. 4. Biosynthesis of nornuciferine-I

tion (DEWAN et al., 1978). Magnoflorine (XXIV) has been shown to be derived from reticuline (X) in *Aquilegia* plants, presumably also via direct ortho-ortho oxidative phenol coupling (BROCHMANN-HANSSEN et al., 1971, 1972).

Several aporphine alkaloids have an unusual hydroxyl substitution pattern which is not to be understood by a straightforward oxidative coupling. For these alkaloids it is supposed that a benzylisoquinoline undergoes an oxidative coupling yielding an intermediate dienone (e.g., XXVI). Rearrangement of such dienones with or without prior reduction may result in a large number of aporphines with different substitution patterns. Cyclohexadienones have a strong tendency to undergo rearrangement. In the acid-catalyzed dienone-phenol rearrangement, the parent dienone is converted to the phenol by migration of a substituent to give a carbonium ion which can aromatize by loss of a proton (e.g., XXVII → XXVIII in Fig. 4).

An alternative, less direct process, the dienol-benzene rearrangement, starts with reduction to the dienol which can now undergo an equivalent rearrangement to give a carbonium ion followed by aromatization to the benzene derivative (e.g., XXVI → XXIX → XXX). Both types of processes are widely used in alkaloid biosynthesis. After the first aromatization the oxygen of the dienone is retained, whereas in the second the oxygen is lost.

Thus isothebaine is formed in *Papaver orientale* by oxidative coupling of orientaline, an isomer of reticuline, via the dienone orientalinone and the dienol orientaline which undergoes a dienol-benzene rearrangement (BATTERSBY et al., 1965a, 1969). Corydine, glaucine, and dicentrine occur in *Dicentra eximia* (Fumariaceae) and their structures can most readily be explained as involving direct coupling of reticuline.

But in this case norprotosinomenine, as the di-O-methylated precursor of the *Dicentra* alkaloids, shows a good incorporation into the three aporphines (BATTERSBY et al., 1971). Here glaucine and dicentrine are formed via boldine. Thus, interestingly, boldine can be formed according to two different pathways, in *Dicentra* via a dienone and in *Litsea* by direct oxidative phenol coupling, as mentioned above.

The so-called "abnormal aporphine" alkaloids lack oxygen function in ring D. Thus nornuciferine-I (XXX) can be biosynthesized in *Croton sparsiflorus* (Euphorbiaceae) from (+)-(S)-methylcoclaurine (XVI) by ortho-para oxidative coupling, yielding the proaporphine (XXVII). Reduction to the dienol (XXIX) followed by dienol-benzene rearrangement finally forms the abnormal aporphine (XXX) (Fig. 4) (BHAKUNI et al., 1979). N-Methylcrotsparine (XXVI), crotsparine (XXVII) and N-methylcrotsparinol (XXIX) are efficiently incorporated intermediates. The results thus strongly supported the following sequence for the biosynthesis of nornuciferine-I in *C. sparsiflorus*: tyrosine → norcoclaurine → coclaurine (XXV) → (+)-(S)-N-methylcoclaurine (XVI) → N-methylcrotsparine (XXVI) → N-methylcrotsparinol-I (XXIX) → nornuciferine-I (XXX).

Besides the proaporphine alkaloids crotsparine (XXVII), N-methylcrotsparine (XXVI) and N,O-dimethylcrotsparine in *Croton sparsiflorus*, the dihydroproaporphine crotsparinine (XXXI) and the aporphine sparsiflorine (XXVIII) occur (BHAKUNI et al., 1970). In contrast to the abnormal aporphine nornuciferine (XXX), sparsiflorine contains one oxygen function in ring D. Nevertheless coclaurine (XXV) was efficiently incorporated into sparsiflorine together with crotsparine and crotsparinine (Fig. 4) (BHAKUNI et al., 1974). Crotsparine (XXVII) and crotsparinine (XXXI) have opposite configurations while sparsiflorine (XXVIII) and crotsparine have the same at the corresponding asymmetric centers. If crotsparine is a biosynthetic precursor of crotsparinine, a change of configuration should occur in the proaporphine during the course of biochemical transformations. Alternately crotsparine and crotsparinine could be biosynthesized by independent routes.

5. Sebiferine

Whereas thebaine, codeine (XXXIV), and morphine (XXXV) are synthesized by an ortho-para oxidative coupling of (-)-(R)-reticuline (XXXIII) via the dienone salutaridine (XXXVI) (Fig. 5) (BARTON et al., 1965), a corresponding para-para coupling of reticuline has been established for the biosynthesis of flavinantine (XXXVII) (STUART et al., 1969; STUART and GRAHAM, 1973), a morphinandienone alkaloid from *Croton flavens* (CHAMBERS and STUART, 1968).

Sebiferine (XXXVIII) is a morphinandienone alkaloid isolated for the first time from the stem bark of *Litsea sebifera* (SIVAKUMARAN and GOPINATH, 1976), and is identical with O-methylflavinantine. Also the biosynthesis of this alkaloid is starting from nor-reticuline and reticuline, giving isosalutaridine by para-para oxidative coupling. In experiments with *Cocculus laurifolius* both enantiomers, (-)- (XXXIII) and (+)-reticuline (X), were incorporated with nearly the same efficiency, establishing the presence of a highly active oxidation-reduction system in *C. laurifolius* by way of the 1,2-didehydroderivative (XXXII) as has been observed in poppy (BATTERSBY et al., 1965b). This supports the following sequence for the biosynthesis of sebiferine in *C. laurifolius*: nor-reticuline → (+)-(S)-reticuline (X) ↔ didehydroreticuline (XXXII) ↔ (-)-(R)-reticuline (XXXIII) → sebiferine (XXXVIII) (Fig. 5) (BHAKUNI et al.,

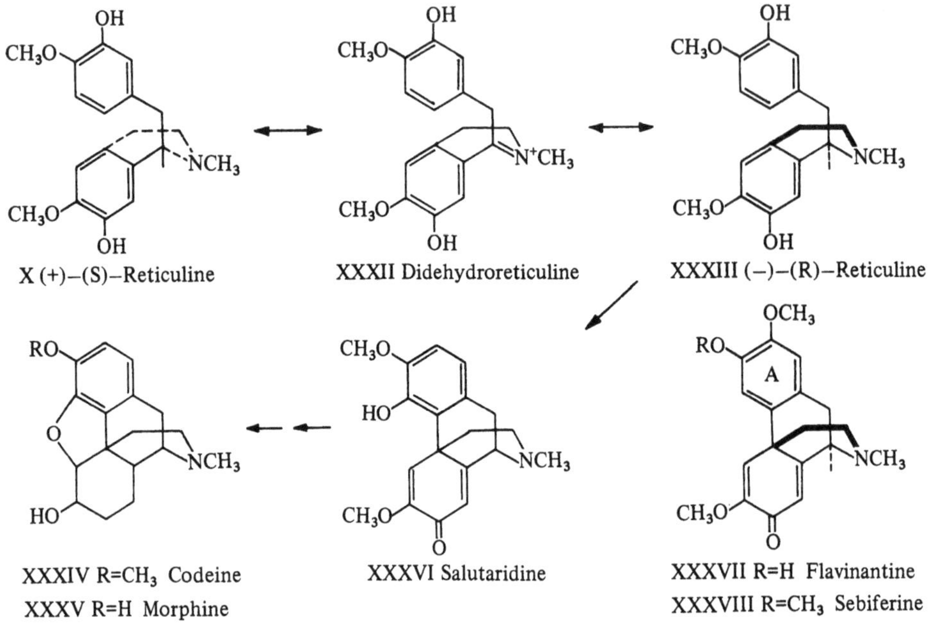

Fig. 5. Biosynthesis of sebiferine

1978c). A similar biosynthetic pathway is suggested for amurine (methylenedioxy group in ring A of sebiferine) and its dienol, nudaurine from *Papaver amurense* Hart (DÖPKE et al., 1968).

6. *Erythrina* Alkaloids

Dienones, established precursors for some aporphines, are also intermediates for another group of bases, the *Erythrina* alkaloids such as erythraline (L), which occur preferably in *Erythrina* sp. (Leguminosae). Here it can be seen that migration is not the only secondary reaction which can lead to modification of the carbon skeleton of a dienone produced by oxidative coupling. In experiments with *Erythrina crista galli* and *E. berteroana* the immediate and also exclusive benzylisoquinoline precursor for the *Erythrina* alkaloids could be found as (+)-(S)-N-nor-protosinomenine (IXL) (BARTON et al., 1968, 1970b, 1974b), although there exists an in vitro synthesis of the *Erythrina*-skeleton starting from N-nor-reticuline (FRANCK and TEETZ, 1971). (+)-(S)-N-Norprotosinomenine is converted to the dienone XL by para-para oxidative phenolic coupling, which is also found as an intermediate in the biosynthetic pathway for the aporphine alkaloids glaucine and dicentrine. This dienone undergoes cleavage to a nine-membered ring system (XLI) rather than migration of a C-C-bond.

The imine thus generated is reduced to give the dibenzazonine intermediate XLII which is efficiently incorporated into erythraline (L) and erythratine (XLVII). The dibenzazonine is subjected to further modification via a recyclization and oxidation leading to erysodienone (XLIII) and especially the (-)-(S)-antipode, which has been related to

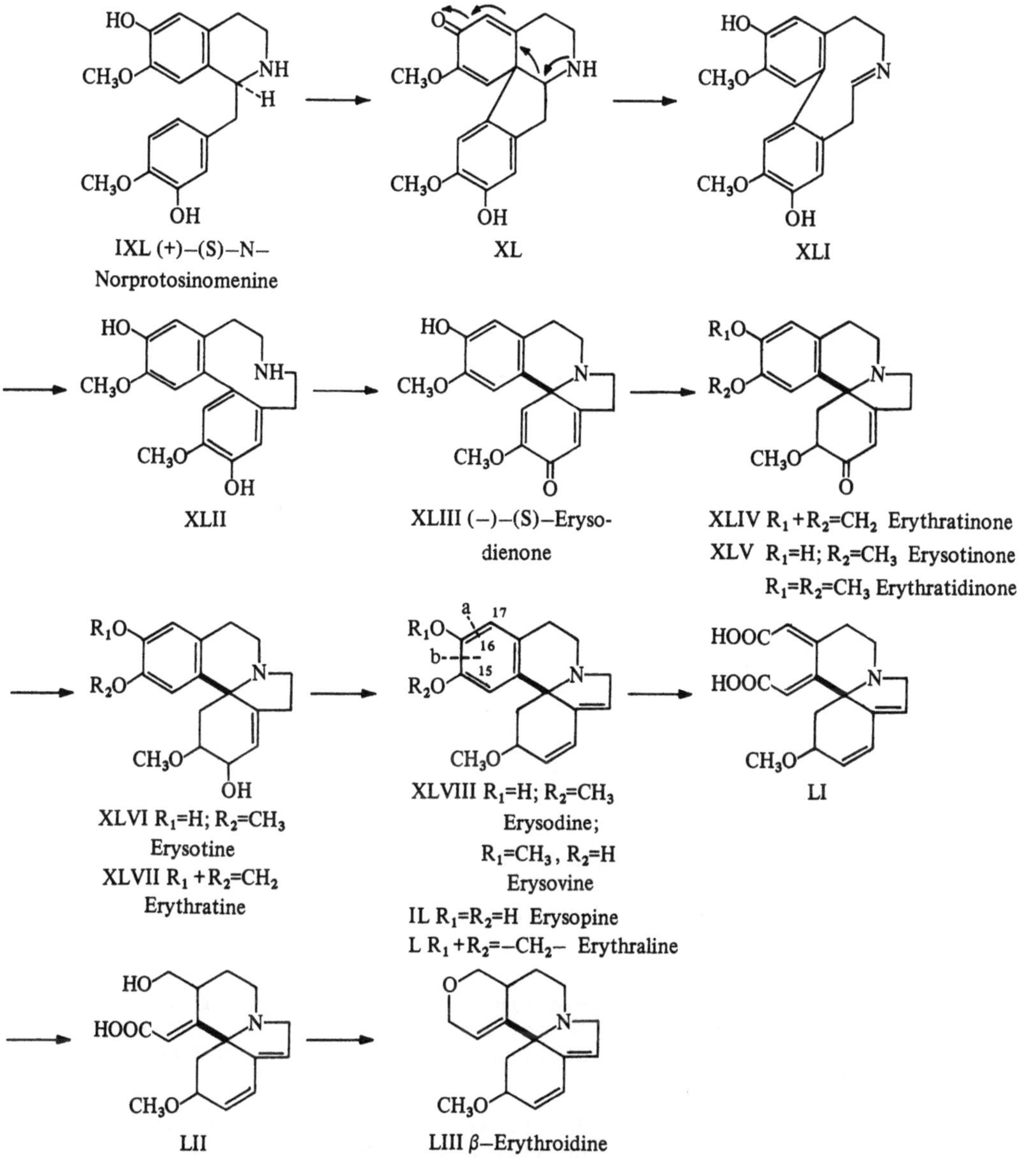

Fig. 6. Biosynthesis of *Erythrina* alkaloids

the natural alkaloids, is the in vivo progenitor of the *Erythrina* alkaloids (Fig. 6) (BARTON et al., 1974a). In vitro this oxidation proceeds via a planar diphenoquinone (the diquinone of XLII), but in vivo the intermediacy of such an intermediate could not been demonstrated (BARTON et al., 1970a). Erythratinone (XLIV), erysodine (XLVIII) and erysopine (IL) are all effectively incorporated into erythraline, suggesting the pathway in Figure 6 for the conversion of erysodienone (XLIII) into erythraline (L). But also the synthetic compounds erysotinone (XLV) and erysotine (XLVI), which have not yet been isolated

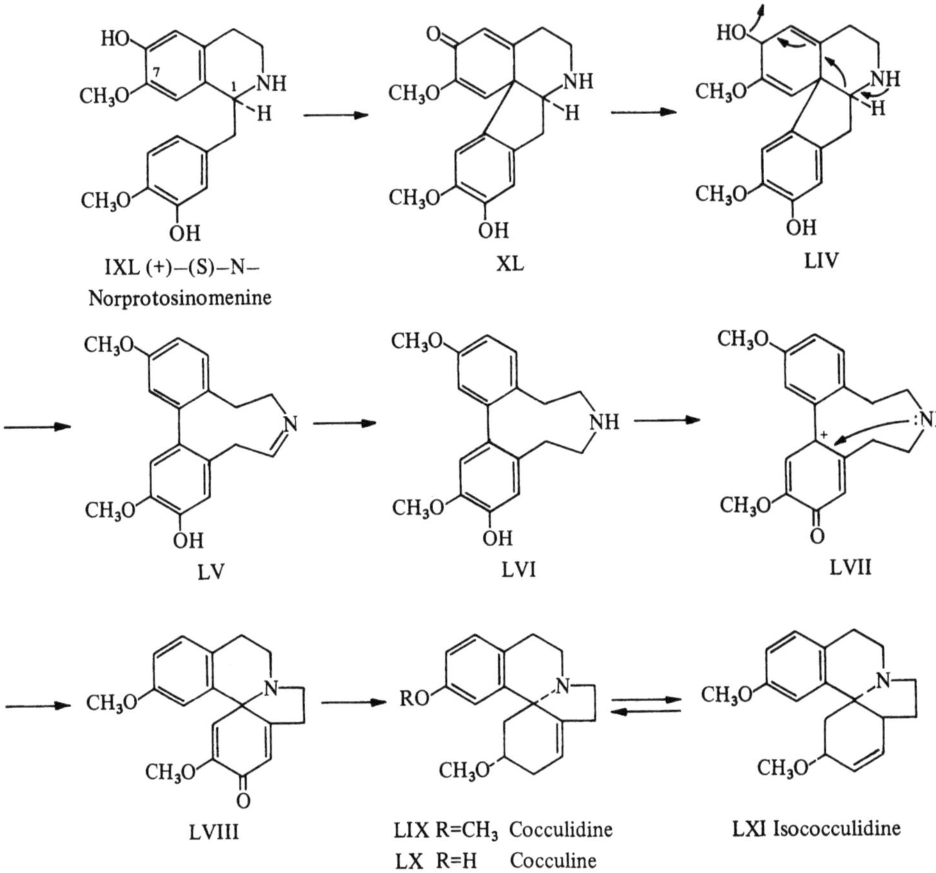

Fig. 7. Formation of "abnormal" *Erythrina* alkaloids

from natural sources, are efficiently converted to erythraline by *E. crista galli*. In a survey of different *Erythrina* species for corresponding alkaloids, some possible intermediate bases have been found (BARTON et al., 1973).

Furthermore the aromatic *Erythrina* alkaloids such as erysodienone (XLIII), erysodine (XLVIII) and erysopine (IL) are shown to be specifically incorporated into the lactone alkaloids such as α-, and β-erythroidines (LIII) with retention of the 17-hydrogen atom probably via the intermediates LI and LII (LEETE and AHMAD, 1966; BARTON et al., 1974a). At the present stage of knowledge, however, it cannot be decided if the lactone alkaloids are formed by an intradiol (C-15-C-16) or extradiol (C-16-C-17) cleavage. Both possibilities would lead to retention of the 17-hydrogen atom and loss of C-16. The cleavage of aromatic rings by higher plants is rare, but has been proven unambiguously for the betalains (MILLER et al., 1968; IMPELLIZZERI and PIATTELLI, 1972).

In addition to the aromatic *Erythrina* alkaloids, which contain the principle oxygen substitution pattern as in the norprotosinomenine, some abnormal *Erythrina* alkaloids are known with more limited oxygen functions, such as cocculidine (LIX) and cocculine (LX), the hypotensive principles, as well as isococculidine (LXI), the neuromuscular blocking principle of *Cocculus laurifolius*, and coccuvine (RAZAKOV et al., 1974; BHAKUNI et al., 1976). But also these structures can be formed in nature from norprotosinomenine (IXL) or its precursors (BHAKUNI et al., 1977a; BHAKUNI and SINGH, 1978b; BHAKUNI and JAIN, 1980) and especially the (+)-(S)-enantiomer is used, yielding at first the same dienone XL as for the normal *Erythrina* alkaloids. But in the bioconversion into the abnormal *Erythrina* alkaloids a reduction of the dienone XL is involved, yielding the dienol LIV which can lose one of the oxygen functions of the precursor by a dienol-benzene rearrangement during the formation of the corresponding imine LV (Fig. 7).

The 7-methoxy group and the C-1 hydrogen atom (at the asymmetric center) in norprotosinomenine (IXL) are retained in the bioconversion into cocculidine (LIX). The next steps are similar to those for the normal *Erythrina* alkaloids.

The keydienone (LVIII), an erysodienone deriviative, from which hitherto known "abnormal" *Erythrina* alkaloids can be formed in nature, can itself be formed from a dibenz|d,f|azonine intermediate (LVI), as shown in (LVII) by recyclization. *C. laurifolius* can convert isococculidine (LXI) into cocculidine (LIX) and cocculidine into cocculine (LX) and vice versa. The occurrence of dibenz|d,f|azonine bases in *C. laurifolius* (PANDE and BHAKUNI, 1976) indirectly supports the intermediacy of these bases in the biosynthesis of the abnormal *Erythrina* alkaloids isococculidine, cocculidine, and cocculine.

7. *Cephalotaxus* and Homoerythrina Alkaloids

Gymnosperms of the genus *Cephalotaxus* (Cephalotaxaceae) contain a group of alkaloids of unique structure. The most abundant member of this group is cephalotaxine (LXXI) (POWELL et al., 1969), which is accompanied in nature by small quantities of related alkaloids (POWELL and MIKOLAJCZAK, 1973; PAUDLER and McKAY, 1973) such as cephalotaxinone (LXXIII), demethylcephalotaxinone (LXXV), and demethylcephalotaxine (LXXIV), as well as by several cephalotaxine esters including deoxyharringtonine (LXXII), isoharringtonine, harringtonine, and homoharringtonine, that possess significant antitumor activity (POWELL et al., 1972b). The alkaloid esters are also known to be potent inhibitors of eukaryotic protein synthesis (FRESNO et al., 1977). In *C. harringtonia*, the alkaloids derived from the cephalotaxine skeleton occur together with a number of bases possessing the homoerythrina skeleton such as 3-epischelhammericine (POWELL, 1972). Otherwise schelhammeridine (LXVIII) and schelhammericine are the major alkaloids of *Schelhammera pedunculata* (Liliaceae) (FITZGERALD, 1969; JOHNS et al., 1969). The occurrence of both structural types (LXVIII and LXXI) together in *Cephalotaxus* species (POWELL et al., 1972a) led to the suggestion that they are divergent products from an initially common biosynthetic pathway.

Schelhammeridine (LXVIII) is structurally related to erythraline (L) which is known to be derived from a 1-benzylisoquinoline as mentioned above. It is thus reasonable to expect schelhammeridine to be biosynthesized (Fig. 8) from a 1-phenethylisoquinoline (as LXV), a system commonly produced by plants of the Liliaceae family of which *Schelhammera* is a member. Consistently with this hypothesis, tyrosine (LXII) is incorporated by *Schelhammera pedunculata* specifically (and solely) into the C_6-C_2 residue (ring A) of schelhammeridine (LXVIII), dopamine (I) follows tyrosine on the biosynthetic pathway and the C_6-C_3 unit of this

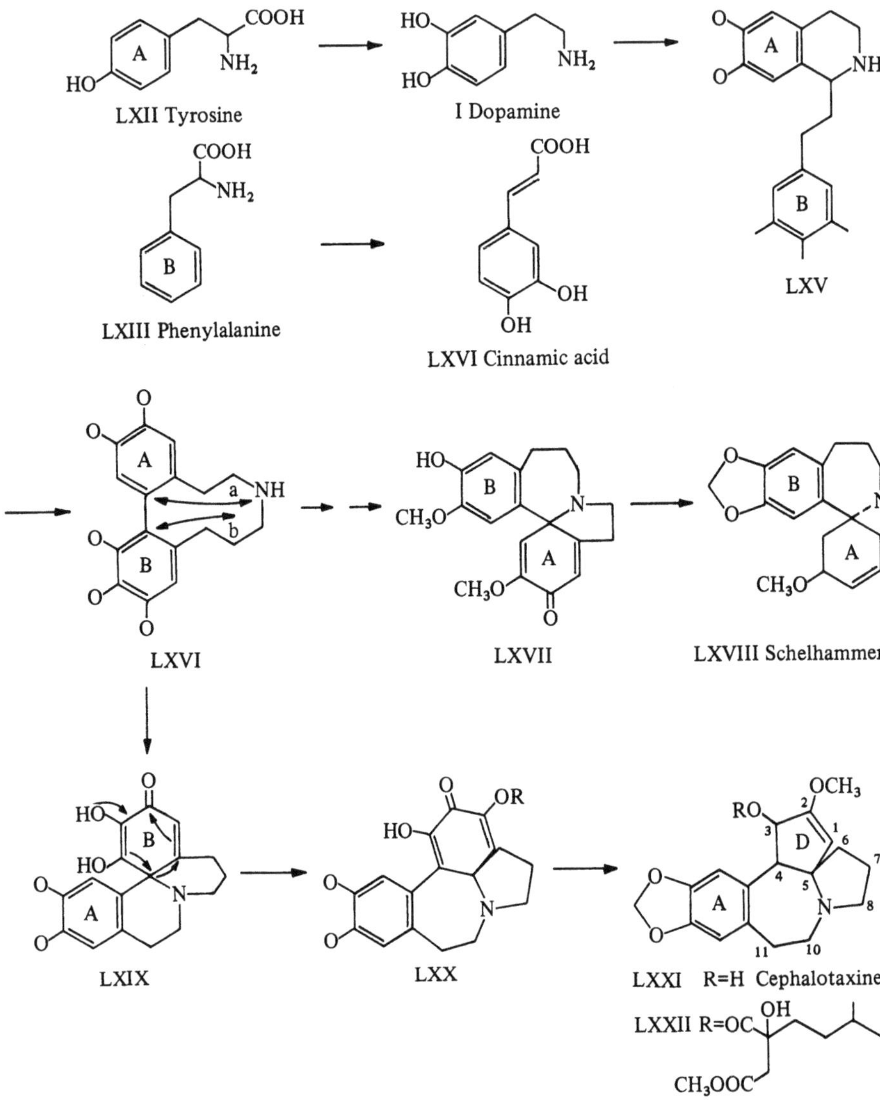

Fig. 8. Biosynthesis of schelhammeridine and cephalotaxine

alkaloid (ring B) is derived from phenylalanine (LXIII) via cinnamic acid (LXIV) (BATTERSBY et al., 1975), as proved for colchicine (BATTERSBY et al., 1972). It is suggested that the phenethylisoquinoline (LXV) is converted in similar reactions as mentioned above for *Erythrina* alkaloids (Fig. 6) into the amine LXVI the homolog of the corresponding *Erythrina* amine (XLII). Probably the dienone LXVII, the homology of erysodienone, is an intermediate formed by ring closure as in LXVI (see Fig. 8).

Also for cephalotaxine it is predicted that this alkaloid might be generated via a 1-phenethyltetrahydroisoquinoline derivative (PARRY,

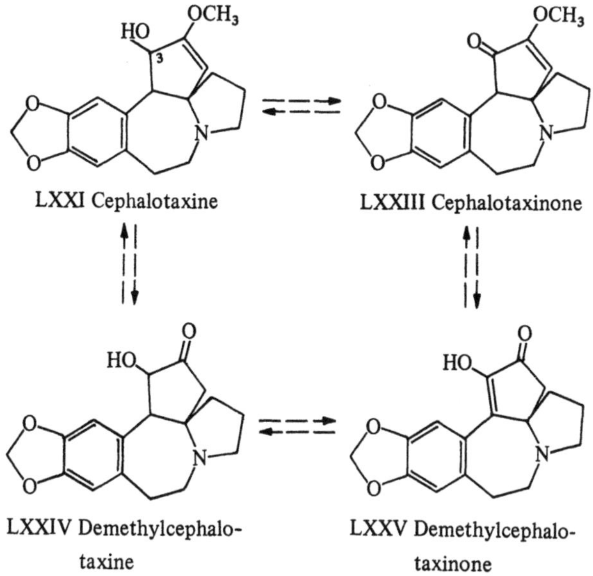

Fig. 9. Metabolic interrelationships between *Cephalotaxus* alkaloids

1979). In experiments with *C. harringtonia* var. *fastigiata* it could be shown that tyrosine (LXII) leads to exclusive labeling of ring A and the C-atoms 10 and 11 of cephalotaxine (LXXI), and that phenylalanine (LXIII) is incorporated via cinnamic acid (LXIV) specifically into the ring D and the C-atoms 6, 7, and 8 of this alkaloid (PARRY and SCHWAB, 1975; SCHWAB et al., 1977; PARRY et al., 1980). The carboxylic group of these acids gives C-8. These results provided substantial evidence that cephalotaxine is a member of the family of phenethylisoquinoline alkaloids which includes the *Schelhammera* and *Colchicum* alkaloids. The suggested intermediate LXVI for the *Schelhammera* alkaloids could undergo a ring closure according to b in LXVI, yielding the dienone LXIX which could be rearranged to the dienone LXX. This could undergo ring contraction to give cephalotaxine (Fig. 8). Thus it is suggested that ring D of this alkaloid is derived from the aromatic ring of phenylalanine by ring contraction and loss of one carbon atom. Incorporation experiments with (p-[^{14}C])- and (m-[^{14}C])-phenylalanine have shown that one of the meta carbon atoms of phenylalanine is lost during the conversion of this amino acid into cephalotaxine. The remaining meta carbon atom is located at C-2 of the alkaloid. These results are consistent with the occurrence of a benzilic acid rearrangement of the hypothetical dienone LXX.

Investigations about the metabolic interrelationships between the *Cephalotaxus* alkaloids cephalotaxine (LXXI), cephalotaxinone (LXXIII), demethylcephalotaxine (LXXIV), and demethylcephalotaxinone (LXXV) have shown that cephalotaxine and cephalotaxinone are interconvertible in *Cephalotaxus* and that demethylation, preferably of cephalotaxinone, takes place readily (Fig. 9).

The key of the elucidation of the biosynthesis of the unusual acyl portions of the antileukemic *Cephalotaxus* ester alkaloids was provided by the recognition that the diacid linked to cephalotaxine in deoxyharringtonine, the deoxyharringtonic acid (LXXXI), bears a close resemblance to a diacid intermediate involved in the biosynthesis of leucine

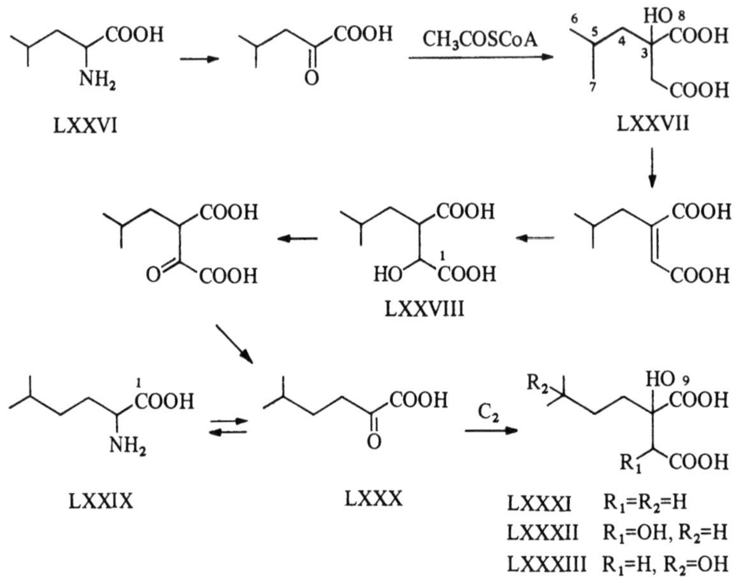

Fig. 10. Biosynthesis of the acids in the cephalotaxine esters

from valine in microorganisms (GROSS et al., 1963). On the basis of this resemblance it is suggested a hypothesis for deoxyharringtonic acid biosynthesis shown in Figure 10 (GITTERMAN et al., 1980). The hypothesis predicts that 3-carboxy-3-hydroxy-5-methylhexanoic acid (LXXVII) should be an intermediate in the biosynthesis of deoxyharringtonic acid (LXXXI) and that carbon atoms 3-8 of LXXVII should be derived from L-leucine (LXXVI). The presence of diacid LXXVII in *Cephalotaxus* plants could be shown by isotopic trapping. Furthermore the hypothesis predicts that 2-hydroxy-3-carboxy-5-methylhexanoic acid (LXXVIII) lies on the biosynthetic pathway to LXXXI. (1-[^{14}C])-LXXVIII could be specifically incorporated into LXXXI. On the basis of the hypothesis suggested in Figure 10, 2-oxo-5-methylhexanoic acid (LXXX), would be expected to be the immediate precursor of LXXXI. If this is the case, then homoleucine (LXXIX) should be specifically incorporated into LXXXI due to the facile interconversion between α-amino acids and the corresponding α-keto acids. It could be demonstrated that homoleucine is a highly efficient and specific precursor of LXXXI. By means of precursor incorporation experiments it could be shown that isoharringtonic acid (LXXXII) and harringtonic acid (LXXXIII) are derived from deoxyharringtonic acid by direct hydroxylation. Homoharringtonic acid could be formed by homologation of deoxyharringtonic acid (LXXXI) in the manner of Figure 10, with subsequent hydroxylation.

The biosynthesis of the acyl portions of deoxyharringtonine, isoharringtonine, and harringtonine having been clarified, attempts were made to determine if deoxyharringtonic acid (LXXXI) is hydroxylated at C-6 before or after it is linked to cephalotaxine. Most probably deoxyharringtonine is directly converted to harringtonine without prior deacylation.

Tissue cultures of *C. harringtonia* synthesize the same alkaloids as intact plants although at much lower concentration, a fact which is important for the production of the antitumor alkaloids (DELFEL and ROTHFUS, 1977). In another callus line a complex mixture of new cephalotaxus alkaloids is produced (DELFEL, 1980b).

The alkaloid composition of *C. harringtonia* is influenced by environmental factors on plant growth (DELFEL, 1980a). Moreover the alkaloids of this plant show a catabolic turnover. In a young plant grown in a controlled environment, the concentration of cephalotaxine esters increases. Physiological stress (pruning) causes hydrolysis of part of the stored alkaloid esters to free cephalotaxine and, in addition the oxidation of cephalotaxine to 11-hydroxycephalotaxine and drupacine (11,2-epoxycephalotaxine) and of homoerythrina alkaloids to its epoxy derivative. It is suggested that, to prevent autotoxicity, high concentrations of alkaloid esters are hydrolyzed to free cephalotaxine, which is only 1/1000 as toxic as its esters, and when the cephalotoxine levels tend to rise too high, further metabolism occur.

References

AKASU, M., ITOKAWA, H., FUJITA, M.: Phytochemistry 15, 471-473 (1976).

BARTON, D.H.R., COHEN, T.: 117-143, in: Festschrift A. STOLL. Basel: Birkhäuser 1957. - BARTON, D.H.R., KIRBY, G.W., STEGLICH, W., THOMAS, G.M., BATTERSBY, A.R., DOBSON, T.A., RAMUZ, H.: J. Chem. Soc. 1965, 2423-2438. - BARTON, D.H.R., KIRBY, G.W., WIECHERS, A.: J. Chem. Soc. 1966, 2313-2319. - BARTON, D.H.R., BHAKUNI, D.S., CHAPMAN, G.M., KIRBY, G.W.: J. Chem. Soc. (C) 1967a, 2134-2140. - BARTON, D.H.R., BHAKUNI, D.S., CHAPMAN, G.M., KIRBY, G.W., HAYNES, L.J., STUART, K.L.: J. Chem. Soc. (C) 1967b, 1295-1298. - BARTON, D.H.R., JAMES, R., KIRBY, G.W., TURNER, D.W., WIDDOWSON, D.A.: J. Chem. Soc. (C) 1968, 1529-1537. - BARTON, D.H.R., BOAR, R.B., WIDDOWSON, D.A.: J. Chem. Soc. (C) 1970a, 1208-1213; - J. Chem. Soc. (C) 1970b, 1213-1218. - BARTON, D.H.R., GUNATILAKA, A.A.L., LETCHER, R.M., LOBO, A.M.F.T., WIDDOWSON, D.A.: J. Chem. Soc. Perkin I 1973, 874-880. - BARTON, D.H.R., BRACHO, R.D., POTTER, C.J., WIDDOWSON, D.A.: J. Chem. Soc. Perkin I 1974a, 2278-2283. - BARTON, D.H.R., POTTER, C.J., WIDDOWSON, D.A.: J. Chem. Soc. Perkin I 1974b, 346-348. - BATTERSBY, A.R.: In: Oxidative Coupling of Phenols, eds. BATTERSBY, A.R., TAYLOYR, W.I. New York: Dekker 1967. - BATTERSBY, A.R., BROWN, R.T., CLEMENTS, J.H., IVERACH, G.G.: Chem. Commun. 1965a, 230-232. - BATTERSBY, A.R., FOULKES, D.M., BINKS, R.: J. Chem. Soc. 1965b, 3323-3332. - BATTERSBY, A.R., BROCKSON, T.J., RAMAGE, R.: Chem. Commun. 1969, 464-465. - BATTERSBY, A.R., McHUGH, J.L., STAUNTON, J., TODD, M.: Chem. Commun 1971, 985-986. - BATTERSBY, A.R., HERBERT, R.B., McDONALD, E., RAMAGE, R., CLEMENTS, J.H.: J. Chem. Soc. Perkin I 1972, 1741-1746. - BATTERSBY, A.R., McDONALD, E., MILNER, J.A.: Tetrahedron Lett. 1975, 3419-3422. - BASMADJIAN, G.P., HUSSAIN, S.F., PAUL, A.G.: Lloydia 41, 375-380 (1978). - BHAKUNI, D.S., JAIN, S.: Tetrahedron 36, 2153-2156 (1980). - BHAKUNI, D.S., JOSHI, P.P.: Tetrahedron 31, 2575-2579 (1975). - BHAKUNI, D.S., SINGH, A.N.: Tetrahedron 34, 1409-1410 (1978a); - J. Chem. Soc. Perkin I 1978b, 618-622. - BHAKUNI, D.S., SATISH, S., DHAR, M.M.: Phytochemistry 9, 2573-2580 (1970). - BHAKUNI, D.S., SATISH, S., UPRETY, H., KAPIL, R.S.: Phytochemistry 13, 2767-2769 (1974). - BHAKUNI, D.S., UPRETY, H., WIDDOWSON, D.A.: Phytochemistry 15, 739-741 (1976). - BHAKUNI, D.S., SINGH, A.N., KAPIL, R.S.: Chem. Commun. 1977a, 211-212. - BHAKUNI, D.S., SINGH, A.N., TEWARI, S., KAPIL, R.S.: J. Chem. Soc. Perkin I 1977b, 1662-1666. - BHAKUNI, D.S., TEWARI, S., KAPIL, R.S.: J. Chem. Soc. Perkin I 1977c, 706-708. - BHAKUNI, D.S., JAIN, S., SINGH, A.N.: J. Chem. Soc. Perkin I 1978a, 380-383. - BHAKUNI, D.S., LABROO, V.M., SINGH, A.N., KAPIL, R.S.: J. Chem. Soc. Perkin I 1978b, 121-125. - BHAKUNI, D.S., MANGLA, V.K., SINGH, A.N., KAPIL, R.S.: J. Chem. Soc. Perkin I 1978c, 267-269. - BHAKUNI, D.S., SINGH, A.N., JAIN, S., KAPIL, R.S.: Chem. Commun. 1978d, 226-228. - BHAKUNI, D.S., JAIN, S., CHATURVEDI, R.: Tetrahedron 35,

2323-2326 (1979). - BHAKUNI, D.S., SINGH, A.N., JAIN, S.: Tetrahedron 36, 2149-1251 (1980). - BLASCHKE, G.: Arch. Pharm. 301, 432-439 (1968); - Arch. Pharm. 303, 358-363 (1970). - BLASCHKE, G., WALDHEIM, G., SCHANTZ, M. v., PEURA, P.: Arch. Pharm. 307, 122-130 (1974). - BROCHMANN-HANSSEN, E., FU, C.-C., MISCONI, L.Y.: J. Pharm. Sci. 60, 1880-1883 (1971). - BROCHMANN-HANSSEN, E., CHEN, C.-H., CHIANG, H.-C., McMURTREY, K.: Chem. Commun. 1972, 1269. - BROCHMANN-HANSSEN, E., CHEN, C.-H., CHEN, C.R., CHIANG, H.-C., LEUNG, A.Y., McMURTREY, K.: J. Chem. Soc. Perkin I 1975, 1531-1537.

CHAMBERS, C., STUART, K.L.: Chem. Commun. 1968, 328-329.

DELFEL, N.E.: Phytochemistry 19, 403-408 (1980a); - Planta Med. 39, 168-179 (1980b). - DELFEL, N.E., ROTHFUS, J.A.: Phytochemistry 16, 1595-1598 (1977). - DEWAN, O.P., BHAKUNI, S., KAPIL, R.S.: J. Chem. Soc. Perkin I 1978, 622-624. - DÖPKE, D., FLENTJE, H., JEFFS, P.W.: Tetrahedron 24, 4459-4476 (1968).

FITZGERALD, J.S., JOHNS, S.R., LAMBERTON, J.A., SIOUMIS, A.A.: Austral. J. Chem. 22, 2187-2201 (1969). - FODOR, G.B., in: Encyclopedia of Plant Physiology, New Series, Vol. 8, eds. E.A. BELL, B.V. CHARLWOOD, p. 92-127. Berlin-Heidelberg-New York: Springer 1980. - FRANK, B., TEETZ, V.: Angew. Chem. 83, 409-410 (1971). - FRESNO, M., JIMENEZ, A., VAZQUEZ, D.: Eur. J. Biochem. 72, 323-330 (1977).

GITTERMAN, A., PARRY, R.J., DUFRESNE, R.F., STERNBACH, D.D., CABELLI, M.D.: J. Am. Chem. Soc. 102, 2074-2081 (1980). - GROSS, S.R., BURNS, R.O., UMBARGER, E.: Biochemistry 2, 1046-1056 (1963).

HERBERT, R.B., in: Comprehensive Organic Chemistry, eds. D.H.R. BARTON, W.D. OLLIS, Vol. 5, p. 1045-1119. Oxford: Pergamon Press 1979.

IMPELLIZZERI, G., PIATTELLI, M.: Phytochemistry 11, 2499-2502 (1972).

JOHNS, S.R., LAMBERTON, J.A., SIOUMIS, A.A.: Aust. J. Chem. 22, 2219-2231 (1969).

LEETE, E., AHMAD, A.: J. Am. Chem. Soc. 88, 4722-4775 (1966). - LIEBISCH, H.W., in: Biosynthese der Alkaloide, eds. K. MOTHES, H.R. SCHÜTTE, p. 101-122. Berlin: VEB Deutscher Verlag der Wissenschaften 1969.

MILLER, H.E., RÖSLER, H., WOHLPART, A., WYLER, H., WILCOX, M.E., FROHOFER, H., MARBY, T.J., DREIDING, A.S.: Helv. Chim. Acta 51, 1470-1474 (1968).

PANDE, H., BHAKUNI, D.S.: J. Chem. Soc. Perkin I 1976, 2197-2202. - PARRY, R.J.: Recent Adv. Phytochem. 13, 55-84 (1979). - PARRY, R.J., SCHWAB, J.M.: J. Am. Chem. Soc. 97, 2555-2557 (1975). - PARRY, R.J., CHANG, M.N.T., SCHWAB, J.M., FOXMAN, B.M.: J. Am. Chem. Soc. 102, 1099-1111 (1980). - PAUDLER, W.W., McKAY, J.: J. Org. Chem. 38, 2100-2112 (1973). - PAUL, A.G.: Lloydia 36, 36-45 (1973). - POWELL, R.G.: Phytochemistry 11, 1467-1478 (1972). - POWELL, R.G., MIKOLAJCZAK, K.L.: Phytochemistry 12, 2987-2991 (1973). - POWELL, R.G., WEISLEDER, D., SMITH, C.R., Jr., WOLFF, I.A.: Tetrahedron Lett. 1969, 4081-4084. - POWELL, R.G., MIKOLAJCZAK, K.L., WEISLEDER, D., SMITH, C.R., Jr.: Phytochemistry 11, 3317-3320 (1972a). - POWELL, R.G., WEISLEDER, D., SMITH, C.R., Jr.: J. Pharm. Sci. 61, 1227-1230 (1972b). - PRAKASH, O., BHAKUNI, D.S., KAPIL, R.S.: J. Chem. Soc. Perkin I 1979, 1515-1518

RAZAKOV, R., YUNUSOV, S.Y., NASYROV, S.M., CHEKHLOV, A.N., ADRIANOV, V.G., STRUCHKOV, Y.T.: Chem. Commun. 1974, 150-151.

SCHÜTTE, H.R., in: Biosynthese der Alkaloide, eds. K. MOTHES, H.R. SCHÜTTE, p. 367-419. Berlin: VEB Deutscher Verlag der Wissenschaften 1969. - SCHÜTTE, H.R.; Biosynthese von niedermolekularen Naturstoffen. Jena: VEB Gustav Fischer 1981. - SCHWAB, J.M., CHANG, M.N.T., PARRY, R.J.: J. Am. Chem. Soc. 99, 2368-2370 (1977). - SHAMMA, M., MONIOT, J.L., YAO, S.Y., MIANA, G.A., IKRAM, M.: J. Am. Chem. Soc. 95, 5742-5747 (1973). - SHAMMA, M., FOY, J.E., MIANA, G.A.: J. Am. Chem. Soc. 96, 7809-7811 (1974). - SHAMMA, M., FOY, J.E., GOVINDACHARI, T.R., VISWANATHAN, N.: J. Org. Chem. 1976, 1293-1294. - SIVAKUMARAN, M., GOPINATH, K.W.: Indian J. Chem. 14B, 150-151 (1976). - SMITH, T.A.: Phytochemistry 16, 9-18 (1977). - STAUNTON, J.: Planta Med. 36, 1-20 (1979). - STUART, K.L., GRAHAM, L.: Phytochemistry 12, 1967-1972 (1973). - STUART, K.L., TEETZ, V., FRANK, B.: Chem. Commun. 1969, 333.

TACKIE, A.N., DWUMA-BADU, D., AYIM, J.S.K., DABRA, T.T., KNAPP, J.E., SLATKIN, D.J., SCHIFF, P.L., Jr.: Lloydia 38, 210-212 (1975). - TEWARI, S., BHAKUNI, D.S., KAPIL, R.S.: Chem. Commun. 1975, 554-555.

UPRETY, H., BHAKUNI, D.S., KAPIL, R.S.: Phytochemistry 14, 1535-1537 (1975).

Professor Dr. HORST-ROBERT SCHÜTTE
Institut für Biochemie der Pflanzen
des Forschungszentrums für Molekularbiologie und Medizin der Akademie
der Wissenschaften der DDR
Weinberg
DDR 401 Halle (Saale)

V. Growth

By NIKOLAUS AMRHEIN

Continuing interest in plant hormones in general is reflected in the monographs by JACOBS (1979) and MOORE (1979), and especially so in the very topical volume edited by MACMILLAN (1980a) on molecular aspects of plant hormones, in which a successful attempt is made to elucidate common principles for the different classes of plant hormones. Conference proceedings were edited by MANDAVA (1979) and SKOOG (1979). With the recent isolation of brassinolide and the determination of its structure (GROVE et al., 1979) the brassins (Progr. Bot. 35, 122) have come of age. The steroidal nature of brassinolide will rekindle the discussion on the possible function of steroidal compounds as plant growth regulators. This topic has, therefore, been included in this review.

1. Gibberellins

Review: GRAEBE and ROPERS (1978); LENTON (1980) edited the proceedings of a meeting on the chemistry, physiology, and use of gibberellins.

a) Chemistry

FUJITA and NODE (1977) reviewed synthetic approaches to gibberellins. Reference to the partial synthesis and isotopic labeling of gibberellins is made by MACMILLAN (1978, 1980b) and by HANSON (1980).

While the total synthesis of GA_3 had been reported in 1978 (Progr. Bot. 41, 109), MANDER and his colleagues presented another approach to the total synthesis of GA_1, GA_3 (LOMBARDO et al., 1980; HOOK et al., 1980), and GA_4 (COSSEY et al., 1980). The brevity and simplicity of these syntheses are emphasized but they, nevertheless, present a serious obstacle to the nonchemist. GA_3 and GA_4/GA_7 from cultures of *Gibberella fujikuroi* are commercially available. Starting from these GA's, BEALE and MACMILLAN (1980) and BEALE et al. (1980) prepared GA_4, GA_5, GA_9, and GA_{20} deuterated or tritiated at specific positions, while AYREY and CHAPMAN (1979) obtained randomly tritium-labeled GA_3 with a specific activity of ca. 7 Ci/mmol. GA_9-17-^{14}C with a specific activity of 4 mCi/mmol was synthesized by LIN and HEFTMANN (1979).

GA_3 derivatives with stepwise deoxygenated position 7 were synthesized by LISCHEWSKI and ADAM (1980a). These and other derivatives, in which the 7-carboxyl group of GA_3 was modified, had very little, if any, activity in a dwarf pea bioassay (ADAM et al., 1979). This group also prepared 6-epi-GA_3 (LISCHEWSKI and ADAM, 1980b). In this compound the carboxyl group is in the 6α- rather than the 6β-position as in GA_3. The biological activity of 6-epi-GA_3 has not yet been reported.

b) Methods of Isolation and Determination

YOKOTA et al. (1980) give detailed information on the extraction, purification, and identification of gibberellins and other plant hormones , while CROZIER (1980) and, in more depth and detail, REEVE and CROZIER (1980), expound theoretical and practical aspects of the quantitative analysis of these substances.

An improvement in the silylation of gibberellins for gas-liquid chromatography was reported by JOLLIFFE et al. (1979). High-performance liquid chromatography (HPLC) without derivatization is gaining ground as the method of choice for the fractionation of gibberellins (BARENDSE et al., 1980; JONES et al., 1980) and conjugated gibberellins (YAMAGUCHI et al., 1979). A recently developed new technique, droplet counter-current chromatography, was adopted by BEARDER and MACMILLAN (1980) for the separation of gibberellins. The potential of the radioimmunoassay for the quantitation of plant hormones had been emphasized in Progr. Bot. 41, 119. WEILER and WIECZOREK (1981) have now introduced a radioimmunoassay, which detects as little as 2 pg of GA_3 in crude extracts. This extreme sensitivity was made possible by the use of [^{125}I]-gibberellic acid-[N-(p-hydroxybenzyl)putrescine]-amide with a specific activity of ca 2300 Ci/mmol as the immunotracer. Of the gibberellins tested only GA_7 cross-reacted strongly.

c) Occurrence

The number of positively identified and fully documented gibberellins has increased to 58. Reference to as yet unpublished results of TAKAHASHI's group on the identification of GA_{59} (2,3-didehydro-GA_{21}) in immature seeds of *Canavalis gladiata* is made by MACMILLAN (1980a). The structures of the first 57 GA's and references to their identification are given by BEARDER (1980); the structure of GA_{58} can be seen in Figure 1. GA_{53} is the name now assigned to 13-hydroxy-GA_{12}, which had previously been identified in *Gibberella fujikuroi* (Progr. Bot. 39, 103). It is one of the six 13-hydroxylated gibberellins identified in immature seeds of *Vicia faba* (SPONSEL et al., 1979) and in spinach shoots (METZGER and ZEEVAART, 1980), and one of the seven GA's in the shoots of *Agrostemma githago* (JONES and ZEEVAART, 1980b). GA_{54} and GA_{55} were isolated from the culture broth of *Gibberella fujikuroi* (MUROFUSHI et al., 1979) as well as from the developing grain of *Triticum aestivum* (GASKIN et al., 1980). Interestingly, TAKAHASHI's group first synthesized the respective gibberellins and then used them as markers in their search for novel minor GA's produced by *Gibberella fujikuroi*. GA_{56} (MUROFUSHI et al., 1979) and GA_{57} (MUROFUSHI et al., 1980) were also detected as minor GA's in the culture broth of the fungus. Finally, GA_{58} (Fig. 1) which is 12α- hydroxy-GA_4, was identified by GRAEBE et al. (1980) as the main native C_{19}-GA in the endosperm of *Cucurbita maxima*, where it seems to occur only in a conjugated form.

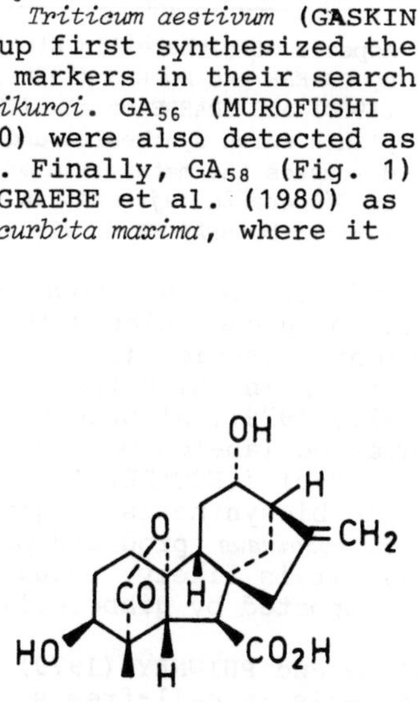

Fig. 1. Structure of GA_{58}

While no microorganisms other than *Gibberella fujikuroi* had previously been shown unequivocally to produce gibberellins, the independent work of two research groups has now established beyond doubt the occurrence of GA_4 in the fungus *Sphaceloma manihoticola* (RADEMACHER and GRAEBE, 1979; ZEIGLER et al., 1980), which is the causal agent of the superelongation disease of cassava *(Manihot esculenta)*. GA_4 appears to be an end-product of gibberellin biosynthesis in *Sphaceloma manihoticola*, while it is an intermediate in the biosynthesis of GA_7, GA_3, and GA_1 in *Gibberella fujikuroi*. GRAEBE et al. (1980) reported the occurrence in small amounts of a number of gibberellins in *Sphaceloma manihoticola*, the nature of which makes it likely that the biosynthetic pathway leading to GA_4 is identical in the two fungi. Considering the problems associated with the separation of the closely related GA_4 and GA_7, *Sphaceloma manihoticola* may be a useful source for pure GA_4 in the future. KUROGOCHI et al. (1979) identified GA_{19} as the principal gibberellin of the rice plant and measured the changes of its level throughout the life cycle of the plant. Drastic changes in the level of GA_{19} were observed at defined developmental stages. While GA_1 is quantitatively a minor gibberellin of the rice plant, it has a far higher growth-promoting activity than GA_{19} in the same plant. Thus, the authors suggest that GA_{19} functions as a "pool gibberellin", from which the active GA_1 is synthesized. Turnover, rather than actual concentration, of GA may therefore be relevant for growth regulation.

A number of reports deals with the identification of gibberellins and estimation of their levels in seeds or embryos (ALPI et al., 1979; BIANCO and BULARD, 1980; BUKOVAC et al., 1979; ISAIA and BULARD, 1978; NAKAGAWA et al., 1979; NOMA et al., 1979; YAMAGUCHI et al., 1980). A controversy has arisen about the nature and concentration of gibberellins in apple seeds (DENNIS et al., 1980; LEWAK and SINSKA, 1980).

GA_9 methyl ester was isolated by YAMANE et al. (1979) from medium in which prothallia of the fern *Lygodium japonicum* had been cultured. The compound acts as an antheridiogen (NÄF et al., 1975) and inhibits the formation of archegonia, but it is unlikely to be the principal endogenous regulator.

d) Biosynthesis and Metabolism

Aspects of gibberellin biosynthesis and metabolism were extensively discussed by GRAEBE and ROPERS (1978), PHINNEY (1979) and SEMBDNER et al. (1980). GRAEBE (1980) and GRAEBE et al. (1980) concentrate on cell-free systems from higher plants which synthesize and transform gibberellins, while SPONSEL (1980a,b) discusses the metabolism of gibberellins in legume seeds. The use of mutants of *Gibberella fujikuroi* in the elucidation of gibberellin biosynthesis was summarized by BEARDER et al. (1979).

In addition to the known cell-free systems from endosperm or whole seeds of higher plants in which gibberellin biosynthesis can be conveniently studied (Progr. Bot. 41, 110), suspensor tissue of *Phaseolus coccineus*, in which the occurrence of GA_1 had been established (ALPI et al., 1979), allowed the preparation of cell-free extracts, which converted labeled mevalonic acid into ent-kaur-16-ene and ent-kaur-16-ene-19β-ol (CECCARELLI et al., 1979) which are central intermediates in the biosynthesis of gibberellins. Extracts from endosperm tissue of *P. coccineus* produced predominantly squalene from mevalonic acid. Thus, it is likely, though not yet proven, that growth of the embryo is supported by gibberellins produced in suspensor tissue.

HEDDEN and PHINNEY (1979) made a very interesting study on ent-kaurene synthesis in cell-free systems from etiolated shoots of either normal

Fig. 2. Structures of ent-kaur-16-ene *(a)* and ent-kaur-15-ene *(b)*

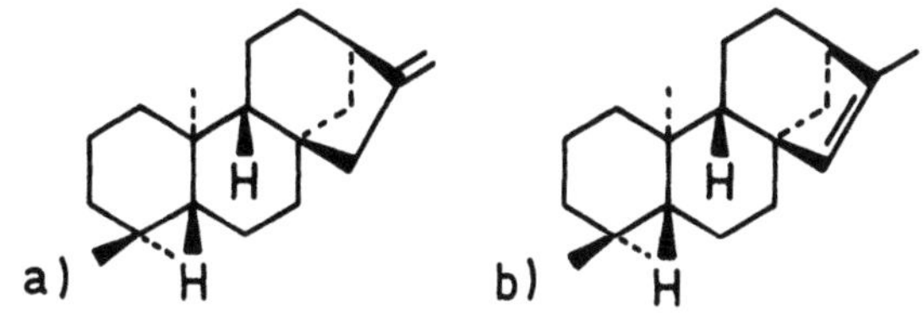

maize seedlings or of the dwarf-5 (d_5) mutant. d_5 is a gibberellin-less mutant, which resulted from a recessive gene mutation. Normal growth is restored by the application of gibberellins, but also of ent-kaurene and its oxidation products ent-kaurenol and ent-kaurenoic acid. Cell-free extracts from normal plants produced the normal intermediate of GA-biosynthesis, ent-kaur-16-ene (Fig. 2a), and ent-isokaurene (ent-kaur-15-ene, Fig. 2b), which does not serve as an intermediate, at a ratio of 8 : 1, while in extracts from d_5 the ratio was 0.13 : 1. Similar ratios were found when copalyl-pyrophosphate was used as substrate for ent-kaurene synthase. Thus, it is apparently the B-activity of this enzyme (Progr. Bot. 41, 110) which is subject to control by the d_5-gene. Evidence has been presented that etioplasts and chloroplasts are compartments of gibberellin biosynthesis (Progr. Bot. 39, 102; 41, 114). In principal support of these findings are the reports on the synthesis of geranylgeraniol derivatives in spinach chloroplasts (BLOCK et al., 1980) and on the concomitant photoinduction of ent-kaurene biosynthesis and of chloroplast development in pea shoots (CHOINSKI and MOORE, 1980).

HILTON and SMITH (1980) reexamined the presence of phytochrome in barley etioplasts and its in vitro regulation of gibberellin levels in etioplasts. SMITH (1980) states definitively that phytochrome is associated with etioplasts and can be shown, in vitro, to control the level of extractable gibberellins in these organelles. Photocontrol of gibberellin levels is not, however, considered to be an obligatory step in all photomorphoses.

In developing pea seeds the greatest capacity for ent-kaurene synthesis from mevalonate in a cell-free system is found shortly before the seeds are fully developed. Thereafter, the synthesizing activity declines rapidly and disappears completely (GRAEBE, 1980). The rate of formation of products from 7β- hydroxykaurenoic acid and from GA_{12}-aldehyde showed similar time courses, but kaurene oxidation to kaurenol and kaurenal was highest in young seeds and then gradually declined. GRAEBE (1980) therefore favors the view that kaurene formation is the rate-limiting step in gibberellin biosynthesis. Cell-free preparations from immature *Cucurbita maxima* seeds have provided further insight in the biosynthetic pathway of gibberellins (GRAEBE et al., 1980). This system provided an important clue to the problem, at which stage of gibberellin biosynthesis the conversion of a C_{20}-GA to the corresponding C_{19}-GA occurs (Progr. Bot. 41, 112). Only $[^{14}C]GA_{36}$, but not $[^{14}C]GA_{13}$ was converted to $[^{14}C]GA_4$ in this system (Fig. 3). Thus, the point of divergence of C_{19}- and C_{20}-gibberellins, which has so long been sought, seems to have been found at last.

In contrast to cucurbitaceous plants, which do not contain gibberellins with a hydroxyl group at C-13, C-13-hydroxylated gibberellins are the rule in legumes. Within the legumes, distinctions with apparent chemotaxonomic significance can be made between the tribes *Phaseolae (Ph. coccineus, Ph. vulgaris, Vigna unguiculata)* and *Viciae (Pisum sativum, V. faba)*: members of the *Phaseolae* contain 3-hydroxylated gibberellins, while these are absent from members of the *Viciae* (SPONSEL et al., 1979; SPONSEL, 1980b). Furthermore, while members of the *Phaseolae* are rich sources of

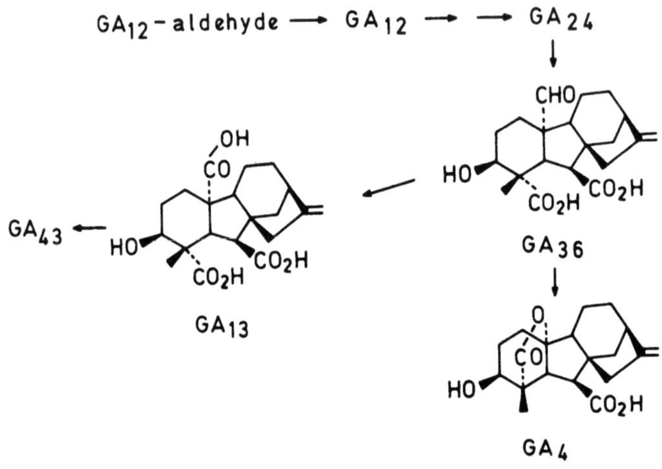

Fig. 3. Scheme of late stages of gibberellin biosynthesis in *Cucurbita maxima* endosperm. Note transition from C_{20}-GA (GA_{36}) to C_{19}-GA (GA_4)

conjugated gibberellins (GA-glucosyl ethers and GA-glucosyl esters), the level of conjugated gibberellins is very low in members of the *Viciae*, which seem to dispose of gibberellins by way of catabolism (Progr. Bot. 41, 112) rather than conjugation.

The formation of a catabolite of GA_{29} in maturing pea seeds (for structure see Progr. Bot. 41, 113) has been further investigated by SPONSEL and MACMILLAN (1980) using $[17\text{-}^{13}C_1]GA_{29}$. The formation of the $[^{13}C_1]$-GA-catabolite from the exogenously added labeled GA_{29} strictly paralleled the accumulation of the native catabolite in the maturing seeds. With $[^{18}O]$-labeled GA-catabolite as internal marker, quantitation of the catabolite was made possible: 25.8 µg per seed were found in 37 days old seeds. GA_{51}, another gibberellin of *P. sativum*, is probably subjected to a similar catabolic reaction during seed maturation. SPONSEL (1980b) presents evidence that germination of *Pisum* and *Phaseolus* seeds proceeds in the absence of gibberellin formation either by de novo synthesis or by release from conjugated forms. Once germination has occurred, de novo synthesis of gibberellins is, however, required for normal development. Thus, the role of gibberellin formed during seed maturation still remains to be clearly defined.

While no correlation had been found between the conversion of the growth-promoting GA_1 to the inactive GA_8 by 2β-hydroxylation in dwarf rice seedlings (Progr. Bot. 41, 113; RAILTON, 1980a) such a positive correlation has been shown for 2β-hydroxylation of GA_{20} to GA_{29} and stem growth in dwarf pea seedlings (RAILTON, 1980b). JONES and ZEEVAART (1980a,b) investigated the role of endogenous gibberellins in the photoperiodic control of stem elongation in the long-day rosette plant *Agrostemma githago*. Control of stem elongation could not be explained on the basis of the observed quantitative or qualitative changes of the seven gibberellins which were identified in this species. It was suggested that the rates of turnover of endogenous gibberellins, as well as the tissue's sensitivity to these gibberellins, may be changed by the photoperiodic induction and thus affect stem elongation.

e) Action

STODDART and VENIS (1980) discuss molecular and subcellular aspects of hormone action, while ZERONI and HALL (1980) discourse about molecular effects of hormone treatment on tissue. PHARIS and KUO (1977) restrict their discussion of the physiology of gibberellins to conifers.

α) *Gibberellin-Induced Elongation.* This topic is discussed by JONES (1980a) and STODDART (1980). Light-induced inhibition of the elongation of excised lettuce hypocotyls is overcome by gibberellins (Progr. Bot. 39, 105; 41, 116). The nature of the light effect was investigated by THOMAS et al. (1980), and it was shown to involve the action of both a blue light receptor and of phytochrome. GA-induced growth is exclusively attributable to cell elongation and is neither mediated by auxin nor dependent on the auxin status of the tissue, as experiments with auxins and antiauxins showed (JONES, 1980a). This is in contrast to earlier results with peas and beans (Progr. Bot. 39, 105). In excised shoot tips of the angiospermic parasite, *Cuscuta chinensis*, GA_3 induced a growth increment which was up to 20 times that of the control (MAHESHWARI et al., 1980).

The growth rate increased within two days of the application of GA_3, and growth was linear for 8 to 10 days before it declined. Involvement of auxin was indicated in this system, and both cell division and cell elongation were thought to be involved in the growth response. In γ-plantlets of peas, in which growth occurs without mitosis and cytokinesis, endomitotic DNA synthesis, epicotyl elongation, and cell elongation are inhibited. Application of GA_3 restores endomitotic DNA synthesis and cell elongation (CALLEBAUT et al., 1980). The authors suggest that γ-irradiation lowers the endogenous gibberellin concentration, but gibberellin levels were not measured. No evidence was found in unirradiated peas for qualitatively different DNA synthesized in GA_3-treated or control epicotyls (MOHAMED and CAPESIUS; 1980; BROEKAERT et al., 1980). The effect of gibberellins on the growth rate of pea stems can be quite rapid: 10 μg of GA_3 applied to a single decapitated stem produced a transient increase in growth rate within 10 min in the dark and within 3 min in red light (NAUNOVIĆ and NESKOVIĆ, 1979). Cucumber hypocotyl sections respond to external application of both auxin and gibberellin, but the responses are qualitatively and quantitatively different under various conditions (KATSUMI and KAZAMA, 1978). Interestingly, GA_3 completely prevents the inhibition of IAA-induced elongation by DCCD (N,N'-dicyclohexylcarbodiimide) in the hypocotyl sections, but involvement of a membrane-bound ATPase was not convincingly demonstrated.

STODDART (1979a, 1979b) and STODDART and WILLIAMS (1979, 1980) have used lettuce hypocotyls as a physiologically well-defined system to study sub-cellular interactions of applied gibberellin. They fed [1,2-^{3}H]GA_1 with a specific activity of 40 Ci/mmol to hypocotyls for various lengths of time and determined the distribution of label in fractions which were obtained by differential centrifugation of buffer extracts. Of the total label more than 95% remained in the 100,000 g supernatant, while 2% - 5% were consistently found in the material sedimenting at 2000 g and which was considered to be cell wall material. Labeling of the 2000 g fraction was highly correlated with growth (STODDART, 1979a). The label was rather tenaciously bound in the fraction, and a nonionic association with, e.g., a carbohydrate fragment was considered possible. In an effort to determine the causal relationship, if there was any, between the labeling of the 2000 g fraction and the elongation response of the hypocotyls STODDART and WILLIAMS (1980) fed [^{3}H]-GA_1 under conditions which allowed uptake, but abolished growth (low temperature), and then transferred the hypocotyls to GA_1-free medium at the usual growth temperature. From the incorporation of label into the 2000 g fraction they concluded that GA_1 incorporation into cell wall (?) material may be a causal event in the tissue's response to GA_1.

Similarly to the approach of STODDART and WILLIAMS (1980) $[^3H]$-GA_1 was fed at low temperature for periods of up to 3 days to barley aleurone layers (KEITH et al., 1980) and to sliced pea epicotyls (KEITH and SRIVASTAVA, 1980), in order to allow uptake and binding to putative receptors without interference of metabolism. Thus the system is comparable to an equilibrium dialysis cell. In the 100,000 g supernatant of the pea epicotyl extracts label presumed to be $[^3H]$-GA_1 was found bound to two protein fractions (m.w. $6 \cdot 10^5$ and 4 to $7 \cdot 10^4$), but the label was not exchangeable. Lack of exchangeability clearly argues against a receptor function of these proteins in barley aleurone layers. Under equilibrium conditions, $[^3H]$-activity per unit volume was higher in the layer than in the medium, but nothing is known about the nature of the binding site(s) in the layers.

Uptake and subcellular compartmentation of $[^3H]$-GA were studied in leaves of barley and cowpea *(Vigna sinensis)* by OHLROGGE et al. (1980). Uptake had a rather narrow pH optimum (5.8) in cowpea leaves and was saturable with an apparent K_m of 15 μM, while no saturation was observed at 300 μM in barley leaves. Surprisingly, 30% to 100% of the label in protoplasts isolated from leaves to which $[^3H]$-GA_1 had been fed were recovered in the vacuolar fraction. It is argued that the accumulation cannot simply be accounted for by the pH of the cell compartments, and a carrier-mediated uptake system was proposed to explain the results.

β) Gibberellin Action in the Aleurone Layer. Review: TOMOS and LAIDMAN (1979) The role of the scutellum versus the role of the aleurone layer in the production of α-amylase and other hydrolases in germinating barley and other cereal seeds has been reinvestigated. In rice (OKAMOTO and AKAZAWA, 1979) barley (GIBBONS, 1979, 1980) and other caryopses (OKAMOTO et al., 1980), it was shown that, at least in the first few days of germination, the scutellar epithelium secretes the major amount of hydrolytic enzymes.

In response to applied GA_3 the translatable mRNA for α-amylase increases in barley aleurone layers (Progr. Bot. 41, 115). In wheat (OKITA et al., 1979) as well as barley (MOZER, 1980a) the in vitro translation product is larger than the α-amylase secreted by the aleurone layers. The purified mRNA for α-amylase from barley aleurone layers is ca. 1650 nucleotides in length, of which about 80% are translated to produce the putative precursor (MOZER, 1980a). While the evidence favors transcriptional control of α-amylase messenger synthesis by GA_3, processing of a precursor RNA cannot be excluded. The production of complementary DNA (cDNA) will allow the quantitation of the α-amylase message by hybridization techniques and thus permit a decision between the two alternatives. MUTUKRISHNAN et al. (1979) found no evidence for activation of preexisting RNA by known mechanisms and thus favor the hypothesis that gibberellin treatment stimulates de novo mRNA synthesis. Clear-cut evidence for translational control by a plant hormone in aleurone layers has, however, been provided (MOZER, 1980b). While abscisic acid inhibits the synthesis of α-amylase in vivo, the in vitro translatable α-amylase mRNA is present in tissue that has been exposed to GA_3 and abscisic acid. Abscisic acid by itself was shown to induce the formation of translatable mRNA and of new protein. Thus, abscisic acid does not merely function as an antagonist of GA_3.

Proliferation of the endomembrane system (rough endoplasmic reticulum) in response to gibberellin has been reported, and enhancement by GA_3 of phospholipid synthesis was suggested by some studies (Progr. Bot. 34, 187; 35, 132; 36, 131; OBATA, 1979), but both the electron microscopic and biochemical evidence put forward in favor of an effect of GA_3 on endomembrane synthesis have been contested, and neither a qualitative nor a quantitative effect of GA_3 on phospholipid metabolism was acknowledged (TOMOS and LAIDMAN, 1979). Requirement for continuous lipid (fatty acid) synthesis for unimpaired α-amylase and phosphatase formation and release was, however, indicated by studies with the antibiotic cerulenin, an inhibitor of de novo fatty acid synthesis (AMRHEIN et al., 1979). JONES (1980b) has recently developed techniques for the purification and quantitation of the endoplasmic reticulum. A reinvestigation of the quantitative and qualitative changes in the endomembranes in the absence and presence of GA_3 (JONES, 1980c) showed mainly a qualitative change (decrease in buoyant density). About 20% of the total extractable α-amylase is localized in the endoplasmic reticulum. Of the remaining 80%, 60% are thought to be in the cell wall and 40% in an undefined cytoplasmic compartment. Whether this latter fraction represents a pool of soluble enzyme in the cytoplasm or has leaked from cisternae of the endoplasmic reticulum during homogenization is unknown. It was shown for acid phosphatase with cytochemical techniques that a presumably newly synthesized fraction of this enzyme is detectable in the lumen of the endoplasmic reticulum in GA_3-treated aleurone layers (PYLIOTIS et al., 1979). The route of secretion of acid phosphatase into the cell wall could not be established.

Lastly, attention was drawn to the involvement of membrane-associated glycosyl transferases in the response of aleurone layers to GA_3 (SCHWAIGER and TANNER, 1979). The activity of these enzymes was enhanced 4 h after addition of GA_3 to these layers, and prevention of endogenous glycosylation reactions by the antibiotic tunicamycin inhibited α-amylase formation.

2. Brassinolide

In 1970 MITCHELL et al. reported the growth-promoting activity of ether extracts of rape (*Brassica napus* L.) pollen. They termed the crude mixture of active principles "brassins" (Progr. Bot. 35, 122) and, with little justification, designated them plant hormones. The growth-promoting activity of the preparations was assayed by their effect on the elongation of the second internode of intact bean plants and was considered to be different from the effect of applied GA_3. Later work (MITCHELL and GREGORY, 1972) showed that brassins enhance the overall growth of bean and elm plants, and the concept of an "alpha hormone" (Progr. Bot. 35, 122) was developed. It was first proposed that brassins have a glyceride structure (MITCHELL et al., 1970), and later that they are 1-β-D-glucopyranosyl esters of various fatty acids (MANDAVA and MITCHELL, 1972). The hormonal nature of the compounds, as well as the elucidation of their structure, was severely criticized and questioned (MILBORROW and PRYCE, 1973). Later work revealed that rape pollen contains 6-D-glucopyranosyl fatty acid esters, but these were found to be inactive in the bean second internode bioassay (GROVE et al., 1978). Synthetic 1-O-glucopyranosyl fatty acid esters were also inactive in the assay (TANAKA et al., 1979), and conflicting results were obtained with a series of O-acyl-D-glucoses and -galactoses, of which only 4-O-acyl-D-glucose seemed to possess activity similar to that of the brassin complex (BEMILLER et al., 1979). Recently, the

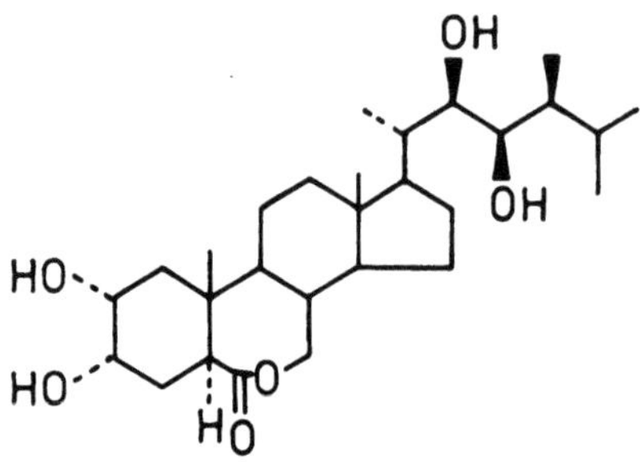

Fig. 4. Structure of brassinolide

isolation and definitive identification of the active principle of the brassin complex was announced by GROVE et al. (1979). From 40 kg of bee-collected rape pollen these authors isolated 4 mg of a compound with the formula $C_{28}H_{46}O_6$, the structure of which was determined by mass spectrometry, [^{1}H]- and [^{13}C]-NMR spectroscopy, and X-ray crystallography to be 2α, 3α, 22(R), 23(R)-tetrahydroxy-24(S)-methyl-B-homo-7-oxa-5α-cholestan-6-one (Fig. 4). The trivial name brassinolide was given to the compound. In the bean second-internode bioassay 10 ng of crystalline brassinolide per plant caused elongation (ca. 200%), swelling, and splitting of the excised internodes. The synthesis of two 22,23 isomeric brassino steroids and a nonlactonic steroid with activity in the bean second-internode assay was reported in the same year (THOMPSON et al., 1979). Homobrassinolide (24-ethyl instead of 24-methyl group) synthesized by MORI (1980) had a high activity in rice plants. The stereoselective synthesis of brassinolide was described by two groups (FUNG and SIDDALL, 1980; ISHIGURO et al., 1980). The synthetic compound was reported to have high biological activity in the wheat coleoptile elongation assay (ISHIGURO et al., 1980), but full documentation has not yet been presented.

According to MEUDT et al. (1979) brassinosteroids act synergistically with auxin. While the growth-promoting activity of brassinolide seems firmly established, its function in the material from which it was isolated is not understood at all, and any growth-regulating (hormonal) function in vivo has as yet to be demonstrated. Nevertheless, the steroidal nature of brassinolide will rekindle the discussion of the involvement of steroids in the regulation of the development of higher plants (for more recent reviews see GEUNS, 1978; HEFTMANN, 1977, HEWITT et al., 1980; SLAMA, 1980; but see also VAN ROMPUY and ZEEVAART, 1979). One must acknowledge, however, that to date, with the exception of antheridiol and the oogoniols – the steroid hormones which control sexual reproduction in the oomycete *Achlya* (McMORRIS, 1978) – unambiguous proof for the function of steroid hormones in plants does not exist.

3. Cytokinins

For extensive reviews see LETHAM (1978) and SKOOG and SCHMITZ (1979).

a) Analytical Methods

Procedures for the isolation and analysis of cytokinins, among other plant hormones, are discussed by YOKOTA et al. (1980) and REEVE and CROZIER (1980).

In addition to the methods listed in the last two reviews (Progr. Bot. 39, 109; 41, 118) new and efficient procedures for high-performance liquid chromatography (HPLC) of cytokinins have been published (ANDERSEN and KEMP, 1979; HORGAN and KRAMERS, 1979).

HPLC and chemical ionization mass spectrometry was used successfully by SUMMONS et al. (1980) to isolate and identify a plethora of cytokinins from immature sweet-corn kernels. WEILER (1980) developed a radioimmunoassay (Progr. Bot. 41, 119) which is highly specific for zeatin and zeatinriboside as well as extremely sensitive (measuring range from 0.06 to 30 pmol), which was made possible by the use of [^{3}H]- and [^{125}J]-labeled immunotracers of very high specific activity. The sensitivity exceeds even that of the most sensitive bioassay and allows the determination of cytokinins in crude extracts without purification. Several hundred analyses per day can be performed.

Stable isotope-labeled internal standards are required for quantification of compounds by mass spectrometry because of losses during extensive purification. Several reports describe the synthesis and application of deuterated (SUMMONS et al., 1979; SUGIYAMA et al., 1980a,b,c) and [^{15}N]-labeled (SCOTT and HORGAN, 1980) cytokinin standards. [^{14}C]-labeled zeatinriboside (spec.act. 5 mCi/mmol) will be of use for metabolic studies (NAKAJIMA et al., 1980), and [^{3}H]-benzyladenine (FOX et al., 1979) with a spec.act. of nearly 27 Ci/mmol will be particularly useful in the search for and study of cytokinin-binding sites.

b) Occurrence

Fully referenced lists of naturally occurring compounds with cytokinin activity are given by SKOOG and SCHMITZ (1979) and BEARDER (1980).

The only previously identified source of the unusual cytokinin 6-(o-hydroxybenzylamino)-9-β-D-ribofuranosylpurine [=N^6-(2-hydroxybenzyl)-adenosine] have been poplar leaves (Progr. Bot. 36, 134; 41, 122). This compound has now been shown to occur in high concentrations (estimated: 0.5 µg/g fresh weight) in fruits of *Zantedeschia aethiopica* (*Araceae*; commonly called Calla) by CHAVES DAS NEVES and PAIS (1980a). A minor cytokinin from the same source is unusual in several respects: the novel compound was identified as 6-(o-hydroxybenzylamino)-2-methylthio-9-β-D-glucofuranosylpurine (CHAVES DAS NEVES and PAIS, 1980b; Fig. 5). Cytokinins with the 2-methylthio group are generally believed in higher plants to be restricted to tRNA (see LETHAM, 1978; for previously reported exception see HASHIZUME et al., 1979). The mass spectral criteria for assigning the furanoside structure to the novel cytokinin is unlikely to withstand the critique of LETHAM et al. (1979a), who have shown that the alleged 7-glucofuranoside of 6-benzylaminopurine that is formed in potato tuber tissue is, in fact, the glucopyranoside. In any case, it appears that *Z. aethiopica*, in which the white spathe regreens by the action of a cytokinin produced by the fruit, is an interesting system to study cytokinin action.

The great advance in our understanding of the molecular biology of crown gall tumors (Progr. Bot. 40, 226) has stimulated research on the relationship between the cytokinin autonomy of the tumor cells and the expression of the Ti plasmid DNA (T-DNA), which is responsible for the transformation of the cells. MESSENS and CLAEYS (1978) provided preliminary evidence that cytokinin production in *Agrobacterium tumefaciens*, the causal agent of the tumor formation, is Ti-plasmid encoded. In this context it is of interest that production of indoleacetic acid has also been proposed to be a function of the Ti-plasmid (LIU and KADO, 1979). The possibility that the Ti-plasmid regulates plant genes of

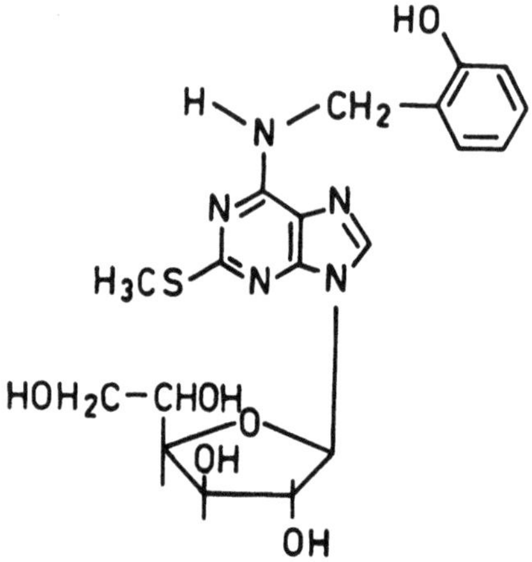

Fig. 5. Structure proposed for new cytokinin from fruits of *Zantedeschia aethiopica*

cytokinin and auxin biosynthesis can, however, not be excluded at present.

The level of zeatin riboside in a cytokinin-requiring callus of *Vinca rosea* growing on a 6-benzylaminopurin containing medium was 2.5 ng per g fresh weight, while it was 402 ng per g fresh weight in a crown gall culture (SCOTT and HORGAN, 1980). The major endogenous cytokinin of these crown gall cultures was identified, however, as zeatin-9-D-glucopyranoside (SCOTT et al., 1980b). The endogenous occurrence of cytokinin-9-glucosides had hitherto not been reported (Progr. Bot. 41, 121). Zeatin riboside is mainly responsible for the cytokinin activity of tobacco crown gall cultures (EINSET, 1980; SCOTT et al., 1980a), and cis- and trans-zeatin riboside were identified in a virulent strain of *Agrobacterium tumefaciens* (MCCLOSKEY et al., 1980). Further work will have to show whether there is a stringent relationship between the induction and maintenance of the tumorous growth and zeatin (riboside) production.

In *Corynebacterium fascians* , the causal agent of the fasciation disease, production of isopentenyladenine, but also of cis- and trans-zeatin in various strains is correlated with their pathogenicity (MURAI et al., 1980). The high production of these cytokinins and the pathogenicity of the bacterium again appear to be related to the presence of plasmids.

Related to the nonrequirement of crown gall tumor cells for an exogenous cytokinin source is the phenomenon of cytokinin habituation, a stable heritable change in the cytokinin requirement of cultured cells. As a result of a series of investigations, MEINS and coworkers (BINNS and MEINS, 1979, 1980; MEINS and LUTZ, 1979, 1980; MEINS et al., 1980a,b) suggest that brief treatment of nonhabituated tissue with exogenous cytokinins at concentrations above a critical threshold causes the shift to the habituated state by establishing a positive feedback loop in the maintenance of the concentration of growth division factors. The induction of cytokinin autonomy by N,N'-diphenylurea in tissue cultures of *Phaseolus lunatus* (MOK et al., 1979) may be a similar phenomenon.

Autonomous growth of cultured cells cannot always be related to elevated hormone levels. This was shown for auxins as well as for cytokinins by a comparison of the hormone content of four different autonomous strains of cultured tobacco cells (NAKAJIMA et al., 1979). The complexity is increased by the fact that cytokinin-degradative activity is higher in cytokinin-autonomous tissue and that exposure to cytokinin will increase cytokinin-degradative activity (TERRINE and LALOUE, 1980). Therefore factors other than, or in addition to, the actual endogenous hormone concentrations must be responsible for the autonomous growth.

Of the numerous reports on the identification of cytokinins and their metabolites only a few can be cited: Isopentenyladenine, previously identified in *Funaria* (Progr. Bot. 41, 122) has been found along with an unidentified cytokinin in the culture media of the moss, *Physcomitrella patens* (WANG et al., 1980). The cytokinin glycosides of lupin pods were quantitated by mass spectrometry with the aid of a series of $[^2H]$-labeled internal standards (SUMMONS et al., 1979). Cytokinins of lupin seeds were investigated by DAVEY and STADEN (1979). Cis-zeatin riboside, which has rarely been found in free form in higher plants, was identified in bamboo shoots (MCCLOSKEY et al., 1979). The identification of cytokinin activity in the suspensor of *Phaseolus coccineus* embryos (LORENZI et al., 1978) underscores the essential role of the suspensor in embryogenesis.

c) Biogenesis and Metabolism

Metabolites of cytokinins are discussed by ENTSCH et al. (1980).

In the culture medium of *Corynebacterium fascians* strains the nature of the cytokinin-active bases and nucleosides suggests degradation of tRNA as their source (MURAI et al., 1979). This is also indicated by the stable relationship between zeatin riboside and 2-methylthio zeatin riboside (and the bases derived from them) in tRNA and in the culture medium. The authors argue, however, that the large amounts of isopentenyladenine which are released by pathogenic strains are probably synthesized via an alternative pathway. The isopentenylation of AMP and subsequent formation of isopentenyladenine in a cell-free system from *Dictyostelium discoideum* had been reported (Progr. Bot. 31, 123) and has now also been found to occur in a cell-free system from cytokinin-autotrophic tobacco tissue cultures (CHEN and MELITZ, 1979). Also working with cultured tobacco cells NISHINARI and SYONO (1980a) observed the formation of isopentenyladenosine, trans-zeatin, and isopentenyladenine with adenosine as substrate, but AMP and adenine served also as acceptors of the isopentenyl group. In short time feeding experiments in vivo, $[^{14}C]$-adenosine was incorporated into zeatin, ribosylzeatin, isopentenyladenosine, and glucosylzeatin (NISHINARI and SYONO, 1980b), providing evidence that the pathway is operative in vivo as well. The Japanese group used partially synchronized cells, in which the endogenous cytokinins increase 5 to 10 times within 3 h (NISHINARI and SYONO, 1980c,d). The time-course of the zeatin content paralleled the mitotic index, while the amount of the cytokinins in the tRNA fraction did not change significantly. These results clearly indicate that de novo synthesis of cytokinins occurs during the cell cycle.

Rate and extent of the conversion of labeled adenine into cytokinins in crown gall tissue of *Vinca rosea* (STUCHBURY et al., 1979) and of mevalonic acid into the cytokinins of cultured potato cells (BARNES et al., 1980) are also compatible with de novo synthesis of cytokinins rather than their formation by catabolism of RNA.

The metabolism of exogenously added cytokinins in plant tissues can be extensive and complex, involving O- and N-glucosylations at various positions of the respective cytokinin (Progr. Bot. 36, 135; 39, 111;

41, 124). Enzymes in extracts from radish cotyledons, which catalyze 7- and 9-glucosylation of benzylaminopurin (Progr. Bot. 41, 124) have been further investigated (ENTSCH et al., 1979). A purified enzyme preparation formed predominantly the 7-glucoside of zeatin which agrees with the known occurrence of zeatin-7-glucoside in radish. UDP-glucose was required as the glucosyl donor, and the enzyme showed a clear preference for an adenine ring with an alkyl or aryl side chain at position N^6, which is in agreement with the structural requirements for cytokinin activity. Some of the substituted pyrrolo[2,3-d]pyrimidines with anticytokinin activity (Progr. Bot. 39, 110) inhibit the glucosylation (ENTSCH et al., 1980). Progress has also been made in the cell-free synthesis of lupinic acid (9-alanyl-zeatin, Progr. Bot. 41, 121) (ENTSCH et al., 1980). The corresponding alanine conjugate of benzylaminopurine (BAP) was found to be the principle BAP metabolite in *Phaseolus vulgaris* seedlings (LETHAM et al., 1979). Studies of RAMINA (1979) with bean plants suggest that benzylaminopurine is translocated in the xylem exclusively as its riboside. Zeatin and zeatin-riboside supplied to poplar leaves are metabolized to a complex of O-glucosides (DUKE et al., 1979).

d) Action

Structure-activity relationships of cytokinins are discussed by IWAMURA et al. (1980) and MATSUBARA (1980). HECHT (1979, 1980) reviews anticytokinins, KULAEVA (1980) summarizes the effects of cytokinins on enzyme activities in plants, and PARTHIER (1979) evaluates the role of cytokinins in chloroplast development.

For a cytokinin (or any hormone) to act, a subcellular recognition site may be assumed to exist (Progr. Bot. 41, 125). Cytokinin-binding proteins have been investigated in more detail, but any relationship between the interaction of a cytokinin with such a protein and the elicited response remains to be established. Wheat germ is a rich source for cytokinin-binding protein(s), which has been exploited by three groups (MOORE, 1979; POLYA and BOWMAN, 1979; ERION and FOX, 1981). The protein, which occurs free in the cytoplasm as well as ribosome-bound, has a molecular weight of ca. 185,000 and appears to consist of four nonidentical subunits. One cytokinin molecule is bound per tetramer, and K_d values for kinetin and benzylaminopurine are in the range of 2 to $5 \cdot 10^{-7}$M. Disturbingly, the highly active cytokinin zeatin is not bound by the protein, and a number of compounds structurally not related to cytokinins displace kinetin from the binding site. The concentration of the protein has been estimated to be $15 \cdot 10^{-6}$ M in wheat germ (ERION and FOX, 1981), which seems quite excessive for a protein with a hormone receptor function. Upon germination, however, the concentration of the protein is greatly diminished, and the authors suggest that the seed is a "reservoir" of the protein, which is diluted by cell division. The biological function of the protein, whether related to cytokinin action or not, remains to be elucidated. A cytokinin-binding protein from cultured tobacco cells is less well characterized (CHEN et al., 1980).

The use of cytokinin photoaffinity probes holds great promise for the identification of the nature and location of a receptor molecule, as well as for the investigation of the topography of the binding site once the receptor has been isolated and identified. MORNET et al. (1979) synthesized a series of cytokinins with an azido group as a photolabile substituent. Of these compounds, 8-azido-trans-zeatin had an exceptionally high cytokinin activity in the tobacco callus bioassay, which showed significant enhancement of growth even at 10^{-11} M. Photolysis of 2-azido-isopentenyladenine in the presence of the cytokinin-binding protein from wheat germ resulted in the inability of the protein to subsequently bind [^{14}C]-kinetin. Using [^{14}C]-labeled 2-azido-benzylaminopurine KEIM and FOX (1980) demonstrated that, after photolysis, the ligand was covalently attached to the protein. All four subunits seemed to participate in the binding.

The association of a cytokinin-binding protein with ribosomes has provoked conjecture that cytokinins affect plant protein synthesis directly, and evidence in favor has been presented (Progr. Bot. 41, 127). Using various viral RNA's and tobacco leaf (poly A)RNA's as templates, RALPH et al. (1980) have reinvestigated the effect of benzylaminopurin and zeatin on in vitro protein synthesis in the wheat germ system. Neither elongation nor initiation steps in protein synthesis were affected by the cytokinins. The authors did notice, however, that changes in ion (K^+) concentration affect the translation of different RNA's preferentially. They propose the attractive hypothesis that cytokinins might affect mRNA selection by altering the ionic milieu in plant cells. There is substantial evidence that responses to cytokinins implicate K^+ and other cations (Progr. Bot. 41, 128). Recent evidence includes betacyanin synthesis (ELLIOTT, 1979a,b), cotyledon expansion (GREEN and MUIR, 1979), membrane potentials in root cells (ABUTALYBOV et al., 1980), and stomatal opening (JEWER and INCOLL, 1980). There is, however, no new evidence for a direct interaction of cytokinins with membranes, which might affect ion compartmentalization.

MILLER (1979, 1980) made the observation that cytokinins at high concentrations (500 μM) inhibit the oxygen consumption of cultured soybean cells and of mitochondria prepared from these cells as well as from tissue of six plant species. The high concentration of cytokinins required to elicit this response, and the inefficacy of zeatin make it unlikely that these effects are physiologically meaningful.

While cytokinins have a pronounced effect on chloroplast development (for review see PARTHIER, 1979), it is still unknown whether cytokinins affect plastogenesis directly (i.e., by control of processes inside the organelle) or indirectly (i.e., by control primarily of extraplastidic events). Preferential stimulation of plastid enzyme synthesis was shown by a comparison of the effect of benzylaminopurine on cytoplasmic and plastid leucyl- and valyl-tRNA synthetases in pumpkin cotyledons (KLYACHKO and PARTHIER, 1980): The activities of the plastid-associated enzymes, which are synthesized on cytoplasmic ribosomes, increased in response to the cytokinin and light, while those of the cytoplasmic enzymes did not. Inhibition of plastid differentiation in cucumber cotyledons by the antibiotic myomycin, which inhibits protein synthesis on 70S chloroplast ribosomes, was counteracted by kinetin (FASULO, 1980). The proposed interaction of kinetin and myomycin at plastid ribosomes remains to be verified and, besides, would not explain the stimulation of the formation of plastid enzymes that are synthesized in the cytoplasm. In excised water melon cotyledons, benzylaminopurine stimulates the development of plastids and microbodies (LONGO et al., 1979a), as well as of mitochondria (LONGO et al., 1979b). On the basis of dosis-effect curves and of short- and long-term treatments with benzylaminopurine, LONGO et al. (1978, 1979c) distinguish between two classes of responses to cytokinins. Growth and mitochondrial development required extended exposure to high cytokinin concentrations in order to be affected, while plastid and microbody development were stimulated by lower concentrations and much lower exposure periods. Only the second class of responses was mimicked by light. Whether or not the two classes of response reflect separate sites and/or mechanisms of actions of the cytokinin is an open question, which is part of the bewildering multiplicity of plant hormone action.

References

ABUTALYBOV, M., MELNIKOV, P., MARDANOV, A., ACHUNDOVA, T., GÖRING, H.: Biochem. Physiol. Pflanz. 175, 529-536 (1980). - ADAM, B., LISCHEWSKI, M., VOIGT, B., MEYER, A., BERGNER, C., SEMBDNER, G.: Modifizierung von Struktur und biologischer Wirkung bei Gibberellinen, 69-75. In: Wirkungsmechanismen von Herbiziden und synthetischen Wachstumsregulatoren, ed. H.R. SCHÜTTE. Jena: VEB Güstav Fischer 1979. - ALPI, A., LORENZI, R. CIONINI, P.G., BENNICI, A., D'AMATO: Planta 147, 225-228 (1979). - AMRHEIN, N., DECKERS, G., FINGERHUT, U.: Biochem. Physiol. Pflanz. 174, 727-737 (1979). - ANDERSEN, R.A., KEMP, T.R.: J. Chromatogr. 172, 509-512 (1979). - AYREY, G., CHAPMAN, J.M.: J. Labelled Comp. Radiopharm. 16, 887-890 (1979).

BARENDSE, G.W.M., VAN DE WERKEN, P.H., TAKAHASHI, N.: J. Chromatogr. 198, 449-455 (1980). - BARNES, M.F., TIEN, Ch.L., GRAY, J.S.: Phytochemistry 19, 409-412 (1980). - BEALE, M.H., MACMILLAN, J.: J. Chem. Soc. Perkin Trans I, 877-884 (1980). - BEALE, M.H., GASKIN, P., KIRKWOOD, P.S., MACMILLAN, J.: J. Chem. Soc. Perkin Trans I, 885-891 (1980). - BEARDER, J.R.: Plant hormones and other growth substances - Their background, structures and occurrence, 9-112; see ref. MACMILLAN, J. (ed.), 1980a. - BEARDER, J.R., MACMILLAN, J.: Separation of gibberellins and related compounds by droplet counter-current chromatography, 25-30; see ref. LENTON, J.R. (ed.), 1980. - BEARDER, J.R., MACMILLAN, J., PHINNEY, B.O.: The use of fungal mutants in the elucidation of gibberellin biosynthesis, 25-35. In: Regulation of Secondary Product and Plant Hormone Metabolism, eds. M. LUCKNER and K. SCHREIBER. FEBS 12th Meeting Dresden 1978, Vol. 55. Oxford: Pergamon Press 1979. - BEMILLER, J.N., LEUNG, G.L.-Y., YOPP, J.E.: Phytochemistry 18, 1283-1287 (1979). - BIANCO, J., BULARD, C.: Z. Pflanzenphysiol. 99, 411-416 (1980). - BINNS, A.N., MEINS, F., Jr.: Planta 145, 365-369 (1979); - Protoplasma 103, 179-187 (1980). - BLOCK, M.A., JOYARD, J., DOUCE, R.: Biochim. Biophys. Acta 631, 210-219 (1980). - BROEKAERT, D., VAN OOSTVELDT, P., BOEKEN, G., VAN PARIJS, R.: Arch. Int. Physiol. Biochim. 88, B121-B156 (1980). - BUKOVAC, M.J., YUDA, E., MUROFUSHI, N., TAKAHASHI, N.: Plant Physiol. 63, 129-132 (1979).

CALLEBAUT, A., VAN OOSTVELDT, P., VAN PARIJS, R.: Plant Physiol. 65, 13-16 (1980). - CECCARELLI, N., LORENZI, R., ALPI, A.: Phytochemistry 18, 1657-1658 (1979). - CHAVES DAS NEVES, H.J., PAIS, M.S.S.: Biochem. Biophys. Res. Commun. 95, 1387-1392 (1980a); - Tetrahedron Lett. 21, 4387-4390 (1980b). - CHEN, Ch., MEUTE, D.K.: FEBS Lett. 107, 15-20 (1979). - CHEN, Ch., MELITZ, D.K., PETSCHOW, B., ECKERT, R.L.: Eur. J. Biochem. 108, 379-387 (1980). - CHOINSKI, J.S., Jr., MOORE, T.C.: Plant Physiol. 65, 1031-1035 (1980). - COSSEY, A.L., LOMBARDO, L., MANDER, L.N.: Tetrahedron Lett. 21, 4383-4386 (1980). - CROZIER, A.: Analysis of gibberellins: verification of accuracy, 17-24; see ref. LENTON, J.R. (ed.), 1980.

DAVEY, J.B., VAN STADEN, J.: Plant Physiol. 63, 873-877 (1979). - DENNIS, F.G., MARTIN, G.C., GASKIN, P., MACMILLAN, J.: Planta 147, 376-377 (1980). - DUKE, C.C., LETHAM, D.S., PARKER, Ch.W., MACLEOD, J.K., SUMMONS, R.E.: Phytochemistry 18, 819-824 (1979).

EINSET, J.W.: Biochem. Biophys. Res. Commun. 93, 510-515 (1980). - ELLIOTT, D.C.: Plant Sci. Lett. 15, 251-264 (1979); - Plant Physiol. 63, 264-268 (1979). - ENTSCH, B., PARKER, Ch.W., LETHAM, D.S., SUMMONS, R.E.: Biochim. Biophys. Acta 570, 124-139 (1979). - ENTSCH, B., LETHAM, D.S., PARKER, C.W., SUMMONS, R.E., GOLLNOW, B.I.: Metabolites of cytokinins, 109-118; see ref. SKOOG, F. (ed.), 1980. - ERION, J.L., FOX, J.E.: Plant Physiol. 67, 156-162 (1981).

FASULO, M.P.: Biochem. Physiol. Pflanz. 175, 322-332 (1980). - FOX, J.E., ERION, J.L., MCCHESNEY, J.D.: Phytochemistry 18, 1055-1056 (1979). - FUJITA, E., NODE, M.: Heterocycles 7, 709-752 (1977). - FUNG, S., SIDDALL, J.B.: J. Am. Chem. Soc. 102, 6580-6581 (1980).

GASKIN, P., KIRKWOOD, P.S., LENTON, J.R., MACMILLAN, J., RADLEY, M.E.: Agric. Biol. Chem. 44, 1589-1593 (1980). - GEUNS, J.M.C.: Phytochemistry 17, 1-14 (1978). - GIBBONS, G.C.: Carlsberg Res. Commun. 44, 353-366 (1979). - GRAEBE, J.E.: GA-Biosynthesis: The development and application of cell-free systems for biosynthetic studies, 180-187; see ref. SKOOG, F. (ed.), 1980. - GRAEBE, J.E., ROPERS, H.J.: Gibberellins, 107-204; see ref. LETHAM, D.S. GOODWIN, P.B., HIGGINS, T.-J.V. (eds.), 1978. - GRAEBE, J.E., HEDDEN, P., RADEMACHER, W.: Gibberellin biosynthesis, 31-47; see ref. LENTON, J.R. (ed.), 1980. - GREEN, J.F., MUIR, R.M.: Physiol. Plant. 43, 213-218 (1978). - GROVE, M.D., SPENCER, G.F., PFEFFER, P.E., MANDAVA, N., WARTHEN, J.D., Jr., WORLEY, J.F.: Phytochemistry 17, 1187-1189 (1978). - GROVE, M.D., SPENCER, G.F., ROHWEDDER, W.K.: Nature (London) 281, 216-217 (1979).

HANSON, J.R.: The partial synthesis and labelling of some gibberellins, 5-16; see ref. LENTON, J.R. (ed.), 1980. - HASHIZUME, T., SUGIYAMA, T., IMURA, M., CORY, H.T., SCOTT, M.F., MCCLOSKEY, J.A.: Anal. Biochem. 92, 111-122 (1979). - HECHT, S.M.: Anticytokinins as probes of cytokinin utilization, 79-98; see ref. MANDAVA, N.B. (ed.), 1979; - Probing the cytokinin receptor site(s), 144-158; see ref. SKOOG, F. (ed.), 1980. - HEDDEN, P., PHINNEY, B.O.: Phytochemistry 18, 1475-1479 (1979). - HEFTMANN, E.: Progr. Phytochemistry 4, 257-276 (1977). - HEWITT, S., HILLMAN, J.R., KNIGHTS, B.A.: New Phytol. 85, 329-350 (1980). - HILTON, J.R., SMITH, H.: Planta 148, 312-318 (1980). - HOOK, J.M., MANDER, L.N., URECH, R.: J. Am. Chem. Soc. 102, 6628-6629 (1980). - HORGAN, R., KRAMERS, M.R.: J. Chromatogr. 173, 263-270 (1979).

ISAIA, A., BULARD, C.: Z. Pflanzenphysiol. 90, 409-414 (1978). - ISHIGURO, M., TAKATSUTO, S., MORISAKAI, M., IKEWANA, N.: J. Chem. Soc. Chem. 20, 962-964 (1980). - IWAMURA, H., FUJITA, T., KOYAMA, S., KOSHIMIZU, K., KUMAZAWA, Z.: Phytochemistry 19, 1309-1319 (1980).

JACOBS, W.P.: Plant Hormones and Plant Development, 339 p. Cambridge: Cambridge University Press 1979. - JEWER, P.C., INCOLL, L.D.: Planta 150, 218-221 (1980). - JOLLIFE, V.A., COGGINS, C.W., Jr., JONES, W.W.: J. Chromatogr. 179, 333-336 (1979). - JONES, M.G., ZEEVAART, J.A.D., Planta 149, 269-273 (1980a); - Planta 149, 274-279 (1980b). - JONES, M.G., METZGER, J.D., ZEEVAART, J.A.D.: Plant Physiol. 65, 218-221 (1980). - JONES, R.L.: The physiology of gibberellin-induced elongation, 188-195; see ref. SKOOG, E. (ed.), 1980a; - Planta 150, 58-69 (1980b); - Planta 150, 70-81 (1980c).

KATSUMI, M., KAZAMA, H.: Bot. Mag. Tokyo, Special Issue I., 141-158 (1978). - KEITH, B., BOAL, R., SRIVASTAVA, L.M.: Plant Physiol. 66, 956-961 (1980). - KEITH, B., SRIVASTAVA, L.M.: Plant Physiol. 66, 962-967 (1980). - KEIM, P., FOX, J.E.: Biochem. Biophys. Res. Commun. 96, 1325-1334 (1980). - KLYACHKO, N.K., PARTHIER, B.: Biochem. Physiol. Pflanz. 175, 333-345 (1980). - KULAEVA, O.N.: Cytokinin action on enzyme activities in plants, 118-128; see ref. SKOOG, F. (ed.), 1980. - KUROGOCHI, S., MUROFUSHI, N., OTA, Y., TAKAHASHI, N.: Planta 146, 185-191 (1979).

LENTON, J.R. (ed.): Gibberellins - Chemistry, Physiology and Use. Monograph No. 5, 143 pp. Wantage: British Plant Growth Regulator Group 1980. - LETHAM, D.S.: Cytokinins, 205-263; see ref. LETHAM, D.S., GOODWIN, P.B., HIGGINS, T.J.V. (eds.), 1978. - LETHAM, D.S., GOODWIN,

P.B., HIGGINS, T.J.V. (eds.): Phytohormones and Related Compounds - a Comprehensive Treatise. Vol. I.: The Biochemistry of Phytohormones and Related Compounds. 641 pp. Amsterdam: Elsevier/North Holland 1978. - LETHAM, D.S., GOLLNOW, B.I., PARKER, C.W.: Plant Sci. Lett. 15, 217-223 (1979a). - LETHAM, D.S., SUMMONS, R.E.: PARKER, C.W., MACLEOD, J.K.: Planta 146, 71-74 (1979b). - LEWAK, S., SINSKA, I.: Planta 147, 378 (1980). - LIN, J.-T., HEFTMANN, E.: J. Labelled Comp. Radiopharm. 16, 537-545 (1978). - LISCHEWSKI, M., ADAM, G.: Tetrahedron 36, 1237-1244 (1980a); - Tetrahedron Lett. 21, 1627-1628 (1980b). - LIU, S.-T., KADO, C.I.: Biochem. Biophys. Res. Commun. 90, 171-178 (1979). - LOMBARDO, L., MANDER, L.N., TURNER, J.V.: J. Am. Chem. Soc. 102, 6626 (1980). - LONGO, G.P., OLGINATI, M., ROSSI, G., VALENTE, M., LONGO, C.P.: Plant, Cell Environ. 1, 39-43 (1978). - LONGO, G.P., PEDRETTI, M., ROSSI, G., LONGO, C.P.: Planta 145, 209-217 (1979a); - Plant Sci. Lett. 14, 213-223 (1979b). - LONGO, C.P., LAMPUGNANI, M.G., SERVETTAZ, O., ROSSI, G., LONGO, C.P.: Plant Sci. Lett. 16, 51-57 (1979c). - LORENZI, R., BENNICI, A., CIONINI, P.G., ALPI, A., D'AMATO, F.: Planta 143, 59-62 (1978).

MACMILLAN, J.: Pure Appl. Chem. 50, 995-1004 (1978); - (ed): Hormonal Regulation of Development I. Molecular Aspects of Plant Hormones, 681 pp. Encyclopedia of Plant Physiology, New Series, Vol. 9. Berlin-Heidelberg-New York: Springer 1980a; - Partial synthesis of isotopically labelled gibberellins, 161-169; see ref. SKOOG, F. (ed.), 1980b. - MAHESHWARI, R., SHAILINI, C., VELUTHAMBI, K., MAHADEVAN, S.: Plant Physiol. 65(2), 186-192 (1980). - MANDAVA, N.B. (ed.): Plant Growth Substances, 310 pp. ACS Symposium Series 111. Washington: American Chemical Society 1979. - MANDAVA, N., MITCHELL, J.W.: Chem. Ind. 930 (1972). - MATSUBARA, S.: Phytochemistry 19, 2239-2253 (1980). - MC-CLOSKEY, J.A., HASHIZUME, T., BASILE, B., SUGIYAMA, T., SEKUKUCHI, S.: Proc. Japan Acad. 55, Ser. B, 445-450 (1979). - MCCLOSKEY, J.A., HASHIZUME, T., BASILE, B., OHNO, Y., SONOKI, S.: FEBS Lett. 111, 181-183 (1980). - MCMORRIS, T.C.: Phil. Trans. R. Soc. Lond. B. 284, 459-470 (1978). - MEINS, F., Jr., LUTZ, J.: Differentiation 15, 1-6 (1979); - Planta 149, 402-407 (1980). - MEINS, F., Jr., LUTZ, J., BINNS, A.N.: Differentiation 16, 71-75 (1980a). - MEINS, F., Jr., LUTZ, J., FOSTER, R.: Planta 150, 264-268 (1980b). - MESSENS, E., CLAEYS, M.: Proc. 4th Int. Conf. Plant Path. Bact., Angers 169-170 (1978). - METZGER, J.D., ZEEVAART, J.A.D.: Plant Physiol. 65, 623-626 (1980). - MEUDT, W.J., WORLEY, J.F., GREGORY, L.E., MANDAVA, N., BUTA, J.G., STEFFENS, G.L., THOMPSON, M.J.: 10th Int. Conf. Plant Growth Substances, Abstr.Nr. 502, Madison 1979. - MILBORROW, B.V., PRYCE, R.J.: Nature (London) 243, 46 (1973). - MILLER, C.O.: Planta 146, 503-511 (1979); - Proc. Natl. Acad. Sci. USA 77, 4731-4735 (1980). - MITCHELL, J.W., GREGORY, L.F.: Nature New Biol. 239, 253-254 (1972). - MITCHELL, J.W., MANDAVA, N., WORLEY, J.F., PLIMMER, J.R.: Nature (London) 225, 1065-1066 (1970). - MOHAMED, Y., CAPESIUS, I.: Z. Pflanzenphysiol. 98, 15-23 (1980). - MOK, M.C., KIM, S.-G., ARMSTRONG, D.J., MOK, D.W.S.: Proc. Natl. Acad. Sci. USA 76, 3880-3884 (1979). - MOORE, F.H., III: Plant Physiol. 64, 594-599 (1979). - MOORE, T.C.: Biochemistry and Physiology of Plant Hormones. 274 pp. Berlin-Heidelberg-New York: Springer 1979. - MORI, K.: Agric. Biol. Chem. 44, 1211-1212 (1980). - MORNET, R., THEILER, J.B., LEONARD, N.J.: Plant Physiol. 64, 600-610 (1979). - MOZER, T.J.: Plant Physiol. 65, 834-837 (1980a); - Cell 20, 479-485 (1980b). - MURAI, N., SKOOG, F., DOYLE, M.E., HANSON, R.S.: Proc. Natl. Acad. Sci. USA 77, 619-623 (1980). - MUROFUSHI, N., SUGIMOTO, M., ITOH, K., TAKAHASHI, N.: Agric. Biol. Chem. 43, 2179-2185 (1979); - Agric. Biol. Chem. 44, 1583-1587 (1980). - MUTHUKRISHNAN, S., CHANDRA, G.R., MAXWELL, E.: Proc. Natl. Acad. Sci. USA 76, 6181-6185 (1979).

NÄF, U.: Bot. Rev. 41, 315-355 (1975). - NAKAGAWA, S., MATSUI, H., YUDA, E., MUROFUSHI, N., TAKAHASHI, N., AKIMORI, N., HISHIDA, S.:

Phytochemistry 18, 1695-1697 (1979). - NAKAJIMA, H., YOKOTA, T., MATSUMOTO, T., NOGUCHI, M., TAKAHASHI, N.: Plant Cell Physiol. 29, 1489-1499 (1979). - NAKAJIMA, H., YOKOTA, T., WATANABE, N., TAKAHASHI, N.: Agric. Biol. Chem. 44, 2717-2719 (1980). - NAUNOVIĆ, G., NESKOVIĆ, M.: Photochem. Photobiol. 29, 1173-1175 (1979). - NISHINARI, N., SYONO, K.: Z. Pflanzenphysiol. 99, 383-392 (1980a); - Plant Cell Physiol. 21, 1143-1150 (1980b); - Plant Physiol. 65, 437-441 (1980c); - Plant Cell Physiol. 21, 383-393 (1980d). - NOMA, M., HUBER, J., PHARIS, R.P.: Agric. Biol. Chem. 43, 1793-1794 (1979).

OBATA, T.: Ann. Bot. 44, 333-337 (1979). - OHLROGGE, J.B., GARCIA-MARTINEZ, J.L., ADAMS, D., RAPPAPORT, L.: Plant Physiol. 66, 422-427 (1980). - OKAMOTO, K., AKAZAWA, T.: Plant Physiol. 63, 336-340 (1979). - OKAMOTO, K., KITANO, H., AKAZAWA, T.: Plant Cell Physiol. 21, 201-204 (1980). - OKITA, Th.W., DE CALEYA, R., RAPPAPORT, L.: Plant Physiol. 63(1), 195-200 (1979).

PARTHIER, B.: Biochem. Physiol. Pflanz. 174, 173-214 (1979). - PHARIS, R.P., KUO, C.G.: Can. J. Forest Res. 7, 299-325 (1977). - PHINNEY, B. O.: Gibberellin biosynthesis in the fungus *Gibberella fujikuroi* and in higher plants, 57-78; see ref. MANDAVA, N.B. (ed.), 1979. - POLYA, G. M., BOWMAN, J.A.: Plant Physiol. 64, 387-392 (1979). - PYLIOTIS, N.A., ASHFORD, A.E., WHITECROSS, M.I., JACOBSEN, J.V.: Planta 147, 134-140 (1979).

RADEMACHER, W., GRAEBE, J.E.: Biochem. Biophys. Res. Commun. 91, 35-40 (1979). - RAILTON, I.D.: Z. Pflanzenphysiol. 96, 103-114 (1980a); - S. Afr. J. Science 76, 33-35 (1980b). - RALPH, R.K., WOJCIK, S.J., AIREY, P.: Plant Sci. Lett. 18, 237-247 (1980). - RAMINA, A.: Plant Physiol. 63, 298-300 (1979). - REEVE, D.R., CROZIER, A.: Quantitative analysis of plant hormones, 203-280; see ref. MACMILLAN, J. (ed.), 1980a.

SCHWAIGER, H., TANNER, W.: Eur. J. Biochem. 102, 375-381 (1979). - SCOTT, I.M., HORGAN, R.: Biomed. Mass Spectrom. 7, 446-449 (1980). - SCOTT, I.M., BROWNING, G., EAGLES, J.: Planta 147, 269-273 (1980a). - SCOTT, I.M., HORGAN, R., MACGAW, B.A.: Planta 149, 472-475 (1980b). - SEMBDNER, G., GROSS, D., LIEBISCH, H.-W., SCHNEIDER, B.: Biosynthesis and metabolism of plant hormones, 281-444; see ref. MACMILLAN, J. (ed.), 1980a - SKOOG, F. (ed.). Plant Growth Substances 1979. 527 pp. Proc. 10th Int. Conf. Plant Growth Substances, Madison 1979. Berlin-Heidelberg-New York: Springer 1980. - SKOOG, F., SCHMITZ, R.Y.: Biochemistry and physiology of cytokinins, 335-413. In: Biochemical Action of Hormones, Vol. VI, ed. G. LITWACK. New York: Academic Press 1979. - SLAMA, K. Biochem. Physiol. Pflanz. 175, 177-193 (1980). - SPONSEL, V.M.: Metabolism of gibberellins in immature seeds of *Pisum sativum*, 170-179; see ref. SKOOG, F. (ed.), 1980a; - Gibberellin metabolism in legume seeds, 49-62; see ref. LENTON, J.R. (ed.), 1980b. - SPONSEL, V.M., MACMILLAN, J.: Planta 150, 46-52 (1980). - SPONSEL, V.M., GASKIN, P., MACMILLAN, J.: Planta 146, 101-105 (1979). - SMITH, H.: Gibberellins in photomorphogenesis, 95-109; see ref. LENTON, J.R. (ed.), 1980. - STODDART, J.L.: Planta 146, 353-361 (1979a); - Planta 146, 363-368 (1979b); - The primary cellular action of gibberellins in the control of elongation, 63-75; see ref. LENTON, J.R. (ed.), 1980. - STODDART, J.L., VENIS, M.A.: Molecular and subcellular aspects of hormone action, 445-510; see ref. MACMILLAN, J. (ed.), 1980. - STODDART, J.L., WILLIAMS, P.D.: Planta 147, 264-268 (1979); - Planta 148, 485-490 (1980). - STUCHBURY, T., PALNI, L.M., HORGAN, R., WAREING, P.F.: Planta 147, 97-102 (1979). - SUGIYAMA, T., IWASAWA, H., HASHIZUME, T.: Agric. Biol. Chem. 44, 1057-1060 (1980a). - SUGIYAMA, T., TATEBA, H., HASHIZUME, T.: Agric. Biol. Chem. 44, 1673-1674 (1980b); - Agric. Biol. Chem. 44, 2755-2756 (1980c). - SUMMONS, R.E., DUKE, C.C., EICHHOLZER, J.V., ENTSCH, B., LETHAM, D.S., MACLEOD, J.K., PARKER, Ch.W.: Biomed. Mass

Spectrom. 6, 407-413 (1979). - SUMMONS, R.E., ENTSCH, B., LETHAM, D.S., GOLLNOW, B.I., MACLEOD, J.K.: Planta 147, 422-434 (1980).

TANAKA, A.K., KOBAYASHI, A., YAMASHITA, K.: Agric. Biol. Chem. 43, 2537-2542 (1979). - TERRINE, C., LALOUE, M.: Plant Physiol. 65, 1090-1095 (1980). - THOMAS, B., TULL, S.E., WARNER, T.J.: Plant Sci. Lett. 19, 355-362 (1980). - THOMPSON, M.J., MANDAVA, N., FLIPPEN-ANDERSON, J.L., WORLEY, J.F., DUTKY, S.R., ROBBINS, W.E., LUSBY, W.: J. Org. Chem. 21, 5002-5004 (1979). - TOMOS, A.D., LAIDMAN: The control of mobilization and metabolism in the aleurone tissue during germination, 119-146. In: Recent Advances in the Biochemistry of Cereals, eds. D.L. LAIDMAN, R.G. WYN JONES. London: Academic Press 1979.

VAN ROMPUY, L.L.L., ZEEVAART, J.A.D.: Phytochemistry 18, 863-865 (1979).

WANG, T.L., COVE, D.J., BEUTELMANN, P., HARTMANN, E.: Phytochemistry 19, 1103-1105 (1980). - WEILER, E.W.: Planta 149, 155-162 (1980). - WEILER, E.W., WIECZOREK, U.: Planta 152, 159-167 (1981).

YAMAGUCHI, I., KOBAYASHI, M., TAKAHASHI, N.: Agric. Biol. Chem. 44, 1975-1977 (1980). - YAMAGUCHI, I., YOKOTA, T., YOSHIDA, S., TAKAHASHI, N.: Phytochemistry 18, 1699-1702 (1979). - YAMANE, H., TAKAHASHI, N., TAKENO, K., FURUYA, M.: Planta 147, 251-256 (1979). - YOKOTA, T., MUROFUSHI, N., TAKAHASHI, N.: Extraction, purification, and identification, 113-201; see ref. MACMILLAN (ed.), 1980a.

ZEIGLER, R.S., POWELL, L.E., THURSTON, H.D.: Phytopathology 70, 589-593 (1980). - ZERONI, M., HALL, M.A.: Molecular effects of hormone treatment on tissue, 511-586; see ref. MACMILLAN, J. (ed.), 1980.

Professor Dr. Nikolaus AMRHEIN
Lehrstuhl für Pflanzenphysiologie
Ruhr-Universität Bochum
D 4630 Bochum

C. Genetics

I. Replication

By WALTER NAGL

1. Reviews and Methods

Recently, several reviews have been published concerning DNA replication as a whole (Cold Spring Harbor Symposium 1978, WINTERSBERGER, 1978; HURWITZ, 1980; KORNBERG, 1980a,b), or certain aspects such as DNA replication in plants (BRYANT, 1980), proteins and enzymes involved (TAYLOR, 1979; COLEMAN and OAKLEY, 1980; FUJIMURA and DAS, 1980; MECHALL, 1980), the replication fork (DE PAMPHILIS and WASSERMANN, 1980), the regulation of initiation (HAND, 1979, KOLTER and HELINSKI, 1979) and the discontinuity of replication (OGAWA and OKAZAKI, 1980).

ROMAN et al. (1980) have shown that isolated plant nuclei are capable of DNA replication. PROBST et al. (1979) developed a method for the isolation of replication forks (from Ehrlich ascites cells), ERICKSON et al. (1979) could isolate and purify replication intermediates (Okazaki fragments), and MURPHY et al. (1980) used metrizamide gradients to isolated newly replicated chromatin (for *Physarum* see BEACH et al., 1980).

The present review deals mainly with aspects of the regulation of initiation of DNA replication, the replicon pattern and rate of elongation, newly detected characters of enzymes inovlved in replication, and new findings on differential DNA replication. The central role of DNA conformational changes in a thermodynamically favored way is emphasized for DNA replication, cell cycles, and cell differentiation.

2. Regulation of Initiation of DNA Synthesis

DNA replication is restricted in most eukaryotes to a phase of the cell cycle, the S period, situated between two "gaps" (G_1 and G_2 periods) of the interphase. Experiments with wild-type and mutant mammalian cell cultures indicate that the arch-type of the cell cycle was a sequence of DNA replication and mitosis, without G phases. Mutations affecting the rate of protein synthesis, and thus of the initiator of DNA synthesis, have introduced the necessity of a G_1 period (PRESCOTT, in press). Similarly, the G_2 period may have evolved. Moreover, recent understanding of the mitotic cycle has led to the conclusion that one must distinguish between the chromosome cycle and the cell growth cycle, because these two cycles are independently regulated and can be uncoupled (e.g., during polyploidization, during the growth of the egg cell and the cleavage divisions without accompanying cell growth). Therefore, MAZIA (in press) called the mitotic cycle a "bicycle" (Fig. 1).

The generation time of individual cells (both eu- and prokaryotic) are highly variable, during both steady-state and exponential growth. In other words, the time at which the S phase begins varies with cell age. The kinetics with which

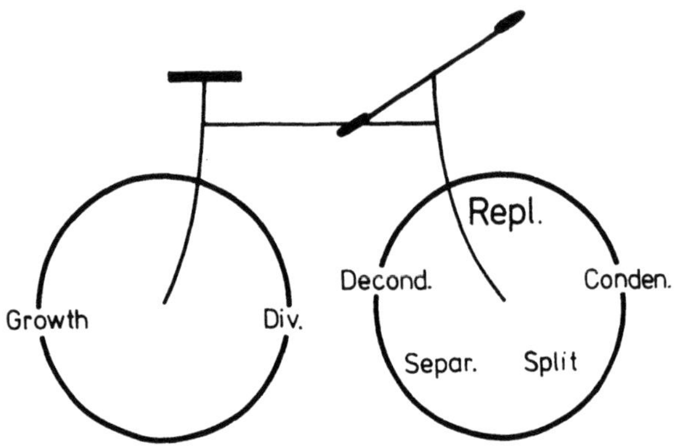

Fig. 1. The cell cycle seen as a bicycle according to suggestions of MAZIA (in press). Only the major events of the growth-division cycle and the chromosome reduplication cycle are indicated

mono- and binucleate cells enter the S phase indicate that the signal responsible for starting S must appear at random in the cytoplasm and may be transient, at least in mammalian cells (BROOKS, 1979). In *Physarum polycephalum*, the start of DNA replication is characterized by a single initiation event simultaneously at approx. 3000 origins. Little or no further initiation occurs for at least 15 min (BEACH et al., 1980). The pattern of DNA replication, i.e., the pattern of replicon initiation throughout S phase, will be discussed in Section 3.

Initiation of DNA replication requires the transfer of a multienzyme complex from the cytoplasm into the nucleus. The complex evidently consists also of DNA precursor-synthesizing enzymes, and its assembly may signal the initiation of the S phase of the cell cycle; REDDY and PARDEE (1980) termed this multienzyme complex "replicase". Also still a subject of controversy, a role of the nuclear membranes (WILLE and STEFFENS, 1979) and the nuclear matrix (PARDOLL et al., 1980) was suggested. The latter authors obtained evidence for fixed sites of DNA replication ("replisomes"), which are consistent with the radial loop model of the eukaryotic chromosome (Fig. 2). Originally, KORNBERG (e.g., 1980b) used this term for the *E. coli* replication complex, which is too unstable to be isolated in toto.

The control of initiation is certainly very complex, and no single "cause" can be given. Besides the well-known factors, such as cell size, auxin, oxygen, ions, etc., recently an essential role for mevalonate synthesis was found in cultured mammalian cells (QUESNEY-HUNEES et al., 1979). Promotion of DNA synthesis by absisic acid was reported by MINOCHA (1979) for Jerusalem artichoke tuber tissue culture in vitro.

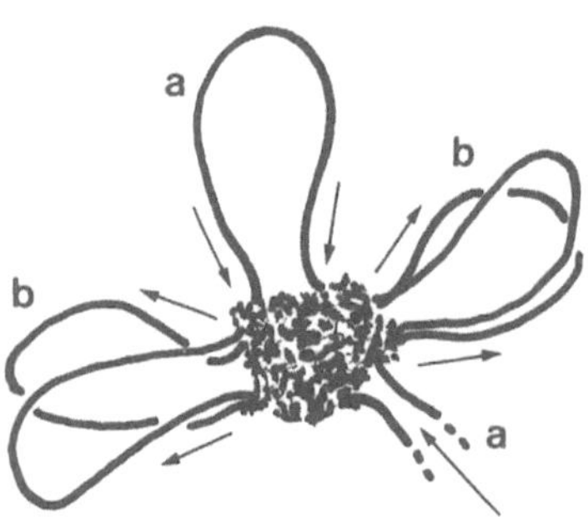

Fig. 2. The nuclear-matrix-bound replisome (model): a cluster of fixed replication complexes allowing bidirectional DNA replication. *Arrows* indicate direction of replication; *a* DNA loops before replication; *b* DNA loops after replication. (Redrawn and modified from PARDOLL et al., 1980)

The promotion occurs when abscisic acid is used in combination with auxin and affects a certain sensitive population of cells only. Several publications suggest the important role of ADP-ribosylation of nuclear proteins in the regulation of DNA synthesis. For instance, in liver nuclei of the chicken *embryo* poly(ADP-ribose) formation stimulates DNA synthesis, probably due to an increase in accessibility of nuclease to DNA as a result of template activation (TANIGAWA et al., 1978a). In the liver of *old* hens, poly(ADP-ribose) polymerization exerts the opposite effect and inhibits DNA replication (TANIGAWA et al., 1978b, see also BERGER et al., 1979). Similar contradicting reports have been published by other authors and may be explained by the multiple functions which can be ascribed to the involved enzyme (see also Progr. Bot. 41, 1979, p. 169).

All functions revolve around specific structural perturbations of chromatin. This and other findings indicate that the biochemistry of initiator synthesis and replicative enzyme activities do not control DNA synthesis per se, but via biophysical changes in the configuration of chromatin and DNA. Modifications (particularly phosphorylation and dephosphorylation) of the histone and nonhistone proteins alter the electro-dynamic situation within the nucleus. As a consequence the conformation of DNA and proteins are changed in a thermodynamically favored way, resulting in the formation and exposure of recognition sites for the enzyme machinery.

The structural changes in chromatin during the cell cycle are actually much more complex than thought previously, when a continuous decondensation from anaphase to late S phase, and a continuous condensation from early G_2 to mitosis was suggested (e.g., MAZIA, 1963). In accordance with the "bicycle model" (MAZIA, in press), careful electron microscopic, circular dichroism, geometric and densitometric texture analyses, and the digitation of nuclear images, are suggestive for two chromatin condensation cycles during one cell cycle (NICOLINI et al., 1975, 1977; KENDALL et al., 1977; NICOLINI, 1979, 1980; see also SETTERFIELD et al., 1978; KWIATKOWSKA and MASZEWSKI, 1979; MOSER and MÜLLER, 1979).

These results falsified the widely accepted hypothesis that chromatin decondensation is maximal during the late S period. The stimulatory effect of caffeine on DNA synthesis, and its preventive effect on inhibition of DNA synthesis by X radiation etc., has also been ascribed to alterations of the conformation of chromatin (PAINTER, 1980). Finally, EREMENKO et al. (1980), studying methylation of human DNA fractions, reached the conclusion that DNA organization, modification, and replication are severely subordinated one to another. Although all these findings were made in mammalian systems, they may be valid also for plant nuclei: despite the fact that the nuclear structure in plants is species-specific and does not show significant changes during differentiation as do mammalian nuclei, chromatin organization undergoes distinct changes during the cell cycle (NAGL, 1977, 1979a).

In conclusion, the recent view of the control of the cell cycle, i.e., mainly the control of the start of DNA replication and mitosis, can be summarized as follows. *The basic mechanisms underlying any biochemical control of the cell cycle are conformational changes of the chromatin in a thermodynamically directed way, which are initiated and maintained through the interrelationship of chromatin with its electrodynamic environment and the position of the cell in various (e.g., hormonal) gradients.*

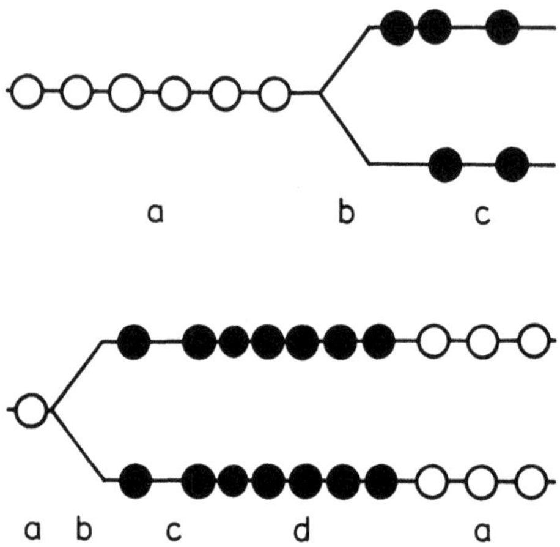

Fig. 3. Changes in nucleosome arrangement during chromatin replication. *All circles* represent 140 base pair nucleosome cores. *Open circles* and *a* nucleosomes with normal, approx. 200 base pair spacing; *filled circles* nucleosomes with abnormal spacing (*b* replication fork; *c* irregularly spaced nucleosomes; *d* short spacing, approx. 180 base pairs long). In reality, the number of nucleosomes between the passage of the replication fork and regaining of normal spacing is certainly much larger. (Redrawn and modified from MURPHY et al., 1978)

3. Rate and Fidelity of Elongation

As initiation of DNA replication, the rate of chain elongation is dependent on structural changes of chromatin. MATTERN and PAINTER (1979) could show in experiments using ethidium bromide that relaxation of DNA supercoiled structures ahead of the replication fork generates enhancement of DNA synthesis, and that the rate of chain elongation is limited by supercoiled regions ahead of the growing point. This does not, however, mean that the conformational alterations are visible at the microscopic level in any case. The rate of DNA replication is not constant during the S period: An initial slow rate is followed by a maximal rate and then by a slow rate (COLLINS, 1978). In mammals the rate is 0.2 - 1.2 µm/min/fork (YUROV, 1979, 1980). KHAITOVA et al. (1980) and KUMURA et al. (1980) found a higher rate in replicons from heterochromatin than from euchromatin. This is consistent with earlier suggestions in annual plants (Anthemideae), where species with high amounts of heterochromatin exhibited faster cell cycles than species with little heterochromatin (NAGL, 1974). In plants the rate of chain elongation is in general slower than in mammals and bacteria, e.g., 0.15 µm/min/fork in *Crepis capillaris* (the number of replicons in this species is about 60,000, the mean size 24 µm; VAN'T HOF and BJERKNES, 1979). Higher plants appear to have a common replicon size and similar rate of elongation (6 to 12 µm/h/fork; in mammals: 30 - 60 µm/h, in bacteria 800 µm/h). Thus, higher plants share the characteristic of a relatively low replication fork rate with cold-blooded animals.

The intermediates of DNA replication (Okazaki fragments) of *Chlorella* have been recently characterized by SSYMANK (1979). They have a sedimentation coefficient of 7S and are, therefore, somewhat smaller than the prokaryotic Okazaki fragments, and they contain single-stranded regions.

The fidelity of DNA replication was reviewed in the last issue of Progr. Bot. 42 (1980), p. 164. Additional aspects were published by BERNARDI and NINIO (1978), BRUSKOV and POLTEV (1979) and KUNKEL and LOEB (1980).

The question as to how the histones (nucleosomes) behave during DNA replication is still under research. SCHLAEGER and KNIPPERS (1979) showed that a fraction of nuclease-resistant nascent DNA appears to be differently bound to nucleosome-like structures than mature DNA, since the 140 base pair fragments of pulse-labeled DNA can be released by 0.5 M NaCl as a free-sedimenting 5S DNA fragment. Fragments of mature DNA (185 base pairs long) remain stably associated with the histones under the same conditions. There is increasing evidence that the internucleosomal spacer of newly replicated chromatin is approximately 20 base pairs shorter than that of total chromatin (MURPHY et al., 1978). Normal spacing is then slowly regained by nucleosome sliding during 20 - 30 min after the passage of the replication fork (Fig. 3). The distribution of the old nucleosomes (randomly, semiconservatively, conservatively) is still not known.

4. Pattern and Termination of Replication

The number of reports that DNA in heterochromatin is late-replicating has raised this claim to a dogma in the mind of many cell biologists. However, in addition to earlier findings (e.g., TANAKA, 1965; NAGL, 1977b) recent studies in the Indian muntjac revealed that heterochromatin does not necessarily replicate late in the S period (KIMURA et al., 1980). KONGSUWAN and SMYTH (1980) found in chromosomes of *Lilium longiflorum* and *L. Pardalinum* that the bright Q-bands are not consistently late-replicating as they apparently are in mammals. Centromeric C bands and those at most nucleolar organizers were not late-replicating. In *Luzula purpurea*, on the other hand, late-replicating regions coincide with C bands, which have been shown to be rich in repetitive DNA sequences (Cot 1 fraction; RAY and VENKETESWARAN, 1979). Also in *Zea mays* heterochromatin is late-replicating (PRYOR et al., 1980).

Special mechanisms are necessary to explain the replication of the ends of linear DNA molecules (and chromosomes), because the requirement of every known DNA polymerase for an oligonucleotide primer ending in a 3'OH group would otherwise result in the production of DNA molecules in which the 5'ends of the progeny strands are missing. Several mechanisms have been proposed to account for completion of the 5'ends of the progeny strands (summarized by BELLETT, 1980; see also Progr. Bot. 37 (1975), p. 196).

Recently, DANCIS and HOLMQUIST (1977, 1979) and HOLMQUIST and DANCIS (1979, 1980) proposed a new model of telomere replication. The authors suggest that telomeres fuse before replication and separate afterwards, thereby eliminating termini at the time of replication. All telomeres require a common palindromic sequence, and any two can recognize each other, whereafter they base pair, covalently fuse end-to-end and form a continuous, endless DNA molecule just before replication. After replication, all fused telomere pairs are recognized by a restriction nuclease and cleaved. The authors state that a fusion model alone is most consistent with the following data: telomeres are often attached to the nuclear envelope and appear clustered or aggregated; end-to-end fusion of nonhomologous telomeres is frequently seen and often so complete that the whole genome appears to be included in a giant ring of fused chromosomes; all *Drosophila* telomeres have the same large (10 kilo

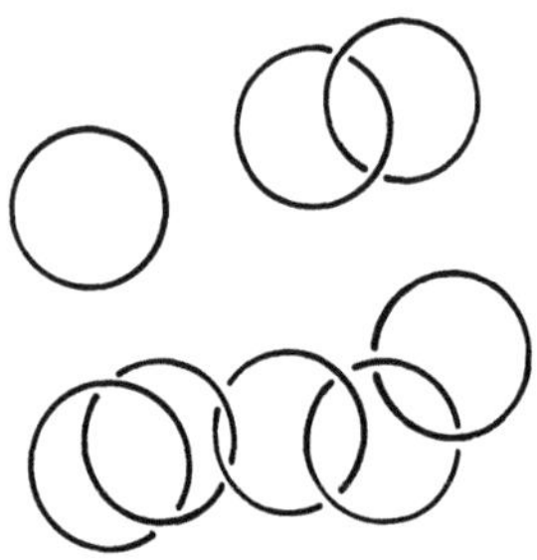

Fig. 4. Catenanes of DNA as they can be seen among free DNA rings

bases) telomere-specific DNA sequence; in the final passage of senescing human primary cell lines, up to 2.5% of the chromosomes are involved in telomere-telomere fusion, consistent with a dysfunctional fission process; further evidence is cited by HOLMQUIST and DANCIS (1980). JOHNSON (1980) detected a family of inverted repeat sequences at the termini of the *Physarum* rDNA palindrome.

5. Enzymes Involved in Replication

The present knowledge about the various types of enzymes involved in DNA replication was discussed in the last issue of Progr. Bot. 42 (1980) p. 161. Therefore, only some new aspects and some references will be inlcuded in this essay.

Several studies were made to elucidate the function of DNA topoisomerases in more details. DNA topoisomerases (or nicking-closing enzymes) introduce transient negative supercoils into DNA of prokaryotes and eukaryotes. This class of enzymes is not only essential for replication, but influences all metabolic processes involving DNA. The well-studied DNA gyrase supercoils DNA by a mechanism called sign inversion (BROWN and COZZARELLI, 1979), whereby a positive supercoil is directly inverted to a negative one by passing a DNA segment through a transient double-strand break (e.g., MIZUUCHI et al., 1980). Reversal of this scheme relaxes DNA. Each round of supercoiling is driven by a conformational change induced by ATP binding. The enzyme is composed of two or four subunits (A_2B_2), where A is particularly associated with the concerted breakage and rejoining of DNA, and B mediates energy transduction from ATP (e.g., KLEVAN and WANG, 1980; reviewed by COZZARELLI, 1980a,b). The binding of DNA gyrase to a specific DNA sequence leads to a covalent complex (e.g., MORRISON et al., 1980; PRELL and VOSBERG, 1980; SUGINO et al., 1980; TSE et al., 1980). Hydrolysis of ATP permits fresh cycles.

Recently, in addition to the already known catalytic reactions of topoisomerase type I of bacteria (i.e., relaxation of superhelical DNA, interconversion between single-stranded DNA rings with and without knots and the interwining of single-stranded DNA rings of complementary sequences into a covalently closed duplex ring with a high linking number), a fourth type of reaction was detected: the enzyme promotes catenation and decatenation of double-stranded DNA rings (TSE and WANG, 1980; the cited authors also describe the differences between type I and type II enzymes). Also DNA gyrase was found to catalyze the formation of DNA catenanes (KREUZER and COZZARELLI, 1980). Catenanes (interlocked duplex DNA rings) were demonstrated as products of DNA replication and recombination in vitro, and are a constant accompaniment of circular DNA molecules in prokaryotes and eukaryotes (Fig. 4). Decatenation of ki-

Fig. 5. Some functions of nucleic acid-dependent ATPases according to WHITEHEAD et al. (1979)

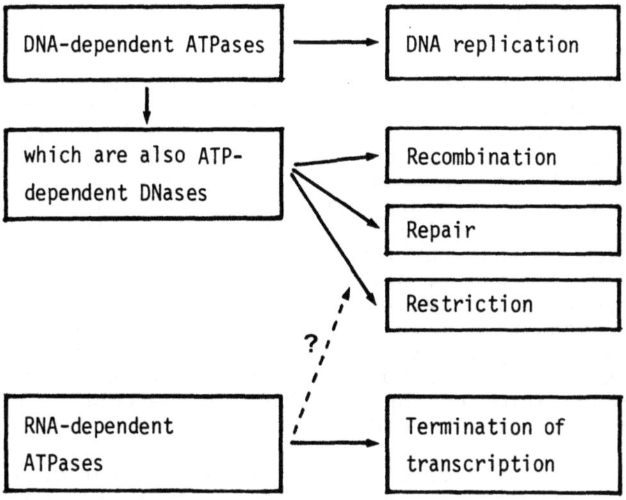

netoplast DNA (the mitochondrial DNA of trypanosomes which has the form of a network of thousands of interlocked DNA rings) was shown by MARINI et al. (1980). Further aspects are discussed and reviewed by WANG and LIU (1979), LIU et al. (1980), and WANG (1980).

Besides *DNA-binding proteins* (e.g., HOTTA and STERN, 1979; REINHARDT and MÜLLER, 1980), *DNA-unwinding enzymes* or DNA helicases are involved in DNA replication of both prokaryotes and eukaryotes (e.g., ABDEL-MONEM and HOFFMANN-BERLING, 1980; BOXER and KORN, 1980; FALASCI et al., 1980; KUHN et al., 1979). These chemomechanically active enzymes use the energy released by ATP hydrolysis for separating the strands of a DNA double helix. WHITEHEAD et al. (1979) reviewed the function of these and other types of DNA-dependent ATPases (some of which are ATP-dependent DNases; Fig. 5). The enzymes are involved in many vital processes such as recombination, replication, restriction, and termination of transcription. They achieve separation of nucleic acid douplex partners and other energetically unfavorable changes of nucleic acid conformation such as supercoiling (= DNA topoisomerase, which can also be seen as an ATPase), and unidirectional processive actions. JAVAHERIAN et al. (1979) reported that nonhistone proteins HMG1 and HMG2 also unwind DNA double helix.

The properties and functional roles of the eukaryotic DNA polymerases were further analyzed, and earlier suggestions could be confirmed. Polymerase α seems to be mainly involved in DNA replication, DNA polymerase β in repair processes of damaged DNA, and polymerase γ is very likely the mitochondrial DNA polymerase (DUBE et al., 1979; HÜBSCHER et al., 1979; WEISSBACH, 1979; WIST, 1979). The eukaryotic DNA polymerases are generally believed to lack an exonuclease activity and hence proof-reading ability. Recently, however, an α-like polymerase δ with associated exonuclease activity has been discovered (SCOVASSI et al., 1980). EHRENBERG and BLOMBERG (1980) and HOPFIELD (1980) discussed the thermodynamic constraints of kinetic proof-reading. AMILENI et al. (1979) purified the major DNA polymerase from cultured plant cells.

An enzyme that fuses DNA molecules at regions of homology was described by POTTER and DRESSLER (1980) and termed DNA synaptase. It is suggestive that this enzyme plays a role in recombination and/or repair.

6. Endoreplication, Meiotic, and Differential DNA Replication

DNA endoreplication (i.e., DNA replication in an endo-cycle such as the enodmitotic and the endoreduplication cycle) occurs in nearly all higher plants and most other plant and animal taxa (reviewed by NAGL, 1978). A systematic scanning cytophotometric study in *Scilla* revealed that 71% of the cells exhibited polyploid nuclei, the highest degrees occurred in the elaiosomes of the seed (512 C) and the antipodal cells of the ovule (1024 C; FRISCH and NAGL, 1979). There is increasing indication that the functional significance of endopolyploidy is twofold: one may be associated with cell differentiation and function (synthetic activity), the other with morphogenesis (organ modeling through cell size). Details have been discussed in recent reviews (NAGL, in press a,b). For an animal system (the silkgland of *Bombyx mori*), PERDRIX-GILLOT (1979) has shown that at the first symptoms of organization of the silkgland in the embryo, mitosis stops and nuclei start to grow by endo-cycles.

From the studies of meiotic pollen mother cells of lily it is known that part of the nuclear DNA is replicated not earlier than the zygotene stage, and an unscheduled DNA synthesis occurs during pachytene (reviewed by HOTTA and STERN, 1975). In an electron microscopic autoradiographic study KURATA and ITO (1978) could demonstrate that the zygotene-DNA is synthesized in the axial cores prior to formation of the synaptonemal complex, and later the silver grains are found over that pairing complex. This is again suggestive for a role of the delayed DNA replication for meiotic chromosome pairing. BOUCHARD and STERN (1980) characterized the DNA synthesized at pachytene and found that it is a nondivergent subclass of moderately repetitive sequences, which might be recombination-specific.

Differential DNA replication (i.e., DNA underreplication in an endocycle, and DNA amplification or extra synthesis in meiotic and somatic cells) was recently reviewed for plants (NAGL, 1979b; NAGL et al., 1979). In the meantime, differential DNA replication was reported to occur in a number of plants and animals. BROEKAERT et al. (1979) found differential DNA replication in *Pisum sativum* seedlings at the onset of germination; the authors used reassociation kinetics to compare the DNA's from various stages. Differential replication of main band and satellite DNA was reported for tissue cultures of *Cucumis melo* (GRISVARD and TUFFET-ANGHILERI, 1980) and *Daucus carota* (HASE et al., 1979; see also SCHÄFER and NEUMANN, 1978). The amplification of rearranged repeated DNA sequences in cereal plants was detected by BEDBROOK et al. (1980). Examples of DNA amplification and underreplication in various animals are listed in Table 1. LAIRD (1980) pointed out to the structural paradox of polytene chromosomes, which exhibit linear continuity, although their segments, perhaps bands and interbands, are differentially replicated and therefore composed of segments with different numbers of chromatids.

One example of amplification of a mRNA gene should be presented in more detail: the extra synthesis and chromosomal incorporation of the dihydrofolate reductase gene as the basis of evolution of resistance against methotrexate (which binds and hence inactivates the coded enzyme). Amplification of this gene in resistant cells was shown for a CHO (Chinese hamster ovary) cell line (SCHIMKE et al., 1978; MELERA et al., 1980; WOHLGEMUTH et al., 1980) and a mouse melanoma cell line (BOSTOCK et al., 1979; BOSTOCK and CLARK, 1980) by various methods. When the resistance is stable, the multiplied genes are incorporated into the chromosome and appear as long "R-band" region, when the re-

Table 1. Recent reports on differential DNA replication in various animals

	Sequence	Event	References
Tetrahymena (ciliate)	genes	A	IWAMURA et al. (1979)
Loxodes (ciliate)	genes	A	BOBYLEVA et al. (1980)
Drosophila (insect)	genes for chorion proteins	A	SPRADLING and MAHOWALD (1980)
	rDNA	A	GRIMM and KUNZ (1980)
	other DNA sequences	A,U	MAHOWALD et al. (1979), COWELL and HARTMANN-GOLDSTEIN (1980a,b), MEER (1980)
Sciara (insect)	various	A,U	EASTMAN et al. (1980)
Prodiamesa (insect)	?	U	ZACHARIAS (1979)
Bacillus (stick insect)	?	U	SCALI et al. (1980)
Geryon (crab)	repetitive DNA	A	CHRISTIE and SKINNER (1980)
Triturus (newt)	satellite DNA	A	LOHMANN (1979), LOHMANN and SCHUBERT (1980)
Rattus (rat)	rDNA	A	MILLER et al. (1979)
Rat,Cat,Chicken,Man	rDNA	?	BRODSKY et al. (1979)

sistance is unstable the genes evidently exist as double minutes (NUNBERG et al., 1978; KAUFMAN et al., 1979). The system was reviewed by SCHIMKE (1980, 1981).

One severe conclusion must be drawn from the drug-induced amplification: mutations are not always a random event, but can be the controlled answer to a disadvantageous environmental factor. This throws a new evolutionary mechanisms to the properties of living matter. Moreover, it demonstrates how the understanding of evolution and differentiation are connected to each other at the level of DNA replication. A second phenomenon is worthwhile to be cited in this connection: The insertion of forein DNA into host DNA (e.g., during transformation, virus insertion, translocation of transposons) often leads to amplification of the foreign DNA (e.g., INNIS et al., 1979; SCHMITT et al., 1979; COLANTUONI et al., 1980; HISCOTT et al., 1980; WIGLER et al., 1980). It seems that transposon-related events are self-regulated in some way (cf. CHOU et al., 1979), in other words, they represent examples of the self-informating and autocatalyzing properties of bio-matter.

Actually, there is less and less space for stochastic events in living organisms, and all the diversity which arises during phylogenesis and ontogenesis of organisms and cells may be attributed to biophysical laws, mainly to the electro-dynamic and thermodynamic laws (PRIGOGINE, 1955, 1979; POPP, 1979a,b). As the increasing complexity of organisms is paralleled by an increasing amount of repetitive and other noncoding DNA sequences (which arise and change by amplification, dispersion within the genome, and diversification), a role of this noncoding sequences for the evolution and somatic realization of organismic complexity can be assumed. NAGL and FUHRMANN (in press) suggested, therefore, the term "thermodynamic code" for the information content of the noncoding

DNA, which is located in the sequence arrangement and the conformation of DNA within chromatin. Differential DNA replication of noncoding sequences during evolution and somatogenesis may be seen in this light as the critical switch in the feed-back loops underlying self-organization of bio-matter. Moreover, recent findings on the coexistence of nucleosomal and various other chromatin configurations (MÜLLER et al., 1980) and on autonomously replicating sequences in eukaryotes (*Saccharomyces*; CHAN and TYE, 1980) stress the highly dynamic nature of the genetic material.

References

ABDEL-MONEM, M., HOFFMANN-BERLING, H.: Trends Biochem. Sci. 5, 128-130 (1980). - AMILENI, A., SALA, F., CELLA, R., SPADARI, S.: Planta 146, 521-527 (1979).

BEACH, D., PIPER, M., SHALL, S.: Exp. Cell Res. 129, 211-221 (1980). - BEDBROOK, J.R., O'DELL, M., FLAVELL, R.B.: Nature 288, 133-137 (1980). - BELLETT, A.J.D.: J. Theor. Biol. 85, 549-551 (1980). - BERGER, N.A., PETZOLD, S.J., BERGER, S.J.: Biochim. Biophys. Acta 564, 90-104 (1979). - BERNARDI, F., NINIO, J.: Biochimie 60, 1083-1095 (1978). - BOBYLEVA, N.N., KUDRJAVTSEV, B.N., RAIKOV, I.B.: J. Cell Sci. 44, 375-394 (1980). - BOSTOCK, C.J., CLARK, E.M.: Cell 19, 709-715 (1980). - BOSTOCK, C.J., CLARK, E.M., HARDING, N.G.L., MOUNTS, V., WALKER, P.M.P.: Chromosoma 74, 153-177 (1979). - BOUCHARD, R.A., STERN, H.: Chromosoma 81, 349-363 (1980). - BOXER, L.M., KORN, D.: Biochemistry 19, 2623-2633 (1980). - BRODSKY, V.Ja., MARSHAK, T.L., MARES, V., LODON, Z., FÜLÖP, Z.: Histochemistry 59, 233-248 (1979). - BROEKAERT, D., VAN OOSTVELDT, P., VAN PARIJS, R.: Biochem. Physiol. Pflanzen 174, 629-640 (1979). - BROOKS, R.F.: Cell Biol. Int. Rep. 3, 707-716 (1979). - BROWN, P.O., COZZARELLI, R.: Science 206, 1081-1083 (1979). - BRUSKOV, V.I., POLTEV, V.I.: J. Theor. Biol. 78, 29-41 (1979). - BRYANT, J.A.: Biol. Rev. (Cambridge) 55, 237-284 (1980).

CHAN, C.S.M., TYE, B.-K.: Proc. Natl. Acad. Sci. USA 77, 6329-6333 (1980). - CHOU, J., CASADABAN, M.J., LEMAUX, P.G., COHEN, S.N.: Proc. Natl. Acad. Sci. USA 76, 4020-4024 (1979). - CHRISTIE, N.T., SKINNER, D.M.: Proc. Natl. Acad. Sci. USA 77, 2786-2790 (1980). - COLANTUONI, V., DAILEY, L., BASILICO, C.: Proc. Natl. Acad. Sci. USA 77, 3850-3854 (1980). - Cold Spring Harbor Symp. Quant. Biol. 43: Replication and Recombination (1978). - COLEMAN, J.E., OAKLEY, J.L.: CRC Crit. Rev. Biochem. 7, 247-289 (1980). - COLLINS, J.M.: J. Biol. Chem. 253, 8570-8577 (1978). - COWELL, J.K., HARTMANN-GOLDSTEIN, I.J.: Chromosoma 79, 329-340 (1980a); - Chromosoma 81, 55-64 (1980b). - COZZARELLI, N.R.: Cell 22, 327-328 (1980a); - Science 207, 953-960 (1980b).

DANCIS, B.M., HOLMQUIST, G.P.: Chrom. Today 6, 95-104 (1977); - J. Theor. Biol. 78, 211-224 (1979). - DE PAMPHILIS, M., WASSERMANN, P.M.: Annu. Rev. Biochem. 49, 627-666 (1980). - DUBE, D.K., KUNKEL, T.A., SEAL, G., LOEB, L.A.: Biochim. Biophys. Acta 561, 369-382 (1979).

EASTMAN, E.M., GOODMAN, R.M., ERLANGER, B.F., MILLER, O.J.: Chromosoma 79, 293-314 (1980). - EHRENBERG, M., BLOMBERG, C.: Biophys. J. 31, 333-358 (1980). - EREMENKO, T., TIMOFEVA, M.Y., VOLPE, P.: Mol. Biol. Rep. 6, 131-136 (1980). - ERICKSON, L.C., ROSS, W.E., KOHN, K.W.: Chromosoma 74, 125-139 (1979).

FALASCHI, A., COBIANCHI, F., RIVA, S.: Trends Biochem. Sci. 5, 154-157 (1980). - FRISCH, B., NAGL, W.: Plant Syst. Evol. 131, 261-276 (1979). -

FUJIMURA, R.K., DAS, S.K.: Progr. Nucl. Acids Res. Mol. Biol. 24, 87-109 (1980).

GRIMM, C., KUNZ, W.: Mol. Gen. Genet. 180, 23-26 (1980). - GRISVARD, J., TUFFET-ANGHILERI, A.: Nucl. Acids. Res. 8, 2843-2858 (1980).

HAND, R.: Eukaryotic chromosome replication and its regulation, p. 389-439. In: Structure and Replication of the Genetic Material, eds. D.M. PRESCOTT, L. GOLDSTEIN. New York: Academic Press 1979. - HASE, Y., YAKURA, K., TANIFUJI, S.: Plant Cell Physiol. 20, 1461-1469 (1979). - HISCOTT, J., MURPHY, D., DEFENDI, V.: Cell 22, 535-543 (1980). - HOLMQUIST, G.P., DANCIS, B.M.: Proc. Natl. Acad. Sci. USA 76, 4566-4570 (1979); - Genetica 52, 151-163 (1980). - HOPFIELD, J.J.: Proc. Natl. Acad. Sci. USA 77, 5248-5252 (1980). - HOTTA, Y., STERN, H.: Zygotene and pachytene-labeled sequences in the meiotic organization of chromosomes, 283-300. In: The Eukaryotic Chromosome, eds. W.J. PEACOCK, R.D. BROCK. Canberra: Austral. Natl. Univ. Press 1975; - Eur. J. Biochem. 95, 31-38 (1979). - HÜBSCHER, U., KUENZLE, C., SPADARI, S.: Proc. Natl. Acad. Sci. USA 76, 2316-2320 (1979). - HURWITZ, J.: CRC Crit. Rev. Biochem. 7, 45-74 (1980).

INNIS, M.A., HARPOLD, M.M., MILLER, D.L.: Arch. Biochem. Biophys. 195, 128-135 (1979). - IWAMURA, Y., SAKAI, M., MITA, T., MURAMATSU, M.: Biochemistry 18, 5289-5294 (1979).

JAVAHERIAN, K., SADEGHI, M., LIN, L.F.: Nucl. Acids Res. 6, 3569-3580 (1979). - JOHNSON, E.M.: Cell 22, 875-886 (1980).

KAUFMAN, R.J., BROWN, P.C., SCHIMKE, R.T.: Proc. Natl. Acad. Sci. USA 76, 5669-5673 (1979). - KENDALL, F., SWENSON, R., BORUN, T., ROWINSKI, J., NICOLINI, C.: Science 196, 1106-1109 (1977). - KHAITOVA, N.M., IL´YNA, G.S., LYAPUNOVA, N.A.: Tsitologiya 22, 640-645 (1980). - KIMURA, S., YAMAZAKI, K., KATO, Y.: Chromosoma 77, 309-323 (1980). - KLEVAN, L., WANG, J.C.: Biochemistry 19, 5229-5234 (1980). - KOLTER, R., HELINSKI, D.R.: Annu. Rev. Genet. 13, 355-391 (1979). - KONGSUWAN, K., SMYTH, D.R.: Chromosoma 76, 151-164 (1980). - KORNBERG, A.: DNA Replication. Oxford: Freeman 1980a; - CRC Crit. Rev. Biochem. 7, 23-43 (1980b). - KREUZER, K.N., COZZARELLI, N.R.: Cell 20, 245-254 (1980). - KUHN, B., ABDEL-MONEM, M., KRELL, H., HOFFMANN-BERLING, H.: J. Biol. Chem. 254, 11343-11350 (1979). - KUNKEL, T.A., LOEB, L.A.: J. Biol. Chem. 255, 9961-9966 (1980). - KURATA, N., ITO, M.: Cell Struct. Funct. 3, 349-356 (1978). - KWIATKOWSKA, M., MASZEWSKI, J.: Protoplasma 98, 363-367 (1979).

LAIRD, C.D.: Cell 22, 869-874 (1980). - LIU, L.F., LIU, C.-C., ALBERTS, B.M.: Cell 19, 697-707 (1980). - LOHMANN, K.: Histochemistry 63, 47-56 (1979). - LOHMANN, K., SCHUBERT, L.: J. Embryol. Exp. Morphol. 57, 61-70 (1980).

MAHOWALD, A.P., CAULTON, J.H., EDWARDS, M.K., FLOYD, A.D.: Exptl. Cell Res. 118, 404-410 (1979). - MARINI, J.C., MILLER, K.G., ENGLUND, P.T.: J. Biol. Chem. 255, 4976-4979 (1980). - MATTERN, M.R., PAINTER, R.B.: Biochim. Biophys. Acta 563, 293-305 (1979). - MAZIA, D.: J. Cell. Comp. Physiol. 62, Suppl. 1, 123-140 (1963); - A critical perspective on the cell cycle and growth. In: Cell Growth, eds. C. NICOLINI. New York: Plenum Press (in press). - MECHALI, M.: Bull. Inst. Pasteur 78, 139-174 (1980). - MEER, B.: Roux's Arch. Devel. Biol. 189, 83-90 (1980). - MELERA, P.W., LEWIS, J.A., BIEDLER, J.L., HESSION, C.: J. Biol. Chem. 255, 7024-7028 (1980). - MILLER, O.J., TANTRAVAHI, R., MILLER, D.A., YU, L.-C., SZABO, P., PRENSKY, W.: Chromosoma 71, 183-195 (1979). - MINOCHA, S.C.: Z. Pflanzenphysiol. 92, 327-339 (1979). - MIZUUCHI, K.,

FISHER, L.M., O'DEA, M.H., GELLERT, M.: Proc. Natl. Acad. Sci. USA 77, 1847-1851 (1980). - MORRISON, A., HIGGINS, N.P., COZZARELLI, N.R.: J. Biol. Chem. 255, 2211-2219 (1980). - MOSER, G., MÜLLER, H.: Eur. J. Cell Biol. 19, 116-119 (1979). - MÜLLER, U., SCHRÖDER, C.H., ZENTGRAF, H., FRANKE, W.W.: Eur. J. Cell Biol. 23, 197-203 (1980). - MURPHY, R.F., WALLACE, R.B., BONNER, J.: Proc. Natl. Acad. Sci. USA 75, 5903-5907 (1978); - Proc. Natl. Acad. Sci. USA 77, 3336-3340 (1980).

NAGL, W.; Nature 249, 53-54 (1974); - Nuclear structures during cell cycles, 147-193. In: Mechanisms and Control of Cell Division, eds. T.L. ROST, E.M. GIFFORD, Jr. Stroudsburg, PA: Dowden, Hutchison & Ross 1977a; - Protoplasma 91, 389-407 (1977b); - Endopolyploidy and Polyteny in Differentiation and Evolution. Amsterdam: North-Holland 1978; - Protoplasma 100, 53-71 (1979a); - Z. Pflanzenphysiol. 95, 283-314 (1979b); - Cell growth and nuclear DNA increase by endoreduplication and differential DNA replication. In: Cell Growth, ed. C. NICOLINI. New York: Plenum (in press, a); - Endoreduplication and differential replication of DNA. In: Encyclopedia of Plant Physiology, New Series, Vol. 14B, eds. B. PARTHIER, D. BOULTER. Berlin-Heidelberg-New York: Springer (in press, b). - NAGL, W., FUHRMANN, B.: The Embryo-Suspenso and Trophoblast as model systems of differentiation. Berlin-Heidelberg-New York: Springer (in press). - NAGL, W., HEMLEBEN, V., EHRENDORFER, F. (eds.): Genome and Chromatin - Organization, Evolution, Function. Vienna-New York: Springer 1979. - NICOLINI, C.: Chromatin structure, from ångstrom to micron levels, and its relationship to mammalian cell proliferation, 613-666. In: Chromatin Structure and Function, ed. C. NICOLINI. New York: Plenum 1979; - J. Submicr. Cytol. 12, 475-505 (1980). - NICOLINI, C., AJIRO, K., BORUN, T.W., BASERGA, R.: J. Biol. Chem. 250, 3381-3385 (1975). - NICOLINI, C., KENDALL, F., GIARETTI, W.: Biophys. J. 19, 163-176 (1977). - NUNBERG, J.H., KAUFMAN, R.J., SCHIMKE, R.T., URLAUB, G., CHASIN, L.A.: Proc. Natl. Acad. Sci. USA 75, 5563-5556 (1978).

OGAWA, T., OKAZAKI, T.: Annu. Rev. Biochem. 49, 421-457 (1980).

PAINTER, R.B.: J. Mol. Biol. 143, 289-301 (1980). - PARDOLL, D.M., VOGELSTEIN, B., COFFEY, D.S.: Cell 19, 527-536 (1980). - PERDRIX-GILLOT, S.: Biochimie 61, 171-204 (1979). - POPP, F.A.: Photon storage in biological systems, 123-149. In: Electromagnetic Bio-Information, ed. F.A. POPP. München, Wien, Baltimore: Urban & Schwarzenberg 1979a; - Umschau Wiss. Techn. 79, 235-239 (1979b). - POTTER, H., DRESSLER, D.: Proc. Natl. Acad. Sci. USA 77, 2390-2394 (1980). - PRELL, B., VOSBERG, H.-P.: Eur. J. Biochem. 108, 389-398 (1980). - PRESCOTT, D.: The genetic basis of eukaryotic cell life cycle: GO and Gl periods. In: Cell Growth, ed. C. NICOLINI. New York: Plenum (in press). - PRIGOGINE, I.: Thermodynamics of Irreversible Processes. New York: Wiley 1955; - Vom Werden zum Sein. Zeit und Komplexität in den Naturwissenschaften. München: Piper 1979. - PROBST, H., HOFSTAETTER, T., STUMPF, C., FRATERMAN, A.: Biochim. Biophys. Acta 561, 132-146 (1979). - PRYOR, A., FAULKNER, K., RHOADES, M.M., PEACOCK, W.J.: Proc. Natl. Acad. Sci. USA 77, 6705-6709 (1980).

QUESNEY-HUNEEUS, V., WILEY, M.H., SIPERSTEIN, M.D.: Proc. Natl. Acad. Sci. USA 76, 6065-5060 (1979).

RAY, J.H., VENKETESWARAN, S.: Chromosoma 74, 337-346 (1979). - REDDY, G.P.v., PARDEE, A.B.: Proc. Natl. Acad. Sci. USA 77, 3312-3316 (1980). - REINHARD, P., MÜLLER, B.: FEBS Lett. 113, 61-64 (1980). - ROMAN, R., CABOCHE, M., LARK, K.G.: Plant Physiol. 66, 726-730 (1980).

SCALI, V., PELLICCIARI, C., MONTANELLI, E., BERNOCCHI, G.: Biol. Cell. 37, 5-12 (1980). - SCHÄFER, A., NEUMANN, K.-H.: Planta 143, 1-4 (1978). - SCHIMKE, R.T.: Sci. Am. 243, 50-59 (1980); - Spektrum Wiss. (Jan.) 99-109 (1981). - SCHIMKE; R.T., KAUFMAN, R.J., ALT, F.W., KELLEMS, R. F.: Science 202, 1051-1055 (1978). - SCHLAEGER, E.-J., KNIPPERS, R.: Nucleic Acids Res. 6, 645-656 (1979). - SCHMITT, R., BERNHARD, E., MATTES, R.: Mol. Gen. Genet. 172, 53-65 (1979). - SCOVASSI, A.I., PLEVANI, P., BERTAZZONI, U.: Trends Biochem. Sci. 335-337 (1980). - SPRADLING, A.C., MAHOWALD, A.P.: Proc. Natl. Acad. Sci. USA 77, 1996-1100 (1980). - SSYMANK, V.: Biochem. Physiol. Pflanzen 174, 215-222 (1979). - SUGINO, A., HIGGINS, N.P., COZZARELLI, N.R.: Nucleic Acids Res. 8, 3865-3874 (1980).

TANAKA, R.: Bot. Mag. Tokyo 78, 50-62 (1965). - TANIGAWA, Y., KITAMURA, A., KAWAMURA, M., SHIMOYAMA, M.: Eur. J. Biochem. 92, 261-269 (1978a). - TANIGAWA, Y., KAWAMURA, M., KITAMURA, A., SHIMOYAMA, M.: Biochem. Biophys. Res. Commun. 81, 1278-1285 (1978b). - TAYLOR, J.H.: Chromosome Structure. New York: Academic Press 1979. - TSE, Y., WANG, J.C.: Cell 22, 269-276 (1980). - TSE, Y., KIRKEGAARD, K., WANG, J.C.: J. Biol. Chem. 255, 5560-5565 (1980).

VAN'T HOF, J., BJERKNES, C.A.: Bioscience 29, 18-22 (1979).

WANG, J.C.: Trends Biochem. Sci. 5, 219-221 (1980). - WANG, J.C., LIU, L.F.: DNA topoisomerases, 65-88. In: Molecular Genetics, Vol. 3, ed. J.H. TAYLOR. New York: Academic Press 1979. - WEISSBACH, A.: Arch. Biochem. Biophys. 198, 386-396 (1979). - WHITEHEAD, E.P.: DNA-dependent ATPases and vice versa, 159-185. In: Macromolecules in the Functioning Cell, eds. S. SALVATORE, G. MARINO, P. VOLPE. New York: Plenum 1979. - WIGLER, M., PERUCHO, M., KURTZ, D., DANA, S., PELLICER, A., AXEL, R., SILVERSTEIN, S.: Proc. Natl. Acad. Sci. USA 77, 3567-3570 (1980). - WILLE, J.J., STEFFENS, W.L.: Nucleic Acids Res. 6, 3323-3339 (1979). - WINTERSBERGER, E.: Rev. Physiol. Biochem. Pharmacol. 84, 93-142 (1978). - WIST, E.: Biochim. Biophys. Acta 562, 62-69 (1979). - WOHLGEMUTH, D.J., BIEDLER, J.L., MELERA, P.W.: Biochem. Genet. 18, 655-667 (1980).

YUROV, Yu.B.: Chromosoma 74, 347-353 (1979); - J. Mol. Biol. 136, 339-342 (1980).

ZACHARIAS, H.: Chromosoma 72, 23-51 (1979).

Professor Dr. WALTER NAGL
Zell-Biologie
Fachbereich Biologie der
Universität
Postfach 3029
D 6750 Kaiserslautern

II. Recombination – Recombination in Higher Plants

By HORST BINDING and REINHARD NEHLS

1. Introduction

Recombination in higher plants is based on events which are coupled to the decisive steps of the sexual cycle, the fusion of the gametes and meiosis; which are introduced by experiments on asexual combinations, e.g., protoplast fusion, organelle transplantation and genetic engineering; and which occur in somatic cells, e.g., somatic crossing-over and sister strand exchange.

Surveys on meiotic and mitotic recombination processes have been given in this series by ABEL (1975), on protoplast fusion and organelle transplantation by BINDING and NEHLS (1979a) and on genetic engineering by HOLLENBERG (1980). The last two topics will not be considered in this article.

2. Sexual Recombination

Experiments on sexual hybridization have been mainly carried out with respect to applied plant breeding. However, several results can be found which are appropriate to elucidate fundamental genetic processes, especially in meiosis.

a) Structural Control of Meiotic Processes

Well-known factors influencing the course and the products of meiosis are the numbers of homologous chromosomes, degree of their homologies, and chromosomal aberrations.

A marked reduction of chiasma frequencies in sunflower plants grown from X-irradiated seeds may probably be explained by reduced homologies of chromosomes (AL-ALLAF and GODWARD, 1979). The authors pointed out that this finding of increased chiasma frequencies is in agreement with other reports on the consequences of X-irradiation of dry seeds, but that irradiation of flower buds brought about rather contradictory results.

The processes of chromosome pairing and chiasma formation are still far from being understood. The roles which euchromatin and heterochromatin play in the formation of the synaptic complex have often been discussed. MAY and APPLES (1980) described observations in rye × wheat hybrids, indicating that telomeric heterochromatin is not decisive in this process.

They analyzed two translocactions in which the chromosomes 2R of rye and 2B of wheat were involved. It appeared by in situ hybridization with DNA recognizing highly repetitive DNA sequences that telomeric heterochromatin was lost by the translocations. However, synapsis was apparently not significantly affected.

EGEL (1979) considered the problem of how terminalization of chiasmata leads to the tight connections at the ends of homologous chromosomes. It seems evident that end-to-end attachment of the sister strands is responsible for this phenomenon.

This may be either caused by covalent linkage of the ends of the DNA, or it may be brought about by the heterochromatic telomeres probably held together by remains of the nuclear envelope.

b) Physiological Processes in Chiasma Formation

SINGH and JAIN (1979) used a number of antimetabolites to trace out metabolic steps involved in chiasma formation. Reduction of chiasma frequencies was observed in *Delphinium ajacis* after application of actinomycin D and mitomycin C, both inhibiting DNA synthesis, and by cycloheximid, known as a blocker of protein formation.

Interestingly, low concentrations of mitomycin increased the chiasma frequencies. This was attributed by the authors to the formation of cross-links between complementary strands of DNA.

c) Genetic Control of Meiotic Processes

Disturbances of the meiotic division is in many cases due to asynapsis or desynapsis either based on low degrees of homologies between pairs of chromosomes or caused by genetic factors. NEL (1979) investigated the gene As/as in maize and confirmed the common feature that the offspring of plants homozygous as/as exhibited low recombination rates.

NEL obtained this result by investigating recombination in the G16-Lg2-A region of chromosome 3. Previous investigations on the As/as gene revealed rather high recombination rates.

The fertility of autopolyploid plants is normally reduced by the formation of multivalents and univalents during the meiotic prophase and consequently aneuploidy of the gonospores. In a number of cases, fertility and the frequencies of euploid plants in the progeny increased with time. FRIEDT (1979) investigated this phenomenon, called phenotypic diploidization, in autotetraploid barley. He observed increased alternate orientation of the centromeres in multivalents.

A similar situation in a quite different system has been described by LELLEY and GUSTAFSON (1979). In a translocation hybrid of rye the centromeres of the multivalents showed controlled alternate orientation.

Both the observations of FRIEDT and of LELLEY and GUSTAFSON indicate a mechanism or mechanisms of centromere orientation which are completely not understood and remind of the controlled translocation heterozygocity in *Oenothera*.

d) Analyses of Linkage Groups

Physical maps of chromosomes constructed by the aid of Giemsa banding techniques have already been used for the location of linkage groups in animals and man. LINDE-LAURSEN (1979) demonstrated that this method is also appropriate for cytogenetic investigations in barley.

The author analyzed the offspring of plants which carried heterozygous markers and characteristic Giemsa C bands in chromosomes 3, 6, and 7. Significant parallelity was found in recombination of both the alleles and the C bands.

BIRCHLER (1980) used another type of a physical chromosome map established by 13 translocation lines of maize with overlapping translocations in the long arm of chromosome 1. He was able to localize the alcohol dehydrogenase-1 gene by the aid of these lines.

Two co-dominantly expressed alleles of the enzyme were taken for the identification in the offspring.

Isoenzyme patterns have also been used for the detection of nucleus substitution lines in *Brassica* by KANAZAWA et al. (1980).

3. Asexual Recombination

a) Somatic Crossing-Over

The classical method of double spot analysis for the detection of somatic crossing over has been applied by EVANS (1980) to investigations on the influence of X-irradiation on crossing-over frequencies. EVANS found a linear dose-effect dependence in soybean after treatment of dry seeds.

The plants were light green by the heterozygous constitution Y/y. Sizes of the spots on leaves of different position at the plants were taken to extrapolate back to their developmental stages in the dormant seeds.

b) Sister Chromatid Exchange

Investigations on sister chromatid exchange (SCE) have been treated in several reports stimulated by the development of an appropriate method for their detection by PERRY and WOLFFS (1974; cf. ABEL, 1975). The method has been modified in various ways (e.g., SCHVARTZMAN, 1979). The frequencies of SCE's are known to be correlated to DNA lesions as demonstrated by the use of mutagens (e.g., BASLER et al., 1979; NAKANISHI and SCHNEIDER, 1979; SCHVARTZMAN et al., 1979; SCHVARTZMAN and HERNANDEZ, 1980).

SCE's are not distributed over the chromosomes at random. A hot spot was found at the nucleolus-organizing region of the M chromosome of *Vicia faba* (RIEGER et al., 1979; SCHUBERT et al., 1979; SCHUBERT and RIEGER, 1980).

SCHUBERT and RIEGER pointed to the undissolved question if the exchange points were located in the euchromatic region or if the heterochromatin was involved in the process. SCHUBERT et al. (1980) concluded from observations in barley that SCE frequencies were independent of the amount of heterochromatin within the exchanged segments of the chromosomes.

c) Protoplast Fusion Technique

Several reports on protoplast fusion have appeared since the last survey in this series (BINDING and NEHLS, 1979a). They were mainly concerned with species of the Solanaceae family.

One reason for the restriction to a few taxonomic groups can be seen in problems of protoplast regeneration. The number of appropriate plant species in the class of Magnoliatae is steadily increasing (cf. BINDING et al., 1981); however, plant regeneration failed or was insufficient in many species. The situation in the class of Liliatae is still rather unsatisfactory. Members of the family of Poaceae are extremely difficult to handle (cf. POTRYKUS et al., 1979). The regeneration of plants from protoplasts of *Pennisetum americanum* is so far the only ray of hope (VASIL and VASIL, 1980).

All regenerated fusion products were derived from fusion bodies formed by the action of PEG, calcium ions, and high pH, singly or in combination. DOUGLAS et al. (1981a) confirmed the fusion-enhancing effect of additional dimethylsulfoxide. The method of protoplast fusion in electric fields was improved by ZIMMERMANN and SCHEURICH (1981) so that it seems likely that it will become a tool in cell hybridization experiments.

d) Markers in Protoplast Fusion Experiments

Mainly the markers already mentioned by BINDING and NEHLS (1979a) have been applied to selection, identification, and analysis of fusion products. GALBRAITH and MAUCH (1980) proposed the detection of fusion bodies by characteristic fluorescence after mutual labeling of the parental protoplast types by different fluorescent dyes. POULSEN et al. (1980) improved the fraction I protein analysis and identified somatic hybrids of potato + tomato by comparison of peptide maps of the small subunits obtained by chymotryptic digestion. DUDITS et al. (1979) utilized DNA-RNA hybridization for the detection of nuclear material of *Aegopodium* in *Daucus* + *Aegopodium* hybrids.

e) Regeneration of Plants from Fusion Bodies

Most of the regenerated plants derived from hybridization experiments within the genus *Nicotiana* (WHITE and VASIL, 1979; AVIV et al., 1980; AVIV and GALUN, 1980; IWAI et al., 1980; DOUGLAS et al., 1980a,b; GRESSEL et al., 1981). Further recombinants in the family of Solanaceae have been obtained in *Petunia* by POWER et al. (1979) and IZHAR and TABIB (1980), in *Datura* by SCHIEDER (1980), of *Datura innoxia* + *Atropa belladonna* by KRUMBIEGEL and SCHIEDER (1979), of potato + tomato by POULSEN et al. (1980) and of *Nicotiana sylvestris* + *Solanum nigrum* by GRESSEL et al. (1981). Intrageneric somatic hybrids in other plant families were formed of *Daucus carota* + *Aegopodium podagraria* (DUDITS et al., 1979) and of *Arabidopsis thaliana* + *Brassica campestris* (GLEBA and HOFFMANN, 1980).

The last five cases of intergeneric combinations are especially interesting with respect to how far the degree of relationships of the parents interfere with the organogenic potencies. Combinations of species within a family do allow – at least in the investigated cases – organized growth. So far, no shoot formation has been observed in interfamiliar hybrid calluses (cf. BINDING and NEHLS, 1979b; BRAR et al., 1980; WETTER and KAO, 1980). It is, however, not considered advisable to draw conclusions from these observations with a view to remote relationships because there was at least one parent involved in each case which itself could not be regenerated under the applied conditions.

f) Chromosomal Behavior in Hybrid Cell Lines and Plants

Karylogical instability proved to be a common phenomenon in interspecific somatic hybrids (KRUMBIEGEL and SCHIEDER, 1979; WHITE and VASIL, 1979; IWAI et al., 1980; LÖNNENDONKER and SCHIEDER, 1979; NAGL and HOFFMANN, 1980; POULSEN et al., 1980; DOUGLAS et al., 1981b). Unilateral elimination of *Nicotiana* chromosomes had been found in soybean + *Nicotiana glauca* calluses. A higher stability of the *Nicotiana* chromosomes was observed by WETTER and KAO (1980) in tissue after backfusion of the hybrid callus + *Nicotiana glauca*.

It has already been pointed out (BINDING and NEHLS, 1979a) that the interpretation of cytological instabilities, especially with respect to incompatibilities, should be made very cautiously because of similar events in normal cell cultures. This emerged, for instance, from comparison of findings in callus of *Vicia faba* and of *Vicia faba* + *Petunia hybrida* (BINDING and NEHLS, 1979b).

DUDITS et al. (1979) obtained a plant of *Daucus* + *Aegopodium* carrying mainly the genetic information of the carrot, besides apparently just a small portion of nuclear material from *Aegopodium*.

The cells contained 18 chromosomes, the diploid number of *Daucus*. The authors concluded the presence of chromosomal material of *Aegopodium* from genetic markers concerning chlorophyll synthesis, root morphology and pigmentation and from DNA-RNA hybridization. One of the possible explanations is that the parental carrot protoplast had been in a mitotic state and stimulated the interphase nucleus of *Aegopodium* to a premature mitotic condensation initiating degradation of the *Aegopodium* chromosomes followed by the integration of fragments into the carrot genome. If this series of processes really occurs, fusion of mitotic protoplasts to those in interphase (SZABADOS and DUDITS, 1980) would be an appropriate method to achieve gene transfer.

SZABADOS et al. (1981) reported on chromosome transplantation experiments finally also aiming at gene transfer.

g) Fates of Extrakaryotic Genophores

The heteroplasmic states of fusion bodies have been found to be carried to the regenerated plants as concluded from fraction I protein analysis by IWAI et al. (1980) and from segregation of cytoplasmic male sterility in the progeny of somatic hybrids by IZHAR and TABIB (1980). Recombinant organelles have been identified in one case. BELLIARD et al. (1979) analyzed restrictase digest patterns of mitochondrial DNA and detected a recombined fragment additionally to others specific for either parents.

Organelle segregation has been found in some experiments. Potato + tomato plants contained hybrid nuclei in combination with either one or the other plastid types (POULSEN et al., 1980). It may be concluded that in these cases the plastids segregated independently of the nuclear constitution. Independent segregation of different extrakaryotic traits has been observed by AVIV and GALUN (1980) and AVIV et al. (1980).

Organelle segregation is a basic mechanism in experiments aiming at the production of organelle substitution lines. GRESSEL et al. (1981) obtained plants containing the nuclei of *Nicotiana sylvestris* and the plastids of *Solanum nigrum* as presumed from the tolerance to atrazin.

The authors fused *Nicotiana* protoplasts to those of *S. nigrum* which had been X-irradiated to inactivate the nuclei. The plastids of *S. nigrum* carried a herbicide resistance.

The fates of plastids in fusion bodies of extremely remote species (carrot + *Chlamydomonas*) have been followed by FOWKE et al. (1979) over a period of just a few days.

h) Asexual Recombination in Plant Breeding

Most of the experiments on asexual recombination have been carried out with economically important plant species as at least one parent. However, just a few papers were concerned with really applied aspects. WENZEL et al. (1979) developed a promising breeding system in potato, making use of somatic hybridization. SMILLIE et al. (1980) detected intermediate chilling resistance of potato + tomato hybrids. It now seems feasible to transfer the higher chilling tolerance of potato to tomato – probably by backfusion of the hybrids to tomato. The transfer of herbicide resistence of *Solanum nigrum* to *Nicotiana sylvestris* by the formation of a plastid substitution hybrid may also be mentioned in this context (GRESSEL et al., 1981).

References

ABEL, W.O.: Progr. Bot. 37, 211-218 (1975). - AL-ALLAF, S.M., GODWARD, M.B.E.: Cytologia 44, 821-833 (1979). - AVIV, D., FLUHR, R., EDELMAN, M., GALUN, E.: Theor. Appl. Genet. 56, 145-150 (1980). - AVIV, D., GALUN, E.: Theor. Appl. Genet. 58, 121-127 (1980).

BASLER, A., BACHMANN, U., ROSZINSKY-KOCHER, G., ROHRBORN, G.: Mut. Res. 59, 209-214 (1979). - BELLIARD, G., VEDEL, F., PELLETIER, G.: Nature 281, 401-403 (1979). - BINDING, H., NEHLS, R.: Progr. Bot. 41, 173-184 (1979a); - Transfer of genetic information in higher plants via protoplast fusion, 315-319. In: Advances in Protoplast Research. Budapest: Akadémiai Kiadó 1979b. - BINDING, H., NEHLS, R., KOCK, R., FINGER, J., MORDHORST, G.: Z. Pflanzenphysiol. 101, 119-130 (1981). - BIRCHLER, J.A.: Genetics 94, 687-700 (1980). - BRAR, D.S., RAMBOLD, S., CONSTABEL, F., GAMBORG, O.L.: Z. Pflanzenphysiol. 96, 269-275 (1980).

DAVEY, M.R., COCKING, E.C., FREEMAN, J., PEARCE, N., TUDOR, I.: Plant Sci. Lett. 18, 307-313 (1980). - DOUGLAS, G.C., KELLER, W.A., SETTERFIELD, G.: Can. J. Bot. 59, 220-227 (1981a). - DOUGLAS, G.C., WETTER, L.R., NAKAMURA, C., KELLER, W.-A., SETTERFIELD, G.: Can. J. Bot. 59, 228-237 (1981b). - DUDITS, D., HADLACZKY, G., BAJSZAR, G., KONCZ, C., LÁZÁR, G., HORVATH, G.: Plant Sci. Lett. 15, 101-112 (1979).

EGEL, R.: Hereditas 91, 138-140 (1979). - EVANS, D.A.: Theor. Appl. Genet. 56, 245-251 (1980).

FOWKE, L.C., GRESSHOFF, P.M., MARCHANT, H.J.: Planta 144, 341-347 (1979). - FRIEDT, W.: Z. Pflanzenzüchtg. 82, 311-339 (1979).

GALBRAITH, D.W., MAUCH, T.J.: Z. Pflanzenphysiol. 98, 129-140 (1980). - GATENBY, A.A., ZAPATA, F.J., COCKING, E.C.: Z. Pflanzenzüchtg. 84, 1-8 (1980). - GLEBA, Y.Y., HOFFMANN, F.: Planta 149, 112-117 (1980). - GRESSEL, J., EZRA, G., JAIN, S.M.: Genetic and chemical manipulation of crops to confer tolerance to chemicals, 1-13. In: Chemical Manipulation of Crop Growth and Development, ed. J.S. Mc LAREN. London: Butterworth 1981.

HOLLENBERG, C.P.: Progr. Bot. 42, 171-185 (1980).

IWAI, S., NAGAO, T., NAKATA, K., KAWASHIMA, N., MATSUYAMA, S.: Planta 147, 414-417 (1980). - IZHAR, S., TABIB, Y.: Theor. Appl. Genet. 57, 241-246 (1980).

KANAZAWA, K., EGUCHI, H., IWASA, S., UEMOTO, S.: J. Fac. Agric. Kyushu Univ. 25, 25-32 (1980). - KRUMBIEGEL, G., SCHIEDER, O.: Planta 145, 371-375 (1979).

LELLEY, T., GUSTAFSON, J.P.: Z. Pflanzenzüchtg. 82, 306-310 (1979). - LINDE-LAURSEN, I.: Hereditas 91, 73-77 (1979). - LÖNNENDONKER, N., SCHIEDER, O.: Plant Sci. Lett. 17, 135-139 (1980).

MAY, C.E., APPELS, R.: Theor. Appl. Genet. 56, 17-23 (1980).

NAGL, W., HOFFMANN, F.: Eur. J. Cell Biol. 21, 227-228 (1980). - NAKANISHI, Y., SCHNEIDER, E.L.: Mutat. Res. 60, 329-337 (1979). - NEL, P.M.: J. Heredity 70, 400-406 (1979).

PERRY, P., WOLFF, S.: Nature 251, 156-158 (1974). - POULSEN, C., PORATH, D., SACRISTÁN, M.D., MELCHERS, G.: Carlsberg Res. Commun. 45, 249-267 (1980). - POTRYKUS, I., HARMS, C.T., LÖRZ, H.: Theor. Appl. Genet. 54, 209-214 (1979). - POWER, J.B., CHAPMAN, J.V., COCKING, E.C., SINK, K.C.: Theor. Appl. Genet. 55, 97-99 (1979).

RIEGER, R., NICOLOFF, H., ANASTASSOVA-KRISTEVA, M.: Biol. Zbl. 98, 385-398 (1979).

SCHIEDER, O.: Z. Pflanzenphysiol. 98, 119-127 (1980). - SCHUBERT, J., RIEGER, R.: Chromosoma 70, 385-391 (1979); - Biol. Zbl. 99, 65-72 (1980). - SCHUBERT, J., STURELID, S., DOBEL, P., RIEGER, R.: Mutat. Res. 59, 27-38 (1979). - SCHUBERT, J., KÜNZEL, G., BRETSCHNEIDER, H., RIEGER, R., NICOLOFF, H.: Theor. Appl. Genet. 56, 1-4 (1980). - SCHVARTZMAN, J.B.: J. Heredity 70, 423-424 (1979). - SCHVARTZMAN, J.B., HERNANDEZ, P.: Theor. Appl. Genet. 57, 221-224 (1980). - SCHVARTZMAN, J.B., POSTIGO, R., GUTIERREZ, C.: Chromosoma 74, 317-328 (1979). - SINGH, U., JAIN, H.K.: Cytologia 44, 123-133 (1979). - SMILLIE, R.M., MELCHERS, G., von WETTSTEIN, D.: Carlsberg Res. Commun. 44, 127-132 (1979). - SZABADOS, L., DUDITS, D.: Exp. Cell Res. 127, 442-446 (1980). - SZABADOS, L., HADLACZKY, G., DUDITS, D.: Planta 151, 141-145 (1981).

VASIL, V., VASIL, I.K.: Theor. Appl. Genet. 56, 97-100 (1980).

WENZEL, G., SCHIEDER, O., PRZEWOZNY, T., SOPORY, S.K., MELCHERS, G.: Theor. Appl. Genet. 55, 49-56 (1979). - WETTER, L.R., KAO, K.N.: Theor. Appl. Genet. 57, 273-276 (1980). - WHITE, D.W.R., VASIL, I.K.: Theor. Appl. Genet. 55, 107-112 (1979).

ZIMMERMANN, U., SCHEURICH, P.: Planta 151, 26-32 (1981).

Professor Dr. HORST BINDING
Dr. REINHARD NEHLS
Botanisches Institut der
Universität
Biologiezentrum
Olshausenstr. 40-60
D 2300 Kiel 1

III. Mutation: Higher Plants

By WERNER GOTTSCHALK

1. Methods for Inducing Gene and Chromosome Mutations

The utilization of chemicals for inducing mutations increases strongly as compared to the use of physical mutagens in higher plants. Some new chemicals, hitherto not used in experimental mutation research, were found to have a high mutagenic efficiency. Moreover, a rapid increase of our knowledge on the mutagenic action of distinct substances used in daily life can be stated. The utilization of cell and suspension cultures becomes of increasing importance in mutation research. A review was given by THOMAS et al. (1979). By means of these methods, mutants have been isolated in *Datura innoxia* (KRUMBIEGEL, 1979), *Petunia hybrida* (COLIJN et al, 1979), in soybean (WEBER and LARK, 1979) and tobacco (CHALEFF, 1980).

a) Physical Mutagens

The mutagenic efficiency of gamma rays can be increased by post-treatment with chloral hydrate, methanol, and aqueous mixtures of these substances with ethanol (REDDY and VAIDYANATH, 1979; rice). A similar effect is reached by using the rays at extremely low temperatures (-196°C) which reduce the M_1 damage. Thus, higher doses of the rays can be applied resulting in higher mutation rates (NAKAI and SAITO, 1979; rice). The influence of oxygen concentration and water content with regard to the gamma ray-induced damage of barley seeds was studied by DONALDSON et al. (1979a,b). Japanese soybean varieties show clear differences in the frequency of chromosome aberrations induced by gamma rays. Groups of resistant and sensitive varieties were isolated, the difference being obviously due to a distinct recessive gene (UKAI and YAMASHITA, 1980b). High-energy protons given to isolated eyes of potato clones, were found to be as effective as fast neutrons and more effective than gamma rays with regard to the frequency of chromosomal rearrangements induced (TARASENKO, 1977).

b) Chemical Mutagens[1]

The mutagenic effects of some compounds widely used in experimental mutation research was comparatively studied in barley. DES was found to be less efficient than NEU and NMU. Interestingly, HA did not produce any chromosome aberrations, whereas it was effective in inducing gene mutations of the waxy type (SEN, 1979). NMU proved to be about three times more effective than gamma rays in *Lens culinaris* (SHARMA and SHARMA, 1979). A review on the mutagenic action of DES, DMS,

---

[1]Abbreviations used in the paper: EMS ethyl methanesulphonate, MMS methyl methanesulphonate, DES diethyl sulphate, DMS dimethyl sulphate, EI ethyleneimine, HA hydroxylamine, MH maleic hydrazide, NMU nitrosomethylurea, NEU nitrosoethylurea, MNNG N-methyl-N'nitro-N-nitrosoguanidine.

and related compounds was given by G. HOFFMANN (1980). N-methyl-N'-nitro-N-nitroso-guanidine (MNNG), a seldom used compound, was successfully applied in cell suspension cultures of *Petunia hybrida* (COLIJN et al., 1979). A particularly potent mutagen is sodium azide. Its efficiency was reviewed by KLEINHOFS et al. (1978). Further results by using NaN_3 have been obtained in rice (SARMA et al., 1979) and in *Petunia hybrida* (KHALATKAR and KASHIKAR, 1980). Propane sultone is not only a potent carcinogen but also an effective mutagenic agent (C. SINGH, 1979; barley). Mutagenic efficiency was furthermore found or confirmed for 5-amino-uracil (SCHVARTZMAN and HERNANDEZ, 1980; *Allium cepa*), for trenimon, a trifunctional alkylating compound (OBE and BEEK, 1979) and for some nitrous esters such as sodium and ethyl nitrite (EHRENBERG et al., 1980). Nitrogen mustard and mustard gas derivatives, on the other hand, proved to be only weakly mutagenic (FOX and SCOTT, 1980).

Studies on the enhancement of the efficiency of mutagenic chemicals were carried out in *Triticum aestivum*. Dioxane, dimethylformamide, and dimethylsulphoxide were found to increase the frequency of chromosome aberrations induced by NMU (SALNIKOVA and DOSMAILOVA, 1979). Similar effects were observed in barley after DES and MMS application when the seeds were stored for 1 to 4 weeks after the treatment at 20% water content. Storage at 30%, however, gave a reduction of the frequency of chromosomal damage (GICHNER and VELEMÎNSKÝ, 1979a). An increase of the frequency of chromatid aberrations, induced by MH and MNNG, was reached by posttreatment with caffeine. Moreover, caffeine influences the distribution of the breakpoints along the chromosomes (HARTLEY-ASP et al., 1980; *Vicia faba*; NICOLOFF et al., 1980; *Hordeum vulgare*). An alteration of the distribution pattern of the breakpoints under the influence of caffeine following treatment with chemical mutagens was also observed in *Crepis capillaris* (DIMITROV, 1979).

The list of pesticides which are found to cause mitotic and meiotic irregularities is continuously increasing. This holds particularly true for organophosphorous insecticides such as Leptophos (AMER and FARAH, 1979; *Vicia faba*), Trichlorfon, Dichlorvos, and similar compounds (PANDA and SHARMA, 1979, 1980; barley). The active ingredients of some systemic and nonsystemic insecticides induced a significantly higher rate of cytological aberrations than EMS (SINGH et al., 1979; barley). Further findings in this field are available in barley following application of the pesticides Thiodan, Folithion, Labaycid, and Kitazin (GROVER and TYAGI, 1980). A review on the effects of some herbicides has been given by SEILER (1978).

Not only new pesticides, but an increasing number of compounds belonging to our daily life are every year added to the lists of substances having cytological or genetic effects. Thioridazine hydrochloride is a widely used antipsycholic drug; it causes chromosome fragmentations (SUBRAMANYAM et al., 1978; *Allium cepa*). Similar effects were observed after application of p-methoxyphenol used for preserving and stabilizing distinct food products (KAUL, 1979). New findings concerning the problem of the genetic risk of cigarette smoking were published by BRIDGES et al. (1979). Undoubtedly, cigarette smoke contains several mutagens, some of which seem to be distributed systematically within the body and may thus reach the gonads. Empirical data on the extent of the genetic hazard for the children of smokers are not yet available; the problem, however, should be considered seriously. Chromosome aberrations and mitotic irregularities in root tips of *Allium cepa* were found after treatment with extracts of green chillies and of papaverine hydrochloride, an alkaloid present in opium (ABRAHAM and KOSHY, 1979; EL-BAYOUMI et al., 1979). Moreover, certain plant flavonols (quercetin, kaempferol), anthraquinones, and hydrazines proved to be mutagenic in the *Salmonella*/mammalian microsome test (TOSK et al., 1979; BROWN and DIETRICH, 1979; BROWN, 1980). The genetic effects of naturally occurring agents derived from fungi and higher plants have been reviewed by NAGAO and SUGIMURA (1978).

c) Protection Against Genetic Damage and Repair of Breaks

A protecting effect against the genetic damage induced by X-ray and neutron irradiation was found for dithiothreital (DELLAERT, 1980; *Arabidopsis*). The mutagenic efficiency of NMU is decreased under the influence of sulphur ether (SALNIKOVA and DOSMAILOVA, 1979). The storage of barley seeds, treated with DES or MMS, at 30% water content for 3 - 28 days leads to a marked decrease of the frequency of chromosome aberrations as compared to the unstored material. The period between mutagenic treatment and the onset of DNA synthesis is prolonged under these storage conditions, resulting in the repair of a part of the breaks induced. Chromatidal breaks were found to recover to a higher extent than chromosomal breaks (GICHNER and VELEMÍNSKÝ, 1979b).

2. Gene Mutations

It is not possible to review the large number of publications dealing with the genetic basis of the anomalies observed in the respective mutants. In almost all the cases studied, they are due to the action of single recessive genes. A mutator system was found in maize causing an approximately 30-fold increase of the seedling mutation rate. It proved to be stable in subsequent generations and is obviously not based on a single Mendelian factor (ROBERTSON, 1978). The distribution of mutated sectors was studied in 58 gamma ray-induced apple mutants (LACEY and CAMPBELL, 1979). In a spontaneously arisen leaf mutant of *Gossypium barbadense*, gene expression is limited to a specific portion of the plant. This unusual behavior is not due to any environmental influences (TURCOTTE and FEASTER, 1980). In barley, mutants have been selected showing certain deviations with regard to the formation of their root system. Number and length of roots, as well as the rate of root system development during ontogenesis, are involved. As expected, the various characters of the shoot and the root system are controlled by independently functioning genes (GORNY, 1978). So far, only very little information on genes influencing the root system is available. Gene-ecological problems have been studied in *Pisum* mutants and recombinants comparatively grown in West Germany and at different locations of India. Strong differences in the behavior of specific genes or gene combinations were observed referring not only to quantitative but also to qualitative traits. Most of the genotypes tested were found to have an adaptational optimum to the ecological conditions tested different from that of the mother variety (GOTTSCHALK and MÜLLER, 1979; GOTTSCHALK and KAUL, 1980). A mutant *Pisum* gene, which causes the degeneration of leaflets and stipules, is unable to express its action under the climatic conditions of Egypt and India as well as in a warm greenhouse. The negative selection value of this gene can be overcome completely by combining it with specific other mutant genes of the genome (GOTTSCHALK, 1979).

In the following, some groups of mutant genes are reviewed, which are not only of genetic, but of general botanical interest or of importance in applied genetics.

a) Genes Controlling Meiosis and Mitosis

A "synaptic" mutant was isolated in the progeny of a hybrid between two diploid tuberous *Solanum* species. Univalents were present at diakinesis and metaphase I, resulting in the later meiotic irregularities generally known for asynaptic and desynaptic mutants. The peculiarity of the recessive gene involved consists in the fact that it only becomes effective in macrosporogenesis. In microsporogenesis, normal synapsis occurs (IWANAGA and PELOQUIN, 1979). This behavior is in so far unusual as all the *as*- and *ds*-genes studied so far influence micro- and macrosporogenesis in a fundamentally similar way. Desynaptic plants of *Corchorus olitorius* were used for obtaining trisomics. All possible primary trisomics of the species except trisomic 6 were produced in this way (BASAK and PARIA, 1980).

A new recessive gene for male sterility (m_3) was found in soybean causing microspore degeneration between tetrad formation and tapetal breakdown. The gene is nonallelic to the genes ms_1, ms_2, and ms_4 of the genome. In the ms_1 locus, four independently arisen spontaneous mutations have been detected (PALMER et al., 1978, 1980). The recessive gene *msp* of the soybean causes partial male sterility. The action of this gene is markedly influenced by both environmental factors and the genotypic background of the plants (STELLY and PALMER, 1980). Male sterility in *Phaseolus vulgaris* was found to be due to the action of a spontaneously arisen recessive gene (VAN RHEENEN et al., 1979). Microsporogenesis has not been investigated. Some mutant genes of *Zea mays* cause irregular distribution of the chromosomes in the first meiotic division resulting in male sterility. In one of three genotypes studied, cytokinesis is partially or completely blocked (GOLUBOVSKAYA and SITNIKOVA, 1980). At least the last-mentioned genotype belongs obviously to the group of typical *ms*-mutants known in many plant species. A review paper on the genetic control of meiosis was published by GOLUBOVSKAYA (1979).

Only little information on genes controlling mitosis is available. Gene *vrg* of the tomato genome is not only responsible for a specific type of leaf variegation, but it causes further the tendency for the delay of the disappearance of the nucleolus in mitoses of early stages of ontogenetic development. A secondary effect of the gene consists in causing the loss of one or both chromosomes No. 2 (the satellite chromosomes). This happens in the earliest mitoses during seedling development in the epidermal and subepidermal cells of highly variegated lines (LESLEY et al., 1979).

b) Mutant Genes of Interest in Applied Genetics

Plenty of work is being done for selecting mutants of a large number of different crops, some of which might be of interest for direct or indirect utilization in plant breeding. Many details on findings in this field are given in the proceedings of three symposia organized by the Institute of Radiation Breeding in Ohmiya, Japan (1977, 1978, 1979). Some relevant problems of mutation breeding have been discussed and reviewed by MICKE (1978, 1979). The present situation of the possibilities for the seed protein improvement of cereals, legumes, and other crops by means of mutant genes becomes clear from the proceedings of a Eucarpia meeting (1976) and of two symposium volumes of the International Atomic Energy Agency (1979). The high-lysine gene of the intensively studied Danish barley mutant 1508 has been located. In the Risø material, three allelic high-lysine genes are now known (JENSEN, 1979).

Of particular interest are new mutant genes causing resistance or tolerance to plant diseases. Problems related to this field have been discussed in a symposium of the Institute of Radiation Breeding in Ohmiya, Japan (1978). After EI treatment of barley seeds, 7 powdery resistant mutants were screened in an M_2 population of about 1,200,000 plants. Interestingly, no mutants of this category were found in gamma ray-treated material (YAMAGUCHI and YAMASHITA, 1979). An early flowering barley mutant proved to be resistant to the barley yellow mosaic virus (UKAI and YAMASHITA, 1980a). Resistance or tolerance of mutant bread wheat lines were found in relatively old mutant material (BOROJEVIĆ, 1978, 1979). A neutron-induced mutant, homozygous for a dominant gene, shows an intermediate level of resistance against bacterial blight in cotton (BRINKERHOFF et al., 1978). In *Nicotiana*, a recessive gene causing resistance to the tobacco vein-mottling virus was found (LEGG et al., 1980).

Some other mutants show specific characters which might be of interest for realizing distinct aims of breeding. This holds true for extreme earliness in jute. An X-ray-induced photoperiod-insensitive mutant of *Corchorus capsularis* flowers in 60 days, whereas the control material starts flowering 120 - 127 days after sowing. A chlorophyll mutant of *Corchorus olitorius* shows an increased fiber length, an improved fiber quality and an approximately 13% increase of fiber yield as related to the control values (HOSSAIN and SEN, 1978a,b). The flowers of okra are normally formed singly in leaf axils. In a spontaneous mutant, a supernumerary inflorescence with several flowers is developed instead of a single flower, resulting in an increased number of fruits per plant. The improvement is caused by a dominant gene (FATOKUN et al., 1979). Experimentally obtained maize mutants with an increased number of ears per stalk are available in Hungary. Moreover, narrow-leaved corn-grass mutants were found to express a high degree of heterosis after crossing with normal inbred maize (PASZTOR, 1979). Coastcross 1 bermudagrass is a hybrid between *Cynodon dactylon* × *C. nlemfuensis*. Improvement by means of the conventional breeding methods is difficult because of the sterility of the plants. Gamma ray treatment of more than 500,000 stems resulted in the selection of a valuable mutant with a high degree of winter hardiness. The character seems to be controlled by several genes derived from *Cynodon dactylon* (BURTON et al., 1980). A group of mutant genes was found to extend the storage life of the tomato fruit considerably (KOPELIOVITCH et al., 1979).

3. Genome Mutations

a) Natural and Experimental Possibilities for Obtaining Polyploid Plants

The most important process for increasing the number of genomes in nature consists in producing unreduced gametes following specific meiotic disturbances. This was observed in tulips (KROON and VAN EIJK, 1977), in *Medicago sativa* (VORSA and BINGHAM, 1979) and in many diploid, triploid, and tetraploid wild potato species (DEN NIJS and PELOQUIN, 1977, CAMADRO and PELOQUIN, 1980). The fusion of spindles in the second meiotic metaphase causes dyad formation and the production of unreduced gametes in diploid *Solanum* clones (RAMANNA, 1979). Similar processes were observed in interspecific or intergeneric hybrids resulting in amphiploid plants (F_1/*Solanum phureja* × *S. tuberosum*; MENDIBURU and PELOQUIN, 1977; F_1/*Aegilops heldreichii* × *Triticum durum*; MAAN and SASAKUMA, 1977).

Some natural compounds were found to induce mitotic anomalies similar to the colchicine action. This holds true for pantopon hydrochloride, an opium alkaloid (KABARITY et al., 1979) and for griseofulvin, a fungicidal substance produced by various *Penicillium* species (SCHIAVO et al., 1980). Mycotoxins of distinct strains of the grain fungi *Fusarium tricinctum* and *Stachybotrys alternans* proved to be strong mitotic inhibitors (LINNAINMAA et al., 1979). A pronounced increase in the frequency of triploid and aneuploid seeds was found in barley infested with all six strains of barley stripe mosaic virus (SANDFAER, 1979b). The action of gene *tri* of the barley genome is of particular interest in this respect. It has no effect on microsporogenesis, but it suppresses the second meiotic division in about half the embryosac mother cells resulting in diploid female gametes. In this way, triploid plants arise on a genetic basis (FINCH and BENNETT, 1979b). The level of ploidy can be discerned easily and reliably in different Gramineae by counting the number of nucleoli in resting nuclei of root tips, leaves, and in distinct parts of the coleoptile and flower (REITBERGER, 1977).

Modern methods of callus culture are being used for producing polyploids. Triploid rice plants were obtained from endosperm cultures (BAJAJ et al., 1980). Diploid callus from *Hemerocallis flava*, treated with colchicine, gave rise to tetra- and octoploid cells from which tetraploid plants in high frequency regenerated (CHEN and GOEDEN-KALLEMEYN, 1979). A fascinating method for the experimental production of amphiploids consists in somatic hybridization by means of protoplast fusion. Four somatic hybrids between the tomato and dihaploids of the potato were obtained in this way by MELCHERS et al. (1978). Three of them had chromosome numbers around 50, thus being close to the expected number of 2n = 4x = 48; the fourth had a chromosome number close to 72. The "*Arabidobrassica*" arose in the same way. Isolated protoplasts of *Arabidopsis thaliana* and *Brassica campestris* were fused and five of the regenerated plants represented true amphiploids. They were highly abnormal and differed drastically from each other in morphology, chromosome numbers, and esterase isozyme patterns (GLEBA and HOFFMANN, 1979). The authors are sceptical with regard to the prospects of this method because of the monstrosity of the products obtained. But it is certainly too early to make any generalizations in this field. For basic research, it is undoubtedly a method by means of which very interesting results can be expected in the near future.

b) Autopolyploids

In cultivated oranges, polyploidy frequently occurs. The tetraploids are without any practical value; they were, however, used for producing triploids with improved growth rate, fruit quality, and with frost resistance (MAISURADZE, 1979). The etheric oils of *Carum carvi* are widely used in the processing of pharmaceutic products, perfume and liqueur carvone being the essential component of the oil. Its content in the seeds of experimentally produced tetraploid plants is considerably higher than in the diploid controls (DIJKSTRA and SPECKMANN, 1980).

The problem of the diploidization of autotetraploid plants was intensively studied by FRIEDT (1978, 1979) in barley. By means of X-rays, the cytogenetic and physiological behavior of the autotetraploids should be altered in the direction of a diploid-like behavior. In F_9 to F_{11} generation, selected lines showed an increased seed set as compared to the productivity of the original tetraploid material. Interestingly, also the course of meiosis was essentially more regular and the frequency of bivalents was higher. It was, however, not possible to detect any kind of preferential chromosome pairing. Moreover, only a weak correlation between meiotic behavior and seed production was found. It is therefore concluded that the meiotic improvement is not due to a diploidization effect, but to a genetic control of quadrivalent orientation. Similar studies were made in autotetraploid maize. Different pairing affinities between normal and restructured chromosomes have been observed due to naturally occurring chromosome aberrations in exotic races and in distinct inbred lines (DOYLE; 1979).

c) Experimentally Produced Allopolyploid Plants

The number of experimentally produced allopolyploids is rapidly increasing. I can only cite a few papers which may be of broader interest. The significance of amphiploids in the Triticeae in basic research and plant breeding was discussed by KIMBER (1979). The pairing of the homoeologous chromosomes of 4n *Briza media/elatior* amphidiploids is negatively influenced by the presence of B chromosomes. The effect is the stronger the higher the number of B chromosomes (MURRAY, 1978). Amphiploids from gooseberries and black currants are fully fertile and represent a prospective starting material for further breeding work (MURAWSKI, 1977). Allopolyploid plants from *Nicotiana goodspeedii* × *N. tabacum* (2n = 88) are partially sterile, but they were found to be highly resistant to the main tobacco diseases and to aphids (PALAKARCHEVA et al., 1978). RAMA VARMA et al. (1977) tried to unite the genomes of *Nicotiana benthamiana* and *glutinosa* with *tabacum*. The chromosome number of the respective allopolyploid plants ranged between 106 and 110. They were almost exclusively present as bivalents in the first pro- and metaphase. None of the plants was self-fertile. Plenty of work has been done during the past years in the synthetic genus *Triticale* with regard to both problems of basic research and utilization in plant breeding. I cannot give the detailed findings; a survey on relevant problems in *Triticale* was given by RAJHATHY (1977).

d) The Genomic Constitution of Natural Polyploid Species

Biochemical methods are being used to an increasing extent for investigating the phylogenetic relationships of plant species in general and the putative origin of allopolyploid species in particular. A summarizing paper on the methods used and the problems studied has been published by LADIZINSKY and HYMOWITZ (1979). In the genus *Brassica* and some other Cruciferae, the biochemical findings were found to be not in full concordance with the cytogenetic findings available. Results in this field have been obtained by UCHIMIYA and WILDMAN (1978) and by YADAVA et al. (1979) using protein and esterase isozyme patterns. The experimentally produced amphidiploid *Brassicoraphanus*, however, has esterase and peroxydase isozyme patterns which can in general be regarded as the summation of the bands of the two parental species (KATO and TOKUMASU, 1979). Electrophoretic analysis of the tuber proteins of some cultivated potatoes confirms the close relationships of the groups *Tuberosum* and *Andigena* as well as the groups *Phureja* and *Stenotomum* (RICKEMAN and DESBOROUGH, 1978). The relations between the genomes of diploid, tetra-, octo-, and decaploid races of *Agropyron elongatum* were studied by means of immunochemical analysis of the grain proteins (KONAREV, 1979).

Cytogenetic investigations were used for genome analysis in *Paspalum*. The tetraploid species *P. conspersum* has common genomes with the diploid species *P. dilatatum* and *jurgensii* (BURSON and BENNETT, 1976; BURSON, 1978). The composite *Emilia praetermissa* is an amphidiploid hybrid derived from *E. sonchifolia* and *coccinea* (OLORODE and OLOPADE, 1978). *Arachis hypogaea*, an allotetraploid species, has most likely derived from the interspecific hybrid *A. cardenasii* × *batizocoi* (SMARTT et al., 1978). *Lilium tigrinum* is an allotriploid species derived from *L. maximowiczii* and *L. tigrinum* var. *flaviflorum* (NODA, 1974). Karyotype analyses and investigations on the pairing behavior show that the tetraploid cytotypes of *Bromus inermis* are allotetraploids of the constitution AABB, whereas the octoploids are auto-allo-polyploids of the constitution AAAABBBB. The A genome could derive from *Bromus erectus*, the B genome from *B. arvensis* (ARMSTRONG, 1977, 1979).

The derivation of the tetra- and hexaploid wheats is not yet completely clear. According to YAMASHITA (1978), the origin of the tetraploid wheats may date back more than 7000 years, the region could be Egypt. Already 6000 to 8000 years ago, hexaploid wheats were cultivated in the area of Iraq and Afghanistan. The N-banded karyotypes of *Triticum* species were used by GERLACH (1977) for discerning their phylogenetic relationships. The banded chromosomes of *Triticum dicoccoides* (AABB) are very similar to corresponding chromosomes of *T. aestivum* (AABBDD). The diploid species *Triticum monococcum, boeoticum, urartu* and *Aegilops squarrosa*, however, do not show a banding which should be expected for the donors of the A and D genomes. *Aegilops speltoides*, on the other side, has a certain degree of banding expected for the donor of the B genome. Similar conclusions with regard to the B genome were drawn by CALDWELL and KASARDA (1978), using the seed albumin and globulin patterns. According to these patterns, *Triticum boeoticum* is most likely the donor of the A genome, a hypothesis which is widely accepted. Further biochemical studies in this field have been carried out by HOLM and FRÖST (1979) using the flavonoid patterns of 24 *Triticum* and 2 *Aegilops* species. The problems becomes even more difficult as *Triticum aestivum* has not a uniform karyotype. On the contrary, striking differences between the bread wheat varieties "Chinese Spring" and "Diamant" were found by means of the Giemsa banding technique (JORDANSKY et al., 1978). The genomic constitution of the hexaploid species *Aegilops crassa* ($M^{cr}M^{cr}$ DD D^2D^2 according to KIHARA, 1957) was studied by CHAPMAN and MILLER (1978). The two D genomes were found to be not homologous with each other and not completely homologous with the D genome of *Aegilops squarrosa*.

New findings on the fern *Ophioglossum reticulatum* – the plant species which has the highest chromosome numbers known so far – have been obtained by SINHA et al. (1979). By means of morphological differences, five subgroups of the species were found. The lowest chromosome number was 2n = 232 - 240, the highest varied between 1270 and 1292. Only 6 to 12 chromosomes were present as univalents in meiosis, whereas all the others formed bivalents.

e) Aneuploids

A large number of aneuploids was obtained in the offspring of 68 autotriploid barley plants ranging between 2n = 14 and 39 in their chromosome numbers. The findings demonstrate that the diploid species *Hordeum vulgare* has a high degree of tolerance against extra chromosomes (SANDFAER, 1979a).

The heterogeneous group of trisomics has been studied so intensively during the past years that I cannot give detailed information with regard to morphology, vitality, fertility, meiotic behavior, and transmission rates because of lack of space. They have been obtained by means of the conventional methods, i.e., by crosses of the types 2n x 3n, 2n x 4n, by selfing triploids or trisomics or by crossing trisomics with diploids. In soybean, male-sterile haploid plants were crossed with diploid ones giving rise to 4 trisomics in the F_1 generation (SORRELLS and BINGHAM, 1979). Moreover, peanut trisomics were found in the M_2 generation following acriflavine treatment (ASHRI et al., 1977).

Complete series of primary trisomics are available in *Pennisetum typhoides* (n = 7; MINOCHA et al., 1976), *Secale cereale* (n = 7; PILCH, 1978), *Trigonella corniculata* (n = 8; SINGH et al., 1978; SINGH and SINGH, 1980), and *Chorchorus olitorius* (n = 7; PARIA and BASAK, 1979a,b). Eleven of the 12 possible types of primary trisomics have been identified in dihaploids of *Solanum tuberosum* (WAGENVOORT and RAMANNA, 1979). Single trisomic types were isolated in *Pennisetum arundinaceae* (JAUHAR, 1978), *Vicia faba* (MARTIN, 1978) and *Lycopersicon peruvianum* (RAMULU et al., 1978). The identification of the extra chromosomes was carried out by pachytene analysis *(Lycopersicon, Solanum)*, by mitotic karyotype analysis *(Trigonella)* or by using the Giemsa C-banding technique (*Secale cereale*, PILCH, 1978; *Hordeum vulgare*, LINDE-LAURSEN, 1978). In *Triticum durum*, the trisomic lines were crossed with aneuploid lines of *Triticum aestivum* ditelocentric for the chromosomes of the A and B genomes. In the hybrids, the supernumerary chromosomes can be reliably discerned from the pairing patterns (BLANCO et al., 1977, 1978). The size of seeds produced in trisomics can be used for selecting trisomics in the offspring. The trisomic frequency was found to be very low from large and very high from small seeds (PILCH, 1979, rye; MINOCHA et al., 1976, pearl millet).

Primary trisomics were used for locating mutant genes in *Corchorus olitorius* (PARIA and BASAK, 1976), *Hordeum vulgare* (SHAHLA and TSUCHIYA, 1980) and *Lolium perenne* (LEWIS et al., 1980).

Double trisomics were isolated in *Vicia faba* (MARTIN, 1978) and *Solanum tuberosum* yielding less seeds than most of the single trisomics (WAGENVOORT and LANGE, 1980). In the wild tomato *Lycopersicon peruvianum*, about 40% of the plants arisen by 3n x 2n crosses were aneuploids. Eighteen percent of them proved to be single trisomics whereas the remaining ones had 2, 3 or 4 supernumerary chromosomes (SREE RAMULU et al., 1977).

Tertiary trisomics of *Pennisetum typhoides*, arisen in combination with reciprocal translocations, appeared in the offspring of selfed triploid plants, as well as in that of radiation-induced interchange heterozygotes (VENKATESWARLU and MANI, 1978). Moreover, a single plant of this structural type was selected in the progeny of a primary trisomic tomato. The translocated extra chromosome was identified in pachytene; the breakpoints could be localized (GILL, 1978). In low frequencies, plants of this group having a more complicated genomic constitution were studied in detail. This holds true for barley monotelotrisomics (SINGH and TSUCHIYA, 1977, LINDE-LAURSEN, 1978), for a ditelosomic plant in a tetraploid broad bean strain (MARTIN, 1978) and for ditelotetrasomics in barley (SINGH and TSUCHIYA, 1977).

Monosomics of polyploids represent important material for gene localization. In *Triticum aestivum*, they have been used for locating further resistance genes (BARTOŠ et al., 1977; KLEIJER et al., 1977; MEYER, 1977; ZELLER and OPPITZ, 1977; REITER and EVDOKIMOV, 1979), as well as genes for seed protein content and tyrosinase activity (BHAT and GOUD, 1978). Similar investigations were carried out in *Gossypium hirsutum* (KOHEL, 1978). The linkage relationships of two mutant genes of Upland cotton were studied by means of monotelodisomics, monoisodisomics, and disomics (ENDRIZZI and BRAY, 1980). Eighteen of the 26 chromosomes of *Gossypium hirsutum* have now been marked by monosomics and monotelodisomics (ENDRIZZI and RAMSAY, 1979, 1980). A whole set of monosomics is available in *Avena sativa* (LARIK, 1978). Two monosomics of *Phaseolus vulgaris* could be identified with regard to the lacking chromosome (MOK and MOK, 1977).

Nullisomics besides mono- and disomics were isolated in the progeny of selfed wheat monosomics (JOSHI et al., 1979). In the offspring of an oat monosomic, a nulli-haploid plant with n = 20 chromosomes was found. A double nulli-haploid plant with 19 chromosomes arose in the progeny of a monosomic-tetrasomic oats hybrid (LEGGETT, 1977).

f) Haploids and Polyhaploids

The production of haploid plants via anther cultures seems to be a promising tool which has been successfully used in a number of species.

Surveys have been given by COLLINS (1977) and by MAHESHWARI et al. (1980). New haploid material obtained by means of androgenesis is available in *Hordeum vulgare* (MIX et al., 1978), *Solanum verrucosum* (WEATHERHEAD and HENSHAW, 1979) and *Nicotiana sylvestris* (BUTTERFASS and KOHLENBACH, 1979). This also holds true for the production of amphihaploid plants of *Brassica napus*, which were obtained by anther culture techniques. Some of them developed spontaneously into diploids (F. HOFFMANN, 1980). Callus cultures of interspecific hybrids between *Hordeum vulgare* × *H. jubatum* (2n = 3x = 21), obtained from immature ovary tissue, gave rise to haploid regenerates in high frequency (ORTON, 1980). On the other hand, haploid plants of *Triticum aestivum* were used for anther culture, giving rise nôt only to haploid but also to diploid regenerates (DE BUYSER and HENRY, 1980).

In the progenies of EMS induced barley mutants, haploids were found in high frequencies, whereas spontaneous haploids are very rare in this species. They were obviously due to the action of a "factor for haploid initiation". The genotypic constitution of the tissue surrounding the embryo sacs seems to determine whether haploids arise or not (HAGBERG and HAGBERG, 1980). A similar case is known in bread wheat. A specific gene or a small group of genes, located on the 1B/1R chromosome, induces obviously haploid parthenogenesis gametophytically (KOBAYASHI and TSUNEWAKI, 1980). Spontaneous haploids of sugarbeet were meiotically analyzed by YU (1980). After application of colchicine, dihaploid barley plants with normal meiotic behavior arose (FINCH and BENNETT, 1979a). The proportion of surviving barley dihaploids can be considerably increased by means of an improved technique adding dimethyl sulfoxide, polyoxyethylene sorbitan monolaurate, and gibberellic acid to the colchicine solution (THIEBAUT and KASHA, 1978).

Polyhaploids are a very heterogeneous group depending on the genomic constitution of the respective parental plants. Interspecific crosses between *Hordeum brachyantherum* (2n = 4x = 28) x *H. bulbosum* (2n = 2x = 14) resulted not only in the triploid hybrids expected but also in dihaploid plants (2n = 14) derived from *H. brachyantherum*. Principally the same result was obtained by crossing *Hordeum depressum* (2n = 4x = 28) x *H. bulbosum*. The reduced chromosome number is obviously due to chromosome elimination during the earliest stages of embryonic development. Trihaploids arose by crossing the hexaploid species *Hordeum arizonicum* and *lechleri* with diploid lines of *Hordeum bulbosum* and *vulgare*. In these cases, the trihaploids derived from the hexaploid parent (SUBRAHMANYAM, 1979, 1980). Polyhaploids (= amphihaploids) of *Brassica napus* — an allotetraploid species — were used for producing homozygous diploid lines, some of which outyielded the original mother varieties with regard to oil production per hectare (THOMPSON, 1979).

Haploid and polyhaploid plants should be an ideal material in experimental mutagenesis. Surprisingly, only few results are available in this field. A survey has been given by KASHA (1976). Androgenetically arisen amphihaploids of *Brassica napus* were treated with mutagenic agents and subsequently diploidized. In this way, a large number of mutants, homozygous for recessive genes, were produced (F. HOFFMANN, 1980). Pollen mother cells of dihaploid potatoes, treated with mutagenic rays and chemicals, were used for anther cultures giving rise to seven mutants (PRZEWOŹNY et al., 1980). Haploid as well as homozygous diploid mutants were isolated in *Nicotiana sylvestris* after having treated callus tissue from haploid leaf veins with EMS and EI (MALEPSZY et al., 1977). Isolated haploid protoplasts of *Datura innoxia*, finally, gave rise to mutants after having applicated MNNG (KRUMBIEGEL, 1979).

4. Chromosome Mutations

Are being considered in Volume 45 of Progress of Botany.

References

ABRAHAM, S., KOSHY, M.P.: Cytologia 44, 221-225 (1979). - AMER, S.M., FARAH, O.R.: Cytologia 44, 907-913 (1979). - ARMSTRONG, K.C.: Z. Pflanzenzücht. 78, 244-252 (1977); - Can. J. Genet. Cytol. 21, 65-71 (1979). - ASHRI, A., OFFENBACH, R., CAHANER, A., LEVY, A.: Z. Pflanzenzücht. 79, 210-218 (1977).

BAJAJ, Y.P.S., SAINI, S.S., BIDANI, M.: Theor. Appl. Genet. 58, 17-18 (1980). - BARTOS, P., KOSNER, J., VALKOUN, J.: Genet. Slecht. 13, 131-136 (1977). - BASAK, S.L., PARIA, P.: Theor. Appl. Genet. 56, 253-256 (1980). - BHAT, S.R., GOUD, J.V.: Euphytica 27, 805-810 (1978). - BLANCO, A., ZITO, C., GIORGI, B.: Caryologia 30, 217-224 (1977). - BLANCO, A., SIMEONE, R., TANZARELLA, O.A., GIORGI, B.: Caryologia 31, 487-491 (1978). - BOROJEVIĆ, K.: Proc. 5th Int. Wheat Genet. Symp., New Delhi, 559-564 (1978); - Gamma Field Symposia, Vol. 18, 25-32. Ohmiya (Japan): Institute of Radiation Breeding, 1979. - BRIDGES, B.A., CLEMMESEN, J., SUGIMURA, T.: Mutation Res. 65, 71-81 (1979). - BRINKERHOFF, L.A., VERHALEN, L.M., MAMAGHANI, R., JOHNSON, W.M.: Crop Sci. 18, 901-903 (1978). - BROWN, J.P.: Mutation Res. 75, 243-277 (1980). - BROWN, J.P., DIETRICH, P.S.: Mutation Res. 66, 223-240 (1979). - BURSON, B.L.: Can. J. Genet. Cytol. 20, 365-372 (1978). - BURSON, B.L., BENNETT, H.W.: Can. J. Genet. Cytol. 18, 701-708 (1976). - BURTON, G.W.: Envir. Exp. Bot. 20, 115-117 (1980). - BUTTERFASS, T., KOHLENBACH, H.W.: Naturwissenschaften 66, 162 (1979). - BUYSER, J.DE, HENRY, Y.: Theor. Appl. Genet. 57, 57-58 (1980).

CALDWELL, K.A., KASARDA, D.D.: Theor. Appl. Genet. 52, 273-280 (1978). - CAMADRO, E.L., PELOQUIN, S.J.: Theor. Appl. Genet. 56, 11-15 (1980). - CHALOFF, R.S.: Theor. Appl. Genet. 58, 91-95 (1980). - CHAPMAN, V., MILLER, T.E.: Wheat Inform. Serv. 47-48, 17-20 (1978). - CHEN, C.H., GOEDEN-KALLEMEYN, Y.: Euphytica 28, 705-709 (1979). - COLIJN, C.M., KOOL, A.J., NIJKAMP, H.J.J.: Theor. Appl. Genet. 55, 101-106 (1979). - COLLINS, G.B.: Crop. Sci. 17, 583-586 (1977).

DELLAERT, L.M.W.: Mutation Res. 71, 109-126 (1980). - DIJKSTRA, H., SPECKMANN, G.J.: Euphytica 29, 89-96 (1980). - DIMITROV, B.: Mutation Res. 62, 267-277 (1979). - DONALDSON, E., NILAN, R.A., KONZAK, C.F.: Envir. Exp. Bot. 19, 153-164 (1979a); - Envir. Exp. Bot. 19, 165-173 (1979b). - DOYLE, G.G.: Theor. Appl. Genet. 54, 103-112 (1979).

EHRENBERG, L., HUSSAIN, S., NOOR SALEH, M., LUNDQVIST, U.: Hereditas 92, 127-130 (1980). - EL-BAYOUMI, A.S., KABARITY, A., HABIB, A.: Cytologia 44, 745-755 (1979). - ENDRIZZI, J.E., BRAY, R.: Genetics 94, 979-988 (1980). - ENDRIZZI, J.E., RAMSAY, G.: Can. J. Genet. Cytol. 21, 531-536 (1979); - J. Hered. 71, 45-48 (1980). - EUCARPIA: Genetika (Beograd) 8, 101-260 (1976).

FATOKUN, C.A., AKEN'OVA, M.E., CHHEDA, H.R.: J. Hered. 70, 270-271 (1979). - FINCH, R.A., BENNETT, M.D.: Can. J. Genet. Cytol. 21, 33-35 (1979a); - Heredity 43, 87-93 (1979b). - FOX, M., SCOTT, D.: Mutation Res. 75, 131-168 (1980). - FRIEDT, W.: Z. Pflanzenzücht. 81, 118-139 (1978); - Z. Pflanzenzücht. 82, 311-339 (1979).

GERLACH, W.L.: Chromosoma 62, 49-56 (1977). - GICHNER, T., VELEMÍNSKÝ, J.: Mutation Res. 60, 181-187 (1979a); - Mutation Res. 66, 135-142 (1979b). - GILL, B.S.: Caryologia 31, 257-269 (1978). - GLEBA, Y.Y., HOFFMANN, F.: Naturwissenschaften 66, 547-554 (1979). - GOLUBOVSKAYA, I.N.: Int. Rev. Cytol. 58, 247-286 (1979). - GOLUBOVSKAYA, I.N., SITNIKOVA, D.V.: Genetika USSR 16, 656-666 (1980). - GÓRNY, A.: Genet.

Pol. 19, 447-456 (1978). - GOTTSCHALK, W.: Genetika (Beograd) 11, 15-28 (1979). - GOTTSCHALK, W., KAUL, M.L.H.: Theor. Appl. Genet. 56, 71-79 (1980). - GOTTSCHALK, W., MÜLLER, H.P.: Seed Protein Improvement in Cereals and Legumes I, 259-272; IAEA Vienna (1979). - GROVER, I.S., TYAGI, P.S.: Can. J. Genet. Cytol. 22, 260 (1980).

HAGBERG, A., HAGBERG, G.: Hereditas 93, 341-343 (1980). - HARTLEY-ASP, B., ANDERSSON, H.C., STURELID, S., KIHLMAN, B.A.: Envir. Exp. Bot. 20, 119-129 (1980). - HOFFMANN, F.: Naturwissenschaften 67, 301-306 (1980). - HOFFMANN, G.R.: Mutation Res. 75, 63-129 (1980). - HOLM, G., FRÖST, S.: Hereditas 91, 295-296 (1979). - HOSSAIN, M., SEN, S.: Indian J. Genet. Plant Breed. 38, 179-181 (1978a); - Z. Pflanzenzücht. 81, 77-79 (1978b).

INSTITUTE OF RADIATION BREEDING: Gamma Field Symposia, Vol. 16, 109 pp. Ohmiya (Japan) 1977; - Gamma Field Symposia, Vol. 17, 91 pp. Ohmiya (Japan) 1978; - Gamma Field Symposia, Vol. 18, 154 pp. Ohmiya (Japan) 1979. - INTERNATIONAL ATOMIC ENERGY AGENCY: Seed Protein Improvement in Cereals and Grain Legumes I, II, 421 and 472 pp. Vienna 1979. - IWANAGA, M., PELOQUIN, S.J.: J. Hered. 70, 385-389 (1979).

JAUHAR, P.P.: J. Hered. 69, 217-223 (1978). - JENSEN, J.: Euphytica 28, 47-56 (1979). - JORDANSKY, A.B., ZURABISHVILI, T.B., BADAEV, N.S.: Theor. Appl. Genet. 51, 145-152 (1978). - JOSHI, B.C., SINGH, D., LAL, B., RAM, D.: Wheat Inform. Serv. 50, 12-13 (1979).

KABARITY, A., EL-BAYOUMI, A., HABIB, A.A.: Mutation Res. 66, 143-148 (1979). - KASHA, K.J.: Genetika (Beograd) 8, 101-110 (1976). - KATO, M., TOKUMASU, S.: Euphytica 28, 339-349 (1979). - KAUL, B.L.: Mutation Res. 67, 239-247 (1979). - KHALATKAR, A.S., KASHIKAR, S.G.: Mutation Res. 79, 81-85 (1980). - KIHARA, H.: Seiken Zihô 8, 3 (1957). - KIMBER, G.: Indian J. Genet. Plant Breed. 39, 133-137 (1979). - KLEIJER, G., BRÖNNIMANN, A., FOSSATI, A.: Z. Pflanzenzücht. 78, 170-173 (1977). - KLEINHOFS, A., OWAIS, W.W., NILAN, R.A.: Mutation Res. 55, 165-195 (1978). - KOBAYASHI, M., TSUNEWAKI, K.: J. Hered. 71, 9-14 (1980). - KOHEL, R.J.: J. Hered. 69, 275-276 (1978). - KONAREV, A.V.: Genetika USSR 15, 510-517 (1978). - KOPELIOVITCH, E., RABINOWITCH, H.D., MIZRAHI, Y., KEDAR, N.: Euphytica 28, 99-104 (1979). - KROON, G.H., VAN EIJK, J.P.: Euphytica 26, 63-66 (1977). - KRUMBIEGEL, G.: Environ. Exp. Bot. 19, 99-103 (1979).

LACEY, C.N.D., CAMPBELL, A.I.: Environ. Exp. Bot. 19, 145-152 (1979). - LADIZINSKY, G., HYMOWITZ, T.: Theor. Appl. Genet. 54, 145-151 (1979). - LARIK, A.S.: Cytologia 43, 695-704 (1978). - LEGG, P.D., LITTON, C.C., COLLINS, G.B.: Can. J. Genet. Cytol. 22, 21-26 (1980). - LEGGETT, J.M.: Can. J. Genet. Cytol. 19, 651-656 (1977). - LESLEY, J.W., LESLEY, M.M., SOOST, R.K.: J. Hered. 70, 103-108 (1979). - LEWIS, E.J., HUMPHREYS, M.W., CATON, M.P.: Theor. Appl. Genet. 57, 237-239 (1980). - LINDE-LAURSEN, I.: Hereditas 89, 37-41 (1978). - LINNAINMAA, K., SORSA, M., ILUS, T.: Hereditas 90, 151-156 (1979).

MAAN, S.S., SASAKUMA, T.: J. Hered. 68, 87-94 (1977). - MAHESHWARI, S.C., TYAGI, A.K., MALHOTRA, K.: Theor. Appl. Genet. 58, 193-206 (1980). - MAISURADZE, N.I.: Genetika USSR 15, 2195-2203 (1979). - MALEPSZY, S., GRUNEWALDT, J., MALUSZYNSKI, M.: Z. Pflanzenzücht. 79, 160-166 (1977). - MARTIN, A.: J. Hered. 69, 421-423 (1978). - MELCHERS, G., SACRISTAN, M.D., HOLDER, A.A.: Carlsberg Res. Commun. 43, 203-218 (1978). - MENDIBURU, A.O., PELOQUIN, S.J.: Euphytica 26, 573-583 (1977). - MEYER, H.: Arch. Züchtungsforsch. 7, 203-210 (1977). - MICKE, A.: Ber. Arbeitstagung Gumpenstein 1978, 239-265 (1978); - Gamma Field Symposia, Vol. 18, 1-23. Ohmiya (Japan): Institute of Radiation Breeding 1979. - MINOCHA, J.L., SHARMA, H.L., SIDHU, J.S., GILL, B.S.: Indian J. Genet.

Plant Breed. 36, 38-43 (1976). - MIX, G., WILSON, H.M., FOROUGHI-WEHR, B.: Z. Pflanzenzücht. 80, 89-99 (1978). - MOK, D.W.S., MOK, M.C.: Theor. Appl. Genet. 49, 145-149 (1977). - MURAWSKI, H.: Arch. Züchtungsforsch. 7, 299-304 (1977). - MURRAY, B.G.: Heredity 41, 227-231 (1978).

NAGAO, M., SUGIMURA, T.: Annu. Rev. Genet. 12, 117-159 (1978). - NAKAI, H., SAITO, M.: Euphytica 28, 697-704 (1979). - NICOLOFF, H., GECHEFF, K.I., STOILOV, L.: Mutation Res. 70, 193-201 (1980). - NIJS, DEN, T.P.M., PELOQUIN, S.J.: Euphytica 26, 585-600 (1977). - NODA, S.: The Lily Yearbook 27, 98-111 (1974).

OBE, G., BEEK, B.: Mutation Res. 65, 21-70 (1979). - OLORODE, O., OLOPADE, D.J.: Cytologia 43, 497-504 (1978). - ORTON, T.J.: J. Hered. 71, 280-282 (1980).

PALAKARCHEVA, M., EDREVA, A., CHOLAKOVA, N., NOVEVA, S.: Z. Pflanzenzücht. 80, 49-63 (1978). - PALMER, R.G., WINGER, C.L., ALBERTSEN, M.C.: Crop Sci. 18, 727-729 (1978). - PALMER, R.G., WINGER JOHNS, C., MUIR, P.S.: J. Hered. 71, 343-348 (1980). - PANDA, B.B., SHARMA, C.B.S.R.: Theor. Appl. Genet. 55, 253-255 (1979); - Mutation Res. 78, 341-345 (1980). - PARIA, P., BASAK, S.L.: Indian J. Genet. Plant Breed. 36, 289-291 (1976); - Indian J. Genet. Plant Breed. 39, 178-187 (1979a); - Indian J. Genet. Plant Breed. 39, 471-479 (1979b). - PASZTOR, K.: Acta Agron. Acad. Sci. Hung. 28, 452-458 (1979). - PILCH, J.: Genet. Pol. 19, 137-152 (1978); - Genet. Pol. 20, 25-29 (1979). - PRZEWÓZNY, T., SCHIEDER, O., WENZEL, G.: Theor. Appl. Genet. 58, 145-148 (1980).

RAJHATHY, T.: Can. J. Genet. Cytol. 19, 595-602 (1977). - RAMANNA, M.S.: Euphytica 28, 537-561 (1979). - RAMA VARMA, K.T., APPAROA, K., NARAYANAN, A.I.: Cytologia 42, 703-709 (1977). - RAMULU, K.S., DEVREUX, M., CARLUCCIO, F., DE NETTANCOURT, D.: Theor. Appl. Genet. 51, 271-275 (1978). - REDDY, T.A., VAIDYANATH, K.: Environ. Exp. Bot. 19, 27-32 (1979). - REITBERGER, A.: Z. Pflanzenzücht. 79, 14-25 (1977). - REITER, B.G., EDVOKIMOV, M.G.: Genetika USSR 15, 1067-1075 (1979). - RHEENEN, H.A. VAN, MUIGAI, S.G.S., KITIVO, D.K.: Euphytica 28, 761-763 (1979). - RICKEMAN, V.S., DESBOROUGH, S.L.: Theor. Appl. Genet. 52, 217-220 (1978). - ROBERTSON, D.S.: Mutation Res. 51, 21-28 (1978).

SALNIKOVA, T.V., DOSMAILOVA, O.I.: Genetika USSR 15, 1469-1479 (1979). - SANDFAER, J.: Hereditas 90, 213-217 (1979a); - Phytopath. Z. 95, 97-104 (1979b). - SARMA, N.P., PATNAIK, A., JACHUCK, P.J.: Environ. Exp. Bot. 19, 117-121 (1979). - SCHIAVO, F.L., RONCHI, V.N., TERZI, M.: Theor. Appl. Genet. 58, 43-47 (1980). - SCHVARTZMAN, J.B., HERNANDEZ, P.: Theor. Appl. Genet. 57, 221-224 (1980). - SEILER, J.P.: Mutation Res. 55, 197-226 (1978). - SEN, L.A.: Genetika USSR 15, 470-477 (1979). - SHAHLA, A., TSUCHIYA, T.: J. Hered. 71, 359-361 (1980). - SHARMA, S.K., SHARMA, B.: Indian J. Genet. Plant Breed. 39, 516-520 (1979). - SINGH, B.D., SINGH, R.B., SINGH, R.M., SINGH, Y., SINGH, J.: Environ. Exp. Bot. 19, 127-132 (1979). - SINGH, C.: Z.Pflanzenzücht. 82, 278-280 (1979). - SINGH, D., SINGH, A.: Caryologia 33, 33-39 (1980). - SINGH, D., SINGH, A., ROY, R.P.: Caryologia 31, 335-361 (1978). - SINGH, R.J., TSUCHIYA, T.: Z. Pflanzenzücht. 78, 327-340 (1977). - SINHA, B.M.B., SRIVASTAVA, D.P., JAYAKAR, JHA: Caryologia 32, 135-146 (1979). - SMARTT, J., GREGORY, W.C., PFLUGE-GREGORY, M.: Euphytica 27, 665-675 (1978). - SORRELLS, M.E., BINGHAM, E.T.: Can. J. Genet. Cytol. 21, 449-455 (1979). - SREE RAMULU, K., CARLUCCIO, F., DE NETTANCOURT, D., DEVREUX, M.: Theor. Appl. Genet. 50, 105-119 (1977). - STELLY, D.M., PALMER, R.G.: Euphytica 29, 295-303 (1980). - SUBRAHMANYAM, N.C.: Theor. Appl. Genet. 55, 139-144 (1979); - Theor. Appl. Genet. 56, 257-263 (1980). - SUBRAMANYAM, S., MURTHY, D.K., LAXMINARAYANA, D.: Egypt. J. Genet. Cytol. 7, 324-330 (1978).

TARASENKO, N.D.: Z. Pflanzenzücht. 79, 79-81 (1977). - THIEBAUT, J., KASHA, K.J.: Can. J. Genet. Cytol. 20, 513-521 (1978). - THOMAS, E., KING, P.J., POTRYKUS, I.: Z. Pflanzenzücht. 82, 1-30 (1979). - THOMPSON, K.F.: Euphytica 28, 127-135 (1979). - TOSK, J., SCHMELTZ, I., HOFFMANN, D.: Mutation Res. 66, 247-252 (1979). - TURCOTTE, E.L., FEASTER, C.V.: J. Hered. 71, 134-135 (1980).

UCHIMIYA, H., WILDMAN, S.G.: J. Hered. 69, 299-303 (1978). - UKAI, Y., YAMASHITA, A.: Japan. J. Breed. 30, 125-130 (1980a); - Japan. J. Genet. 55, 225-234 (1980b).

VENKATESWARLU, J., MANI, J.N.R.: Genetica 48, 145-149 (1978). - VORSA, N., BINGHAM, E.T.: Can. J. Genet. Cytol. 21, 525-530 (1979).

WAGENVOORT, M., LANGE, W.: Euphytica 29, 281-293 (1980). - WAGENVOORT, M., RAMANNA, M.S.: Euphytica 28, 633-642 (1979). - WEATHERHEAD, M.A., HENSHAW, G.G.: Euphytica 28, 765-768 (1979). - WEBER, G., LARK, K.G.: Theor. Appl. Genet. 55, 81-86 (1979).

YADAVA, J.S., CHOWDHURY, J.B., KAKAR, S.N., NAINAWATEE, H.S.: Theor. Appl. Genet. 54, 89-91 (1979). - YAMAGUCHI, I., YAMASHITA, A.: Technical News 20. Ohmiya-machi (Japan): Institute of Radiation Breeding 1979. - YAMASHITA, K.: Wheat Inform. Serv. 47,48, 1-6 (1978). - YU, M.H.: Can. J. Genet. Cytol. 22, 375-380 (1980).

ZELLER, F.J., OPPITZ, K.: Z. Pflanzenzücht. 78, 79-82 (1977).

Professor Dr. WERNER GOTTSCHALK
Institut für Genetik
der Universität Bonn
Kirschallee 1
D 5300 Bonn

IV. Function of Genetic Material

Genetics of Storage Proteins and Gene-Enzymes Relationships in Higher Plants

R. Blaich

Our current knowledge of the function of genetic material has its origin in genetics and enzymology of prokaryotic organisms and fungi. The function of eukaryotic genes has been studied mainly in animals. Compared to the results in these fields the work on higher plants is still at its beginning. Numerous studies on flowering plants are to a great extent descriptive rather than functional, adding further examples to well-known facts, without allowing a deeper insight in gene function on a molecular level. Though descriptive work may be a valuable basis for advanced studies, the following review will consider only those results which consider both genes and the consequences of their action. Well-known gene products, whose genetic basis is also fairly well analyzed, are the storage proteins of some cereals and enzymes. These topics are discussed here in some detail. Literature cited is restricted to recent papers, which should be consulted for earlier work and further information. Subjects discussed earlier in this series are not considered, or only some important new results are mentioned, as in Section 3 (gene expression).

Literature on the function of fungal genes may be found in ZIMMERMANN (1975, 1978, 1980). The molecular structure of the eukaryotic genome is discussed by HERZFELD (1976), HERZFELD and KIPER (1979) and information on nucleic acid synthesis and related processes is given by BINDING (1977).

1. Genetics of Storage Proteins

One of the goals of breeding is to improve the biological value of storage substances. The clarification of synthesis and regulation mechanisms would allow the genetic manipulation of storage systems. The genetics of zein proteins in maize will serve here as an example of the current knowledge, though some work with other storage systems has also been done.

As a basis for genetic engineering MOZER (1980) isolated α-amylase mRNA from barley aleurone and BRANDT (1979) was successful in cloning double-stranded DNA coding for hordein polypeptides. BOYER et al. (1980) analyzed the action of the *ae* locus in maize by gene dosage techniques in the endosperm. The mutation causes an increasing chain length of starch molecules, shifting the balance between amylopectin and amylose, and is therefore called amylose extender.

a) The Zein-Biosynthesis in Maize

Zein, the prolamin fraction of maize, constitutes 50% of the total protein in a kernel. It is localized in membrane-bound structures of the endosperm, called protein bodies. The high concentration of intermediates of the protein synthesis during the accumulation of storage proteins allows the isolation of zein-mRNA in the form of polyribosomes from the surfaces of the protein bodies. WIENAND and FEIX (1978) separated two types of zein-mRNA, coding for the two main components of zein (Z1 and Z2 with molecular weights of 22,000 and 19,000, respectively). Studies of MELCHER (1980) corroborate these results. Transcription of the mRNA by reverse transcriptase into labeled complementary DNA (cDNA) yielded two components with distinct hybridization kinetics.

LARKINS et al. (1979) demonstrated that maize storage proteins are synthesized very efficiently in oocytes of the frog *Xenopus laevis* injected with zein-mRNA.

The synthesis continued for several days and the zein sometimes amounted to 10% of the total protein synthesized in the oocyte. In cell-free systems of wheat germ or reticulocytes only the so-called pre-zeins are formed, which have a molecular weight 2000 larger than the zein proteins synthesized both by the oocytes and by the maize endosperm.

Amino terminal sequence analysis eventually confirmed that *the oocyte system is able to process pre-zein polypeptides* into native zeins. This fact again points to the universality of genetic mechanisms of plants and animals.

Processing of enzymes is also described by RIEZMAN et al. (1980) in cucumber. In this case the malate-dehydrogenase isolated from glyoxysomes is smaller than the in vitro product of a wheat germ system programmed with cotyledonary RNA from cucumber seedlings.

b) Structural Genes Coding for Zein

A general impression on the structure of the zein system is given by VIOTTI et al. (1979), whose studies suggest a tenfold reiteration of each mRNA sequence coding for zein polypeptides. The total number of zein genes per haploid maize genome is thus estimated to be about 120 copies. According to PEDERSEN et al. (1980) the zein genes are, however, present only in 1 to 5 copies per gene, or a maximum of 30 copies per genome.

By electrophoretic methods the two major components of the zein complex (Z1 and Z2) are separated into numerous compounds |for literature see SOAVE et al. (1978); PARK et al. (1980)|, and it is not clear how many of them correspond to structural genes.

LARKINS et al. (1979) submitted the product of his artificial zein synthesis system to cyanogen cleavage and demonstrated the real existence of several distinct proteins by analysis of the cleavage products.

Although the charge heterogeneity of zeins does not result from the presence of a large number of totally different mRNA, a number of at least six loci seem to code for these storage proteins (PARK et al., 1980).

SOAVE et al. (1978) reported five structural loci (*Zp1*, *Zp2*, *Zp3*, *Zp6*, *Zp12*) showing normal Mendelian segregation. *Zp1*, *Zp2* and *Zp3* are closely linked and located near the *opaque-2* locus on chromosome 7.

Studies of VALENTINI et al. (1979) using translocation techniques with trisomic genotypes, located other factors near the R-locus on chromosome 10, on the short arms of chromosome 9 and 4, and on the long arm of chromosome 4.

VIOTTI et al. (1979) used in situ hybridization of nucleic acids to locate zein genes. Microsporic chromosomes of the KYS-strain (with specific heterozygous interchanges and paracentric inversions) were labeled with |^{125}I|-zein-mRNA and |^{3}H|-cDNA. Hybridization sites were found on diplotene-metaphase chromosomes on the long arm of chromosome 4, 5, and 10, and on the short arm of chromosome 7.

Since different methods yield similar results, the number of structural genes for zeins may well be estimated at five or six.

c) Genes Influencing Zein-Synthesis

A number of genes have been described which modify the zein accumulation in the endosperm and which therefore were called regulatory genes. However, the underlying mechanisms are analyzed in only a few cases.

JONES (1978) studied the function of the *floury-2* locus. He introduced an increasing number (zero through three) of *floury-2* alleles into the endosperm and obtained a stepwise decrease of zein accumulation. SDS-polyacrylamide gel electrophoresis revealed a proportionate reduction of the two major zein components, whereas the relative proportions of the minor ones were altered. These changes were accompanied by a decreasing recovery of membrane-bound polysome material, the amount of which was correlated again with reductions on zein accumulation and also in pre-zein synthesis by in vitro systems.

The *opaque*-loci are also considered to be regulatory genes because — according to spectra obtained by isoelectric focusing — they reduce zein storage without influenc.ng the relative quantities of the components (SOAVE et al., 1978).

Results of PEDERSEN et al. (1980) indicate, however, that the mRNA for the 22,000 m.w. components has a reduced sequence complexity in *opaque-2* endosperms, whereas LARKINS et al. (1979) reported that *Xenopus* oocytes injected with mRNA from this mutant did not produce the Z2 component at all, while oocytes injected with wild-type RNA did (for more details see Sect. 1a).

A combination of *opaque-2* with the starch-deficient mutation *brittle-2* leads to an almost complete inhibition of the synthesis of Z1 and Z2 components in the double mutant. This effect seems to be caused by a high RNase activity (TSAI et al., 1978).

2. Regulation of Gene Expression

a) Gene Number Estimates

Since this subject and the restrictions of estimation methods used has been discussed in detail by HERZFELD and KIPER (1979) in Vol. 41, Progr. Bot., only one new paper will be mentioned here. KAMALAY and GOLDBERG (1980) tried to measure the extent to which structural gene expression is regulated in an entire tobacco plant.

Leaves, root, stem, petals, anthers, and ovaries were analyzed separately by excess RNS/single-copy DNA hybridization. At least 11% of the single-copy DNA or 60,000 diverse structural genes are expressed in the entire plant during its generative life cycle. This amount of genetic information constitutes only about 5% of the whole genome (these estimates lie well beyond earlier reports). Approximately 25,000 average-sized mRNA's are present in the polysomes of all organs. Each organ has at least 6000 diverse specific gene transcripts which are not present in other organs. Hybridization experiments with nuclear RNA indicated, however, that structural genes not utilized in a given organ are nevertheless transcribed.

Thus, post-transcriptional selection mechanisms seem to play an important role in the regulation of plant gene expression.

b) Transcription

The assumption that a tripartite transcription system is present in all eukaryotes was further corroborated by HAHN and SERVOS (1980), who found and characterized all three classes of DNA-dependent RNA-polymerases in quiescent wheat embryos and by HAHN and SIEMES (1978), who isolated these enzymes from *Zea mays* seedlings.

New results support the current theory on the prokaryotic origin of some cell organelles. McINTOSH et al. (1980) analyzed the gene sequence for the large subunit of the ribulose phosphate carboxylase of maize chloroplasts. There is a typical prokaryotic ribosome-binding site close to the site at which translation is initiated. The gene, however, exhibits no prokaryotic promoter sequence.

Premature transcription in pea seedlings is probably controlled by a transcription-inhibiting factor. WIELGAT et al. (1979) demonstrated that chromating from 5-day-old seedlings exhibits no template activity for chromatin-bound RNA polymerase and inhibited transcription of *E. coli* polymerase with both calf thymus DNA and pea DNA as a template.
The inhibitor seems to be an enzyme of the nonhistone protein fraction with UTPase and ATPase activity.

A direct photoregulation of transcription is described by BEDBROOK et al. (1978) from maize chloroplasts. Developing plastids and mature chloroplasts, but not etioplasts from dark-grown seedlings, contain mRNA for a 34,000 m.w. polypeptide. This mRNA constitutes the major nonribosomal RNA of developing plastids and was mapped by hybridization on the maize plastid chromosome. It is translated in a rabbit reticulocyte system to the same polypeptide as was isolated from the plastids.

c) Double-Stranded RNA in a Higher Plant

IKEGAMI and FRAENKEL-CONRAT (1979) characterized double-stranded RNA in tobacco leaves. It seems to be the product of the RNA-dependent RNA-polymerase recently described from this and other plants. This new type of nucleic acid seems to be an intermediate of transcription and/or has a regulatory role.

3. Genetics and Enzymes

The first step to an understanding of the function of genetic material was the "one-gene-one-enzyme hypothesis". In contrast to microorganisms in higher plants it is more difficult to establish this relationship definitely. At present the main emphasis of research lies on the clarification of the genetic basis of isoenzymes.

ROOSE and GOTTLIEB (1980) investigated the ADH-1 alcohol dehydrogenase isoenzymes of *Stephanomeria exigua* (Compositae). They made test crosses between 59 homo- and heterozygous plants with different ADH-equipment. ADH-1 is dimeric and forms homo- and heterodimers. They are *specified by a composed locus* consisting of one to several tightly linked structural genes which are expressed coordinately and show no recombination among more than 500 test-cross progeny.

According to TANSKLEY (1980) a number of phosphoglucose isomerase bands (B through D in the zymogram) are *controlled by a single gene PG-1*, though they exhibit changes during seed germination.

GOODMAN and STUBER (1980) and at the same time NEWTON and SCHWARZ (1980) studied the genetic basis of the major malate dehydrogenase (MDH) isoenzymes in maize. The mitochondrial enzymes are *encoded by three independently inherited nuclear genes. Mdh1, Mdh2 and Mdh3* , which could be located exactly on the chromosomes 8, 6 and 3 respectively. In addition a modifier locus *(mmn)* was mapped on chromosome 1.

The soluble MDH-isoenzymes are *coded by two unlinked genes, Mdh4* on chromosome 1 and *Mdh5* on chromosome 5.

All Mdh-loci exhibit multiple alleles, but the *formation of intergenic and interallelic heterodimers is restricted to gene products that occupy the same subcellular compartment.*

SCANDALIOS et al. (1980a,b) found a third locus for a catalase in maize and a regulatory gene near the structural locus for catalase 2. Catalase 3 was characterized and its expression could be shown to be tissue- and time-dependent.

Two seed proteins of soybean, coded for by the locus *Sp1* and hitherto known in electrophoresis as Sp1a and Sp1b-protein bands were identified as β-amylase isoenzymes by HILDEBRAND and HYMOWITH (1980). BEDBROOK et al. (1979) mapped the gene for the large subunit of ribulosephosphate carboxylase on the maize chloroplast chromosome and BIRCHLER (1980) localized the alcohol-dehydrogenase-1 locus in maize on the long arm of chromosome 1.

These examples show that no universal genetic basis exists for isoenzymes, but that their behavior reflects the multiplicity of their evolutionary origin. Evidently changes in isoenzymic banding pattern are not necessarily due to a change in structural gene expression; there exists epigenetic formation of different enzymes from one locus, as well as separate structural loci for all or for part of the bands of an isoenzyme pattern.

References

BEDBROOK, J.R., LINK, G., COEN, D.M., BOGORAD, L., RICH, A.: Proc. Natl. Acad. Sci. USA 75, 3060-3064 (1978). - BEDBROOK, J.R., COEN, D.M., BEATON, A.R., BOGORAD, L., RICH, A.: J. Biol. Chem. 254, 905-910 (1979). - BINDING, H.: Progr. Bot. 39, 173-181 (1977). - BIRCHLER, J.A.: Genetics 94, 687-700 (1980). - BRANDT, A.: Carlsberg Res. Commun. 44, 255-268 (1979). - BOYER, C.D., DAMEWOOD, P.A., MATTERS, G.L.: Starch Staerke 32, 217-222 (1980).

GOODMAN, M.M., STUBER, C.W., LEE, C.-N., JOHNSON, F.M.: Genetics 94, 153-168 (1980).

HAHN, H., SERVOS, D.: Z. Pflanzenphysiol. 97, 43-58 (1980). - HAHN, H., SIEMES, L.: Ber. Dtsch. Bot. Ges. 91, 361-368 (1978). - HERZFELD, F.: Progr. Bot. 38, 196-204 (1976). - HERZFELD, F., KIPER, M.: Progr. Bot. 41, 198-211 (1979). - HILDEBRAND, D.F., HYMOWITZ, T.: Crop. Sci. 20, 165-168 (1980).

IKEGAMI, M., FRAENKEL-CONRAT, H.: Proc. Natl. Acad. Sci. USA 76, 3637-3640 (1979).

JONES, R.: Biochem. Genet. 16, 27-38 (1978).

KAMALAY, J., GOLDBERG, R.B.: Cell 19, 935-946 (1980).

LARKINS, B.A., PEDERSEN, K., HANDA, A.K., HURKMAN, W.J., SMITH, D.L.: Proc. Natl. Acad. Sci. USA 76, 6448-6452 (1979).

McINTOSH, L., POULSEN, C., BOGORAD, L.: Nature 288, 556-560 (1980). - MELCHER, U.: Plant Sci. Lett. 18, 133-142 (1980). - MOZER, T.: Plant Physiol. 65, 834-837 (1980).

NEWTON, K., SCHWARTZ, D.: Genetics 95, 425-442 (1980).

PARK, W.D., LEWIS, E.D., RUBENSTEIN, I.: Plant Physiol. 65, 98-106 (1980). - PEDERSEN, K., BLOOM, K.S., ANDERSEN, J.N., GLOVER, D.V., LARKINS, B.A.: Biochemistry 19, 1644-1650 (1980).

RIEZMAN, H., WEIR, E.M., LEAVER, C.J., TITUS, D.E., BECKER, W.M.: Plant Physiol. 65, 40-46 (1980). - ROOSE, M.L., GOTTLIEB, L.D.: Genetics 95, 171-186 (1980).

SCANDALIOS, J.G., TONG, W.-F., ROUPAKIAS, D.G.: Mol. Gen. Genet. 179, 33-42 (1980a). - SCANDALIOS, J.G., CHANG, D.Y., McMILLIN, D.E., TSAFTARIS, A., MOLL, R.H.: Proc. Natl. Acad. Sci. USA 77, 5360-5364 (1980b). - SOAVE, C., SUMAN, N., VIOTTI, A., SALAMINI, F.: Theor. Appl. Genet. 52, 263-268 (1978). - SULLIVAN, D., BRISSON, N., VERMA, D.P.D.: Gene (Amst.) 7, 317-334 (1979).

TANSKLEY, S.D.: Can. J. Genet. Cytol. 22, 271-278 (1980). - TSAI, C.Y., LARKINS, B.A., GLOVER, D.V.: Biochem. Genet. 16, 883-896 (1978).

VALENTINI, G., SOAVE, C., OTTAVIANO, E.: Heredity 42, 33-40 (1979). - VIOTTI, A., SALA, E., MAROTTA, R., ALBERI, P., BALDUCCI, C., SOAVE, C.: Eur. J. Biochem. 102, 211-222 (1979). - VIOTTI, A., POGNA, N.E., BALDUCCI, C., DURANTE, M.: Mol. Gen. Genet. 178 (1980).

WIELGAT, B., JANKOWSKI, J.M., KLECZKOWSKI, K.: Acta Biochim. Pol. 26, 423-430 (1979). - WIENAND, U., FEIX, G.: Eur. J. Biochem. 92, 605-612 (1978).

ZIMMERMANN, F.K.: Progr. Bot. 37, 247-258 (19785); - Progr. Bot. 40, 247-260 (1978); - Progr. Bot. 42, 197-213 (1980).

Priv.Doz. Dr. R. BLAICH
Bundesforschungsanstalt für
Rebenzüchtung
Geilweilerhof
D 6741 Siebeldingen b.Landau/Pfalz

V. Extranuclear Inheritance

Plastid Genetics

By RUDOLF HAGEMANN and THOMAS BÖRNER

1. Introduction

The chapters on plastid inheritance of the preceding volumes of Progress in Botany dealt with evolutionary and morphological aspects (ARNOLD, 1977; ARNOLD and GAFFAL, 1979). In this volume we review papers concerning the very intensely developing field of research on *transcription and organization of chloroplast DNA* and concerning the related problems of chloroplast DNA *replication and repair*.

Other aspects of plastid genetics and biogenesis have been reviewed recently (chloroplast protein synthesis: BÖRNER, 1980; ELLIS, 1980; HAGEMANN and METZLAFF, 1980; nucleo-cytoplasmic and chloroplast interactions: HAGEMANN 1980; structure of chloroplast DNA: TEWARI, 1979; TEWARI and MEEKER, 1979). In addition, the reader is referred to the reviews contained in two recently published volumes (REINERT, 1980; SCHWEMMLER and SCHENK, 1980).

2. The Replication of Plastid DNA Molecules

The *plastid DNA (chloroplast DNA, ptDNA) molecule is a circular, covalently closed molecule*, isolated as a supercoiled structure or as an open circle.

The contour lengths and molecular weights of ctDNA are in angiosperms and ferns 40 - 45 µm and 80 - 90 • 10^6 (GC% 37 - 39), in *Chlamydomonas reinhardii* 62 µm and 128 • 10^6 (GC% 37); in some algae (e.g., *Vaucheria*) and mosses (e.g., *Sphaerocarpos*) the values are smaller than in higher plants. *Euglena gracilis* ptDNA has 44 µm, 92 • 10^6 Dal, but a distinctly different GC% of only 28 (KIRK and TILNEY-BASSETT, 1978; HERRMANN and POSSINGHAM, 1980; HERRMANN et al., 1980).

The *plastid DNA molecule represents one replicon*. The replication is semiconservative. It starts at unique origins on opposite strands. In *Pisum sativum* and *Zea mays* the ptDNA contains two displacement (D) loops at two adjacent sites of opposite strands.

In *Pisum* the smallest size of the D-loops (so far found) is 820 base pairs and the average distance between the outside edges of the two D-loops is 7.15 kilobase pairs. In *Zea* the size of the D-loop corresponds to 860 base pairs and the outer distance between the two D-loops is 7.06 kilobase pairs (KOLODNER and TEWARI, 1975; TEWARI and MEEKER, 1979; TEWARI, 1979).

The replication proceeds by expansion of the D-loops bidirectionally to complete a Cairns round of replication (TEWARI, 1979).

Both in *Pisum* and *Zea* circular ptDNA molecules have been found with tails that were longer than the attached monomer length circular molecule. This finding eliminates the possibility that these tails arose by break-

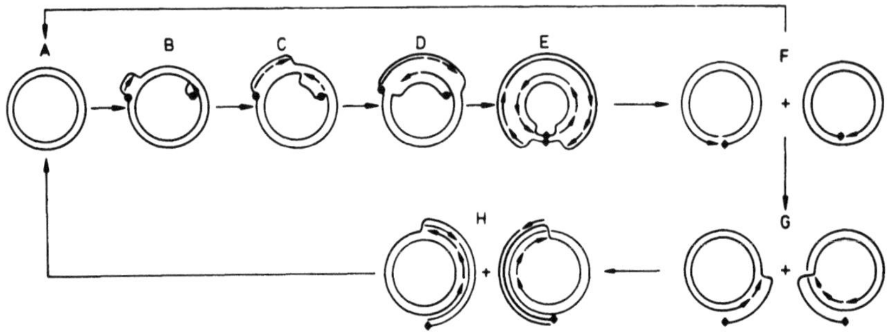

Fig. 1. A model for the replication of plastid DNA. *A* closed circular parental molecule; *B* molecule containing D loops; *C* molecule containing expanded D loops; *D*, *E* Cairns type of replicative intermediate; *F* nicked progeny molecules; *G*, *H* rolling circles.
The *thin* and *thick lines* mark the opposite strands of a molecule. The *lines with the arrows* are the daughter strands. The ● indicates the positions of the two origins of D loop synthesis. The ◆ indicates the terminus of the Cairns round of replication. (KOLODNER and TEWARI, 1975b; TEWARI and MEEKER, 1979)

age of a Cairns forked structure at a replicative fork. It rather indicates that – in addition to the *Cairns type of replication* – a *rolling circle replication* takes place (KOLODNER and TEWARI, 1975b; TEWARI, 1979).

These findings led KOLODNER and TEWARI (1975b) to a *model for the replication of plastid DNA* (Fig. 1): As a first step a full Cairns round of replication of ptDNA takes place, which may be followed, as a second step, by rolling circle replication.

Studies on ptDNA synthesis in vitro have been started to identify the proteins necessary for the replication of ptDNA (TEWARI and MEEKER, 1979; TEWARI, 1979).

Studies on ptDNA synthesis in isolated plastids have so far led only to few insights. There is incorporation of externally added precursor nucleosides. But it is so far not known whether this incorporation is due to replication, repair, or terminal addition of nucleotides to ptDNA (cf. review of HERRMANN and POSSINGHAM, 1980).

The replication of ptDNA is performed by the *ptDNA polymerase*, which is *coded for by nuclear DNA*, transcribed, and synthesized in the nucleo-cytoplasmic compartment and afterwards transported into the plastids. This was shown by the use of inhibitors of transcription and translation (cf. BÖRNER, 1973; BEDBROOK and KOLODNER, 1979). The same result was obtained from our studies using plastid-ribosome-deficient mutants of barley and *Pelargonium*; these plastome mutants with a blocked plastidal protein-synthesizing system nevertheless contain ptDNA and regularly replicate it (KNOTH et al., 1974; HAGEMANN and BÖRNER, 1978). The plastidal ptDNA polymerase seems to be tightly bound to plastidal membranes (HOWELL and WALKER, 1972).

In *Euglena gracilis* it was found that in a cell the ratio of ptDNA to nuclear DNA is always constant; nevertheless every time the nuclear DNA is replicated once, the ptDNA is replicated 1.5 times. Therefore a regular degradation of ptDNA has to be assumed. Indications have been found for the involvement of specific nucleases which are red-light-induced and degrade ptDNA molecules and thus keep the regular quantitative relation between the amount of nuclear and ptDNA constant (LYMAN and SRINIVAS, 1978; SRINIVAS and LYMAN, 1980).

Replication of ptDNA is in its time course independent of the replication of nuclear DNA. In some cases nuclear and ptDNA replicate at the same stage of the cell cycle (*Euglena:* RICHARDS and MANNING, 1975), in other cases ptDNA replicates at a quite different time (*Chlamydomonas:* CHIANG and SUEOKA, 1967; LEMIEUX et al., 1980).

3. The Number of Plastid DNA Molecules per Plastid

A plastid (proplastid, leucoplast, chloroplast, chromoplast) *contains more than one ptDNA molecule.* The ptDNA molecules within the plastid are aggregated and form ptDNA regions (= plastid nucleoids) consisting of many ptDNA molecules.

The *number of DNA regions* per plastid characterized in electron microscopical investigations is due to significant variations. In *Beta vulgaris* their number varies between 10 and 100. The same order of magnitude was found also for other species. These differences are dependent upon the nuclear-controlled general pattern of cell differentiation and the state of development and differentiation of the cells and tissues. Small plastids in meristematic cells contain only a few DNA regions, whereas large chloroplasts, for example of the spongy parenchyma, have many more DNA regions, which have been studied in detail in electron microscopic studies using serial sections (cf. HERRMANN et al., 1974; BUTTERFASS, 1979).

For some time it seemed as if there is a close correlation between the number of DNA regions, identified in electron microscopial studies, and the *number of ptDNA molecules* per plastid. Several recent publications call this correlation into question.

In *Pisum sativum* it was found that during greening and expansion of the first true foliage leaves the average number of chloroplasts per cell increased from 24 ± 10 to 64 ± 20 (without a parallel increase of the amount of nuclear DNA). During this increase of chloroplast number per cell there was a decrease of the number of ptDNA copies per chloroplast from 272 to 102; that is a 2.7-fold dilution of the number of ptDNA molecules per plastid during leaf expansion (LAMPPA et al., 1980). Similar results have been reported for expanding spinach leaves (SCOTT and POSSINGHAM, 1980). Cells from the basal half of 2 cm leaves contained 20 chloroplasts, whereas cells from larger leaves contained 130 - 170 chloroplasts. However, the small chloroplasts in the base of the 2 cm leaves contained about 200 ptDNA copies (= 32×10^{-15} g DNA), while the chloroplasts in the large leaves had only 30 ptDNA copies. Staining with the DNA fluorochrome DAPI allowed the detection of 10 - 15 plastid nucleoid areas (= ptDNA regions) in chloroplasts of larger leaves, suggesting the presence of 2 - 3 ptDNA molecules per plastid nucleoid (SCOTT and POSSINGHAM, 1980). The number of ptDNA copies per plastid nucleoids of chloroplasts of smaller leaves is obviously much higher.

The number of ptDNA copies per plastid can also be changed under the influence of the plastome. In *Hordeum vulgare* plastome mutations, induced by the gene mutation *albostrians*, lead to a plastid ribosome deficiency and thus to a block of the plastidal protein-synthesizing system. Autoradiographic studies revealed striking differences in $|^3H|$-thymidine incorporation in ptDNA between the mutant plastids and normal chloroplasts; the plastome mutation seems to be connected with a marked reduction of ptDNA (KRAHNERT cit. in HAGEMANN and BÖRNER, 1978).

4. Repair of Damage of Plastid DNA

Treatment with *physical and chemical agents* does not only *cause damages* in the nuclear DNA, but also *in the plastid DNA* (and in the mitochondrial DNA) of the cell. Intense studies on repair of ptDNA have only been performed in *Euglena gracilis* and *Chlamydomonas reinhardii*.

After UV-irradiation of *Euglena* cells there is a very *efficient photoreactivation* both at the plastidal and the nuclear level (SCHIFF and EPSTEIN, 1966; NICOLAS et al., 1980). SCHIFF and EPSTEIN (1966) and HILL et al. (1966) analyzed the action spectrum for photoreactivation of plastid DNA in *Euglena* and found a great similarity with the curve for the photoreactivation of viability of *Escherichia coli* as well as for the phage T4 in *E. coli*. The photoreactivating enzyme of *Euglena* has been characterized by DIAMOND et al. (1975). The photoreactivating enzyme of *Euglena* is most probably coded for by nuclear DNA.

NICOLAS et al. (1980) have confirmed the efficiency of photoreactivation of plastid DNA damage in *Euglena*. They could demonstrate dark repair only for nuclear DNA, although they could not exclude plastidal dark repair of a very limited extent.

In *Chlamydomonas reinhardii*, too, efficient photoreactivation of plastid DNA damage after UV-irradiation has been proved; moreover some indications for a plastidal dark repair system were found (SMALL and GREINMANN, 1977).

5. Transcription of Plastid DNA and Its Regulation

The *plastid* has a *specific transcription and translation system* which is the basis of the semiautonomous state of the chloroplast. The transcription within the plastid is perforemd by a ptDNA-dependent RNA polymerase which is firmly bound to membrane structures (cf. WOLLGIEHN and PARTHIER, 1980). There is until now only one report indicating that there might be a second RNA polymerase in the stroma of the chloroplast with properties different from the membrane-bound form (JOUSSAUME, 1973).

The *plastid RNA polymerase* is *encoded in nuclear DNA*, synthesized on 80S cytoplasmic ribosomes, and *afterwards transported into the plastids*. This conclusion is based on several lines of evidence: The enzyme is active in the presence of inhibitors of translation on chloroplast 70S ribosomes (chloramphenicol, spectinomycin) and of rifampicin which inhibits under certain conditions (see below) the transcription of plastid DNA (SURZYCKI, 1969; GOODENOUGH, 1971; ELLIS and HARTLEY, 1971). Plastids lacking ribosomes due to a heat treatment of the plants or to a mutation of plastid DNA nevertheless are able to transcribe their DNA (BÜNGER and FEIERABEND, 1980; SIEMENROTH, 1981). Among the transcripts of ribosome-deficient plastids are the mature high molecular-weight rRNA's, i.e., not only the chloroplast RNA polymerase, but also the enzyme(s) for rRNA processing are made in the nucleo-cytoplasm (SIEMENROTH, 1981).

The chloroplast-specific RNA polymerase has been partially characterized in green algae and in higher plants. RNA polymerase activity was found in DNA-protein complexes isolated from *Euglena* chloroplasts (HALLICK et al., 1976; SCHIEMANN et al., 1977) and from spinach chloroplasts (BRIAT et al., 1979). The transcription in these complexes is insensitive to rifampicin or rifamycin SV (HALLICK et al., 1976; BRIAT et al., 1979).

Among the products of transcription chloroplast rRNA could be detected (HALLICK et al., 1976; BRIAT et al., 1979). A purified RNA polymerase was solubilized from isolated maize chloroplasts (SMITH and BOGORAD, 1974). It consists of different subunits with molecular weights of 180, 140, 100, 95, 85, 42 kD. Maize nuclear RNA polymerase II has subunits of similar molecular weights (180, 160, 43), which, however, could clearly be separated from the respective polypeptides of the chloroplast enzyme by two-dimensional mapping of tryptic digests (KIDD and BOGORAD, 1979).

Spinach chloroplast RNA polymerase isolated from a transcriptionally active DNA-protein complex was found to consist of subunits with molecular weights quite different from maize RNA polymerase. Five subunits with molecular weights of 69, 60, 55, 34, and 15 kD were observed in the active enzyme (BRIAT and MACHE, 1980).

As the DNA-protein complex the spinach enzyme is insensitive to rifampicin, an antibiotic which inhibits transcription in bacteria by binding to the β subunit of *Escherichia coli* RNA polymerase. In contrast, crude preparations of corn chloroplast RNA polymerase (BOTTOMLEY et al., 1971) and purified chloroplast RNA polymerase from *Chlamydomonas* (SURZYCKI and SHELLENBARGER, 1976) are inhibited by rifamycins. Whether this reported different sensitivity to rifamycins actually reflects true species-specific differences or is due to different treatment conditions, still remains a matter of controversy (cf. WOLLGIEHN and PARTHIER, 1980).

A putative δ factor for chloroplast RNA-polymerase has been described in *Chlamydomonas* (SURZYCKI and SHELLENBARGER, 1976). This protein with a molecular weight of 51 kD stimulates the transcription by chloroplast RNA polymerase and, to a lesser extent, by *Escherichia coli* RNA polymerase (core enzyme). It does not stimulate the activity of nuclear RNA polymerases. In maize, a 27.5 kD protein was isolated from chloroplast RNA polymerase preparations (JOLLY and BOGORAD, 1980). This polypeptide, called *S factor*, was shown to be necessary for a *preferential transcription of chloroplast DNA sequences* (the gene for the large subunit of ribulose 1,5-bisphosphate-carboxylase incorporated in cloned bacterial plasmids).

The amount of RNA (mRNA, rRNA, tRNA) transcribed from chloroplast DNA varies during chloroplast differentiation and development and is stimulated by light. The literature in this field is summarized and discussed in a recent review by WOLLGIEHN and PARTHIER (1980). We shall therefore deal in the following only with data concerning differential expression of chloroplast genes published since 1978.

After illumination of maize leaves there is during greening a specific accumulation of a 32 kD thylakoid protein and its translatable mRNA in plastids. This mRNA hybridizes with the "plastid photogene 32", coding this photoregulated thylakoid protein. This gene is located in the Bam H1 fragment 8 of maize plastid DNA (BEDBROOK et al., 1978; BOGORAD et al., 1978). This gene is not expressed in proplastids and etioplasts, but it is transcribed into large amounts of mRNA following light induction in developing and green chloroplasts. During the light induction of the expression of this gene there is a sharp rise of the transcriptional activity of the plastid RNA polymerase (APEL and BOGORAD, 1976). Initially a 33.5 kD precursor is synthesized, which then is processed to the 32 kD thylakoid protein. In the Lemnaceae species *Spirodela oligorrhiza* illumination was similarly found to induce the synthesis of a 32 kD polypeptide as a membrane protein; it is first formed as a 33.5 kD precursor which is rapidly processed to 32 kD (EDELMAN and REISFELD, 1980). The 32 kD protein in maize, in *Spirodela* and proteins of similar molecular weight and behavior in other plants have identical fingerprint patterns as compared after partial proteolytic digestion (EDELMAN et al., 1980).

Research work in the laboratories of SCHIFF (1978) and HEIZMANN et al. (1978) has characterized the sequence of developmental processes connected with the morphogenesis of the proplastids in dark-grown *Euglena* cells into differentiated chloroplasts following illumination. In the course of these processes a light activation of nuclear and plastid genes takes place. In *Euglena* the direct photocontrol of the plastid seems to act via a red-blue system based on protochlorophyll(ide). It has been shown that Eco RI fragments of *Euglena* chloroplast DNA differ from each other with respect to their transcriptional activity during light-induced greening (HEIZMANN et al., 1978). The transcription of restriction fragments of *Chlamydomonas* chloroplast DNA was followed during the cell cycle of this alga. The genes coding for chloroplast rRNA, the large subunit of ribulose-1,5-bisphosphate carboxylase and several others were found to be transcriptionally active throughout the cell cycle (ROCHAIX et al., 1980).

The *plastid gene for the large subunit of RuBPCase* has been intensively studied with regard to the *regulation of its expression*. This was especially successful in maize. *Zea mays* is a C_4 plant, which has two types of photosynthetic cell types: *Bundle-sheath cells* contain ribulose-1,5-bisphosphate carboxylase/oxygenase and perform the dark reactions of photosynthesis via the Calvin cycle; the *LS gene is fully expressed*, transcribed into messenger RNA and translated within the plastids. However, in *mesophyll cells* the *LS gene is not transcribed*; no mRNA from this gene is present, and the RuBPCase is missing in mesophyll cells. The fixation of CO_2 is performed in these cells by the "C_4 pathway": CO_2 is fixed by phosphoenol-pyruvate which is transformed into oxalacetate and then into malate; malate is transported from the mesophyll cells into the bundle-sheath cells where it is cleaved into pyruvate and CO_2, which then is fixed by RuBP and introduced into the Calvin cycle.

The differentiation of the maize leaf into bundle-sheath and mesophyll cells is obviously determined by nuclear genes (DAY, 1977), and thus presumably also the transcription of this plastid gene in bundle-sheath cells and the block of transcription in mesophyll cells. Of particular interest is the fact that this regulation is very specific and concerns only this particular plastid gene. The "2.2 kb plastid gene" which is located closely by the side of the LS gene (only separated by about 200 base pairs) is expressed in both tissues (BOGORAD, pers. comm.).

The kind of the regulatory signal from the nucleocytoplasm to the plastid is still unknown at present, but the analysis of the transcription of plastid DNA by the plastidal RNA polymerase has given an interesting hint. In the process of isolation of the polypeptides comprising the RNA polymerase a 27.5 kD S factor was analyzed. In a chimerical bacterial plasmid containing plastid DNA sequences this S factor caused the preferential transcription of some plastid DNA sequences, especially that of the LS gene (JOLLY and BOGORAD, 1980). The characterization of the role of this S factor in the preferential transcription of plastid DNA sequences may be an important means in the analysis of the regulatory signals for the plastid DNA.

6. The Localization of Plastid Genes

a) Restriction Maps and Physical Maps of the Plastid Chromosome

Restriction endonucleases of type II cut double-stranded DNA at definite sites which have specific base sequences. Thus, the action of par-

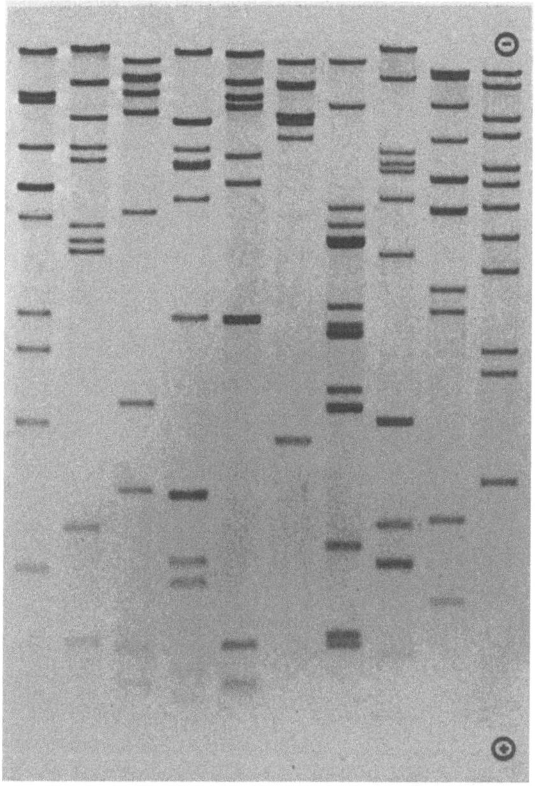

Fig. 2. Digestion of spinach ptDNA with ten different restriction endonucleases from left: Sal I, Pst I, Kpn I, Xho I, Pvu I, Bgl I, Xma I, Pvu II, Hpa I and Sac I (fragments separated on a 0.5% agarose gel). The molecular weights range from about 34 Md down to 0.8 Md. Smaller fragments have run off the gel. The sizes of the fragments obtained in the individual digestions add up to the molecular weight of the circle within the limits of errors of the method (CROUSE et al., 1978). The stoichiometry of the fragments is not fully apparent since the lower parts of the gel have been overexposed to show the small fragments. Low backgrounds, especially between the largest fragments amounting to about 1/3 of the circle, indicate that DNA was almost undegraded. (HERRMANN and POSSINGHAM, 1980)

ticular restriction endonucleases causes specific DNA fragmentation patterns (Fig. 2). The combination of the restriction patterns of several restriction enzymes enables the construction of "restriction maps" of the plastid chromosome. The application of the restriction enzyme techniques to the research of plastid inheritance was a key step for the great progress in this field. Such restriction maps of the plastid chromosome have so far been constructed for *Zea mays*, *Spinacia oleraceae*, several species of *Oenothera*, *Nicotiana tabacum*, *Chlamydomonas reinhardii* and *Euglena gracilis* (cf. reviews by HERRMANN and POSSINGHAM, 1980; HERRMANN et al., 1980; TEWARI, 1979). A general result of these studies is the finding that ptDNA is segmentally organized into nonrepetitive (unique) and repetitive regions (which are present two, three, and partially even four times).

The use of the restriction fragments of ptDNA for DNA-RNA hybridization allowed the localization of specific genes for tRNA, rRNA's and mRNA's

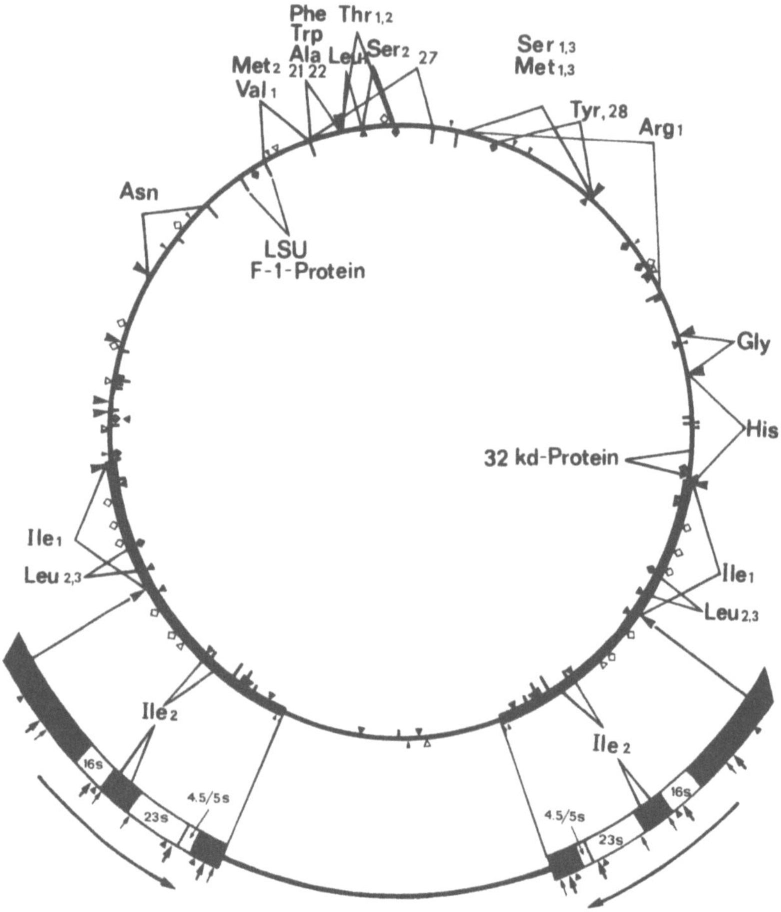

Fig. 3. The circular physical map of the *Spinacia oleracea* plastid chromosome. The chromosome accounts for about 95 MD or 140 kbp; the inverted repeat is demonstrated by thick arc regions. The map order of the various DNA fragments was determined by reciprocal digestion with several restriction endonucleases. The relative cut positions of endonucleases are drawn to scale and indicated by the following symbols: Sal I = ▼, Pst I = ◆ , Kpn I = T , Xma I = ▾, Xho I = ı , Bgl I = ▿, Pvu I = ◊, Sac I = ▾. The location of the two sets of genes for ribosomal RNA within the inverted repeat is indicated in the expanded drawing. The symbols specify the cleavage sites of EcoRI = ⬇, Bam HI = ↓, Sma I = ▼. The polarity of transcription in the rDNA region preceeds from the 16S gene toward that of the 23S *(arrow)*. For the mapping of tRNA and polypeptide genes the smallest DNA segment to which hybridization was observed is indicated for each single component (HERRMANN and POSSINGHAM, 1980)

for particular polypeptides. In this way physical maps of the plastid chromosome have been constructed. Figure 3 shows a recent version of the physical map of the plastid chromosome of *Spinacia oleracea*. Details of the localization of specific genes on the physical map of the plastid chromosome are given in the following paragraphs.

b) Organization of Ribosomal DNA

Chloroplast genes for rRNA of most species studied so far are located in extended repeated regions of ptDNA.

This holds true for *Zea* (BEDBROOK and BOGORAD, 1976; BEDBROOK et al., 1977), *Spinacia* (HOBOM et al., 1977; CROUSE et al., 1978), *Pisum* (KOLODNER and TEWARI, 1979), *Nicotiana*, *Oenothera* (HERRMANN et al., 1980), *Euglena* (GRAY and HALLICK, 1978; JENNI and STUTZ, 1978; RAWSON et al., 1978), and *Chlamydomonas* (ROCHAIX, 1978; ROCHAIX and MALNOE, 1978). There are two inverted repeats each containing one copy of the genes coding for 16S, 23S and 5SrRNA in *Zea*, *Spinacia*, *Nicotiana*, *Oenothera*, and *Chlamydomonas*, two tandem repeats in *Pisum* and three tandem repeats in *Euglena*. A fourth 16SrRNA sequence was found in *Euglena gracilis*, strain Z, about 6 kb pairs away in 5'-direction from the 3'end of the next 16SrRNA gene (JENNI and STUTZ, 1979).

Vicia faba ptDNA is the only hitherto known exception in having only one set of rRNA genes (KOLLER and DELIUS, 1980).

The genes seem to be generally arranged in the order 16S - 23S - 5S and are encoded in the same strand (BEDBROOK et al., 1977; ROCHAIX and MALNOE, 1978; DELIUS and KOLLER, 1980; HERRMANN and POSSINGHAM, 1980; KUSUDA et al., 1980; TAKAIWA and SUGIURA, 1980; OROZCO et al., 1980a). In higher plants there is a sequence between the 23S rRNA gene and the 5S rRNA gene coding for a 4.5S RNA of still unknown function (BOHNERT et al., 1974; DYER et al., 1977; WHITFELD et al., 1978; TAKAIWA and SUGIURA, 1980). A 7S and a 3S RNA which originate from the large subunit of chloroplast ribosomes are transcribed from the region between the 16S rRNA gene and the 23S rRNA gene in *Chlamydomonas* (ROCHAIX and MALNOE, 1978). A further unusual feature of *Chlamydomonas* ptDNA is that the 23S rRNA genes are interrupted by a 870 bp *intervening sequence* (ROCHAIX and MALNOE, 1978). The flanking regions and the ends of the intervening sequence have been sequenced (ALLET and ROCHAIX, 1979). The ends of the intervening sequence show a partial complementarity. The flanking regions have sequences in common with the *E. coli* 23S rRNA gene.

The genes for 16S rRNA and 23S rRNA are separated by a spacer sequence which differs in length from species to species. This spacer contains (besides the 3S and 7S RNA genes in *Chlamydomonas*) in analogy to *E. coli* sequences hybridizing with tRNA. By hybridization of the respective tRNA to ptDNA, a gene for a $tRNA^{Ile}$ was localized in the spacer region of *Spinacia* (BOHNERT et al., 1979) and a gene for a $tRNA^{Ile}$ and a gene for a $tRNA^{Ala}$ in the spacer of *Euglena* (KELLER et al., 1980). Sequencing of cloned 16S and 23S regions of *Zea* ptDNA revealed the presence of two split tRNA genes in the spacer coding for $tRNA^{Ile}$ and $tRNA^{Ala}$, respectively (KÖSSEL et al., 1980). The genes contain intervening sequences of 949 bp ($tRNA^{Ile}$) and approx. 850 bp ($tRNA^{Ala}$). The intervening sequences start one or two bases distal to the anticodon, as was observed for yeast cytoplasmic tRNA genes (KNAPP et al., 1979). The existence of one ($tRNA^{Glu}$) or two ($tRNA^{Ile}$ and $tRNA^{Ala}$) tRNA genes in the spacer of rDNA is also known from *E. coli* (LUND et al., 1976; MORGAN et al., 1978). Sequencing of the spacer in *Euglena* revealed in accordance with the hybridization studies the existence of genes for $tRNA^{Ile}$ and $tRNA^{Ala}$. The sequence of the leader region of the 16S rRNA gene was found to show 68% homology with the spacer and to contain a complete "pseudo" $tRNA^{Ile}$ gene (OROZCO et al., 1980b).

c) Transfer RNA Genes and Transfer RNA Sequences

Transfer RNA genes are located on the chloroplast DNA in the spacer between the genes for 16S rRNA and 23S rRNA (see 6.b), in the repeated region of the DNA (but outside the rRNA sequence) and, in majority, in the larger of the two nonrepeated regions (HALLICK et al., 1978; MALNOE and ROCHAIX, 1978; STEINMETZ et al., 1978; DRIESEL et al., 1979).

Fractionation of chloroplast tRNA's from spinach by two-dimensional polyacrylamide gel electrophoresis yielded in 35 spots. Twenty seven of them (corresponding to 16 amino acids) could be identified by aminoacylation with the respective aminoacyl-tRNA synthetase. Until now 21 genes coding for tRNA's for 14 different amino acids were localized on the map of spinach chloroplast DNA (DRIESEL et al., 1979). Out of the 21 genes, 15 genes corresponding to 12 different amino acids have been detected in the large single copy region (Fig. 3). In addition to the gene for $tRNA_2^{Ile}$ in the spacer between the genes for 16S and 23S rRNA the repeats contain outside the rRNA unit the genes for $tRNA_1^{Ile}$, $tRNA_1^{Leu}$, and $tRNA_2^{Leu}$. In maize 13 genes for identified tRNA's could be localized on chloroplast DNA (BURKARD et al., 1980a,b). Nine of the tRNA genes are located on the large single copy region. The inverted repeats contain outside the rDNA the genes for $tRNA_2^{Leu}$, $tRNA_1^{Ile}$, $tRNA^{His}$ (BURKARD et al., 1980b) and $tRNA^{Val}$ (KÖSSEL et al., 1980). Ten tRNA genes have been localized so far on *Euglena* chloroplast DNA, of these seven are found in the single copy region (BURKARD et al., 1980b). Three tRNA genes could be localized in the tandem repeats, the genes for $tRNA^{Ile}$ and $tRNA^{Ala}$ in the spacer between 16S and 23S rDNA and a gene for $tRNA^{Trp}$ or $tRNA^{Glu}$ outside the spacer (KELLER et al., 1980).

The nucleotide sequences of a number of chloroplast tRNA's has been determined: $tRNA^{Phe}$ from *Euglena* (CHANG et al., 1976) and *Phaseolus* (GUILLEMAUT and KEITH, 1977), initiator $tRNA_F^{Met}$ from *Spinacia* (CALAGAN et al., 1980) and *Phaseolus* (CANADAY et al., 1980), elongator $tRNA^{Met}$ from *Spinacia* (OLINS and JONES, 1980), *Spinacia* $tRNA^{Thr}$ (KASHDAN et al., 1980) and three isoaccepting $tRNA^{Leu}$ from *Phaseolus* (BURKARD et al., 1980a). Some features seem to be characteristic for chloroplast tRNA's: There is a methylated G-G sequence in the dihydrouridine loop and a predominance of T-Ψ-C-A in loop IV as compared with T-Ψ-C-G found in nonchloroplast tRNA's (cf. CALAGAN et al., 1980). Generally there is more homology to prokaryotic tRNA sequences than to eukaryotic tRNA sequences. An exception is the chloroplast $tRNA^{Thr}$ from spinach which shows equal homology to both prokaryotic and eukaryotic threonine tRNA's (KASHDAN et al., 1980). There is 92% homology between the spinach and the bean chloroplast $tRNA_F^{Met}$ sequences (CALAGAN et al., 1980; CANADAY et al., 1980). Both show the characteristics of prokaryotic initiator tRNA's in having a 5' terminal nucleotide (C) which is not base-paired and a T-Ψ-C-A sequence in loop IV. The sequence homology with prokaryotic initiator tRNA's (80%-84%) is much higher than with eukaryotic $tRNA's^{Met}$ (64% - 69%) and with mitochondrial initiator $tRNA's_F^{Met}$ (55% - 62%).

The sequences of *Euglena* and bean $tRNA^{Phe}$ only differ by five nucleotides apart from small differences in post-transcriptional modifications (CHANG et al., 1976; GIULLEMAUT and KEITH, 1977). This suggests a stability of the primary structure of tRNA's during evolution. However, this similarity between a *Euglena* and a higher plant chloroplast tRNA seems to be an exception. By hybridization of higher plant chloroplast and *Euglena* chloroplast tRNA's to homologous and heterologous chloroplast DNA's a high degree of homology between higher plant chloroplast tRNA's of different species (*Spinacia*, *Phaseolus*, *Zea*) was observed, but there was only a very poor hybridization of *Euglena* chloroplast tRNA's to higher plant chloroplast DNA. From 14 individual *Euglena* chloroplast tRNA's studied only one, $tRNA^{Phe}$, hybridized to chloroplast DNA from *Zea* and *Phaseolus* which is in agreement with the similar nucleotide sequences of chloroplast tRNA^{Phe} from *Euglena* and bean (MUBUMBILA et al., 1980).

d) The Gene for the Large Subunit of RuBP-Carboxylase

The enzyme responsible for the first step in the Calvin cycle is the ribulose-1,5-bisphosphate carboxylase (RuBPCase). It catalyzes the carboxylation of ribulosebisphosphate with CO_2 to synthesize 3-phosphoglycerid acid. This enzyme is in fact identical with the Fraction I Protein (F-I-P) which is present in very high amount in leaves of green plants. RuBPCase is in vivo an octomer consisting of four (identical) small subunits (MW = 12,000) and four (identical) large subunits (MW = 56,000). The small subunit is coded for by the nuclear DNA, whereas the large subunit is coded for by the plastid DNA (CHAN and WILDMAN, 1972; KAWASHIMA and WILDMAN, 1972; KUNG, 1976; WILDMAN, 1979; BOTTOMLEY, 1980).

The plastid gene for the large subunit of RuBPCase *(= LS gene) is today the best studied plastid gene*. In *Zea mays* the LS gene has been localized on the physical map of plastid DNA: it is contained in a 2500 base pair long sequence of the Bam HI fragment 9, which is part of the Sal I A fragment (BEDBROOK et al., 1979). A 4000 base pair long fragment containing this sequence has been cloned and used in a linked transcription-translation system; it directed the synthesis of a polypeptide practically identical with the large subunit (COEN et al., 1977). The use of several restriction enzymes made it possible to define the LS sequence within this Bam HI fragment 9. *In 1980 the nucleotide sequence of the LS plastid gene of maize was determined* by McINTOSH, POULSEN and BOGORAD.

The protein-coding sequence consists of 1425 nucleotide pairs. From the DNA-sequence the LS amino-acid sequence was predicted. The predicted sequence was compared with the amino acid sequences of cyanogen bromide cleavage fragments from the LS of barley and of LS fragments of spinach. The alignment of these sequences with the predicted maize LS sequence revealed a very great homology; there are only a few differences. The analysis of the DNA sequence upstream of the start codon for the N-terminal methionine showed two interesting features: it contains neither a "classical" prokaryotic promoter sequence nor its proposed eukaryotic counterpart. However, the sequence has a typical prokaryotic ribosome-binding site five nucleotides upstream from the N-terminal methionine codon (5'-GGAGG-3') that is complementary to the "Shine-Dalgarno" sequence at the 3' terminus of the maize chloroplast 16S rRNA.

The analysis of the codon usage in the maize LS gene has shown that almost all amino acid coding triplets are used (except one for Thr, one for Arg, and one for Ser). *The codons are obviously translated by the "universal code"* rather than by the variation of it found in mitochondria. The natural stop codon terminating the translation of the LS protein is ocher. There is a strict colinearity between the LS DNA sequence, and the LS mRNA sequence, which is translated into the LS polypeptide.

In *Chlamydomonas reinhardii* the LS gene has been localized on the physical map of the ptDNA in the Eco RI fragment R15 (ROCHAIX and MALNOE, 1978). In *Spinacia oleraceae* the LS gene has also been localized on the physical ptDNA map (HERRMANN and POSSINGHAM, 1980).

e) Genes for Thylakoid Polypeptides

Besides the gene for the large subunit of the ribulose-1,5-bisphosphate carboxylase a number of thylakoid polypeptides have been localized on the map of *Chlamydomonas* chloroplast DNA (ROCHAIX; 1981). This concerns the polypeptides D1 and D2. Moreover a tentative assignment is reported for the polypeptides M 4.1, 4.2, 5, and 6 (nomenclature according to CHUA and GILLHAM, 1977). The position of the gene for the 32 kd protein which shows a light-stimulated transcription (cf. 6.4. 5), has been determined in *Zea* (BEDBROOK et al., 1978), in *Spinacia* (DRIESEL et al., 1980) and in *Chlamydomonas* (MALNOE et al., 1979). Chloroplast DNA seems to contain only one copy of this gene. It is located in the chloroplast DNA of the three species studied in the larger single copy region near to one of the two inverted repeats (Fig. 3).

References

ALLET, B., ROCHAIX, J.-D.: Cell 18, 55-60 (1979). - APEL, K., BOGORAD, L.: Eur. J. Biochem. 67, 615-620 (1976). - ARNOLD, C.-G.: Progr. Bot. 39, 182-191 (1977). - ARNOLD, C.-G., GAFFAL, K.P.: Progr. Bot. 41, 212-220 (1979).

BEDBROOK, J.R., BOGORAD, L.: 369-373 in: Genetics and Biogenesis of Chloroplasts and Mitochondria, ed. Th. BÜCHER et al. Amsterdam: North Holland 1976. - BEDBROOK, J.R., KOLODNER, R.: Annu. Rev. Plant Physiol. 30, 593-620 (1979). - BEDBROOK, J.R., KOLODNER, R., BOGORAD, L.: Cell 11, 739-749 (1977). - BEDBROOK, J.R., LINK, G., COEN, D.M., BOGORAD, L., RICH, A.: Proc. Natl. Acad. Sci. USA 75, 3060-3064 (1978). - BEDBROOK, J.R., COEN, D.M., BEATON, A.R., BOGORAD, L., RICH, A.: J. Biol. Chem. 254, 905-910 (1979). - BÖRNER, Th.: Biol. Zbl. 92, 545-561 (1973); - Biol. Zbl. 99, 1-11 (1980). - BOGORAD, L., BEDBROOK, J.R., COEN, D.M., KOLODNER, R., LINK, G.: 541-551 in: Chloroplast Development, eds. G. AKOYUNOGLOU, J.H. ARGYROUDI-AKOYUNOGLOU. Amsterdam: North Holland 1978. - BOHNERT, H.J., SCHMITT, J.M., HERRMANN, R.G.: Port. Acta Biol. Ser. A, 14, 71-90 (1974). - BOHNERT, H.J., DRIESEL, A.J., CROUSE, E., GORDON, K., HERRMANN, R.G., STEINMETZ, A., MUBUMBILA, M., KELLER, M., BURKARD, G., WEIL, J.H.: FEBS Lett. 103, 51-56 (1979). - BOTTOMLEY, W.: 179-199 in: Chloroplasts (Results and Problems in Cell Differentiation, Vol.10), ed. J. REINERT. Berlin-Heidelberg-New York: Springer 1980. - BOTTOMLEY, W., SPENCER, D., WHEELER, A.M., WHITFELD, P.R.: Arch. Biochem. Biophys. 143, 296-275 (1971). - BRIAT, J.-F., LAULHERE, J.-P., MACHE, R.: Eur. J. Biochem. 98, 285-292 (1979). - BRIAT, J.-F., MACHE, R.: Eur. J. Biochem. 111, 503-509 (1980). - BÜNGER, W., FEIERABEND, J.: Planta 149, 163-169 (1980). - BURKARD, G., CANADAY, J., CROUSE, E., GUILLEMAUT, P., IMBAULT, P., KEITH, G., KELLER, M., MUBUMBILA, M., OSORIO, L., SARANTOGLOU, W., STEINMETZ, A., WEIL, J.-H.: 313-320 in: Genome Organization and Expression in Plants, ed. C.J. LEAVER. New York: Plenum Press 1980a. - BURKARD, G., CROUSE, E.J., KELLER, M., MUBUMBILA, M., STEINMETZ, A., WEIL, J.-H., SELDON, R., McINTOSH, L., BOGORAD, L.: 34 in: Abstracts 6th EMBO Annual Symp.: Molecular Biologists Look at Green Plants, Heidelberg: EMBL 1980b. - BUTTERFASS, Th.: Patterns of Chloroplast Reproduction. Wien-New York: Springer 1979.

CALAGAN, J.L., PIRTLE, R.M., PIRTLE, I.L., KASHDAN, M.A., VREMAN, H.J., DUDOCK, B.S.: J. Biol. Chem. 255, 9981-9984 (1980). - CANADAY, J., GUILLEMAUT, P., WEIL, J.-H.: Nucleic Acid Res. 8, 999-1008 (1980). - CHAN, P.-H., WILDMAN, S.G.: Biochim. Biophys. Acta 277, 677-680 (1972). - CHANG, S.H., BRUM, C.K., SILBERKLANG, M., RAJBHANDARY, U.L., HECKER,

L.I., BARNETT, W.E.: Cell 9, 717-723 (1976). - CHIANG, K.-S., SUEOKA, N.: Proc. Natl. Acad. Sci. USA 57, 1506-1513 (1967). - CHUA, N.-H., GILLHAM, N.W.: J. Cell Biol. 74, 441-452 (1977). - COEN, D.M., BEDBROOK, J.R., BOGORAD, L., RICH, A.: Proc. Natl. Acad. Sci. USA 74, 5487-5491 (1977). - CROUSE, E.J., SCHMITT, J.M., BOHNERT, H.-J., GORDON, K., DRIESEL, A.J., HERRMANN, R.G.: 565-572 in: Chloroplast Development, eds. G. AKOYUNOGLOU, J.H. ARGYROUDI-AKOYUNOGLOU. Amsterdam: North Hollan 1978.

DAY, P.R.: Science 197, 1334-1339 (1977). - DELIUS, H., KOLLER, B.: 36 in: Abstracts 6th EMBO Annual Symp.: Molecular Biologists Look at Green Plants. Heidelberg: EMBL 1980. - DIAMOND, J., SCHIFF, J.A., KELNER, A.: Arch. Biochem. Biophys. 167, 603-614 (1975). - DRIESEL, A.J., CROUSE, E.J., GORDON, K., BOHNERT, H.J., HERRMANN, R.G., STEINMETZ, A., MUBUMBILA, M., KELLER, M., BURKARD, G., WEIL, J.H.: Gene 6, 285-306 (1979). - DRIESEL, A.J., SPEIRS, J., BOHNERT, H.J.: BBA 610, 297-310 (1980). - DYER, T.A., BOWMAN, C.M., PAYNE, P.I.: 121-133 in: Nucleic Acids and Protein Synthesis in Plants, eds. L. BOGORAD, J.H. WEIL. New York-London: Plenum Press 1977.

EDELMAN, M., REISFELD, A.: 353-362 in: Genome Organization and Expression in Plants, ed. C.J. LEAVER. New York: Plenum Press 1980. - EDELMAN, M., MATTOO, A.K., HOFFMAN-FALK, H.: 5th Internat. Congr. Photosynth. Abstract, 159 (1980). - ELLIS, R.J., HARTLEY, M.R.: Nature New Biol. 233, 193-196 (1971). - ELLIS, R.J., SMITH, S.M., BARRACLOUGH, R.: 321-336 in: Genome Organization and Expression in Plants, ed. C.J. LEAVER. New York, London: Plenum Press 1980.

GOODENOUGH, U.W.: J. Cell Biol. 50, 35-49 (1971). - GRAY, P.W., HALLICK, R.B.: Biochemistry 17, 284-289 (1978). - GUILLEMAUT, P., KEITH, G.: FEBS Lett. 84, 351-356 (1977).

HAGEMANN, R.: Biol. Rdsch. 18, 329-348 (1980). - HAGEMANN, R., BÖRNER, Th.: 709-720 in: Chloroplast Development, eds. G. AKOYUNOGLOU, J.H. ARGYROUDI-AKOYUNOGLOU. Amsterdam: North Holland 1978. - HAGEMANN, R., METZLAFF, M.: Biol. Rdsch. 18, 121-132 (1980). - HALLICK, R.B., LIPPER, C., RICHARDS, O.C., RUTTER, W.J.: Biochemistry 15, 3039-3045 (1976). - HALLICK, R.B., GRAY, P.W., CHELM, B.K., RUSHLOW, K.E., OROZCO, E.M.: 619-622 in: Chloroplast Development, eds. G. AKOYUNOGLOU, J.H. ARGYROUDI-AKOYUNOGLOU. Amsterdam: North Holland 1978. - HEIZMANN, P., VERDIER, G., YOUNIS, H.: 623-628 in: Chloroplast Development, eds. G. AKOYUNOGLOU, J.H. ARGYROUDI-AKOYUNOGLOU. Amsterdam: North Holland 1978. - HERRMANN, F.H., YUSUPOVA, G.A., GILLER, Yu.E., BÖRNER, Th.: Stud. Biophys. 46, 9-12 (1974). - HERRMANN, R.G., POSSINGHAM, J.V.: 45-96 in: Chloroplasts (Results and Problems in Cell Differentiation, Vol. 10), ed. J. REINERT. Berlin-Heidelberg-New York: Springer 1980. - HERRMANN, R.G., SEYER, P., SCHEDEL, R., GORDON, K., BISANZ, C., WINTER, P., HILDEBRANDT, J.W., WLASCHEK, M., ALT, J., DRIESEL, A.J., SEARS, B.B.: 97-112 in: Biological Chemistry of Organelle Formation (31. Colloq. Ges. Biol. Chemie, Mosbach), eds. T. BÜCHER, W. SEBALD, H. WEISS. Berlin-Heidelberg-New York: Springer 1980. - HILL, H.Z., SCHIFF, J.A., EPSTEIN, H.T.: Biophys. J. 6, 125-133 (1966). - HOBOM, G., BOHNERT, H.J., DRIESEL, A., HERRMANN, R.G.: 63-69 in: Acides Nucleiques et Synthese de Protein chez les Vegetaux, eds. L. BOGORAD, J.H. WEIL. Paris: Ed. CNRS 1977. - HOWELL, S.H., WALKER, L.L.: Proc. Natl. Acad. Sci. USA 69, 490-494 (1972)

JENNI, B., STUTZ, E.: Eur. J. Biochem. 88, 127-134 (1978); - 142 in: Genome Organization and Expression in Plants, ed. Ch. LEAVER. Edingburg: FEBS Advanced Course 1979. - JOLLY, S.O., BOGORAD, L.: Proc. Natl. Acad. Sci. USA 77, 822-826 (1980). - JOUSSAUME, M.: Physiol. Veget. 11, 69-82 (1973).

KASHDAN, M.A., PIRTLE, R.M., PIRTLE, I.L., CALAGAN, J.L., VREMAN, H.J. DUDOCK, R.S.: J. Biol. Chem. 255, 8831-8835 (1980). - KAWASHIMA, N., WILDMAN, S.G.: Biochim. Biophys. Acta 262, 42-49 (1972). - KELLER, M., BURKARD, G., BOHNERT, H.J., MUBUMBILA, M., GORDON, K., STEINMETZ, A., HEISER, D., CROUSE, E.J., WEIL, J.H.: Biochem. Biophys. Res. Comm. 95, 47-54 (1980). - KIDD, G.H., BOGORAD, L.: Proc. Natl. Acad. Sci. USA 76, 4890-4892 (1979). - KIRK, J.T.O., TILNEY-BASSETT, R.: The Plastids, Their Chemistry, Structure, Growth and Inheritance. Amsterdam: North Holland 1978. - KNAPP, G., OGDEN, R.C., PEEBLES, C.L., ABELSON, J.: Cell 18, 37-45 (1979). - KNOTH, R., HERRMANN, F.H., BÖTTGER, M., BÖRNER, T.: Biochem. Physiol. Pflanzen 166, 129-148 (1974). - KÖSSEL, H., EDWARDS, K., KOCH, W., SCHWARZ, Zs.: 32-33 in: Abstracts 6th EMBO Annual Symp.: Molecular Biologists Look at Green Plants. Heidelberg: EMBL 1980. - KOLLER, B., DELIUS, H.: Mol. Gen. Genet. 178, 261-269 (1980). - KOLODNER, R.D., TEWARI, K.K.: J. Biol. Chem. 250, 8840-8847 (1975a); - Nature 265, 708-711 (1975b); - Proc. Natl. Acad. Sci. USA 76, 41-45 (1979). - KUNG, S.-D.: Science 191, 429-434 (1976). - KUSUDA, J., SHINOZAKI, K., TAKAIWA, F., SUGIURA, M.: Mol. Gen. Genet. 178, 1-7 (1980).

LAMPPA, G.K., ELLIOT, L.V., BENDICH, A.J.: Planta 148, 437-443 (1980). - LEMIEUX, C., TURMEL, M., LEE, R.W.: J. Cell Biol. 87, 185a (1980). - LUND, E., DAHLBERG, J.E., LINDAHL, L., JAKUNAS, S.R., DENNIS, P.P., NOMURA, M.: Cell 7, 165-177 (1976). - LYMAN, H., SRINIVAS, U.K.: 593-607 in: Chloroplast Development, eds. G. AKOYUNOGLOU, J.H. ARGYROUDI-AKOYUNOGLOU. Amsterdam: North Holland 1978.

MALNOE, P., ROCHAIX, J.-D.: Mol. Gen. Genet. 166, 269-275 (1978). - MALNOE, P., ROCHAIX, J.-D., CHUA, N.H., SPAHR, P.-F.: J. Mol. Biol. 133, 417-434 (1979). - McINTOSH, L., POULSEN, C., BOGORAD, L.: Nature 288, 556-560 (1980). - MORGAN, E.A., IKEMURA, T., LINDAHL, L., FALLON, A.M., NOMURA, M.: Cell 13, 335-344 (1978). - MUBUMBILA, M., BURKARD, G., KELLER, M., STEINMETZ, A., CROUSE, E., WEIL, J.H.: Biochim. Biophys. Acta 609, 31-39 (1980).

NICOLAS, P., HUSSEIN, Y., HEIZMANN, Ph., NIGON V.: Mol. Gen. Genet. 178, 567-572 (1980).

OLINS, P.O., JONES, D.S.: Nucleic Acid Res. 8, 715-729 (1980). - OROZCO, E.M.,Jr., GRAY, P.W., HALLICK, R.B.: J. Biol. Chem. 255, 10991-10996 (1980a). - OROZCO, E.M.,Jr., RUSHLOW, K.E., DODD, J.B., HALLICK, R.B.: J. Biol. Chem. 255, 10997-11003 (1980b).

RAWSON, J.R.J., KUSHNER, S.R., VAPUNEK, D., ALTON, N.K., BOERMA, C.L.: Gene 3, 191-209 (1978). - REINERT, J. (ed.): Chloroplasts (Results and Problems in Cell Differentiation, Vol. 10). Berlin-Heidelberg-New York: Springer 1980. - RICHARDS, O.C., MANNING, J.E.: 213-224 in: Les Cycles Cellulaires et Leur Blocage, eds. M. LEFORT-TRAN, R. VALENGIA. Paris: C.N.R.S. No. 240, 1975. - ROCHAIX, J.D.: J. Mol. Biol. 126, 597-617 (1978). - ROCHAIX, J.D.: Experientia 37, 323-332 (1981). - ROCHAIX, J.D., MALNOE, P.: Cell 16, 661-670 (1978). - ROCHAIX, J.D., MALNOE, P., DARLIX, SCHNEIDER, M., DRON, M., ALLET, B., SPAHR, P.-F.: 27-29 in: Abstracts 6th EMBO Annual Symp.: Molecular Biologists Look at Green Plants. Heidelberg: EMBL 1980.

SCHIEMANN, J., WOLLGIEHN, R., PARTHIER, B.: Biochem. Physiol. Pflanzen 171, 474-478 (1977). - SCHIFF, J.A.: 747-767 in: Chloroplast Development, eds. G. AKOYUNOGLOU, J.H. ARGYROUDI-AKOYUNOGLOU. Amsterdam: North Holland 1978. - SCHIFF, J.A., EPSTEIN, H.T.: 341-353 in: Biochemistry of Chloroplasts I, ed. T.W. GOODWIN. London-New York: Academic Press 1966. - SCHWEMMLER, W., SCHENK, H.E.A.: Endocytobiology, Endosymbiosis and Cell Biology, Vol. I. Berlin-New York: de Gruyter 1980. - SCOTT, N.St., POSSINGHAM, J.V.: J. Exp. Bot. 31, 1081-1092 (1980). - SMALL, G.D., GREINMANN, C.S.: Nucleic Acids. Res. 4, 2893-2902 (1977). - SMITH, H.J., BOGORAD, L.: Proc. Natl. Acad. Sci. USA 71, 4839-4842 (1974). - SIEMENROTH, A.: Dissertation Fak. Naturwiss. Martin-Luther-Universität Halle/S. 1981. - SRINIVAS, U., LYMAN, H.: Plant Physiol. 66, 295-301 (1980). - STEINMETZ, A., MUBUMBILA, M., KELLER, M., BURKARD, G., WEIL, J.H., DRIESEL, A.J., CROUSE, E.J., GORDON, K., BOHNERT, H.J., HERRMANN, R.G.: 573-580 in: Chloroplast Development, eds. G. AKOYONOGLOU, J.H. ARGYROUDI-AKOYUNOGLOU. Amsterdam: North Holland 1978. - SURZYCKI, S.J.: Proc. Natl. Acad. Sci. USA 63, 1327-1334 (1969). - SURZYCKI, S.J., SHELLENBARGER, D.L.: Proc. Natl. Acad. Sci. USA 73, 3961-3965 (1976)

TAKAIWA, F., SUGIURA, M.: Gene 10, 95-103 (1980); - Mol. Gen. Genet. 180, 1-4 (1980). - TEWARI, K.K.: 41-108 in: Nucleic Acids in Plants, Vol. 1, eds. T.C. HALL, J.W. DAVIES. Boca Raton/FL, USA: CRC Press 1979. - TEWARI, K.K., MEEKER, R.: 93-137 in: Molecular Biology of Plants, eds. I. RUBENSTEIN, R.L. PHILLIPS, C.E. GREEN, B.G. GENGENBACH. New York-London: Academic Press 1979.

WHITFELD, P.R., LEAVER, Ch.J., BOTTOMLEY, W., ATCHISON, B.A.: Biochim. J. 175, 1103-1112 (1978). - WILDMAN, S.G.: Arch. Biochem. Biophys. 196, 598-610 (1979). - WOLLGIEHN, R., PARTHIER, B.: 97-145 in: Chloroplasts (Results and Problems in Cell Differentiation, Vol. 10), ed. J. REINERT. Berlin-Heidelberg-New York: Springer 1980.

Professor Dr. RUDOLF HAGEMANN
Dr. Sc. THOMAS BÖRNER
Department of Genetics
Section Biosciences
Martin-Luther-University
Domplatz 1
DDR 402 Halle/S.

VI. Population Genetics

By KLAUS WÖHRMANN and JÜRGEN TOMIUK

1. Introduction

During the period of report no spectacular discussion has been induced comparable, for example, with the formulation of the neutrality hypothesis by KIMURA. However, attention should be called to two interesting considerations concerning the existence of genetic variability and its maintenance. Firstly, it is assumed that variation in regulatory genes, which may overlap with minor genes or polygenes as understood in quantitative genetics, is more important for the variation of fitness or quantitative traits than structural genes (HEDRICK and McDONALD, 1980).

AYALA and McDONALD (1980) showed in alcohol tolerance in *Drosophila melanogaster* that the observed variation is largely determined by variation of regulatory genes modifying the activity of major genes. Gene regulation may therefore be of great importance in the evolution of organisms.

Secondly, doubt in the assumption of widespread heterozygote advantage as a force to maintain polymorphism led CLARKE (1979) to show that frequency-dependent selection is a powerful and perhaps dominant factor for maintaining genetic diversity. But the question may be raised whether the evolutionist and geneticist should search for "the main force of evolution" or accept a diversity within the explanations of evolutionary change.

Publications revealed that most investigations are theoretical considerations which, however, are of interest for plant as well as for animal populations. Theories deal mainly with multilocus systems, density and frequency-dependent selection, and the evolution of sexuality.

In natural populations of plants questions on the enzyme polymorphism are still of great interest. Such studies have been less a preoccupation with the question of how much variation than emphasis on problem orientation (BROWN, 1979). Particular emphasis was laid on evolutionary and ecogenetic questions and the diversity of the breeding systems.

2. Theoretical Approach

a) The Origin of Genetic Variability

α) Mutation. The accumulation of mutations within populations is a basic theme of population genetics and has long been discussed. In 1964 MULLER described the accumulation of mutations in individuals of asexually reproducing populations known as "Mullers Ratchet". Since then this subject has been investigated in haploid organisms by HAIGH (1978). Estimations of the speed of the "Ratchet" and the change of the frequency distributions for the number of individuals carrying mutations are given. In finite diploid populations the conditions for accumulation were found to be very similar (HELLER and MAYNARD SMITH, 1978).

The distribution of allele frequencies depends on the effective population size and selection differences between alleles. The mean number of alleles is more strongly effected by selection than rare alleles. Even a slight heterozygote advantage increases the mean degree of heterozygosity of a population in comparison to neutral mutations. On this basis the existence of selection can be tested by the variance of heterozygote frequency but only when population size is large (LI, 1978).

The fate of mutations at one of two duplicate loci under selection and random drift was investigated by KIMURA and KING (1979). Assuming an intermediate selection optimum of a number of active gene copies a rapid loss of duplicate gene expressions occur if uncomplete recessiveness of a null allele and linkage exist. A loss may occur already shortly after duplication (ALLENDORF, 1979). The evolutionary importance of such gene duplications and of multigene families was discussed by OHTA (1980).

β) Gene Flow and Dispersal. The frequency of selectively neutral alleles decreases within a small number of generations by migration between populations. A more favorable situation for dispersion over an area is the combination with selective alleles which may raise the frequency of an immigrant allele under special conditions. Sedentary perennial plants with a long generation time are not suitable for gene exchange over a large area (GRANT, 1980). If a neutral locus is closely linked to another a cline as barrier to gene flow induces a frequency change of the neutral allele which is proportional to the gradient of the neutral allele frequency and to the reciprocal recombination rate (BARTON, 1979). MOODY (1979) investigated migration models applicable to hermaphrodite plant populations with mixed selfing and random mating with respect to the maintenance of genetic variation.

γ) Recombination. It is supposed that the process of genetic recombination has been established because it is of advantage to its carrier. Populations may evolve more rapidly and may have better chances in changing environments with recombination than without any rearrangement of the genetic material.

Increased recombination is the consequence of normalizing selection with changing optimum in infinite random mating and diploid populations with additive polygenic determination of the phenotypes. Intragenic recombination in an equilibrium population increases the frequency of the most common allele, the number of alleles which occur only once in the sample and the degree of homozygosity. It may therefore be responsible for the excess of rare and also of common alleles in samples from natural populations (MORGAN and STROBECK, 1979).

D. CHARLESWORTH and B. CHARLESWORTH (1979a) assumed in a finite population 80 selective loci and a modifier gene affecting the map length, and found that selection for reduction in recombination outweighed any selection for increased recombination. Certain types of fitness interactions in partially self-fertilized populations can increase selection for recombination between the loci considered if the recombination rate is controlled by the alleles of a third locus (CHARLESWORTH et al., 1979). CAMPBELL (1980) investigated the effect of migration and recombination on the equilibrium structure in subdivided populations under symmetric selection regimes with the result that in general the less the migration and the recombination rate the greater the possibilities of polymorphic equilibrium. The population is in linkage equilibrium although in individual demes disequilibrium exists.

b) Multilocus Systems

The fitness of single loci is not solely caused by the mode of action of these loci but it is also influenced by exogenic and endogenic factors. The interaction between genes and the linkage of genes are of great importance for the understanding of evolution. Thus, in addition to all attempts to study the genetic variation and microevolution by changes of the gene frequencies and the maintenance of polymorphism of single loci in the recent years, remarkable attention has been paid to multilocus systems and their evolution (HEDRICK et al., 1978). They considered theoretical models and estimation procedures of linkage disequilibrium, and described evidence for disequilibrium in natural populations.

In multiplicative viability models it has been shown by KARLIN and FELDMAN (1978) that the known region of recombination values, for which linkage equilibrium (D=0) is locally stable, does not preclude the local stability of a disequilibrium (D≠0). The analysis of multilocus systems is difficult and despite continuous progress far from being completely understood. GINZBURG and BRAUMANN (1980) proposed a radical change in the point of view. They could show that the likelihood of selection prevailing over recombination grows with the number of loci involved. This means that the average fitness may behave in most cases as if the systems are in complete linkage equilibrium. AVERY and HILL (1979) gave distributions of linkage disequilibrium for populations with finite size and selection. If a substantial perturbation of frequencies occurs by bottleneck, and populations are subsequently very large, the linkage disequilibrium takes hundreds of generations to return to their stable state. The authors questioned the utility of the studies of linkage disequilibrium and stable points within infinite populations.

c) Sexuality

α) The Evolution of Sexuality. Plants have an immense variability in their breeding systems and, therefore, are of great interest for comparative studies of breeding structure and variation within natural populations. It is therefore understandable that extensive investigations have been conducted in order to understand the evolution of such diverse systems.

One of the most comprehensive puplications to this theme is *The Evolution of Sex* by MAYNARD SMITH (1978). CHARNOV (1979) examined models of the evolution of sexual and reproduction with particular implications to sex allocation. Reviewing the evolution of breeding systems in flowering plants D. CHARLESWORTH and B. CHARLESWORTH (1979b) reported that some strong disadvantage to selfing has to be acting during the evolution of outbreeding which has been suggested to be inbreeding depression. A special problem is the evolution of anisogamy. BELL (1978) found conditions for which anisogamy without disassortative fusion was favored by natural selection. These investigations and those of CHARLESWORTH (1978) confirm the conclusions of PARKER et al. (1972) that under the assumption of close linkage between a mating type locus and a locus determining the size of the gametes, nonrandom associations of alleles may lead to disassortative fusion with respect to gamete size. Considering the effect of partial selfing and inbreeding, it is shown that a mutant causing male sterility can be selected in hermaphrodites for the evolution of gynodioecy under stringent conditions (B. CHARLESWORTH and D. CHARLESWORTH, 1978; D. CHARLESWORTH and GANDERS, 1979).

A different approach to this problem has been undertaken by GIVNISH (1980), who found a correlation between animal-dispersed gymnosperms and dioecy as well as between wind-dispersed gymnosperms and monoecy. On the basis of a model a theory has been outlined on the evolution of dioecy, monoecy, and hermaphroditism in terms of the ecology of pollination and seed dispersal. The predictions of the model are supported by the distribution of breeding, pollination and dispersal systems in various groups of plants.

A model for the evolution of distyly has caused the assumption that morphological distyly is unlikely to evolve before a system of incompatibility has been established (D. CHARLESWORTH and B. CHARLESWORTH, 1979c). The maintenance of distyly depends on the opposition of selection and recombination in a heterostyled population (B. CHARLESWORTH and D. CHARLESWORTH, 1979).

β) Mating Systems and Polymorphism. Of particular interest is the continued maintenance of heterozygotes in apomictic populations. Apomixis has no effect on the maintenance of polymorphism in species which otherwise outcross, but in species with partially or completely self the possibility of the selective maintenance of balanced polymorphism is enhanced (MARSHALL and WEIR, 1979).

Selection results may be modified in experimental populations with partial selfing (HOLDEN, 1979). Partial selfing at equilibrium reduces population size and the recombination values (GOLDING and STROBECK, 1980). Breeding systems with an opposite effect on heterozygosity and dioecy have been investigated under certain conditions at equilibrium by HEUCH (1979, 1980) and NAKYLAKI (1979a,b). The fact that population parameters are not constant is considered in the case of outbreeding rates and predominantly self-pollinated plants by BROWN and ALBRECHT (1980).

Statements on the conditions of equilibria and the nature of stability have been made (KARLIN and FARKASH, 1978; RAPER et al., 1979) particularly for an admixture of positive and negative assortative mating (OTTO, 1978) without and under selection (KARLIN and O'DONALD, 1978; KARLIN, 1978) and for the effect of breeding systems at one locus with a second linked locus (GREGORIUS, 1980; HEDRICK, 1980). Assortative mating may have a potential effect on reproductive isolation and speciation (MOORE, 1979).

d) Environment and Polymorphism

In general fitness values are not constant. They may vary with the development of the individual and during the seasons or the individual may experience different environments. One population may be distributed over a range of different territories forming subpopulations by different selection effects. An increasing number of papers are concerned with problems which arise from different selection in subopulations or with temporary varying conditions.

An extensive study by TEMPLETON and ROTHMAN (1978) is concerned with the evolution of homeostasis in a fine-grained environment. A more principal consideration has been made by MAYNARD SMITH and HOEKSTRA (1980) about the robustness of models describing polymorphism in variable environments, showing that in some cases protected polymorphism requires that the relative niche size is within a small range. The authors exclude models relying on habitat selection or restricted migration. The only ones which may sustain criticism are those for diploid populations with partial dominance with respect to fitness. The fundamental assumption is that in all niches the fitness of the heterozygotes is higher than the arithmetic mean of the homozygotes.

SLATKIN and WILSON (1979) developed a model for coevolution in structural demes. They predicted that coevolved relations between species can result from various interactions beside the direct forms of competition and predation. Frequency and density-dependent interactions may be of importance.

The spatial distribution of mutant alleles in a subdivided population has been simulated by SLATKIN (1980), who distinguished two types of quantities: occupancy distribution, which provides information on the absence or presence of a mutant in different demes, and the conditional frequency distribution, which gives information on the extent of local differentiation when the mutant is present in different numbers of demes. Classical overdominance is not needed to maintain polymorphism in models of exploitative competition assuming a gene locus with multiple alleles and additive genotypic values which influence the niche position (CHRISTIANSEN and LOESCHKE, 1980).

e) Frequency and Density-Dependent Selection

Fitness values can be a function of the frequencies of the genotypes or of the number of individuals per plot or space (De BENEDICTIS, 1978). CLARKE (1979) emphasized frequency-dependent selection as an important factor maintaining genetic diversity and listed several examples to prove the limited importance of heterozygote advantage in natural populations.

Under the assumption that average population fitness decreases with increasing density, regions of allelic frequencies should be distinguished which imply limited or unlimited population growth or perhaps extinction. In this context the problem of sudden loss of alleles and population extinction has been discussed by GREGORIUS (1979). The Allee effect (growth rates are low at low density and increase to a maximum if populations reach an intermediate optimum size) has been considered by ASMUSSEN (1979a). LANGE and GLADSTIEN (1980) determined the equilibrium state under these assumptions where the fitness of the heterozygotes was of great importance.

Whereas according to POHLMANN (1980) overdominance is necessary for a polymorphic stable equilibrium at the carrying capacity, ASMUSSEN (1979b) developed a model which maintained genetic variation without overdominance. WOLF and WÖHRMANN (1979) pointed out that in theoretical considerations absolute numbers should be used for viability and fertility because the outcome of competition may be changed by using different absolute numbers of the same relation.

The change of population size has an effect on the correlation between mutation rate and heterozygosity (LI, 1979a). Numerical computations have shown that the correlation is stronger in a population which is expanding or expanded in the recent past but weaker in a population which is decreasing or has experienced a reduction recently compared with an equilibrium population with the same mean degree of heterozygosity. Finally ADAMS and DUNCAN (1979) offered a statistical method for analyzing experiments on frequency-dependent fitness.

f) Random Drift and Neutrality of Mutations

Random genetic drift influences the genotypic frequencies in small populations. It is of most interest with respect to the neutral gene theory and for breeding in animals and plants.

The genetic variability maintained in finite populations under mutation and autocorrelated random fluctuations has been discussed by TAKAHATA and KIMURA (1979) and KIMURA (1980). In diploid populations with overlapping generations fixation probabilities have been given by EMIGH and POLLAK (1979); EMIGH (1979a,b) considered also haploid populations. Approximations have been developed for the frequency spectrum, the mean number of alleles, and the homozygosity to describe the genetic variation under the pressure of neutral and deleterious mutations (LI, 1979b). In addition, STROBECK (1979, 1980) considered partial selfing and self-incompatibility. Fixation time and probabilities for two independent loci under mutation and selection have been calculated by LITTLER and GOOD (1978).

g) The Unit of Selection

What is the unit of selection: the gene, the individual, the kin, or the group? This is still a controversial theme of geneticists and evolutionists. In an extensive paper WRIGHT (1980) reviewed these problems, correcting misinterpretations and defending his shifting balance theory: "In a panmictic population natural selection operates on the field of variability provided by mutation merely according to the average effects of allele differences in all combinations. It is a creative process but one severely limited by the complexity of the relations between primary gene actions and the characters involved directly in the fitness of individuals, a complexity that insures that each allelic difference has pleiotropic effects".

Several hypothesis exist for the evolution of altruism under kin and group selection. The major ones have been reviewed by UYENOYAMA and FELDMAN (1980). Kin selection has been supposed by GRANT (1978) to be probably a rare and exceptional process in nature if it is a real process at all. WADE (1978) argued similarly in his critical review on group selection. He stated that although group selection may be existent, it could not prevail over the effect of individual selection within populations except for a highly restricted set of parameter values. The models have been reexamined in the light of empiric studies with laboratory populations of *Tribolium* and proposals have been made for further theoretical investigations.

3. Investigations on Natural Populations

a) Enzyme Polymorphism

The introduction of enzyme electrophoresis as a population genetic tool in 1964 by HUBBY and LEWONTIN has stimulated a remarkable number of investigations. This method has rendered possible an insight into the hidden genetic variability of natural populations resulting in an unexpected amount of variation of structural genes. A great number of scientists are still engaged in detecting and describing enzyme polymorphism in various species within the kingdom of plants as well as of animals. The observed degree of variation, which would have resulted in an immense genetic load and thereby in the inability of any population to exist, led to the hypothesis of neutrality formulated by KIMURA and OHTA. This theory and the detection of enzyme polymorphism have given new impulses for discussion and till now there is no final decision in favor of the "neutralists" or "selectionists". Certainly the extreme positions have been abandoned and the majority of scientists place a special significance on the selection of the majority of amino acid substitutions. The dilemma to prove one of the theories or to give an estimation of the amount of neutrality is shown by two papers summarizing and evaluating the results of experimental investigations. Whereas NEVO (1978) concluded from the results of 243 species in which 14 or more enzyme loci were tested that "natural selection, in some form, may often be the major determinant of genetic population structure and differentiation", CHAKRABORTY et al. (1980) stated on the basis of investigations of 138 populations in which at least 20 enzyme loci were considered that an agreement "with the theoretical distribution expected under the mutation-drift hypothesis" existed and that "the allele frequency distribution observed was incompatible with the prediction from the overdominance hypothesis".

Nevertheless, extensive investigations have been undertaken on enzyme polymorphism within natural populations of plants of which cultivated plants and their relatives are the most interesting.

FEJER et al. (1979) found six isozymes of alcohol dehydrogenase (ADH) in different hexaploid wheat species and varieties. With respect to ADH isoelectric focusing is useful because six isozymes can be demonstrated instead of three with standard techniques. The enzyme patterns support the hypothesis that in hexaploid wheat the ADH isozymes are dimers composed of six possible combinations of subunits and they are coded by triplicate structural genes.

In ray *(Secale cereale)* a genetic analysis of nonspecific esterase was conducted. In leaves two loci with four or two alleles per locus were found. Varying intensities of the bands were explained to be dose-dependent (SCHMIDT-STOHN, 1979). KIANG and WU (1979) proved in rice that 75% of the investigated loci were polymorphic and an average degree of heterozygosity of H = 0.22 existed. Considering 23 isozyme loci of *Pinus ponderosa* (O'MALLEY et al., 1979) from 10 habitats an average degree of heterozygosity of H = 0.12 was found.

Significant nonrandom segregation was consistingly detected at three of the investigated loci. With progressive age an increasing degree of heterozygosity was observed in *Pinus silvestris* by POLOZOVA et al. (1978). The highest amount found was H = 0.72. The gene expression can depend on the age of the plant (ADAMS and JOLY, 1980). Malate dehydrogenase (MDH) and glutamate oxaloacetate transaminase (GOT), two out of ten investigated enzymes of *Pinus taeda*, demonstrated that a gene must not necessarily be detectable in tissues of embryos and megagametophytes.

Genetic diversity within several enzyme systems existed between three regions in which *Picea glauca* has been sampled (TSAY and TAYLOR, 1978). *Demodium nudiflorum* has been investigated by SMITH and SCHAAL (1979) and SCHAAL and SMITH (1980). Eight enzyme systems encoding 13 genetic loci were considered. Samples taken from five populations showed a proportion of polymorphic loci of 46.2%, whereas the average within one population was 17.5%. There was only a low average degree of individual heterozygosity (2.3%). Significant differences between populations were observed but no population structures were detected. The average genetic distance of the populations was independent of the geographic structure and amounted to 1.4%. The structure of the populations resembled a random patchwork. *Hordem spontaneum* was investigated on 28 enzymes from 28 populations in Israel. *H. spontaneum* is the progenitor of the cultivated barley (BROWN et al., 1978; NEVO et al., 1979). One to 15 alleles per locus were detected. The mean probability was 0.19 for obtaining randomly two gametes which are genetically different at a locus, evidence that this wild species possesses a manifoldness of genetic variability. The total allozyme variance has been apportioned into 17% between regions, 37% between populations and 51% within populations. Most of the investigated morphological variation (spikelets) has occurred between populations. These results support field-sampling strategies emphasizing more sites rather than number per site.

Marginal populations of *Hordeum jubatum* in Utah were less variable in enzymes than central populations. Populations within one marginal group (high elevation or salt flat) were almost identical while populations in different groups were quite different. The authors suggested genetic drift and natural selection as the factors causing these differences (SHUMAKER and BABBLE, 1980).

b) Quantitative Traits

Many investigations have been conducted to estimate genetic variability in natural populations of related and wild forms of cultivated plants. Knowledge of qualitative and quantitative variation is a necessity for the development of breeding strategies. In such investigations mostly variations of the yield and its morphological components are of interest. Most publications have reported and discussed the variation as such and no frequency estimates have been given.

Populations of *Agrostis tenius* and *Fustuca rubra* showed selective adaptation to copper in that tin- and plumbum-tolerant populations had better adaptability to copper than nontolerant ones. The amount of tolerance which was obtained by artificial selection has been compared with that under natural conditions (KARATAGLIS, 1980a,b).

c) Breeding Systems and Polymorphism

It becomes apparent from a review of BROWN (1979) in which the experimental results on plant populations over the last decades have been summarized and evaluated that mating systems occupy a central position. Some generalizations can be made, for example that outbreeding species have a deficit of heterozygotes in comparison with panmictic ones and that inbreeders commonly have an excess of heterozygotes as compared with the neutral inbreeding law of Wright. Short-living outbreeders seem to have a greater deficit than long-living and the deficit of wind-pollinated plants is less than that of animal-pollinated ones. Some forces explain this "heterozygosity paradox", among others the capacity to form very small subpopulations and to differentiate adaptivety with respect to variation in the local environment, the necessity for partial inbreeding due to restricted population sizes and therefore limited gene flow, and the selection for multilocus associations in the inbreeding populations. Consequently more attention should be paid by scientists to the breeding system of plant populations. The simple classification into inbreeders and outbreeders is outdated because of the temporal and spatial flexibility of the mating systems as shown in *Elymus canadensis* by SANDERS and HAMRICK (1980) and SANDERS et al. (1979).

There is a variation in the degree of self-incompatibility from the central part of the population range to the margins in *Lycopersicum hirsutum*. High self-incompatibility in the center is correlated with high variability in allozymic and morphological characters (RICK et al., 1979).

d) Selection Experiments

The performance of experiments is indispensable in the case of plant breeding for studying the nature of selection. Therefore it is not surprising that most of the investigated objects are of agricultural or horticultural importance.

In a bulk experiment over ten generations with cotton *(Gossypium hirsutum)*, natural selection was powerful with respect to some morphological traits such as seed set, lint yield, coarseness of fiber, seed size, and fiber strength, while no differences in fiber length were detected (QUISENBERRY et al., 1978). Similar results were obtained by CHOO et al. (1980 a,b).

Using bulks of *Hordeum vulgare*, they found rough awns, yellow aleuron, etc. favored by natural selection but with grains per spike or grain yield very little effected. The extent of change was influenced by the environment in which the bulks were grown. Propagation under alternating environment influenced the rates of change. The frequency of alternation was of minor significance and influenced only the speed of change.

An important question is the selective influence of one trait on another not artificially selected. In *Secale cereale* populations which were selected for high seed set, the frequency of aneuploids was reduced with increasing seed set (HOSSAIN, 1978). In *Zea mays* the frequency changes of alleles at eight-enzyme loci were monitored, resulting in a greater change than was consistent with the hypothesis of selective neutrality (STUBER et al., 1980).

Since the first attempts of ALLARD and his co-workers several attempts to detect the selective forces and to estimate their intensities have been made. Continuing these efforts CLEGG et al. (1978) estimated selection parameters during the life cycle of *Hordeum sativum* in the "Composite Cross V". The calculations showed that selection is intensive at various stages of the life cycle and that the viability and fertility components often intercompensate. The three enzyme loci observed revealed large differences of selective factors between homozygotes. Fitness parameters of phenotypes with different awn length (controlled by two loci) indicated no strong evidence of heterozygote advantage. The selection values of various alleles were associated with the phenotypic expression of awn length and the initial association influenced the direction and intensity of selection (JAIN et al., 1979).

e) Geographic Differentiation and Adaptation

One attempt to study and to prove selection is to observe populations in different environments and to correlate the genetic variation with environmental factors. Such investigations have been performed on several plant species. Twenty seven loci were the subject of consideration in 31 populations of *Avena barbata* from different habitats in Israel (KAHLER et al., 1980). The array of isozymes found in each population differed distinctly. There was no cline but the correlation of different isozyme phenotypes justified the assumption of the combined effect of directional and diversifying selection.

Similar investigations have been conducted in the same country on *Hordeum spontaneum* (NEVO et al., 1979). In the self-fertilizing species the variation is mainly carried by different homozygotes of the 28 enzyme loci. Likewise spikelet variation could be correlated with environmental factors although there was no correlation between allozymic and spikelet variation. A variation of the distribution of homozygotes in *Picea abies* and *Festuca pratensis* has been described by TIGERSTEDT (1979). With the approach to the northern limit of vegetation the frequencies of homozygotes in both species increase although the different alleles still exist in the population. Similar results have been obtained by VALIZADEH and VALDEYRON (1979) in *Ficus caria* in which the alleles of five enzyme loci are nonrandomly distributed and the frequencies of two can be correlated with the mean temperature of February. However, other correlations might exist not considered in this experiment. *Poa annua* populations on a bowling green containing almost entirely prostrate growth forms differ from those of adjacent flower beds which consist of prostrate and errect forms. No differentiation in the electrophoretic esterase patterns between both forms could be demonstrated (WARWICK and BRIGGS, 1978). The variation in four quantitative traits appear to be proportional to geographic distances and to genetic differentiation. Measurements of the genetic differentiation have revealed that about 74% of the variation is contributed by the environment. This implies the importance of phenotypic plasticity for the adaptation of the species (WU and JAIN, 1978). Some phenotypic plasticity has been also found in *Fragaria vesca* which is assumed to be a specialist. This has been concluded from the aberration of 25 quantitative traits and two enzyme systems from 13 different natural sites in California. A considerable ecological differentiation has been observed in many traits (HANCOCK and BRINGHURST, 1978, 1979).

All these results support the theory of differentiation between and within populations and of the adaptiveness of monomorphic and polygenic traits. However, they give no evidence to what extent a single enzyme locus is involved. The importance of genotypic variation for the adaptability or for the competitive ability is a matter of interest in ecology and population genetics. Competition experiments have proved that generally high polymorphic populations of *Avena fatua* compete better with monomorphic *Avena barbata* and high polymorphism in *A. barbata* is a better basis for competition with monomorphic *A. fatua*. A relative stability over environments seems to favor polymorphism in one or both of the competitors (YAZDI-SAMADI et al., 1978).

Differences in several factors influencing reproduction in plants could be detected in populations of *Trifolium hirtum* from Californian ranges and from seasonal populations established by natural colonization (JAIN and MARTINS, 1979). Local differentiation has also been detected in allele frequencies, where more homozygotes existed in the range populations. The evolution of road-side populations has produced microevolutionary shifts most favorable to selection. Moreover, it can be proved that genetic polymorphism has been irrelevant to colonization ability. The gene flow by seed dispersal and pollination by honey bees is of more significance (MARTINS and JAIN, 1979, 1980).

Plantago cordata (MEAGHER et al., 1978) is a declining species in the eastern U.S.A. Investigations suggest that *P. cordata* is adapted to long-standing stable environment with a life cycle that is not preadapted to shifting man-made unstable environments.

f) Genetic Diversity and Relationship

The degree of similarity in electrophoretic enzyme patterns is used to investigate speciation and phylogeny. Enzyme patterns are regarded as additional morphological and physiological characteristics describing diversities between populations and species. Enzymatical similarities, however, do not necessarily correspond with taxonomic relationships in *Vicia* (YAMAMOTO and PLITMANN, 1980). The advantage and problems of using this method have been discussed with respect to plants by HURKA (1980).

Investigations on the flavonoid patterns of *Hordeum species* (FROST et al., 1979) have shown that cultivation has had a noticeable selective effect on the flavonoid patterns and confirmed the taxonomic separation of *H. vulgare* complex from all other species. SYMEONIDIS et al. (1979) have shown that greater agreement in barley patterns is found between different varieties of *Aegilops caudata* and

A. comosa than between the varieties of the two species. This seems to be a result of breeding.

The electrophoretic results from seedling esterases of wheat support a monophyletic origin of the primitive tetraploid and also of the hexaploid species from which contemporary taxa of polyploid wheats have emerged through recurrent introgressive hybridization and accumulation of mutations (JAASKA, 1980). The relationship and genetic variation of *Triticum* and *Aegilops* has been investigated by BRODY and MENDLINGER (1980).

Speciation in the section *Androcas* in *Solanum*, of which most populations are monomorphic at 7 loci, seems to be accompanied by evolution of internal barriers to hybridization that has caused seeds to abort (WHALEN, 1979).

No indication has been found to include the two complexes, macrosperma and microsperma, in the cultivated *Cicer arietinum*. These forms have evolved recently (MORENO and CUBERO, 1978). It is proposed to include *C. reticulata* as a subspecies of *C. arietinum*.

In *Pinus nigra* (BONNET-MASIMBERT and BIKAY-BIKAY, 1978) it is possible to distinguish the subspecies on the basis of enzyme patterns. Just the opposite can be found in *Chenopodium incanatum* and *C. fremontii* where the genetic distance between the species is much greater than the distance between some varieties of *C. incanum* (CRAWFORD, 1979).

Xanthium strumarium includes four naturalized races (MORAN and MARSHALL, 1978). Investigations have shown that, at least on the enzyme level, genetic polymorphism has been of little significance in the adaptation of the species during its colonization of Australia. Populations of *Typha latifolia* and *T. domingensis* have involved no intraspecific variation on the biochemical level and no support of hybridization exists (MASHBURN et al., 1978).

References

ADAMS, W.T., DUNCAN, G.T.: Behav. Genet. 9, 7-21 (1979). - ADAMS, W.T., JOLY, R.J.: J. Hered. 71, 33-40 (1980). - ALLENDORF, F.W.: Heredity 43, 247-258 (1979). - ASMUSSEN, M.A.: Am. Nat. 114, 796-809 (1979a); - Theor. Popul. Biol. 16, 172-190 (1979b). - AVERY, P.J., HILL, W.G.: Genet. Res. 33, 29-48 (1979). - AYALA, F.J., McDONALD, J.F.: Genetica 52/53, 1-15 (1980).

BARTON, N.H.: Heredity 43, 333-339 (1979). - BELL, G.: J. Theor. Biol. 73, 247-270 (1978). - BONNET-MASIMBERT, M., BIKAY-BIKAY, V.: Silvae Genet. 27, 71-79 (1978). - BRODY, T., MENDLINGER, S.: Plant. Syst. Evol. 136, 247-258 (1980). - BROWN, A.H.D.: Theor. Pop. Biol. 15, 1-42 (1979). - BROWN, A.H.D., ALBRECHT, L.: J. Theor. Biol. 82, 591-606 (1980). - BROWN, A.H.D., NEVO, E., ZOHARY, D., DAGAN, O.: Genetica 49, 97-108 (1978).

CALLOW, R.S., PARKER, J.S.: New Phytol. 83, 538-547 (1979). - CAMPBELL, R.B.: Genet. Res. 36, 29-40 (1980). - CHAKRABORTY, R., FUERST, P.A., NEI, M.: Genetics 94, 1039-1063 (1980). - CHARLESWORTH, B.: J. Theor. Biol. 73, 347-357 (1978). - CHARLESWORTH, B., CHARLESWORTH, D.: Am. Nat. 112, 975-997 (1978); - Am. Nat. 114, 499-513 (1979). - CHARLESWORTH, D., CHARLESWORTH, B.: Genetics 91, 575-580 (1979a); - Proc. R. Soc. London 205, 513-530 (1979b); - Am. Nat. 114, 467-498 (1979c). - CHARLESWORTH, D., GANDERS, F.R.: Heredity 43, 213-218 (1979). - CHARLESWORTH, D., CHARLESWORTH, B., STROBECK, C.: Genetics 93, 237-244

(1979). - CHARNOV, E.L.: Am. Nat. 113, 465-480 (1979). - CHOO, T.M., KLINCK, H.R., ST-PIERRE, C.A.: Can. J. Plant Sci. 60, 31-40 (1980a); - Can. J. Plant Sci. 60, 41-47 (1980b). - CHRISTIANSEN, F.B., LOESCHCKE, V.: Theor. Pop. Biol. 18, 297-313 (1980). - CLARKE, B.C.: Proc. R. Soc. London B 205, 453-474 (1979). - CLEGG, M.T., KAHLER, A.L., ALLARD, R.W.: Genetics 89, 765-792 (1978). - CRAWFORD, D.J.: Bull. Torrey Bot. Club. 106, 257-261 (1979).

DE BENEDICTIS, P.: Evolution 32, 915-916 (1978).

EMIGH, T.H.: Genetics 92, 323-337 (1979a); - Genetics 92, 339-351 (1979b). - EMIGH, T.H., POLLAK, E.: Theor. Popul. Biol. 15, 86-107 (1979).

FEJER, O., OROSZ-FEJER, K., BELEA, A.: Theor. Appl. Genet. 54, 37-39 (1979). - FROST, S., HARBORNE, J.B., ASKER, S., SALEH, N.: Hereditas 90, 251-256 (1979).

GINZBURG, L.R., BRAUMANN, C.A.: Theor. Popul. Biol. 17, 298-320 (1980). - GIVNISH, T.J.: Evolution 34, 959-972 (1980). - GOLDING, G.B., STROBECK, C.: Genetics 94, 777-789 (1980). - GRANT, V.: Biol. Zentralbl. 97, 385-392 (1978); - Biol. Zentralbl. 99, 157-169 (1980). - GREGORIUS, H.R.: J. Math. Biol. 8, 375-391 (1979); - Theor. Appl. Genet. 57, 17-24 (1980).

HAIGH, J.: Theor. Popul. Biol. 14, 251-267 (1978). - HANCOCK, J.F. Jr., BRINGHURST, R.S.: Am. J. Bot. 65, 795-803 (1978); - Am. J. Bot. 66, 367-375 (1979). - HEDRICK, P.W.: Genetics 94, 791-808 (1980). - HEDRICK, P.W., McDONALD, J.F.: Heredity 45, 83-97 (1980). - HEDRICK, P.W., JAIN, S., HOLDEN, L.: Multilocus systems in evolution. In: Evolutionary Biology, Vol. 11. New York-London: Plenum Press 1978. - HELLER, R., MAYNARD, SMITH, J.: Genet. Res. 32, 289-293 (1978). - HEUCH, I.: Theor. Popul. Biol. 15, 43-57 (1979); - Hereditas 92, 53-57 (1980). - HOFF, R.J., McDONALD, G.I.: Res. Note, USDA For. Serv., Intermountain For. and Range Exp. Stn 249, 4 (1978). - HOLDEN, L.R.: Genetics 93, 217-236 (1979). - HONDELMANN, W.: Proc. V. Intern. wheat gen Symp. 149-155 (1979). - HOSSAIN, M.G.: Euphytica 27, 137-143 (1978). - HURKA, H., in: Chemosystematics: Principles and Practice, eds. F.A. BISBY, J.G. VAUGHAN, C.A. WRIGHT. London-New York: Academic Press 1980.

JAASKA, V.: Theor. Appl. Genet. 56, 273-284 (1980). - JAIN, K.B.L., JAIN, S.K., SCHALLER, C.W.: Genetica 51, 115-124 (1979). - JAIN, S.K., MARTINS, P.S.: Am. J. Bot. 66, 361-366 (1979). - JOARDER, O.I., GHOSE, S.K., SALEHUZZAMAN, M.: Z. Pflanzenzücht. 81, 248-257 (1978).

KAHLER, A.L., ALLARD, R.W., KRZAKOWA, M., WEHRHAHN, C.F., NEVO, E.: Theor. Appl. Genet. 56, 31-47 (1980). - KARATAGLIS, S.S.: Pl. Syst. Evol. 134, 215-228 (1980a); - Pl. Syst. Evol. 134, 23-31 (1980b). - KARLIN, S.: Theor. Popul. Biol. 14, 430-445 (1978). - KARLIN, S., FARKASH, S.: Theor. Popul. Biol. 14, 446-470 (1978). - KARLIN, S., FELDMAN, M.W.: Genetics 90, 813-825 (1978). - KARLIN, S., O'DONALD, P.: Heredity 41, 165-174 (1978). - KIANG, Y.T., WU, L.: Bot. Bull. Acad. Sin. 20, 103-115 (1979). - KIMURA, M.: Proc. Natl. Acad. Sci. USA 77, 522-526 (1980). - KIMURA, M., KING, J.L.: Proc. Natl. Acad. Sci. USA 76, 2858-2861 (1979).

LANGE, K., GLADSTIEN, K.: Theor. Popul. Biol. 18, 31-43 (1980). - LEONARD, K.J.: Can. J. Bot. 56, 1809-1815 (1978). - LI, W.H.: Genetics 90, 349-382 (1978); - J. Mol. Evol. 12, 319-329 (1979a); - Genetics 92, 647-667 (1979b). - LITTLER, R.A., GOOD, A.J.: Theor. Popul. Biol. 14, 204-214 (1978).

MARSHALL, D.R., WEIR, B.S.: Heredity 42, 159-172 (1979). - MARTINS, P.S., JAIN, S.K.: Am. Nat. 114, 591-595 (1979); - J. Hered. 71, 29-32 (1980). - MASHBURN, S.J., SHARITZ, R.R., SMITH, M.H.: Evolution 32, 681-685 (1978). - MAYNARD SMITH, J.: The Evolution of Sex. Cambridge: Cambridge University Press 1978; - Genet. Res. 35, 269-277 (1980). - MAYNARD SMITH, J., HOEKSTRA, R.: Genet. Res. 35, 45-57 (1980). - McLAUGHLIN, R.J., CHRISTIE, B.R.: Can. J. Plant Sci. 60, 547-554 (1980). - MEAGHER, T.R., ANTONOVICS, J., PRIMACK, R.: Biol. Conserv. 14, 243-257 (1978). - MEINARTOWICZ, L.E., SZMIDT, A.: Silvae Genet. 27, 59-62 (1978). - MOODY, M.: J. Math. Biol. 8, 73-109 (1979). - MOORE, W.S.: Heredity 42, 173-186 (1979). - MORAN, G.F., MARSHALL, D.R.: Aust. J. Biol. Sci. 31, 283-291 (1978). - MORENO, M.T., CUBERO, J.I.: Euphytica 27, 465-485 (1978). - MORGAN, K., STROBECK, C.: Nature 277, 383-384 (1979).

NAGYLAKI, T.: Ann. Hum. Genet. 43, 143-150 (1979a); - J. Math. Biol. 8, 123-131 (1979b). - NEVO, E.: Theor. Popul. Biol. 13, 121-177 (1978). - NEVO, E., ZOHARY, D., BROWN, A.H.D., HABER, M.: Evolution 33, 815-833 (1979).

OHTA, T.: Evolution and Variation of multigene families. Lecture Notes in Biomathematics, Vol. 32. Berlin-Heidelberg-New York: Springer 1980. - O'MALLEY, D.M., ALLENDORF, F.W., BLAKE, G.M.: Biochem. Genet. 17, 233-250 (1979). - O'TOOLE, J.C., MOYA, T.B.: Crop. Sci. 18, 873-876 (1978). - OTTO, P.A.: J. Hered. 69, 207-209 (1978).

POHLMANN, J.: Biol. Zentralbl. 99, 553-570 (1980). - POLOZOVA, L.Y., PRAVDIN, L.F., DUKHAREV, V.A.: Dokl. Akad. Nauk SSSR 243, 778-781 (1978).

QUISENBERRY, J.E., ROARK, B., BILBRO, J.D., RAY, L.L.: Crop. Sci. 18, 799-801 (1978).

RAPER, J.K., KARLIN, S., O'DONALD, P.: Heredity 43, 27-34 (1979). - RICK, C.M., FOBES, J.F., TANKSLEY, S.D.: Plant Syst. Evol. 132, 279-298 (1979).

SANDERS, T.B., HAMRICK, J.L.: Evolution 34, 117-122 (1980). - SANDERS, T.B., HAMRICK, J.L., HOLDEN, L.R.: Am. Midl. Nat. 101, 1-12 (1979). - SCHAAL, B., SMITH, W.G.: Evolution 34, 214-221 (1980). - SCHMIDT-STOHN, G.: Z. Pflanzenzücht. 83, 155-162 (1979). - SHARMA, J.R., MURTY, B.R.: Genetica 51, 45-53 (1979). - SHUMAKER, K.M., BABBLE, G.R.: Evolution 34, 110-116 (1980). - SLATKIN, M.: Genetics 93, 755-771 (1979); - Genetics 95, 503-523 (1980). - SLATKIN, M., WILSON, D.S.: Proc. Natl. Acad. Sci. USA 76, 2084-2087 (1979). - SMITH, W.G., SCHAAL, B.A.: Biochem. Syst. Ecol. 7, 121-123 (1979). - STROBECK, C.: Genetics 92, 305-315 (1979); - Evolution 34, 779-788 (1980). - STUBER, C.W., MOLL, R.H., GOODMAN, M.M., SCHAFFER, H.E., WEIR, B.S.: Genetics 95, 225-236 (1980). - SYMEONIDIS, L., KARATAGLIS, S., TSEKOS, I.: Plant Syst. Evol. 131, 1-15 (1979).

TAKAHATA, N., KIMURA, M.: Proc. Natl. Acad. Sci. USA 76, 5813-5817 (1979). - TEMPLETON, A.R., ROTHMAN, E.D.: Theor. Popul. Biol. 13, 340-355 (1978). - TIGERSTEDT, P.M.A.: Holartic Ecol. 2, 264-268 (1979). - TSAY, R.C., TAYLOR, I.E.P.: Can. J. Bot. 56, 80-90 (1978)

UYENOYAMA, M.K., FELDMAN, M.W.: Theor. Popul. Biol. 17, 380-414 (1980).

VALIZADEH, M., VALDEYRON, G.: Ann. Amelior. Plant. 29, 213-225 (1979)

WADE, M.J.: Q. Rev. Biol. 53, 101-114 (1978). - WARWICK, S.I., BRIGGS, D.: New Phytol. 81, 711-723 (1978). - WHALEN, M.D.: Syst. Bot. 4, 203-222 (1979). - WOLF, H.G., WÖHRMANN, K.: Oecologia 143, 365-371 (1979). - WRIGHT, S.: Evolution 34, 825-843 (1980). - WU, K.K., JAIN, S.K.: Can. J. Bot. 56, 873-879 (1978).

YAMAMOTO, K., PLITMANN, U.: Jap. J. Genet. 55, 151-164 (1980). - YAZDI-SAMADI, B., WU, K.K., JAIN, S.K.: Genetica 48, 151-159 (1978).

Professor Dr. KLAUS WÖHRMANN
Dr. JÜRGEN TOMIUK
Institut für Biologie
Lehrstuhl für Populationsgenetik
der Universität
Auf der Morgenstelle 28
D 7400 Tübingen

D. Taxonomy

I. Systematics and Evolution of Seed Plants

By HANS-HELMUT POPPENDIECK

With a Contribution by KLAUS KUBITZKI

1. Introduction

The development of plant population biology (SOLBRIG et al., 1979) as a discipline which connects plant systematics with ecology has been one of the most important trends within recent years, though at present more questions have been asked than answered. Likewise, the study of reproductive biology, notably of tropical plants, has turned into a field of tremendous activity; most of it will be reviewed in this volume under pollination ecology, but the results are of course of great interest for the evolution and systematics of seed plants. Methodological progress has been made in the interpretation of phytochemical data. Apart from that, the trends mentioned in the last review (Progr. Bot. 41, 239) are still of interest. As to the problems of classification in general, the number of cladistic studies has increased, but it is still doubtful if they will have a major impact. Comprehensive, integrated approaches to the systematics and biology of major groups include the Solanaceae symposium (HAWKES et al., 1979), and the resumption of the *Natürliche Pflanzenfamilien* series with a multi-authored treatment of the Loganiaceae (LEEUWENBERG, 1980).

Past, present, and future of systematic biology have been reviewed by RAVEN (1977) and HEYWOOD (1980) on occasion of the bicentenary of the U.S.A. and the Linnaeus anniversary, respectively. HEYWOOD (1980) believes taxonomy to be in a highly critical stage; the rapidly progressing destruction of plant life on earth calls for a reappraisal of scientific policies, but despite the vastly improved techniques the goals of systematic biology have become controversial. HEYWOOD (1980) sees the most urgent task of systematic botany today in giving a survey of diversity as representative as possible, and in answering some questions about the biology of plants, especially from the tropics, as long as we are able to study them. This view is shared by RAVEN (1979), perhaps with a slight difference, as he argues especially for multi-disciplinary studies on a few selected species and believes that we may have a fairly representative overview of plant diversity already. The gaps that exist in our knowledge of tropical floras have been pointed out by FOSBERG (1979) for islands and by PRANCE (1979) for the lowland Neotropics; lack of critical collections and revisions render many phytogeographic interpretations conjectural.

As always, revisions and monographic studies unfortunately will not be treated in the fullness which they would rightfully deserve. The reviewer feels that this is due to the fact that their complexity prevents them from being summarized within a few lines. In contrast to these studies which contain a multitude of results in one publication, it is now becoming more and more fashionable to utilize one finding for several papers, and the attention of the reader is drawn to BROAD (1981) for a full discussion of this phenomenon. This short summary would be incomplete without mentioning the recent upswing of plant taxonomy on the Southern hemisphere, especially in Australia (historical review: CARR 1981) and New Zealand

Textbooks etc.: Plant Systematics: JONES and LUCHSINGER (1979); beautifully illustrated account of angiosperm families: HEYWOOD (1978); identification of angiosperm families: DAVIS and CULLEN (1979); new edition of Zander's dictionary: ENCKE et al. (1979).

2. Problems of Classification

a) The Art of Phylogenetic Reconstruction

A small but very prolific group is currently trying to convince the botanical community of the merits of cladistics, a concept which sees the very task of biological systematics in the production of phylogenetic classifications, or at least of phylogenetic reconstructions; this is in marked contrast to the phenetics movement which started some 20 years ago with the aim of producing better classifications without the preoccupation with phylogeny, and whose procedures have by now become something like a standard tool in many monographic studies. Already, several partly conflicting schools of cladism have developed, but, as WAGNER (1980) noted, the differences in the often intricate and to the outsider somewhat confusing terminology must not necessarily reflect different concepts. Prevailing among zoologists, and advocated strongly for botany by BREMER and WANNTORP (1978, 1981), is the method of Hennig; a fine application of this approach can be found in HUMPHRIES (1979). WAGNER's (1980) groundplan-divergence method, developed about 30 years ago, has had a large number of applications within systematic botany, and Wagner's account of his method stands from the rest of many cladistic papers in being lucid and readable. A different concept is used by the so-called clique group which bases its phylogenetic estimates on the largest collection (= "clique") of mutually compatible characters (e.g., CRISCI, 1980; CRISCI et al., 1979; DUNCAN, 1980a; LA DUKE and CRAWFORD, 1979; MEACHAM, 1980; STUESSY, 1979), an approach that has been severely criticized by Hennigian cladists (FARRIS and KLUGE, 1979). Recently, DUNCAN et al. (1980) have compared different numerical and cladistic methods within one group, the hypothetical "Dendrogrammaceae", and concluded that all of them may elucidate evolutionary history, but no-one provides the "best" or "right" answer. This leaves the reviewer somewhat worried about the question whether methods which are capable of producing evolutionary histories in fictitious groups might not produce fictitious phylogenies in real taxa.

The apparent lack of enthusiasm for cladistics among botanists may lie in the fact that the high degree of reticulation caused by polyploidy and the absence of uniquely derived characters in angiosperms impede the constructions of cladograms, and that botanical systematics has always been more pragmatical than zoological systematics (HEYWOOD, 1980). Apart from that, many botanists feel that cladisms just present "The Emperor's New Taxonomic Dress" (FAEGRI, 1979) which merely synonymizes and formalizes well-known concepts. A further objection, also raised by FAEGRI (1979), is that the consistent application of cladistic principles will lead either to a plethora of categories, one for each time the tree is supposed to have branched, or to too comprehensive and thus useless taxa. Recently proposed re-classifications of higher taxa of land plants (BREMER and WANNTORP, 1981; PARENTI, 1980) have aimed at refuting these objections, but will probably not convince many of the advantages of cladistic classifications.

PARENTI (1980) tries to cut the Gordian knot of land plant systematics with the cladistic razor. Current classifications are severely criticized because they contain many nonmonophyletic groups, e.g., dicotyledons. Since the study utilizes the circumscriptions and reasonings of elementary textbooks, the result is far from conveying something excitingly new, though the arguments presented are often, to say the least, bizarre: The Gneticae are a sister group to the angiosperms with which they share among others the presence of a compound pollen cone; the two uniquely derived characters of angiosperms are the presence of an ovary, and the fact that sporophylls are differentiated into sepals and petals (!); uniquely derived characters of monocotyledons are atactostele, one cotyledon, and a reduced number of sporophylls due to fusion (!). BREMER and WANNTORP (1981), in a similar attempt, wisely avoid naming the uniquely derived characters defining their monophyletic groups.

Just as phenetics helped to clarify some concepts about characters, cladistics may be beneficial for the now often obscure lines of argument used in phylogenetic discussions. Repugnant, however, is the loud claim of some cladists to the one and only true scientific method. Heavily basing on the authority of the philosopher Popper, WANNTORP (1980) refutes BURGER's (1979) criticism and maintains that cladistic classifications are falsifiable, and therefore scientific; classifications based on general similarity cannot be falsified, and are thus unscientific. This is a new angle of the apparently unending discussion about phylogenetic and phenetic classification of which McNEILL (1979b) has given a critical review.

A symposium on cladistics and plant systematics gives a good overview about different lines of thinking (Syst. Bot. 5, 109-231, 1980), although the panel discussion suffered somewhat from lack of opposition. Fortunately, there are at least some studies in which new data are presented, as those by MEACHAM (1980), whose cladistic analysis of the Berberidaceae confirms Janchen's classification, and the wood anatomical studies by KOEK-NOORMAN (1980) and BAAS and ZWEYPFENNING (1979), which employ cladistics and are illustrative of the difficulties of assigning evolutionary polarity to character states and decide on mono- or polyphyletic origin of specialized characters.

b) Identification of Primitive Character States

Prerequisite for any cladistic scheme is the assigning of primitive or advanced status to a number of character states. STEVENS (1980) gives a critical review of the different criteria and, not surprisingly, arrives at the conclusion that none is flawless or to be applied in every given case. The most satisfactory principle seems to be outgroup analysis, i.e., a character state shared with a more primitive ancestor or, less rigorously, with most members of the higher taxon, is held to be primitive; unfortunately, this is not possible to apply in many cases as, for example, in isolated taxa. Criteria as ontogeny, teratology, correlations, or associations of characters are considered to be unreliable. Also unadmissible are general ideas about the nature of evolutionary processes, not because they are necessarily wrong as generalizations, but because of the danger of circular reasoning: Cladistic and other evolutionary studies concern particular cases from which this kind of generalizations are hoped to be obtained. The same topic is treated by CRISCI and STUESSY (1980), understandably in a more optimistic way, although it is admitted that absolute statements are impossible. Interestingly, they are at variance with STEVENS (1980) in two points: First, they consider paleobotanical evidence to be most valuable, whereas STEVENS (1980) considers it as not reliable because of the uncertainty of connecting different organs and taxa. Secondly, in-group analysis, i.e., common equals primitive, is ranked very high; this criterion is dismissed by STEVENS (1980), not only because many systematists will hold the opposite to be more probable as generalization, but also because it excludes the possibility of a common character being derived (see also RIEPPEL, 1979).

c) Handling Diversity

In contrast to the cladists, many other systematists would consider the chief aim of biological classification to characterize diversity in a mentally satisfying way. This is, according to LEVIN (1979a), also the only task which the concept of the species can perform. In order to be acceptable, classifications have undergone a process of simplification which is reflected, among others, by the fact that the average number of subordinates to a given taxon is something between 5 and 6, a number which seems to be convenient to memorize (HEYWOOD, 1980). The extent to which classification meets this criterion of simplification is termed structural value by McNEILL (1979a) who also presents a procedure to select the best simplification when the dendrogram obtained is too complicated for good formal taxonomy. It is interesting to see that numerical phenetics which started as a move to abolish intuitively derived classifications, now have gone full circle and incorporate psychological findings for their design. Relevant to this topic is also a study by GILMARTIN (1980) who compared patterns of variation within temperate Asclepiadaceae and Umbelliferae, and tropical Bromeliaceae. The first two families are similar on each level of classification, whereas Bromeliaceae are different on population level, but similar on the generic one. According to GILMARTIN (1980), this reflects the attempt of taxonomists trying to produce the same amount of generic variation in each family.

The reason for the present excessive proliferation of oligotypic sections in *Pandanus* reflects partly the high diversification of the genus and partly the incomplete knowledge of some species (HUYNH, 1980a). A numerical study within a species complex of *Stipa* reveals one species with two subspecies, one of them with two varieties, instead of three species as was initially assumed (BARKWORTH et al., 1979) which again leads to the question of subspecific categories. STACE (1979), in his very readable textbook of plant taxonomy and biosystematics, again makes a plea for the use of a full array of subspecific categories in some extensively treated taxa, e.g., the wild relatives of cultivated plants, and KLOTZ (1980) maintains that infraspecific categories are as real in nature as species, genera, and families. Fortunately, he argues that at least the subforma should not be used any longer. An interesting example of the species problem — the legal one — is illustrated by SMALL (1979) in the context of *Cannabis* cultivation.

The practice of EDP in plant systematics has been dealt with in several publications. Pankhurst's book on biological identification is strongly oriented toward computerized procedures; JOHNSTON (1980) presents a program to construct polyclave keys from data matrices, and ČEŠKA and TRUMPOUR (1979) one for computerized editing of serial and indented keys. A preliminary to the numerical taxonomy of angiosperm leaves has been published by HILL (1980). ESTABROOK (1979) presents the data bank of seed plant types at the University of Michigan herbarium, and BRENAN et al. (1979) have edited a symposium volume on computers in botanical collections, published 6 years after the meeting.

3. Morphology

a) Anatomy and Micromorphology

Metcalfe and Chalk's well-known systematic anatomy of dicotyledons is now in its second edition. The first instalment, a multi-authored volume, contains the general chapters and constitutes something like a textbook of systematic plant anatomy in the widest sense, since it contains also items like the classification of leaf architecture; Takhtajan's system was chosen as a basis which is thought to fit moderately well with the anatomical features of the dicotyledons. It is quite impressive to see the progress which this field of research has made since the publication of the first edition some 30 years ago. A very readable textbook of practical plant anatomy was presented by CUTLER (1978). FAHN (1979) surveyed the different types of secretory tissue in plants, and thus provided a wealth of information, but was rather cautious in interpreting these data. CZAJA's (1978) book on starch and starch accumulation in vascular plants is a rich source of important information for the plant taxonomist, though his conclusions do not appear to be well-founded in all cases.

The occurrence of leaf sclereids in angiosperms has been reviewed in the context of Dahlgren's system (see p. 217) by RAO and DAS (1979); they are apparently absent or rare in some groups often considered to be advanced, such as Compositae, Centrospermae, or Monocotyledons. PAYNE (1979) presents a new classification of stomatal patterns in embryophytes, based on ontogeny, and discusses the evolution of these patterns; interestingly, he considers the diameristic mesoperigenous type (i.e., guard mother cell dividing with a wall at right angles to the preceding wall, with only the proximal cell mesogene) to be truly primitive not only for lower embryophytes, but also for monocots, Centrospermae, and several lines within the dicots.

The wood anatomy of the Myrtales is currently under study by a group of Dutch botanists; groups already investigated include Blakeeae (Melastomataceae) (KOEK-NOORMAN et al., 1979) and Lythraceae (BAAS and ZWEYPFENNING, 1979). The latter work is interesting because, as in KOEK-NOORMAN's (1980) study of some rubiaceous genera, cladistics are used to elucidate wood phylogeny; in both cases several problems emerge. On the basis of wood anatomy, Punica(ceae), *Rhynchocalyx* and *Psiloxylon* are very close to Lythraceae, and the affinity to this family decreases on the line Oliniaceae-*Alzatea*-Sonneratiaceae-Onagraceae-Melastomataceae.

The wood structure of *Archidendron* is advanced, and puts in question the supposedly primitive condition of the pluricarpellate gynoecium in this mimosoid genus (BARETTA-KUIPERS, 1979). CARLQUIST (1980) comments on some concepts of ecological wood anatomy, as fiber dimorphism, adaptive morphology of vessel elements, growth rings, pedomorphosis, and some general problems of interpretation. As regards nodal anatomy, SUGIYAMA (1979) considers the multilacunar type of Magnoliaceae to be the most primitive, and trilacunar and other multilacunar types to be derived. In conifers, the different types of the course of the resin canals seem to have evolved independently of the vascular system, and to reveal certain trends of evolution (SUZUKI, 1979).

Only some of the numerous TEM and SEM studies can be mentioned. SPETA (1979) found protein bodies in cell nuclei to be of prospective taxonomic importance. The notoriously difficult classification of American oak species may be faciliated by using foliar trichome characters (HARDIN, 1979; THOMPSON and MOHLENBROCK,

1979). Epidermal characters of succulent Liliaceae and Aizoaceae may be of systematic value, but caution must prevail as these show infraspecific variation or may be influenced by environmental factors in the latter family (CUTLER and HARTMANN, 1979). The unique hair-bristles at the leaf tip of the S. African succulent *Trichodiadema* (Aizoaceae) are capable of taking up droplets of condensed water from the atmosphere and play an important role for the water balance of these plants (IHLENFELDT, 1980); furthermore, these structures aid infrageneric classification. The surface sculptures of seed have been studied in a systematic context by HESSE and MORAWETZ (1980) for some Bignoniaceae, by BARTHLOTT and VOIT (1979) for Cactaceae, by CAROLIN (1980) for *Goodenia*, by T'HART and BERENDSEN (1980) for *Sedum*, and by HAMANN et al. (1979) for Hydatellaceae. The application of SEM techniques to wood anatomy allows closer inspection of hitherto neglected characters of systematic value as crystals or cystoliths, e.g., of *Sparattanthelium* (Hernandiaceae) (TER WELLE, 1980).

b) Vegetative Structures

The architectural approach to the vegetative morphology of trees by HALLÉ et al. (1978) has stimulated further studies, e.g., of *Araucaria* in New Caledonia (VEILLON 1980), and of herbaceous Zingiberaceae which conform to Tomlinson's model despite their wide range of variation (HALLÉ, 1979); in *Alpinia*, BELL (1979) and BELL and TOMLINSON (1980) detected a regular hexagonal branching pattern. J.B. FISHER and HONDA (1979a,b) presented theoretical studies of branch geometry and effective leaf area size in *Terminalia*. Another theoretical study of interest to the systematist is that of GIVNISH (1979) on the adaptive significance of leaf form, as some of his arguments have been employed by HICKEY and DOYLE (1977) for their reconstruction of the habitat of early angiosperms; he provides, among others, models and equations to cope with the well-known duality of xeromorphism and oligomorphism, as encountered in some bog plants.

HAINES and LYE (1979) are convinced that in both dicotyledons and monocotyledons, monocotyly is always syncotyly, i.e., derived from an ancestral pair of cotyledons. The occurrence of stipules in Irvingioideae and Rectioideae suggests the removal of these groups from the otherwise estipulate Simaroubaceae which is also confirmed by palynology and anatomy (WEBERLING et al., 1980). A detailed and beautifully illustrated account of the blastogeny of woody Malesian plants has been given by DE VOGEL (1980).

Studies of life forms include those on tropical-alpine vascular plants by HEDBERG and HEDBERG (1979), on Alismataceae by LIEU (1979), in *Teucrium* (Labiatae) by KÄSTNER (1979), as well as those by SHULKINA (1980) in Campanulaceae and by AESCHIMANN and BOCQUET (1980) in *Silene vulgaris*, where 15 different growth forms contribute to the polymorphy of this species.

Epiphylly in angiosperms, review: DICKINSON (1978). Symposium on juvenile forms of seed plants: Bull. Soc. Bot. France 126, Actual. Bot. 3 (1979).

c) Floral Structures

BRIGGS and JOHNSON (1979) give a very thorough evolutionary profile of the Myrtaceae, with special reference to the notoriously difficult morphology of the inflorescences. Decisive impulses for the interpretation of these structures have come from Troll's concepts of inflorescence typology; in fact, this seems to be the first attempt at a large-scale application of inflorescence typology outside Central Europe. Several changes and additions to the terminology are suggested, and listed in an annotated glossary. Conspicuous trends within the Myrtaceae are the massing of flowers in order to attract pollinators, and the reduction of the degree of branching correlated with the development of large, fleshy, zoochorous fruits. These obviously adaptive features in relation to pollination and dispersal, and the extensive parallelism observed, render it impossible to determine affinities and subgeneric groupings by using inflorescence morphology.

FROEBE (1980) presents a study of the pseudanthia of Umbelliferae; the enlargement of the phyllomes can effect umbellet or umbel or both, so that many insufficient degrees of synorganization and only few optimal stages can be found; it is maintained that these special structures are probably not directly influenced by pollinators. Flowers and inflorescences of Solanaceae and Nolanaceae have been the topic of a very thorough study by HUBER (1980).

A. WEBER (1980a) surveyed the different modes of origin of the perigone within monocotyledons, and concluded that the variation observed makes it impossible to decide whether it originated from bracts or from stamens. Possible ways leading from a centrifugal complex androecium to a simple one are pointed out by LEINS (1979); a single stamen may be homologous to a whole staminal bundle, or a part of it. Secondary polyandry has arisen separately from basic trimery in several groups of palms, probably correlated with a sequence from simple to specialized cantharophily (UHL and MOORE, 1980). Pseudomonomerous gynoecia with different evolutionary history have been detected by MAGIN (1980) in umbelliferous *Pentagynia* and *Lagoecia*.

d) Palynology and Embryology

Symposium: Ann. Missouri Bot. Gard. 66, 591-838 (1979).

While structural investigation by TEM and SEM and their application to systematics still make up most of the palynological literature, several articles have dealt with various functional aspects which may provide a new background to evolutionary interpretations. MULLER (1979), in a highly readable survey of origin, form, and function in angiosperm pollen, points to three basic functions which are reflected by the structure of the pollen wall, viz. protection, harmomegathic function, i.e., to withstand tension due to desiccation, and reservoir function for storing of incompatibility substances. Numerous examples demonstrate how these functions may influence the exine architecture, as well as shape and size of pollen grains. He describes two new evolutionary trends occurring independently in several families, (1) peristasy, the reduction of colpate ectoapertures with retention of endoapertures which leads to spherical pollen grains where the harmomegathic stresses are distributed equally over the surface, and (2) latistasy which results in a latitudinally oriented harmomegathic stress pattern, and which can be observed, e.g., in *Utricularia* or *Xanthophyllum* where equatorial costae are bordering the encircling endoapertures. In widespread *Peucedanum* species, the architecture of the exine has been found to be very variable and correlated to climate (NIGAUD, 1980). The ontogeny of the exineless pollen of *Heliconia* has been studied by STONE et al. (1979).

The lack of pollenkitt in *Gnetum* and other gymnosperms, and its omnipresence in angiosperms (HESSE, 1980a,c) confirms the coherence of the latter and suggests that the development of pollenkitt was a prerequisite for the switch from anemophily to entomophily which was to become the ancestral pollination mode in angiosperms. Within angiosperms,

HESSE (1979a,b,c,d) was able to demonstrate that changes of this single feature may be crucial for the alteration in the mode of pollination; in Ericaceae, where a highly specialized mechanism for locating pollen on the insect body exists, the stickiness of pollen is low, which enables both wind and insect pollination (HESSE, 1979e); the "secondary" nature of insect pollination in *Euphorbia* seems to be confirmed by the data from pollenkitt studies (HESSE, 1980b). Viscin threads as in Onagraceae and Leguminosae are functionally analogous to pollenkitt (HESSE, 1979f) and bring about a highly effective pollination (CRUDEN and JENSEN, 1979).

According to the nature of their resources, pollen grains can be divided into starchy and starchless ones; starch is easier to metabolize and to store (cf. FREY-WYSSLING, 1980), but bulkier than lipids. Lipids are less space-consuming, more difficult and energetically more costly to metabolize, but of higher nutritive value to Diptera and Hymenoptera. Consequently, BAKER and BAKER (1979) found that fly- and bee-pollinated species, as well as those with small pollen grains, tend to be starchless, whereas pollen is starchy in anemophilous and entomophilous species. As this single trait is influenced by both mode of pollination and nutrient requirements of pollen tube growth, the overall picture is complicated by several findings, e.g., that in bird and lepidopteran flowers where pollen has to traverse long styles, there is a selection toward large oil-containing or very large starchy grains. As BAKER and BAKER (1979) point out, the systematic and phylogenetic implications are not very conspicuous; the majority of angiosperm families are starchless, and many are discordant. Apparently the starch-containing families are concentrated within the "core" of angiosperms, whereas both "advanced" (e.g., Compositae, Labiatae) and "primitive" families (e.g., Magnoliaceae, Winteraceae) possess starchless pollen.

The adaptive traits of the pollen wall have been studied by HESLOP-HARRISON (1979a,b) with regard to the hydrodynamics of pollen. The desiccated pollen grain takes the water necessary for germination from the stigma. This system evolved under progressive co-adaptation of the two partners with regards to control, uptake, use, and loss of water, and can be viewed as yet another evolutionary innovation of the angiosperms.

NOWICKE and SKVARLA (1979) studied pollen from several dicotyledonous orders. Their main results are that the coherence of the Centrospermae is stressed once again, that Plumbaginaceae, Polygonaceae, and Primulaceae are neither closely related to each other nor to Centrospermae, and that the Ranunculales are rather heterogeneous but, because of the rather similar spinulose and punctate/perforate tectum, the Ranunculaceae may well be the closest relatives of the Centrospermae. HIDEUX continued his studies within the Saxifragales; the ontogeny of the sporoderm proved to be helpful for the elucidation of trends and levels of evolution, and suggests that the Saxifragaceae s. str. are highly evolved and isolated within the Rosidae. *Vigna* and *Phaseolus*, closely related, differ strongly with regard to the architecture of the exine (HORVAT and STAINIER, 1980).

Olacaceae and Icacinaceae distinct, with reverse trends of differentiation: Olacaceae with complication and ornamentation of exine, Icacinaceae with loss of endexine (LOBREAU-CALLEN, 1980). 3-zonocolporate pollen of Didymelaceae suggests relation to Buxaceae/Euphorbiaceae (KÖHLER, 1981). Pollen morphology of S. African Malvales, with numerical analysis (COETZEE and VAN DER SCHIJFF, 1979). Pollen morphology in Commelinaceae correlated to structure of androecium (POOLE and HUNT, 1980). Finely reticulate, 3-colpate pollen probably ancestral in Bignoniaceae, otherwise very diverse and useful for generic delimitation but because of many parallel lines of evolution no application for higher levels (GENTRY and TOMBS, 1979). Several unique pollen types in Leguminosae (GRAHAM et al., 1980; FERGUSON and SKVARLA, 1979); peculiar rod-shaped pollen in *Crossandra* (Acanthaceae) (BRUMMITT et al., 1980). Palynology of large-flowered neotropical and Australian Loranthaceae support the idea of a transpacific evolutionary connection (FEUER and KUJIT, 1980). CHANDA et al. (1979) discuss trends of evolution of mono- and diaperturate pollen grains. Pollen dimorphism in Lecythidaceae: MORI et al. (1980). Viscin threads in *Camissonia* (Onagraceae): SCHLISING et al. (1980).

A textbook of the embryology of gymnosperms has been presented by SINGH (1978). YAKOVLEV and ZHUKOVA (1980) have surveyed angiosperm embryos containing chorophyll, and their results were mapped by DAHLGREN (1980b) in his diagram; the occurrence of this character is loosely correlated to the lack of endosperm and seems to have arisen independently in several larger groups. Embryological studies confirm family rank for Eriocaulaceae (AREKAL and RAMASWAMY, 1980). *Corynea* (Balanophoraceae) possesses a crassinucellate, ategmic ovule (ENGELL, 1978).

4. Karyology and Cytogenetics

a) Karyotypes and DNA Content

Symposium on organization, evolution, and function of genome and chromatin: Plant Syst. Evol. Suppl. 2 (1979).

NAGL (1979) summarizes the current ideas about structural organization of chromosomes in relation to evolution and ontogeny. Noncoding repetitive DNA which is variable in quantity and organized in such a way that it determines the macrostructure of the chromosomes, is decisive for the control of karyotype evolution and gene regulation and thus for the control of phylogenetic diversification. Gene mutation, protein diversification, and selection contribute only to the fine-tuning of the "Baupläne". This may explain the observation that molecular evolution seems to be largely independent of rates of organismal evolution, viz. that morphologically conservative organisms have not experienced a slower gene sequence evolution than organisms that have evolved unusually rapidly. As the same mechanisms of control are effective in both ontogenetic and phylogenetic changes, this revives Haeckel's "Biogenetisches Grundgesetz" at least for the field of molecular biology. This is partially supported by L.D. GOTTLIEB's (1979) study of *Stephanomeria* (Compositae), one of the rare cases where progenitor and sympatric derivative species could be identified with certainty; morphological and allozyme analysis show that the differences observed are due to changes in the regulation of genes rather than to the products they specify.

SCHAAN and NAGL (1979) studied the amount of repetitive DNA in magnoliid angiosperms (*Magnolia, Decaisnea, Cinnamomum, Liriodendron*) and found it to be clearly lower than in "higher" angiosperms, the variation between the genera being less than that found within genera of more advanced families. This is taken to confirm the hypotheses that the proportion of repetitive DNA increased during macroevolution and that primitive, or stable, taxa are less variable in their genome composition than more advanced, rapidly evolving ones. A clear correlation between increase of nuclear complexity and increase of nuclear DNA content has been found by NAGL and FUSENIG (1979), both being highest in monocotyledons. LEVIN and FUNDERBUG (1979), in a review from literature, found genome size to be usually twice as large in temperate as in tropical families; in Gramineae, the proportion may rise as high as 3:1, whereas Compositae and Leguminosae show no differences between temperate and tropical taxa. The decrease of nuclear DNA content in *Microseris* is paralleled by a trend from perennial to annual species, and obviously influences the timing of the different ontogenetical steps, as can be induced from changes in capitulum structure (BACHMANN et al., 1979). The nuclear DNA content within the *Scilla bifolia*-group has decreased progressively from primitive yellow-seeded species to more advanced ones with black seeds; the accumulation of heterochromatin in two side lines suggests that this has been an additional element in genome evolution (GREILHUBER, 1979).

b) Cytogenetics

Genetic self-incompatibility which has been postulated to be involved in the origin of complex heterozygotes in *Oenothera* subgen. *Oenothera*, was detected for the first time in *O. grandiflora*. This species is assumed to be one of the parents in the original cross that led to *O. biennis* (STUBBE and RAVEN, 1979a). A multiple independent origin of the permanent translocation heterozygote species *Oenothera laciniata* as an interpopulational hybrid of *O. grandis* is suggested on the basis of allozyme data by ELLSTRAND and LEVIN (1980). STUBBE (1980) summarizes what is known about the evolution of complex heterozygosity in *Oenothera* which can arise either from interspecific hybridization or from structural heterozygotes from a single population which may slowly evolve toward complex heterozygosity. Translocation heterozygosity has played an important part in the evolution of dioecy within *Viscum* (WIENS and BARLOW, 1980).

Little structural variations and almost no failure of chromosome pairing at meiosis have been observed in the tuber-bearing Solanums (RAMANNA and HERMSEN, 1979); the suppression of crossing-over between homoeologous chromosomes may be due to some small observable or postulated cryptic structural changes. A biosystematic study of *Lycopersicon* (RICK, 1979) shows that in divergence from *Solanum* chromosomal differentiation has occurred, while within *Lycopersicon* the species seem to have evolved entirely by gene substitution. In *Dioscorea*, the absence of multivalents suggests allopolyploid origin of some of the species (PAL and SHARMA, 1980). Gene-controlled pairing with regular formation of bivalents favors the formation of polyploid hybrids in *Cymbopogon* (QURAISH and FARUQI, 1979). The genome organization in *Zea mays* and its wild and cultivated relatives was investigated by HAKE and WALBOT (1980), using DNA hybridization, and is illustrative of the problems of the "molecular clock", as genome differences would imply a higher age of the species than is usually estimated for the advent of grass genera.

c) Variation of Chromosome Numbers

The instability of the genome as regards different levels of ploidy in aging clones of Lemnaceae is one of the most conspicuous traits within the family which has been treated by URBANSKA-WORYTKIEWICZ (1980a). Cytological variation occurs on the individual, populational, and racial level. Euploid chromosome numbers range from 2n = 20 to 30, 40, 50, 70, and 80; deviating numbers might be due to selective endoduplication. Peculiarities of the life-cycle, asexual reproduction, and very small chromosomes make the identification of genotypes and the distinction between genets and ramets difficult.

In Elaeagnaceae, chromosome numbers x = 11, 12, 13, and 14 have been reported, but the evolutionary direction is difficult to decide, since the order of decreasing symmetry which is usually taken as a criterion is different when chromosome size or centromeric position are used (AROHONKA and ROUSI, 1980). A previous cladistic analysis of *Bupleurum* suggests that the numbers x = 7 and x = 6 have been derived from x = 8 and a nearly hypothetical x = 4 (CAUWET-MARC, 1979). The most common number in Bignoniaceae, n = 20, is interpreted as being derived from x = 7 as a putative hexaploid with subsequent loss of a pair of chromosomes which, however, seems to have been retained in the primitive genus *Delostoma* which has n = 21 (GOLDBLATT and GENTRY, 1979). The high frequency of polyploidy in the exclusively woody Australian Myrtaceae seems to be connected with apomixis; this is considered by RYE (1979) to be successful in short terms, but of limited evolutionary potential. Several myrtaceous genera are characterized by descending dysploidy. The fact that the large and successful genus *Eucalyptus* completely lacks polyploidy remains to be explained. The Old and New World species of Vernonieae (Compositae) have different base numbers, n = 9 or 10, and n = 17, respectively; the latter may represent ancient polyploids derived by aneuploidy from n = 9 (S.B. JONES, 1979). A hybrid origin seems possible for most Mediterranean species of *Limonium* with numbers higher than 2n = 18 (ERBEN, 1979); the numbers range from 2n = 12 to 54, and the genus has a large percentage of apomictic species. The rare type of an autopolyploid series with 2n = 14, 28, 42 has been found by DEWEY (1979) within the *Hordeum violaceum*-complex. In Stylidiaceae, chromosome numbers range from n = 5 to n = 16, with occasional polyploid series based on n = 13, 14, 15; the latter number is held to be primitive by JAMES (1979).

Allogamic perennials of the grass tribes Airinae and Aristaveninae are highly variable colonizers of open habitats and exhibit high levels of polyploidy; in contrast, annual species are well distinct morphologically, autogamous, and mostly diploid but often with chromosomal variations as decreasing dysploidy. In critical habitats, perennials will be apomictic and annuals obligately selfing (ALBERS, 1980a,b,c). No influence of nuclear DNA content on duration of either life-cycle or microsporo- or microgametogenesis in these tribes could be found (ALBERS, 1980a). Similar correlations between life-cycle and cytological strategies could be detected in Iranian Dipsacaceae, with perennials frequently exhibiting polyploidy and aneuploidy, and annuals aneusomaty and dysploidy (VERLAQUE, 1980). In the presumably primitive outbreeding perennials of *Lathyrus*, heterochromatin bands are more frequent but scattered, whereas in annuals they are either terminal or centromeric; this is interpreted as a trend for reduction in constitutive heterochromatin and C bands from perennials to annuals (LAVANIA and SHARMA, 1980). A correlation between *mean* chromosome number and life form of different groups has been postulated by LAVIE (1979) for some Vitaceae. The highest chromosome number known so far for the Compositae, n = 110-120, has been found in *Montanoa* where polyploidy is correlated with arborescent habit and high altitudinal habitat (FUNK and RAVEN, 1980).

The B-chromosomes in *Astranthium* (Compositae) which apparently reduce pollen fertility, originated by aneuploidy and subsequent fragmentation (CHINAPPA and MORTON, 1979). In some cases, B-chromosomes have been found to bring about detectable additional esterase isozyme (RUIZ REJÓN et al., 1980).

Cytogeographic studies in *Arnica* (WOLF, 1980) and *Festuca* (PILS, 1980) show again polyploids with wider range of distribution and diploids restricted to unglaciated areas. Polyploidy is of minor importance as an adaptive mechanism on the

Canary Islands (BORGEN, 1979). Secondary cycle of polyploidy in Australian *Senecio*: LAWRENCE (1980). Very thorough study on chromosome variability in Genisteae: SAÑUDO (1979). Polytopic origin of polyploidy in *Sempervivum arachnoideum*: WELTER (1979). Karyology of *Iris*: SAUER and STEGMEIER (1979). Cytotaxonomy of Italian *Ornithogalum*: TORNADORE and GARBARI (1979). Chromosome numbers elucidate origin of some *Juniperus* cultivars: HALL et al. (1979).

Genome relation of *Hordeum jubatum*, *Agropyron sericeum*, and their hybrids: MURRAY and TAI (1980). Triploid karyo-ecotypes in Californian *Fritillaria* have arisen by functioning of unreduced gametes in diploids: MARCHANT and MACFAIRLAINE (1980). Puzzling complexity of chromosome number in *Pilocosta* (Melastomataceae-Tibouchineae) of n = 7, 14, 18, 33 in a tribe where n = 9 is the base number: ALMEDA and WHIFFIN (1980).

5. Chemical Systematics*

a) Origin of Secondary Compounds as Allelochemicals

The interaction of herbivores with secondary plant metabolites has been exhaustively treated in a multi-authored book edited by ROSENTHAL and JANZEN (1979); apart from 8 chapters dealing with ecological and evolutionary processes, 12 articles give concise overviews of the different chemical classes of secondary plant substances and their known allelochemic function. The overall impression emerging from this volume is that a vast body of chemical information and ecological and coevolutionary theory have been assembled, but experimental testing of most of the ideas is still in its infancy. Nevertheless, there is a growing consensus about the ecological role of secondary metabolites, especially about their defensive function against herbivores and pathogens (see, for instance, SWAIN, 1979). A little-used argument in favor of this is the fact that secondary compounds have changed in structural complexity and variety during the course of evolution which becomes clear from the increasing diversification of metabolic pathways (GOMES and GOTTLIEB, 1980). If secondary compounds were merely waste products of plant metabolism, such changes would be completely unnecessary. For example, none of the nonflowering plants produce alkaloids from either aromatic amino acids or by incorporation of nitrogen into steroidal nuclei, yet these two routes produce by far the most effective mammalian toxins known (LEVIN and YORK, 1978). Therefore one may conclude that the evolution of these plant groups was directed by their coevolution with mammals (SWAIN, 1979).

While it is clear that at least some natural compounds are defence substances, it is difficult to provide definite proof that this is true for all of them; in fact this is not only unnecessary but also improbable because plants are largely anachronisms, i.e., often the selective pressures responsible for the evolution of a certain trait do not exist any longer (JANZEN, 1979). In this connection it is less important whether secondary metabolites originally (i.e., before there was selection for their production in great quantity) were "normal" metabolites (JONES, 1979), waste products (the old view), or novel substances (SWAIN, 1976).

It has often been noted (recently again: BELL, 1978) that plants, in contrast to animals, cannot rely on physical mobility to escape their predators and have therefore evolved chemical defences against such enemies. The coevolutionary race between plants and their animal herbivores, including overcoming the toxic barrier by the animal and the

*by KLAUS KUBITZKI

need for the plant to evolve new allelochemics, have repeatedly proven to be reasonable postulates. All this is compatible with the growing evidence for the repellent or toxic action of certain plant substances on certain animals. Recent discoveries in this field include the recognition of the effect of prenylation on umbelliferon which brings about insect toxicity of the resulting furanocoumarin (BERENBAUM, 1978); the toxic and deterrent properties of sesquiterpene resins of *Hymenaea* (LANGENHEIM et al., 1978); the resistancy of *Pinus taeda* to bark beetles which is brought about by unusually high values of myrcene (GOLLOB, 1980); the dual activity of isoflavonoid phytoalexins from several legumes against insects and fungi (SUTHERLAND et al., 1980); the detoxification mechanism for L-canavanine which has evolved in the bruchid beetle *Caryedes brasiliensis* which depends on this highly toxic nonproteic amino acid as its sole food source (ROSENTHAL et al., 1979) – all this complements what has been known earlier, but also shows that few generalizations about the toxicity of a given compound – let alone a whole biogenetic group – are possible: one beast's drink is another beast's poison (JANZEN, 1979)! This leaves us with the nearly unsolvable task of investigating millions of relationships.

Feeding experiments with plant populations whose members differ phytochemically provide useful models for showing how chemical differences might have come into being. Recent work involving plant material of different terpene composition of *Satureja douglasii* (RICE et al., 1978) or using artificial diets containing the sesquiterpene resins which naturally occur in the leaves of *Hymenaea* (LANGENHEIM et al., 1980) show that differential herbivory indeed can be a factor determining quantitative variation of chemical composition among plant populations.

The question remains for the magnitude of genetic change necessary for the development of novel secondary compounds. McKEY (1980) has pointed to the interesting possibility of merging of different common pathways that previously have been separated which include only minor genetic changes. He considers complex indole alkaloids and benzylisoquinoline alkaloids as having originated through condensation, possibly catalyzed by a single new enzyme, from precursors present already in the plant taxa where these condensation products occur (indole components plus monoterpenoids in the first case, dopamine plus 3,4-dihydroxyphenylpyruvic acid in the latter).

Sometimes the interpretation of data appears somewhat arbitrary. For instance, petal flavonoids in the Limnanthaceae seem to be more conservative, evolutionarily, than whole plant flavonoids and hence are considered to be better systematic indicators (PARKER and BOHM, 1979) which contradicts the widespread view that traits with an obvious ecological function are of less use in systematics. On the other hand, HARBORNE and SMITH (1978) find the anthocyanin chemistry of autogamous species not different from their animal pollinated relatives.

b) Ecogeographical Differentiation and Community Relationships of Secondary Metabolites

MEARS (1980) points to an interesting relationship between life form, distribution, and chemical differentiation between populations and species in *Parthenium*. Widespread species occurring in a variety of habitat types are found to be characterized by a large number of flavonoid glycosides based on common aglycones. Species which are geographically and ecologically restricted appear to exhibit a tendency toward depauperate flavonoid patterns with major methylated aglycone components. It is assumed that methylation is a less expensive means of autodetoxification than glycosylation, but at the same time also presents a less toxic potential to predators.

FEENY (1976) and RHOADES and CATES (1976) had put forward the hypothesis that plants (or plant parts) which are more readily found by insect predators (apparent, or predictable plant resources) are more likely to have quantitative defences like tannins or similar digestibility-reducing substances than plants (or plant parts) that are rare in space and/or time (unapparent, or ephemeral, plant resources) which put their effort into the production of qualitative defences such as alkaloids and other toxins. RHOADES and CATES (1976) also have proposed that qualitative defences are divergent systems evolved in response to generalist herbivore pressure on ephemeral resources, whereas quantitative defences of which there are few chemical types are a convergent system evolved in response to specialist herbivore pressure on available resources. Consequently, divergence in toxic defences will reduce the resource basis of existing herbivores and can be expected to be traceable in the first instance in closely related sympatric species which basically should be endowed with a similar defence chemistry. Correspondingly, WATERMAN et al. (1978) showed in a study of six species of West African members of the Rutaceae/Toddalioideae that the species with a similar alkaloid and triterpene chemistry have different ecological ranges, while species coexisting with each other differ chemically. Equally, eight sympatric species of Crucifers were found to differ in glucosinolate composition (RODMAN and CHEW, 1980). On the other hand, within one major taxon stepwise chemical variations of adjacent para- and allopatric taxa were detected by SALATINO and GOTTLIEB (1980). Starting from a biogenetic classification of quinolizidine alkaloids in the Papilionoideae, these authors were able to trace the spatial evolution of the subfamily starting from a center in tropical Africa. It is suggested that the gradual extension of the ancestral taxon subsequently has led to new taxa with specific allelochemicals. A southern temperate "pre-cytisine" route, a northern temperate "pre-cytisine" route, and a tropical "ormosanine" route are being distinguished. The argument which is strictly chemical and which ignores, for instance, resorting to paleogeography, is reminiscent of Croizat's concept of "genorheitra" which during their flows left behind them the segregates which have led to present-day taxa (see Fortschr. Bot. 16, 156). The gradual change of secondary metabolites which obviously has occurred after the exposure of dispersing plant populations to new environments gives indirect proof of the allelochemic nature of secondary plant substances.

There have been attempts to test the plant apparency hypothesis on the community level, but neither in the taiga (BRYANT and KUROPAT, 1980) nor in the tundra (JUNG et al., 1979) could the predicted relationship between apparency and chemical defence be verified; this may be due to the diminishing influence of biotic factors toward the poles, while toxicity of alkaloids, as gauged by an index of relative toxicity for mammals, was found to be much greater in tropical plant taxa than in temperate ones (LEVIN and YORK, 1978). In general, however, it appears that it is in the first instance evolutionary history as expressed by systematic position that determines the biosynthetic pathways a plant can utilize for defence (SWAIN, 1979). This is not denying that there are different ways in which plant species in various environments protect themselves from attack.

c) Chemosystematic Principles and Practice

As outlined in the previous sections, chemical differentiation which is being brought about through environmental pressure operates mostly in small steps and thus leads, within the evolutionary/systematic context, to very gradual changes which allow to trace the lines of phylogenetic differentation. A vast body of information on secondary metabolites has been assembled during the last decades, but there is general agreement that methodology for making use of this information in systematic biology is still insufficient. This is well reflected by the contributions to a symposium volume (BISBY et al., 1980) in which several procedures for chemical data analysis are suggested. CRONQUIST (1980) offers the pragmatic view that chemical features will gain momentum in classification if they prove to be useful, i.e., if they are correlated with other features of classificatory importance. However, there is very often a marked incongruence between different sets of data (for instance, morphological and chemical ones) and HARRIS and BISBY (1980) deal with the classificatory problems that result therefrom. In an attempt to replace the "narrative" approach of systematic biology by formalized analytical methods for the evaluation of chemical data, HUMPHRIES and RICHARDSON (1980) try to apply Hennig's cladistic methods, but since everything depends on the distinction between ancestral and derived character states which is often more problematical with chemical characters than with morphological ones, one may regard this undertaking sceptically. ESTABROOK (1980) suggests extending the use of his character compatibility method to the study of biochemical systematics.

With regard to all these numerical methods one is tempted to cite again RAVEN (1974) who had stated that "as long as the chief goal of a particular chemosystematic study is the elucidation of some systematic hypothesis, it will be unlikely that broader relationships will be perceived". It is here that the ambitious, highly dynamic approach of O.R. GOTTLIEB (1980) comes in, which gives due attention to the occurrence of natural products in nature and their biogenesis, and is considered a systematic criterion independent of morphology. In an endeavor to determine the direction of micromolecular evolution, several principles derived from the study of a vast material are put forward. The First Principle postulates that with regard to primary precursors of the shikimate pathway from which biogenetic groups characterizing taxa of vascular plants of high hierarchic level are derived, evolution proceeded by blocking of reaction steps. This principle is found to be operative, for instance, in alkaloid evolution of the angiosperms (GOMES and GOTTLIEB, 1980) (see Fig. 1). Among taxa of ligneous plants of low hierarchic level, biochemical evolution of secondary metabolites involves gradual specialization by substitutional and/or skeletal diversification of compounds within biogenetic groups (Second Principle, GOTTLIEB, 1980). While resemblance between taxa with respect to the structural variation of the metabolites of their chemosystematically meaningful biogenetic groups is a reasonable indication of their affinity, difference in chemical composition says nothing about their lack of affinity, because switch-overs which lead to different metabolites in closely related taxa appear to be not seldom (see neolignan- vs. pyrone-producing species in *Aniba*: GOTTLIEB and KUBITZKI, 1981). This notion is embodied in the Third Principle: The homology of biosynthetic routes, not the substances produced, is a plausible indication of affinity. Recognition of such switch-overs will, however, only be possible through recourse to morphological criteria. A Fourth Principle deals with a supposed equilibrium between the association of macromolecules and micromolecules within an organism which is crucial for its adjustment to the environment.

In a similar way, but with more narrow outlook, CRAWFORD (1978) and GIANNASI (1978) discuss the systematic significance of flavonoids in connection with their biosynthesis and distribution, and GORNALL and BOHM (1978) recognize two controversial trends of flavonoid distribution in angiosperms, one involving reduction in structural complexity, following a reverse direction down the biosynthetic pathway, and a second trend, often superimposed on the first, involving diversifica-

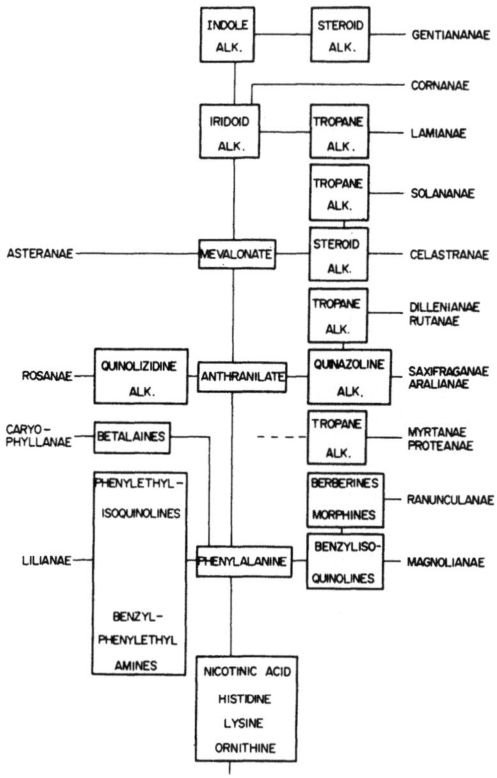

Fig. 1. Distribution of alkaloids in Dahlgren's superorders of angiosperms as an illustration of the First Principle of Biochemical Systematics. (After GOMES and GOTTLIEB, 1980) (see p. 202). Phenylalanine-derived benzylisochinolines abound in Magnolianae, generally considered the most primitive angiosperms, and in biosynthetically modified form in Ranunculanae, whereas other phenylalanine-derived alkaloids characterize Caryophyllanae and Lilianae. By blocking of the shikimate pathway chorismic acid becomes available for the production of anthranilic acid-derived alkaloids characteristic of Rutanae but widespread also in other superorders. In contrast to phenylalanine-derived alkaloids, anthranilic acid derivatives frequently incorporate small terpenoid units to give rise to indole alkaloids. The fullest expression of mevalonate utilization occurs in the total absence of shikimate-derived precursors in the biosynthesis of iridoid and steroid alkaloids, in which nitrogen is linked directly to terpenoid moieties. Mevalonate-derived alkaloids are concentrated in Solananae, Lamianae, Gentiananae, and Cornanae. The superposition on the general use of nicotinic acid, histidine, ornithine, and lysine, considered an ancient heritage of phenylalanine, anthranilic acid, and mevalonate for alkaloid synthesis in three stages is ± consistent with the major division of dicotyledons in Magnoliidae, Rosidae, and Asteridae

tion (such as O-methylation, extra oxygenation, etc.). GORNALL et al. (1979) deal with the systematic distribution of flavonoids in angiosperms.

HARBORNE (1980) gives a useful survey of the newer experimental approaches to plant chemosystematics, including such aspects as UV patterning of flowers, nectar constituents, and phytoalexin induction. In Vicieae, the phytoalexin response leads to pterocarpan formation; in *Vicia* and *Lens*, however, furanoacetylene synthesis has been superimposed on this pattern and has largely replaced it, possibly in response to coevolution with specific damaging insects or fungal parasites (ROBESON and HARBORNE, 1980).

COOPER-DRIVER and BALICK (1978) and CORADIN and GIANNASI (1980) have tested the effects of drying and chemical preservation on plant material later to be used for chemical study.

Papers dealing with the application of chemosystematics to taxonomic problems are mentioned in Section 10. For macromolecular systematics, see Chapter F I.

6. Reproductive Biology

Important symposium volume: New Zealand J. Bot. 17 (4), 425-685 (1979). Reviews: Breeding systems in tropical forests: M.T.K. ARROYO (1979), BAWA (1979), on islands: EHRENDORFER (1979), during secondary succession in tropical forest: OPLER et al. (1980), on tropical barro Colorado Island: CROAT (1979).

a) Costs and Benefits of Reproduction

The recent boom in this field of research is not only reflected by the increasing number of experimental studies, but also by an extension of theory. Ever since the formulation of the synthetic theory of evolution in the late 1930's, it was assumed that the genetic system of plants operates so as to strike a compromise, or attain an optimal balance between the immediate advantage of genetic conformity and the long-range advantage of genetic recombination which allows populations to adapt to a changing environment. Such a long-range advantage, however, can only be an indirect result in the evolution of the reproductive system, and is not under natural selection which operates on the individual and not on the group level (LLOYD, 1979c; SOLBRIG, 1979). Consequently, the various components of the genetic system have to be assessed in terms of their effect on the individual's fitness which has been attempted by SOLBRIG (1979) in a cost-benefit analysis of recombination in plants.

The reproductive costs involve the so-called meiotic costs, i.e., the loss of parental alleles in outcrossing species as compared to selfing ones, and the energetic costs of transporting pollen from flower to flower which can be assessed in terms of the caloric budget. Since flower, nectar, and pollen production is costly, selection will operate toward inbreeding or apomixis if the benefits of out-crossing (e.g., heterosis) do not overcompensate energetic and meiotic costs. A good example is the trend from allogamy to autogamy in colonizing species: the decrease in pollen/ovule ratio (CRUDEN, 1977) reflects the transfer of costs from xenogamy to seed production. The great appeal of cost-benefit analysis lies in the fact that the various aspects of the reproductive biology of plants, such as pollination, breeding system, and dispersal, can be viewed under one unifying aspect. It should be noted, however, that the testing of many of the hypotheses based hereon involves the difficult monitoring of the post-dispersal fate of pollen and seeds (WILLSON, 1979), and that the parental success of a plant cannot be measured directly, but only be estimated on the basis of pollen and seed production (LLOYD, 1979c).

A note of warning is therefore sounded by URBANSKA-WORYTKIEWICZ (1980b) who studied reproductive strategies in *Cardamine*. Though she considers the concept of sexual reproductive effort to be useful in general, she finds it difficult to apply in cases where actual seed production is inadequate to the considerable amount of energy invested in floral structures as a whole. According to the notions of energy partitioning, males should be more vigorous than females which have to bear the cost of producing offspring; this has been confirmed for *Fragaria chiloensis* by HANCOCK and BRINGHURST (1980), but not for *Populus tremuloides* (M.C. GRANT and MITTON, 1979); altogether, this stresses the importance of an unbiased empirical approach.

b) Sex Distribution and Related Topics

Evolution of dioecy, review: BAWA (1980).

Up till now, the evolution of dioecy has been viewed primarily in the context of the regulation of gene flow. At least two additional factors have to be considered, the altered patterns of resource allocation and the spatial changes affecting the behavior of pollinators, seed dispersers, and predators. As regards the first one, male sterile mutants (females) may have an adaptive advantage, as more energy can be devoted to seed production and, in contrast to hermaphrodite flowers, there are no losses owing to the abortion of flowers that receive incompatible pollen (BAWA, 1980). This is in line with experimental studies on apioid Umbelliferae in New Zealand by LLOYD (1979a) and WEBB (1979b) who show that gynodioecy, the most common type of breeding system within this group, apparently has been derived from hermaphroditism via andromonoecy. Dioecy, though rare, is achieved by decreasing fruit set of the pollen parent. Other possible pathways to dioecy are those via monoecy or heterostyly (LLOYD, 1979b). An example is distylous *Linum perenne* where the pollination of pin and thrum flowers is brought about by different insects; if one misses, the plants behave as dioecists (HEITZ, 1980); similar observations have been made by BEACH and BAWA (1980) for *Cordia* and *Coussarea*.

GIVNISH (1980) calls attention to the striking fact that in gymnosperms monoecy and wind-dispersal, and dioecy and animal dispersal are strictly correlated and he sets up a model to account for this phenomenon. It is predicted that in wind-pollinated species a dioecious female is likely to invade a monoecious population if the effort located at the production of ovules and seeds leads to an increase of seeds timely dispersed, since dispersers will prefer to visit trees with the highest concentration of edible diaspores. This model can be extended to, and has partly been confirmed for, angiosperms by BAWA (1980) where dioecy is frequent in animal-dispersed species with unspecialized pollination agents such as wind or small and unfaithful animal vectors. Plants pollinated by specialized insects or birds will tend to be hermaphroditic, as the considerable costs of pollinator attraction has to be borne by both sexes. An interesting inference is that the high prevalence of dioecy on islands may not be a consequence of the selection for outbreeding, but an indirect result of the correlation between dioecy and dispersal by birds over long distances.

A moot point of this and similar hypotheses is that they are based primarily on statistical data and that ecological information about dioecious plants is rather rare; there may be numerous aberrations and exceptions. Another question relevant to Givnish's model is whether the findings from gymnosperms can be transferred to angiosperms without constraint. The crucial point is the attraction of dispersers (which would have been reptiles in early gymnosperms), but this is exclusively determined by female effort in gymnosperms where genera like *Ginkgo* and *Cycas* develop numb yet nourishable seeds irrespective of fertilization, a marked contrast to angiosperms where seed set is fertilization-limited. The value of the resource allocation theory in plants may not lie in the interpretation of well-known data, but in the stimulus it hopefully provides to look for new ones. The genus *Thymelaea* which has been revised by TAN (1980b) may provide a suitable object for further research because of its interesting variation in sex distribution, ranging from monoecy, dioecy, gynodioecy, and androdioecy also to hermaphroditism in an apparently advanced annual.

In *Solanum* subgen. *Leptostemonum*, andromonoecy is common but often remained undetected because of rudimentary gynoecia in male flowers. Without alluding to the ideas presented above, SYMON (1980b) concludes that floral heteromorphism does not have the chief function of ensuring out-crossing, because many species in this group are self-incompatible, but that the excess of pollen parents is necessary to attract and feed pollinating bees, thus ensuring pollination. This is especially important in an arid environment where pollinators are rare and temporarily restricted. An interesting case of floral mimicry has been discovered in *Jacaratia* (Caricaceae) where the stigmatic lobes of the female flowers are of the same shape and size as the corolla lobes of the male one (BAWA, 1980b); this observation is explained by the assumption that two conflicting trends, diverging for functional parsimony and converging for recognition by the pollinator, have been followed successively.

Study of the resource cost of gender in normally selfing Umbelliferae: HARPER and LOVETT DOUST (1980), LOVETT DOUST (1980). Sex ratio in *Thymus*: BONNEMAISON et al. (1979). Gynodioecy in *Stellaria*: PHILIPP (1980), in *Euonymus*: WEBB (1979), in *Ficus*: VALDEYRON and LLOYD (1979). Andromonoecy in *Leptospermum scoparium* (Myrtaceae): PRIMACK and LLOYD (1980). Occasional occurrence of hermaphroditic flowers in *Quercus* indicative of ancestral condition: TUCKER et al. (1980). Differences in spatial distribution of male and female plants in *Chamaelirium* due to ecological differentiation: MEAGHER (1980).

In *Trichostema* (Labiatae), xenogamous species expend more energy per floral unit and are less effective seed producers than closely related autogamous species (SPIRA, 1980). A correlation between woody habit and allogamy, and herbaceous habit with autogamy has been found by COUDERC (1980) for *Anthyllis*, and in *Rhinanthus* CAMPION-BOURGET (1980) demonstrated a trend from allogamy to autogamy. *Anthoxanthum odoratum* is nearly self-incompatible (less than 0.2% self-fertility) and exhibits a strong inbreeding depression in selfed seeds (WU and JAIN, 1980). AUQUIER and STACE (1980) studied cleistogamous species of *Vulpia* and COSTAS-LIPPMANN (1979) agamospermy in weedy *Cortaderia jubata*. The widespread occurrence of sexually reproducing diploids in *Taraxacum* originated partially through chromosome reduction and re-sexualization of agamospermous triploids and hinders a satisfactory microtaxonomic treatment (DEN NIJS and STERK, 1980). PANDEY (1979) has reviewed incompatibility in *Nicotiana*.

c) Dispersal

Tetragastic panamensis (Burseraceae) is characterized by an enormous seed waste. The fruits contain little but sugar and are not depleted by animals and the trees are visited by a variety of vectors with different effects for dissemination. The generalized dispersal strategy of this species has obviously not closely co-evolved with a particular frugivore (HOWE, 1980). The fruiting pattern of *Catalpa speciosa* is obviously influenced by the life history of its principle herbivore *Ceratonia catalpae*. Fruits develop slowly as long as the larvae feed on the leaves of *Catalpa* and mature rapidly when they leave; the number of aborted fruits is influenced by the amount of defoliation (STEPHENSON, 1980).

The nutlets of *Salvia columbariae* match the soil color of the site in small populations and are variable in larger ones which is interpreted by BRAYTON and CAPON (1980) as an adaptation for escape from granivory. Variable fire patterns within the ecosystems led to the evolution of serotinous cones, opening after fire, in *Pinus contorta* var. *latifolia* (PERRY and LOTAN, 1979); a mathematical model which allows for the explanation of the low frequency of this feature within the populations has been devised by HARTL (1979). In a survey of the phenology of mid-latitude ornithochorous fruits, THOMPSON and WILLSON (1979) were able to find three different strategies. The Australian species of *Solanum* show a remarkable diversity in fruit structure and dispersal syndrome, ranging from succulent or hard and dry berries to trample burrs, and from capsules to tumble weeds which should render them an ideal object of study in dissemination biology (SYMON, 1979c).

Aerodynamics of spinning samaras: GREEN (1980). Wind tunnel to study dispersal: EHRENDORFER et al. (1980). Predator satiation hypothesis of mast seeding confirmed statistically: SILVERTON (1980). Ornithochory in *Solanum* makes assessment of geographical origin difficult: NEE (1979). Dual dispersal modes in *Agrostis hiemalis*, barychory, and tumble weed: RABINOWITZ and RAPP (1979). Classification of germination syndromes, avoidance strategy with subdivisions as drought, fire, etc., and tolerance strategy, viz. maternal investment syndrome, the latter typical for climax vegetation: ANGEVINE and CHABOT (1979). Dispersal spectra for different European plant communities: LUFTENSTEINER (1979). Seed appendages, overview: KAPIL et al. (1980). Myrmecochory in *Calathea*: HORVITZ and BEATTIE (1980). Peculiar mechanism of attachment to substrate in seeds of orchidaceous *Chiloschista lunifera*: BARTHLOTT and ZIEGLER (1980). Long-distance dispersal once again dismissed as myth: VAN STEENIS (1979).

7. Evolution

a) General Considerations

A multi-authored volume edited by SOLBRIG et al. (1979) provides a good overview of current approaches and philosophies and is especially valuable for pointing to the many gaps in our knowledge. One inescapable impression of this and other publications in this field is that practical investigations apparently have not kept pace with plethora of theories and hypotheses formulated in recent years which is signified by the high incidence of the phrase "I/theory predict(s)" in many published papers. The interpretation of the adaptive significance of certain characters is especially problematic and it is interesting to note that apparently many of the arguments of the ecological/evolutionary discussion from the beginning of this century have been revived. The search for the adaptive value, e.g., of leaf form (GIVNISH, 1979) has led to the application of optimality models to plant evolution, as has been advocated strongly but somewhat unconvincingly by HORN (1979) who maintains that adaptive arguments need not depend on evolutionary history or genetic machinery. When speaking of a character state as being adaptive, however, we assume two possibilities (ANTONOVICS, 1980): It may be of importance in conferring differential fitness in present-day populations, but it may also be the product of past selection. The second alternative is stressed by JANZEN (1980) who draws attention to the historical context, pleads for caution in using the word "co-evolution", and maintains that plants are anachronisms: many of the characters observed today may have been brought under selection pressures no longer operating. JAIN (1979) critically reviews the concept of adaptive strategies and also sounds some notes of caution.

Analogies to economics have been associated with evolutionary theory since Darwin. It is amusing to see now that the more fashionable point of view expects evolution to operate as a manager of a multi-national enterprise would direct his business: Parsimonously pursuing his strategy, cost-efficient (SOLBRIG, 1979), and with an optimal outcome (HORN, 1979). One has to be thankful for some diverging opinion, such as that of VAN STEENIS (1980) who applied his ideas on autonomous and fortuitous evolution (Progr. Bot. 41, 239) to the study of rheophytes (flood-resistant plants), and who maintains that mutation is the creative agency and the so-called adaptation only a secondary pruning process.

b) Populations and Species

LEVIN (1979) discusses the nature of plant species and argues that the empirical concepts of species, grown out of the attempts to describe diversity, are based on assumption found to be lacking in plants. Plant species are said to be devoid of reality, cohesion, independence, and simple evolutionary and ecological roles. Thus, the concept of the species can only serve as a tool to characterize diversity in a mentally satisfying way. Likewise, RAVEN (1979) once again makes a plea for abandoning the biological species as a unit of evolutionary studies and instead concentrate on populations which are local, integrated, and resistant to change. An integrated multidisciplinary approach to population biology may generate answers to open questions as how and under what conditions morphologically and ecologically distinct units are produced and why species remain coherent units even in the absence of gene flow. Thus it seems that by a shift of emphasis from species to populations the problems of speciation may be attacked, a point of view which is shared by M.J.D. WHITE (1978) in his book on modes of speciation. The mainly zoological material accumulated by the author suggests that the models brought forward in the "New Synthesis" were oversimplified, especially with respect to the decisive role postulated for geographic isolation. It also seems that the importance of hybridization and karyotype changes have been underrated for a long time by zoologists, and that the modes of speciation in animals and plants are not as different as has been assumed.

α) Variation within Populations. Our knowledge of electrophoretic genetic variation in plant populations in relation to longevity is summarized by HAMRICK (1979). Forest trees which have been studied most intensively, are highly variable in this respect except for some conspicuous species as, e.g., *Pinus resinosa*, and generally woody plants contain three times the heterozygosity found in herbaceous plants. Several hypotheses have been brought forward to account for this phenomenon, but the data to support them are still scarce. A crucial problem pertinent to all these studies is that the role of variation at enzyme loci in plant adaptation is as yet not understood, as has been pointed out by G.B. JOHNSON (1979). The issue is not the overall genetic variability, but rather that of the loci which affect the phenotype which is the target of selection. G.B. JOHNSON (1979) gives some examples of physiological processes which could be profitably studied in this respect and points also to the still somewhat inadequate techniques which need refinement.

The decreased genetic variability in annuals from *Solanum* sect. *Androceras* as compared to climax perennials of the same section is explained by WHALEN (1979c) as a consequence of the genetic bottleneck associated with colonization. Natural selection and genetic drift are held to be responsible for the decrease of allozyme variation from central to marginal populations of *Hordeum jubatum* (SHUMAKER and BABBLE, 1979).

A study by HILL et al. (1978) on three neotropical riverine legume species shows a concise correlation between genetic variability and habitat. Populations with a high extinction probability on perturbed habitats are more homozygous than those with a greater chance of survival. The strictly unidirectional migration in riverine habitats makes it easy to assess its affects which in other cases may be responsible for some spurious correlations.

Some systematic applications of allozyme data: Genome relationships in *Avena*, established by cytogenetic and morphological-ecological studies, are largely confirmed despite the highly significant variation within populations (JAIN and SINGH, 1979). In connection with other techniques, EDMONDS (1979) was able to prove allopolyploidy in *Solanum* sect. *Solanum*. Speciation on tropical mountains was investigated by LEE and LOWRY (1980). The possible sample errors in measurements of interspecific genetic variability have been discussed by THORPE (1979).

β) Differentiation and Divergence of Populations. Optimal outcrossing distances in *Delphinium nelsonii* have been investigated by PRICE and WASER (1979); hand pollination experiments revealed that seed set is optimal within 1-10 m distance, with naturally occurring pollination at the lower end. This is in line with the idea that natural selection should favor mating between individuals of intermediate genetic similarity because of the detrimental effects of both inbreeding and outbreeding depression. The actual pollen flow observed is a compromise between many conflicting pressures, e.g., on plants to promote outcrossing, and on pollinators to minimize flight distance; note that KREBS (1979) considers pollinator and inflorescence structure of plants like *Delphinium* to be coevolutionary behavior systems in which neither partner can do any better. A similar phenomenon, differing in magnitude, is to be found in *Stylidium crassifolium* where BANYARD and JAMES (1979) detected infrapopulational inbreeding depression. Recessive lethal genes eliminate most of the products of self-pollination, and inter-populational crosses yield the highest seed set which, however, again decreases with increasing distance of the respective populations. Together with SCHILLING and HEISER's (1979) findings of intraspecific hybrid sterility in the *Solanum nigrum* complex, this is a hard blow to the biological species concept which is based on hybrid fertility within species.

Populations of *Tsuga canadensis* in the Eastern U.S.A. apparently consist mostly of two ecotypes with different environmental responses which are distinct with very little intergradation due to diverging selection by the habitat (KESSELL, 1979). Similarly, the broader ecological amplitude of octoploid *Fragaria* in California is a consequence of ecotypic differentiation and not, as sometimes assumed, of fixed heterozygosity (HANCOCK and BRINGHURST, 1979). Different types of colonizing potential and competetive success of clones in combination with a flexible breeding system contributes to the success of *Spartina patens* (SILANDER, 1979; SILANDER and ANTONOVICZ, 1979). The phenotypic plasticity of *Poa alpina*, especially the relationship between chasmogamic and bulbil-producing flowers, is influenced by several abiotic factors (M.A. BACHMANN, 1980); the evolutionary ecology of vivipary in *Festuca vivipara* has been studied by LEE and HARMER (1980).

LEVIN (1979b) discusses and summarizes the effects of pollinator foraging behavior on plant populations; experimental studies by SCHMITT (1980) show the different effects of butterflies (bypassers) and bumblebees (near-neighbor visitors) on gene flow. Stigma position is decisive for the direction of gene flow within *Rhinanthus*, as was shown from the introgression of *R. minor* into *R. serotinus* (KWAK, 1980). Heavy metal tolerance which can arise within a few generations, provides a good marker for measuring gene flow and has spread from plants on mining sites to surrounding pastures in the direction of prevailing winds (KARATAGLIS, 1980b).

Geographical isolation as well as pre- and postzygotic reproductive barriers have been effective in *Solanum* sect. *Androceras*; hybridization may have happened occasionally, but cannot be proven because of the uniform karyotype (WHALEN, 1979a). DANIN et al. (1979) were able to identify Mexico at least as a center of diversity for the polymorphic cosmopolitan *Portulaca oleracea*-polyploid complex. Divergent evolution of two *Hypericum* species from a common ancestor after the separation of the Mascarenes from Africa has been reported by ROBSON (1979). *Trifolium hirtum* shows significant genetic variation after 30 years since the first introduction to California which may be due to local selective forces and promoted by the retention of outbreeding together with genetic variation and some few and rapid morphological changes (JAIN and MARTINS, 1979; MARTINS and JAIN, 1980). The observed polymorphism of weedy *Vicia sativa* is of recent origin and was enforced by the agricultural revolution in the Old World; colonization is associated with reticulate evolution and recombination of previously separated gene pools by occasional hybridization (ZOHARY and PLITMANN, 1979). Quantum speciation has been postulated by SCHOLZ (1979b) for one *Stipagrostis* species. Clinal variation in seed size within *Capsella* is correlated to climate, but discontinuities occur in disturbed habitats (HURKA and BENNEWEG, 1979).

γ) *Hybridization.* LÜÖND and LÜÖND (1980) detected unidirectional gene flow from *Ajuga reptans* to *A. genevensis* in hybrid swarms from disturbed habitats. Character coherence begins to fade in older populations and is anyway considered to be a variable and mostly overrated feature of naural hybrid populations by V. GRANT (1979) who has studied hybridization in *Opuntia* (V. GRANT and K.A. GRANT, 1979a). In hybrid populations of *Epilobium* in forest clearings, the closing of the habitat favors the return to parental types within the hybrid swarm (FITTER, 1980).

The re-investigation of a hybrid population of *Helianthus divaricatus* and *H. microcephalus* after 22 years by HEISER (1979a) leads him to point out the difficulty of differentiation between introgression (i.e., incorporation of germ plasm) and transitory hybrid swarms. *H. divaricatus*, a species of semi-open habitats, remained strongly in most spots even after closing; it could well be that this is due to genes taken up from the first species. The importance of hybridization for speciation within *Impatiens* is stressed by GREY-WILSON (1980a); in *Atriplex*, introgression of 4n *A. canescens* into 6n *A. tridentata* produces several new adaptive 6n derivatives (STUTZ et al., 1979). Applications and limitations of using a distance coefficient as hybridization index have been discussed by WELLS (1980) and MAZE (1980).

c) Geographical and Ecological Aspects

A symposium on *plants and islands* (BRAMWELL, 1979) focused on biological problems of biota which rank a close second to the tropical rain forest both in species diversity and current rate of extinction, and which have been important for the shaping of evolutionary theories. MABBERLEY (1979b) does not see a need to postulate special mechanisms for island evolution, however, since many of the plants from continental areas grow on ecological islands and may also show "island syndromes". He argues that pachycauly is the primitive condition and therefore rare and localized, thus inverting the notions of Age and Area, and common equals primitive. GREUTER (1979) points to the difficulty of correlating prolonged evolutionary standstill (e.g., of Aegean relicts) with the reality of angiosperms evolution and explains this by the postulation of short dynamic phases (e.g., Messinian changes in the sea level of the Mediterranean Sea) followed by more or less stable periods. The impact of man, undesirable as it may be in destroying the original island vegetation, may provide some excellent opportunities to study factors, such as establishment, competition, and adaptation of colonizing species and their effect on the indigenous flora; especially interesting will be the cases where aliens invade climax communities (MOORE, 1979).

Climatic stability and steep ecological gradients between habitats are among the key factors for speciation on islands (HUMPHRIES, 1979b); in this respect, the situation on island groups like Macaronesia and regions like Western Australia, also more or less isolated, is comparable (HOPPER, 1979). In both cases, species richness is highest in the transitional zone between moist and arid areas where small fluctuations move local conditions beyond the tolerance of one or the other species. Steep ecological gradients also play a role in the evolutionary divergence and endemism of NW. American *Sedum* (DENTON, 1979).

CARDONA and CONTANDRIOPOULOS (1979) discuss endemism and evolution in Western Mediterranean islands. SUNDING (1979) surveys the endemic Macaronesian flora and points to several interesting disjunctions, e.g., to New Zealand, India, S. America, and S. Africa. A symposium on tropical botany (LARSEN and HOLM-NIELSEN, 1979) focuses on some topics of current phytogeographic studies on tropical plants and biota.

In the study of *host-parasite interactions* (symposium: Bull. Soc. Bot. France 126, Actual. Bot. 4, 1979), ecological, evolutionary, and taxonomic aspects are interwoven. The revived interest in parasitical angiosperms is reflected by two reviews by OZENDA and CAPDEPON (1979a) (with a nice diagram) and H.C. WEBER (1980). In parasitical Scrophulariaceae and Orobanchaceae, evolution proceeded from small annual root parasites (e.g., *Euphrasia*) to larger root parasites (e.g., *Rhinanthus*) and finally to forms with large leaf haustoria (some *Orobanche*), corresponding to a shift of the contacting organ from secondary rootlets to hypocotyl (root system reduced) and further to bulky leaf haustoria (H.C. WEBER, 1979, 1980). Host specifity is restricted to plants with primary haustoria but otherwise rare in parasitic angiosperms (KUIJT, 1979). In hemiparasitic *Castilleja*, HANSEN (1979) found that the host species buffer the plant against variations of the physical environment. A short account of neotropical saprophytes is given by MAAS (1979).

The application of parasitological data to the taxonomy of their host plants has always been a challenging area of research, and an highly interesting symposium [Symb. Bot. Upsal. 22 (4) 1979] gives a synopsis of our current knowledge within this field. The general impression is that parasitological data have to be applied with great caution, because of the complex evolutionary interactions and because some of the observed patterns may be due to fortuitous colonization (THORNE, 1979); as EASTOP (1979) has put it nicely for *Stenorrhyncha*: "A similar parasitic fauna is no proof of a common ancestry. Some *Stenorrhyncha* are better physiologists than taxonomists, many have some ability as analytical chemists, but above all they excel as opportunists." Some cases where good correlations between parasite fauna and plant classifications exist include gall midges and *Arabis* (SYLVÉN, 1979) as well as some cases within Leguminosae where the parasitic species of *Meliola* (Plectomycetes) give an indication of tribal relationships within Papilionoideae (HIJWEGEN, 1979) and where the uredinalic *Ravenelia* suggests a closer relationship between Caealpinioideae and Mimosoideae and is correlated with the classification of the latter (EL-GAZZAR, 1979). Also cyst-nematodes seem to have evolved with their hosts in a broadly correlated way, and their distribution is more or less in line with plant classification (STONE, 1979). Though fungi may serve as veritable taxonomists, e.g., when they jump from a native species to a closely related imported one which formerly was free from the disease because of geographical isolation, the distribution of parasitical fungi does not seem to be very helpful for the systematics of higher categories to HIJWEGEN (1980) who mapped their occurrence using Dahlgren's diagram, and DAHLGREN (1979) concludes that these data can be applied here only if correlated with other characters. Nevertheless, the host range of a given parasite is of great heuristic value.

KUKKONEN and TIMONEN (1979) point to the fact that the smut genera *Ustilago* and *Entyloma* occur in primitive orders of flowering plants, but both are highly collective and have to be revised before conclusions can be drawn. Largely based on a study by NANNFELDT (1979), they show that the distribution of the smut *Anthracoidea* within the Cyperaceae provides some evidence that unisexual flowers of the Caricoideae have been derived from the hermaphroditic ones of the Scirpoideae. To SAVILE (1979a,b), the lack of Ustilaginales as parasites in Arecidae suggests a basic dichotomy within the monocotyledons, and the lack of Uredinales in Alismatidae indicates an early differentiation of the group before smuts became available; furthermore, he assumes that the origin of Juncaceae from ancestors of the Cyperaceae provides the only parsimonous explanation of the data obtained from the distribution of rust fungi within these families.

STRONG and LEVIN (1979) surveyed the parasites of different life forms and found, not surprisingly, that trees have more parasites than shrubs or herbs and climax herbs more than weedy annuals; otherwise, there are no differences between angiosperms and gymnosperms, and between monocotyledonous and dicoyledonous herbs. This is partly in line with SIMMONDS' (1980) review of the occurrence of monocarpy where the escape from pest pressure is suggested to be a major factor in its evolution. SYMON (1980) reports that butterfly larvae in Australia feed almost exclusively on supposedly recently immigrated families and concludes that they have invaded Australia together with their host plants of tropical relationships in the Oligocene.

MALLOCH et al. (1980) review the ecological and evolutionary role of *mycorrhizal symbiosis* in vascular plants and present some far-reaching conclusions. Endotrophic mycorrhiza, frequent in tropical forests, involves apparently ubiquituous fungi, so that dispersal is unproblematic. Ectotrophy, though energetically more expensive, enables the phytobiont to colonize extreme environments, e.g., soils with low nutrient content. Even more conspicuous, the altitudinal and latitudinal timberline is almost exclusively formed by trees with ectotrophic mycorrhiza. As ectotrophic fungi are more specialized, ectotrophic plants tend to occupy pure stands. Living in pure stands in places marginal for tree growth and with a low density of pollinators thus may have been responsible for the evolution of anemogamy in temperate tree species. It would be interesting to know if the rosaceous *Polylepis* (revision: SIMPSON, 1979), occurring as high as 5200 m.s.m., has ectotrophic mykorrhiza.

8. Plants and Man

Origin of some cultivated New World plants, review: HEISER (1979b). Symposium covering all aspects of agronomically important legumes: SUMMERFIELD and BUNTING (1980). Taxonomic aspects of African economic botany: KUNKEL (1979). See also Solanaceae symposium: HAWKES et al. (1979).

Although there is some agreement about the evolution of maize following domestication, its progenitor and place of origin are still controversial. MANGELSDORF et al. (1978) plead for a South American origin from a wild maize which seems to be supported by the discovery of the remains of pre-ceramic maize in Peru by GROBMAN and BONAVIA (1978). DOEBLEY and ILTIS (1980), however, dismiss wild maize as a myth and maintain that Mexican annual teosinte was the ancestor of cultivated corn. In a revised classification (ILTIS and DOEBLEY, 1980), both cultigen and putative ancestor are treated as subspecies within *Zea mays*. The recent discovery of *Zea diploperennis* (ILTIS et al., 1979) is of the greatest practical and theoretical value: this diploid perennial shares the chromosome number with cultivated maize ($2n = 20$), with which it is fully interfertile and thus of highest interest for breeding programs; according to DOEBLEY and ILTIS (1980), however, it has played no part in maize evolution.

The wild species of fenugreek (*Trigonella foenum-graecum*) were pre-adapted to cultivation because they do not shatter their seeds. In lentil, pea, and grass pea, the initial stages of domestication were due to a single mutation in a major gene which prevented pod dehiscence (LADIZINSKY, 1979).

The wild progenitor of cultivated sesame arose probably as an intersectional hybrid, as could be demonstrated using cytological and morphological evidence by IHLENFELDT and GRABOW-SEIDENSTICKER (1979); a reduction in chromosome number from $2n = 32$ in *Sesamum latifolium* led to $2n = 26$ in *S. indicum* via $2n = 26 + 6B$ in some primitive cultivars. Intergradation between primitive domesticates and their respective wild relatives have been responsible for the wide range of varieties found today in groups like *Cocos* (review: HARRIES, 1978) or *Capsicum* (numerical study: PICKERSGILL et al., 1979). Here as in the crop-weed complexes in neotropical *Chenopodium* (biosystematic study: WILSON and HEISER, 1979) it is difficult to cope with the observed variation within the existing taxonomic framework because of the probably polytopic domestication.

PICKERSGILL et al. (1979) argue that polytopic origin of cultivation should be less probable in staple foods than in vegetable or spices where yield is not that important and selection consequently weaker. Rye, then, would be an exception since taxonomic and archeological, as well as linguistic evidence, suggests that this cultigen evolved independently in several places from a weed in other cereals to a crop of its own (SENGER and HAWKES, 1980). SCHOLZ (1979a) points to the fact that the wild "mimetic" weeds accompanying the cultivar may not be ancestral but often spontaneously originated off-types of the cultivated species. Interactions of this type seem less probable now, as industrialized agriculture reduces the variability in arable weeds (*Papaver rhoeas*: HAMMER and HANELT, 1980).

Plants to serve energy crisis, diesel fuel from *Copaifera* and *Euphorbia* spp.: MAUGH (1979), CALVIN (1980). Wild relatives of coffee: LEROY (1980a). East African origin of finger millett, with secondary center of diversification India: HILU et al. (1979). *Hordeum spontaneum*, potential use in barley breeding programs: NEVO et al. (1979a,b). Solanaceous crops: cytoplasmic analysis suggests Chilean origin of modern cultivated potatoe: GRUN (1979). Ancestral forms of egg-plant cultivated by Sanskrit-speaking people: KHAN (1979). Spread of *Solanum* sect. *Brevantherum* related to 16th century trading routes: ROE (1979). Self-incompatibility in cultivated *Solanum muricatum* (pepino) and related species: ANDERSON (1979).

SCHULTES (1979) reviewed solanaceous hallucinogens and their cultural impact and pointed to the difficulties of botanical research on medical, toxic, and narcotic plants (SCHULTES, 1978). Acculturation and the disappearance of native cultures with a large and important ethnobotanical heritage press botanists to the hurry. According to GILLETT (1979), this kind of study is the field where native citizens of the developing countries can make most easily valuable contributions to science.

The traditional approach to plant utilization has been criticized by BROCKWAY (1979) in her timely but somewhat overstated study on science and colonial expansion, as being imperialistic; this is a further sign of the growing tendency of considering plants as a national heritage of their country of origin. The conversion of knowledge to profit and power by the British botanical elite of Kew in the last century is held to be responsible for making possible the exploitation of tropical countries. It should seem ironical, then, that contemporary botanists see their role increasingly in healing the wounds of the so-called development, viz. in conservation and conservation-oriented research. This is exemplified by the ever-growing number of symposia and publications on this issue:

Systematic botany, plant utilization, and biosphere conservation: HEDBERG (1979). Rare and endangered plants of New England: Rhodora 82, 1-237 (1980). Practical role of botanic gardens in plant conservation: SYNGE and TOWNSEND (1979). Conservation in the context of island biology: BRAMWELL (1979). Red data book containing detailed accounts of 250 threatened plant species throughout the world: LUCAS and SYNGE (1978).

9. Evolution and Classification of Higher Taxa

a) Early Angiosperms

The last ten years have seen the accumulation of a large body of paleobotanical evidence concerning early stages of angiosperm evolution and the attempts to collate this into one coherent picture, e.g., by DOYLE and HICKEY (1977). Nevertheless, most of the controversial points have remained unsettled, probably because fossil evidence of early angiosperms is largely restricted to pollen and leaf remains. But the evolution of seed plants is largely synonymous with that of their reproductive structures which have been reviewed by several authors.

TAYLOR and MILLAY (1979) survey the reproductive structures of early seed plants which provide a mosaic of different stages from different taxa, though all levels of organization can be found preserved in precise detail. Anemophily was the predominant mode of pollination in the Palaeozoic; cupule lobes and integumentary filaments may have directed pollen to the ovule, as was also shown in an elegant wind tunnel study by NIKLAS (1981). Entomophily is, however, likely to have occurred as well, as is indicated by the exceptionally large pollen of Medullosaceae, coprolites of pollen grains suggesting cantharophily, and perhaps also by the occurrence of stalked capitate glands functioning as nectaries or food tissue. Siphonogamy was present in one family of seed ferns. CREPET (1979) reviews the reproductive structures of Middle Eocene angiosperms, at which time most of the basic insect pollination syndromes had evolved, but the more advanced families as well as the more specialized forms of insect pollination were still missing.

DILCHER's (1979) report of reproductive structures in early angiosperms is interesting in several aspects. First he shows pollen similar to *Retimonocolptites* to be associated with small florets arranged in catkins which may have been anemophilous. Those and related pollen types had been interpreted as entomophilous by DOYLE and HICKEY (1977), the exine structure supposedly indicating both adhesive devices and the presence of incompatibility mechanisms, as inferred from the correlations observed in extant angiosperms.

Dilcher further points to the remarkable differentiation found among the reproductive organs of Lower and Middle Cretaceous angiosperms, e.g., loosely closed follicles helically arranged, weakly or strongly connate carpels, differentiation into style and stigma, and fused perianths, and concludes that both unisexual anemophilous and bisexual entomophilous flowers are ancient for earliest angiosperms and that it is impossible to decide which condition was primitive. Rapid, reticulate evolution and the many possible shifts between anemophily and entomophily render the establishment of homologies of the different floral parts rather difficult. Therefore, the Ranales may represent one but not the only type of a primitive flower, and the simple and so-called reduced flowers of Trochodendrales, Eupteleales, Cercidiphyllales (floral structures interpreted as pseudanthia by LEROY, 1980), Hamamelidales, and Piperales may be considered as initially simple rather than reduced from a monoclinous ancestor.

Not unexpectedly, DILCHER's (1980) findings are grist to Meeuse's mill, as this challenges the so-called "Russo-American Credo" (GUÉDÈS, 1979). MEEUSE (1980) stresses again the significance of the Gnetatae as the closest relatives and surviving representatives of at least some of the pre- and protoangiosperms. The continuity of function of certain structures already found in the progenitors of seed plants requires the perpetuation of the structural element, so that the cupule of seed ferns, the chlamys of *Gnetum*, and the carpels in all magnoliophytic taxa have to be considered as homologs (MEEUSE, 1979). Therefore the pollination droplet, the presentation of which was also taken to confirm the entomophilous character in *Ephedra* by MOUSSEL (1980), is considered to be the original food source for pollinators, and not pollen (but see FAEGRI, 1980, for criticism).

One of the many conceptual problems associated with the origin of angiosperms has been clarified again by FAEGRI (1980) who points to the often neglected fact that the terms polyphyly and monophyly can only be sensibly applied in relation to a given level. As this is practically never done, reasonings vacillate uncontrollably between the first incidence of an angiosperm characteristic and the upper level of the full syndrome of angiospermy.

MULCAHY (1979) presents a new theory for the explanation of dominance of angiosperms, suggesting that the closing of carpels was not only beneficial as protection of the ovules, but had also genetic benefits: It introduced a sort of "test-track" on which unbalanced genomes could be eliminated by competition during the gametophytic phase. As exemplified by Caryophyllaceae, gametophytic competition can modify the quality of the sporophyte. Within the angiosperms, the self-incompatibility system evolved from primitive gametophytic to more advanced sporophytic and finally to more refined systems as floral heteromorphism, as postulated by BEACH and KRESS (1980); the decisive factor was apparently the response of the sprophyte to an autonomous gametophyte.

b) "Living Fossils"

Clearly, the interpretation of fossil evidence must be based on models obtained from the study of living plants; therefore, the investigations into the reproductive biology of supposedly primitive extant taxa of angiosperms are of greatest interest. A study by ENDRESS (1980b) of the flowers of the Monimiaceae points again to the floral diversity within the Polycarpicae. A trend from more or less generalized ranalean flowers to forms with massive mucilage secretion can be observed, ultimately leading to the formation of a so-called hyperstigma, i.e., an extracarpellary stigmatic zone in the narrow entrance of the floral cup, formed by the inner tepals. This is of course a wonderful example of transfer of function; the "angiocarpous" flowers resemble somewhat the inflorescence of *Ficus* and are as far from "angiospermy" as is "angiospermy" from "gymnospermy".

The reproductive biology of *Drimys brasiliensis* was studied by GOTTSBERGER et al. (1980) and led to interesting new interpretations of floral evolution within primitive Magnoliidae. The rather generalized, open and unprotected flowers are visited by thrips and dipters as well as by beetles. Since a similar observation has been made by THIEN (1980) in *Degeneria*, this challenges the presumed central position of cantharophily for the evolution of the angiospermous flower. GOTTSBERGER et al. (1980) consider the condition of *Drimys* to be primitive among extant Magnoliidae, the further floral evolution of which was connected with the elaboration of a more precise beetle pollination as, for example, that of *Calycanthus* (RICKSON, 1979) where a highly specialized food tissue resembling the Beltian bodies if ant-Acacias is offered to the pollinators; more specialized cantharophily led also to the switch from self-compatibility, held to be primitive for extant Magnoliidae, to self-incompatibility. The phytogeography of *Drimys* is discussed by EHRENDORFER et al. (1979) in the context of plate tectonics and is illustrative of the difficulty of correlating the age of a given taxon with its biological behavior: The observed infra-populational variability is remarkable for a relic genus in which conformity would be expected.

Another example of "living fossils" is *Austrobaileya* which has been studied in detail by ENDRESS (1980a) and should be placed close to Magnoliaceae and Annonaceae, and not into Laurales. The carpel is ascidiate, not plicate, and the pollen interestingly resembles that of the oldest (probable) angiosperm fossil, *Clavatipollenites* (ENDRESS and HONEGGER, 1980). (For a discussion of the somewhat irritating fact that most European students find ascidiate and most Americans plicate carpels, see Progr. Bot. 42, 52.)

But are conclusions from the so-called living fossils to the early angiosperms justified? It must be remembered that the assignment of primitive status to an extant plant implies the acceptance of some phylogenetic theory which consequently cannot be proven without bias. GUÉDÈS (1979) maintains that there is no compelling evidence to focus on Ranales, the tropics, or both, when discussing the origin of flowering plants. The extant so-called living fossils, such as those discussed above, are in many cases island plants and the putatively primitive features may be well due to special evolutionary processes on isolated islands; otherwise, these plants are in many respects more advanced than many temperate families.

c) Angiosperm Systematics

DAHLGREN (1980), SPORNE (1980), and TAKHTAJAN (1980) have presented new publications relevant to higher order systematics of angiosperms. SPORNE (1980), in a re-investigation of character correlations among dicotyledons, was able to confirm earlier results and to increase the number of correlated characters from 26 to 39; the most primitive families are Aextoxicaceae, Degeneriaceae, Stachyuraceae, and Magnoliaceae, and the most advanced ones are Dipsacaceae, Valerianaceae, and Hippuridaceae. TAKHTAJAN (1980) gives a modified outline of his system, together with a presentation of his ideas on evolutionary morphology which form the basis of his classification. Some of the problems mentioned include living fossils (surviving by means of "idioadaptation"), heterobathmy, and the relations between the paleosystematist and neosystematist in the construction of phylogenetic systems. TAKHTAJAN's

(1980) hierarchy is more structured than that of DAHLGREN (1980), since he uses subclasses as well as superorders, some of which, however, like Nymphaeanae comprizing both Nymphaeaceae and Nelumbonaceae, seem to be maintained for convention only. With respect to the circumscription of superorders within the monocotyledons, both systems differ only in the size of the *Commelina*-superorder which includes Juncaceae and Cyperaceae in DAHLGREN's (1980), but not in TAKHTAJAN's (1980) system. Within the dicotyledons, the differences are more pronounced.

As to the presentation of DAHLGREN's (1980) system, it is interesting to note that a similar scheme, of course vastly different in details, had already been constructed by Linnaeus' pupil Giseke (JONSELL, 1979); controversial or bridging genera had also been indicated, although Linnaeus would have them kept as well-hidden secrets within the large natural orders, because he feared that they might create too much confusion. One of the problems of presenting multidimensional relationships between several taxa has been pointed out by CLIFFORD and WILLIAMS (1980) for their numerically derived diagram of the monocotyledons: Arecaceae, Pandanaceae, and Cyclanthaceae appear widely separated. The crucial point is that it is often difficult to tell if the position of two groups close to each other in such a scheme is due to distortion, brought about by reducing the multi-dimensional net of relations down to two dimensions, due to convergent evolution, or if it is a reflection of the real relationship. In the case mentioned, CLIFFORD and WILLIAMS (1980) conclude that the respective similarities are due to convergence, and the problem is solved by representing the groups in question at two places in the scheme. Fortunately, large scale convergence appears to be rare otherwise.

In the modified version of his well-known diagram (Progr. Bot. 39, 194), DAHLGREN (1980) has tried to incorporate the data which have accumulated since 1975, as well as to give a synthesis of different ideas about the relationships of families and orders. The number of the superorders (now ending in -florae instead of -anae) has decreased from 34 to 24 and some radical changes in their circumscription have been introduced. Most controversial will probably be the combination of Piperales and Nymphaeales into one superorder, the Nymphaeiflorae which may serve to stress their intermediate position between monocotyledons and ranalean dicotyledons. But both orders are of course highly isolated. TUCKER (1979, 1980) stresses again the distinction of Piperales from the magnolialean line of angiosperms as shown by the development of the inflorescence and the flower, and YI-ZHEN (1980) again points to the primitive condition of their pollen. The unique position of the Nymphaeales was again confirmed by WEIDLICH (1980) for anatomical characters and by COLLINSON (1980) for seed characters, though the relative rank of their component taxa are still somewhat controversial (INAMDAR and ALEYKUTTY, 1979). To DAHLGREN (1980), the rationale for the union of these highly isolated groups lies in the fact that, apart from some other mutual characters, both Piperales and Nymphaeales possess starchy perisperm and well-defined endosperm enclosing a small embryo, characters which he believes cannot have evolved separately. This argument marks, however, a deviation from the original approach which was purely phenetical, and it is perhaps significant that DAHLGREN (1980) seems to be more optimistic now about the applications of his diagram. Starting with a device to map the distribution of characters, he now hopes that it will develop into a system of classification as natural as possible in the sense that, in the position of each family, phylogenetic relationships are reflected.

The Malviflorae (=Dilleniiflorae) are very broadly conceived by DAHLGREN (1980) and include now not only Rhamnales (without Vitaceae), Elaeagnales, and Thymelaeales, but also tentatively Plumbaginales, the embryology of which points to an euphorbiaceous affinity, while the seed coat is reminiscent of Rhamnaceae or Elaeagnaceae. Also included here, within Euphorbiales, are Simmondsiaceae and Didymelaceae. Whereas the position of the former was confirmed by SCOGIN (1980), based on serological investigations, the position of the latter is still uncertain. The trizonocolpate pollen which, according to MULLER (1979), resembles the Albian genus *Hexaporotricolpites*, is taken by KÖHLER (1981) to suggest a relation to Buxaceae or Euphorbiaceae which are, however, widely separated now by both TAKHTAJAN (1980) and DAHLGREN (1980). Within the Malvales, the genus *Pakaraimea* has been controversial between Dipterocarpaceae and Tiliaceae which may stress the close relationship between the two; while KOSTERMANS (1978) pleads for the inclusion within the latter family, additional material is presented by MAGUIRE and ASHTON (1980) to confirm the position within the former. *Plagiopteron* should be placed into the Malvales as a monotypic family which is suggested from palynological and anatomical findings by BAAS et al. (1979), although there are also close links to the Flacourtiaceae (Violales). The allocation of Urticales to Malviflorae is questioned by LEINS and ORTH (1979) who maintain that the striking similarity in floral ontogeny to Fagales and Polygonales supports the traditional concept of relationship (!).

Another very large superorder are the Rosiflorae which contain now the Hamamelidales, Trochodendrales, Buxales, a large proportion of the Amentiferae, together with Rosales, Cunoniales, and Saxifragales. The Hamamelidales occupy a critical position, as they link the wind-pollinated orders to those with insect pollination. Probably closely related to Rosiflorae are the fairly natural Myrtiflorae. DAHLGREN (1980) distinguishes the Araliiflorae from the Corniflorae which are placed close to each other by TAKHTAJAN (1980). The multidisciplinary study of *Aralidium* (Taxon 29, 391-416, 1980) is therefore very interesting as this genus is held to bridge the two groups. Some characters point to a relationship with Araliales, e.g., wood anatomy (PHILIPSON and BUTTERFIELD, 1980), others suggest connections with Cornales, e.g., serology and the presence of iridoid glycosides (FAIRBROTHERS, 1980; JENSEN and NIELSEN, 1980), and still other characters, e.g., pollen morphology (TSENG, 1980), are neutral. After a critical selection and weighting of characters, PHILIPSON and STONE (1980) conclude that *Aralidium* is close to Cornaceae, especially *Griselinia*, but sufficiently distinct to form a monotypic family.

Retzia, a monotypic sympetalous genus from S. Africa which has been included in the Loganiaceae in the latest issue of the *Natürliche Pflanzenfamilien* by LEEUWENBERG (1980), was placed into the Lamiiflorae by DAHLGREN (1980), based on chemical (carbocyclic iridoids) and morphological evidence from DAHLGREN et al. (1979); the closest relative may be the endemic Cape family Stilbaceae. Other controversial dicotyledonous genera which have recently been studied include *Dipterygium*, an intermediate between Cruciferae and Capparidaceae, assigned to the latter family because of chemical evidence (methylglucosinolates), but for which no generic ally could be named (HEDGE et al., 1980); *Sphenoclea*, the close relation of which to Campanulaceae was confirmed by MONOD (1980); and *Idiospermum* which, according to C.L. WILSON (1979), is sufficiently distinct from Calycanthaceae in its vegetative anatomy to revive family rank. The assignment of family status in cases like the latter has been criticized by VAN STEENIS (1978) who maintains that the segregation of families can only have scientific value if the segregate does not prove to be the closest ally to the family from which it is split.

A study of androecial evolution within Velloziaceae leads MENEZES (1980) to assume that multistaminate flowers, often with united filaments, are ancestral for this family from which flowers with six shortened filaments and eventually corollas similar to that of the Amaryllidaceae may have evolved. In extending her findings to further groups of petaloid monocotyledons, MENEZES (1980) postulates Vellloziaceae to be ancestral to Hypoxidaceae, Amaryllidaceae, Bromeliaceae, Liliaceae, etc. which will certainly meet with objections.

10. Seed Plant Systematics on the Family Level and Below

Gymnosperms

Araucariaceae. *Agathis*, monograph: WHITMORE (1980), leaf anatomy: PAGE (1980); the genus has probably invaded the Malayan archipelago from the old Gondwana centers New Caledonia and northern Queensland: WHITMORE and PAGE (1980). Gnetaceae. Rev. for Amazonia: CAVALCANTE (1978). Pinaceae. Classification, determination of fossil material and detection of evolutionary lines within *Pinus* using umbo (i.e., central field of apophyse of cone scale) characters: KLAUS (1980). Terpenoids and relationships in E. Asiatic *Abies*: ZAVARIN et al. (1978). Biosystematics of Mexican *Abies* spp.: TAYLOR and PATTERSON (1980). Embryology in *Pinus* and *Abies*: KONAR and NAGMANI (1980).

Monocotyledons

Alismataceae. Rev. for Brazil: RATAJ (1978). Amaryllidaceae. *Crinum* in Cameroun: NORDAL and WAHLSTRÖM (1980). Araceae. *Anubias*, rev.: CRUSIO (1978). *Arisaema* in Japan: OHASHI and MURATA (1980). Cyperaceae. *Ascolepis*: GOETGHEBEUR (1980). *Cyperus*, rev. of *Luzulae* groups: DENTON (1978). Gramineae. Chorology of Old World grasses used to define major phytochoria: CLAYTON and COPE (1980); none of these regions has less than 64 grass species, those poor in grasses lie at either end of the rainfall spectrum: CROSS (1980). Leaf anatomy of Chlorideae: VIGNAL (1979, 1980). Cuticles of W. African grasses allow determination down to species level, potentially useful for study of herbivore preferences: BARTHLOTT and MARTENS (1979). Genera of Bambusoideae in the Americas, conspectus with keys: CALDERÔN and SODERSTROM (1980). Pooideae, circumscription, with discussion of controversial genera: MACFARLANE and WATSON (1980). Key to Olyreae: SODERSTROM and CALDERÓN (1980). *Bouteloua*, monogr.: GOULD (1979). *Dichanthelium* in U.S.A. and Canada: GOULD and CLARK (1978). *Hordeum* sect. *Stenostachys*, part. rev.: VON BOTHMER (1979). *Hordeum secalinum* and *H. capense*, biosystematics: VON BOTHMER and JACOBSEN (1979). *Lasiacis*, rev.; all spp. self-compatible, C_3-plants: DAVIDSE (1978). *Pennisetum* in trop. Africa, agamospermy and probable hybrid origin of some spp.: BRUNKEN (1979). Hydrocharitaceae. Syst. anatomy: ANCIBOR (1979). Iridaceae. Cytology of Australasian spp.: GOLDBLATT (1979a). Cytology of *Crocus* very polymorphic and highly variable within spp.: BRIGHTON (1980). *Ferraria*, monogr.: DE VOS (1979). *Galaxia*, monogr., aneuploidy and autogamy: GOLDBLATT (1979b). *Gynandiris*, disjunct Mediterranean/S. African distr.: GOLDBLATT (1980a). *Homeria*, *Moraea*, *Rheome*, tax., cytology and evol. trends: GOLDBLATT (1980b). Lemnaceae. Multiauthored volume incl. key to all spp., new spp., cytology, physiology, and chemistry: LANDOLT (1980). Liliaceae. *Aloë* in E. Africa, evolution: CUTLER et al. (1980). *Asparagus* in Macaronesia: VALDÈS (1979). *Polygonatum* in E. Asia, rev.: JEFFREY (1980a). Marantaceae. *Thalia geniculata*-complex, rev.: ANDERSSON (1981). Orchidaceae. *Aerangis* in Africa, rev.: STEWART (1979). *Brownleea*, rev.: LINDER (1981). Palynology of Neottieae confirms homogeneity of the tribe except for *Lecanorchis*, trend from single grains to pollen tetrads: ACKERMAN and WILLIAMS (1980). *Polystachya*, part. rev.: GEERINCK (1979), PODZORSKI and CRIBB (1979). *Schizochilus*, rev.: LINDER (1980). *Stelis*, synopsis and key: GARAY (1979). Epidermal characters aid identification in diandrous orchids: ATWOOD and WILLIAMS (1979). Palmae. Syst. and ecology, review: TOMLINSON (1979). *Ceratolobus*, monogr.: DRANSFIELD (1979). *Salaca*, part. rev.: MOGEA (1980). Pandanaceae. Palynology of *Pandanus* subg. *Vinsonia*: HUYNH (1980b). Leaf anatomy: HUYNH (1979a,b). Restionaceae. Significant differences in flavonoid patterns between Australasian and S. African spp., the former containing gossypetin or a related 8-hydroxyflavonoid and little proanthocyanidins, the latter possess flavonols, flavones, and glycosylflavones plus proanthocyanidins: HARBORNE (1979). Zingiberaceae. *Aframomum* in W. Africa, rev.: LOCK (1980).

Dicotyledons

Acanthaceae. Epidermal characters in tribal classification: KARLSTRÖM (1980). Actinidiaceae. *Saurauia* in S. America: SOEJARTO (1980). Amaranthaceae. Betalains in *Bosea*: KULL and KÜHN (1979). Apocynaceae. Exceptional valvate estivation in *Parsonia* and *Artia*: ALLORGE (1979). *Adenium*, *Diplorhynchus*: PLAIZIER (1980). Araliaceae. Phytogeography of New Caledonian genera: BERNARDI (1979a). Asclepiadaceae. *Dischidia* in Malesia: RINTZ (1980). Balsaminaceae. Very fine monograph of African *Impatiens*; West Gondwanaland origin, with subsequent splitting in northern group with linear and southern with fusiform fruits; Lake Victoria region is center for Africa; hybridization frequent in disturbed habitats: GREY-WILSON (1980b). Betulaceae. *Alnus* in N. America: FURLOW (1979). Bignoniaceae. Dispersal and distribution, anemochorous spp. with larger areas than mammal-dispersed ones: GENTRY (1979). Bonnetiaceae. Xanthones in *Bonnetia* and *Archytaea* suggest strong affinity to Guttiferae, from which this family has been split off: KUBITZKI et al. (1978). Boraginaceae. Karyosystematics of *Onosma alboroseum*-group: TEPPNER (1980). Burseraceae. Evolutionary tendencies in *Bursera*: RZEDOWSKI (1979). *Bursera leptophloeos* from Brazil, included in this genus for purely geographical reasons, has to be placed into the otherwise African *Commiphora*: GILLET (1980). Cactaceae. Study of Mexican columnar cacti using a wide array of evidence from different fields reveals S. Mexico as center of origin for Pachycereae, from where lines of speciation led to other regions; the high number of endemics in the Andean cordillera is taken as an indication of a relatively recent origin of the group: GIBSON and HORAK (1978). Different types of pollination indicate geographical lines of relationship in cacti from U.S.A.: V. GRANT and K.A. GRANT (1979b). Caryophyllaceae. Biosystematics within *Silene*: BOCK (1980), MELZHEIMER (1980), PRENTICE (1979). Chenopodiaceae. *Chenopodium*, numerical study: REYNOLDS and CRAWFORD (1980), artificial hybridization: H.D. WILSON (1980). Salicornieae in Australia: P.G. WILSON (1980). Chloanthaceae. Physopsideae: MUNIR (1978). *Pityrodia*: MUNIR (1979). Cochlospermaceae. Monograph, related to but distinct from Bixaceae, convergent evolution of geoxylic suffrutices in America and Africa: POPPENDIECK (1980). Compositae. Cladistic study of *Anacyclus* incl. cytology, nuclear DNA content, and artificial hybridization: HUMPHRIES (1981). Chorology and cytology of *Amellus*: ROMMEL (1979). *Rothmaleria*, primitive genus of Lactuceae from S. Spain: LACK et al. (1980). Biology, dispersal etc. of *Scalesia pedunculata*: HAMANN (1979). Evolutionary trends in Macaronesian *Sonchus*: ALDRIDGE (1979). Morphological studies: Testa characters in Mutisieae: GRAU (1980). Palynology of Calenduleae: PRAGLOWSKI and GRAFTSTRÖM (1980); of *Urospermum*: LACK and LEUENBERGER (1979); of *Vernonia*: KEELY and JONES (1979). Embryology of Arctoteae-Arctotinae: AHLSTRAND (1979), of Inuleae: PULLAIAH (1979). Phytochemistry: Sesquiterpene lactones and systematics of *Artemisia*: KELSEY and SHAFIZADEK (1979), of *Ambrosia*: SEAMAN and MABRY (1979), and of Melampodiinae: SEAMAN et al. (1980). Revisions, synopses, etc.: *Athrixia* and related genera: KRONER (1980). *Aster* and *Lasallea* in N. America: SEMPLE and BROUILLET (1980). *Brachycome lineariloba*-complex: CARTER (1978). Grangeinae: FAYED (1979). *Hazardia*: CLARK (1979). *Lasianthea*: BECKER (1979). *Lipochaeta*: GARDNER (1979). *Pterocaulon*: CABRERA and RAGONESE (1978). *Saussurea*: LIPSCHITZ (1979). *Schistocarpha*: ROBINSON (1979). *Senecio* in Argentina: CABRERA and ZARDINI (1980). *Senecio* sect. *Senecio* in the Mediterranean: ALEXANDER (1979). *Zaluzania*: OLSEN (1979). Connaraceae. Palynology: DICKISON (1979). Convolvulaceae. Revision for Lesser Antilles: POWELL (1979). Cornaceae. Herbaceous spp. of *Cornus* in Pazific N. America: BAIN and DENFORD (1979). Crassulaceae. Chromosome numbers and classification of Mexican *Sedum*: UHL (1980). *Rosularia* and *Rhodiola* subgen. *Primuloides* in Asia: OHBA (1980). Cruciferae. Seed characters of *Brassica* sect. *Brassica*: STORK et al. (1980). Cytogeography of *Erysimum*: FAVARGER (1978), revision for SW. Europe: POLATSCHEK (1979). *Heliophila* in SW. Africa: SCHREIBER (1979). *Smelowskia* originated probably in Central Asia, with secondary center in Western N. America: VELICZKIN (1979). Cucurbitaceae. Review: JEFFREY (1980b). Cunoniaceae. Nodal anatomy: DICKISON (1980). *Caldcluvia*, *Pullea*, *Acsmithia*, *Spiraeanthemum*: HOOGLAND (1979). Dichapetalaceae. Ovule and seed-coat of *Dichapetalum*: BOESEWINKEL and BOUMAN (1980). Dipterocarpaceae. Evolutionary trends and patterns: ASHTON (1979). Stamen morphology and evolution: WOON and KENG (1979). Ebenaceae. Revision for Thailand: PHENGKAI (1978). *Diospyros* in Africa: F. WHITE (1978, 1980). Ericaceae. Phenetics and cladistics used for study of Andromedeae: JUDD (1979). *Grisebachia*: OLIVER (1980). *Rhododendron*, synopsis: CULLEN and CHAMBERLAIN (1979), CULLEN (1980). Euphorbiaceae. *Argythamnia*: INGRAM (1980a,b). Leaf epidermis of *Jatropha*: DEHGAN (1980). *Macaranga*: WHITMORE (1978). Pantropic *Margaritaria*: WEBSTER (1979). Northern S. America, the Antilles, and Malesia centers of diversity for *Phyllanthus*: HOLM-NIELSEN (1979). Platylobeae of Australasia: AIRY-SHAW (1980). Flacourtiaceae. Primitive position of *Berberidopsis* and relation of this family with Dilleniaceae: VAN HEEL (1979). Geraniaceae. *Geranium* in Malesia: VELDKAMP and MOERMAN (1978). *Monsonia*: VENTER (1979). *Sarcocaulon*: MOFFETT (1979). Gesneriaceae. Segregation of *Schizoboea* from *Didymocarpus* supported by floral morphology: A. WEBER (1980b). Goodeniaceae. *Calogyne*: CAROLIN (1979). Guttiferae. Anatomy of *Calophyllum*: D'ARCY and KEATING (1979). *Rheedia* in Brazil: M.E. VAN DEN BERG (1979). Hydrophyllaceae. Biosystematics of *Hydrophyllum*: BECKMANN (1979). Juglandaceae. Generic monograph, list of all spp., new interpretation of floral structures: MANNING (1978). Krameriaceae. Lipids of floral glands similar to those found in Malpighiaceae: SIMPSON et al. (1979). Labiatae. *Hedeoma*: IRVING (1980). Leaf anatomy of *Hyptidinae*: RUDALL (1980). *Leucas*: SEBALD (1980). *Origanum*: IETSWART (1980). *Salvia* in SW Europe: LIPPERT (1979). New infrageneric classification of *Stachys*: BHATTACHARJEE (1980). Lauraceae. Re-instatement of *Clinostemon*, based especially on xylotomy: KUBITZKI et al. (1979). Fruit and seed morphology confirm position of *Cassytha*: MOHANA RAO (1979/80). Lecythidaceae. First part of Flora Neotropica monograph, covering all biological aspects of the family and utilizing extensive field studies: PRANCE and MORI (1979). Leguminosae. Caesalpinioideae. Palynology of Madagascan spp.: SENESSE (1980). *Bauhinia* sect. *Tylotea* in Brazil: VAZ (1979). Mimosoideae. Reaction of *Acacia* seeds to heavy compression as taxonomic marker: TRAN and CAVANAGH (1980). *Acacia* in Queensland, part. rev.: PEDLEY (1978, 1979); *Uninerves-Triangularis* group: MASLIN (1978); series *Pulchellae*: GUINET et al. (1980); distribution in W. Australia: HNATIUK and MASLIN (1979). *Albizia* in mainland SE Asia: NIELSEN (1979a). Realignements in SE Asian Ingeae: NIELSEN (1979b). Papilionoideae. Flavonoids in *Anthyllis vulneraria*: GONNET (1980). *Andira*, rev.: MATTOS (1979). *Coelidium*: GRANBY (1980). *Coronilla*: SCHMIDT (1979). Classification and phytogeography of *Crotalaria* discussed in context of flavonoid distribution: PILBEAM et al. (1979). Leaflet anatomy supports new infratribal classification of Phaseoleae: LACKEY (1978). Classification of *Phaseolus* and related genera: VERDCOURT (1980b). Canavanine, a nitrogen-rich nonprotein amino acid supposedly acting as both nitrogen storage and defense in seeds and seedlings, is absent from basal tribes and appears in Tephrosieae and more advanced tribes: BELL et al. (1978). Loganiaceae. Full-scale treatise for the Natürliche Pflanzenfamilien indicates continuation of the series after 25 years with a change of format: LEEUWENBERG (1980). *Geniostoma* subgen. *Geniostoma*, monogr.: CONN (1980). Loranthaceae. Pollen morphology of *Psittacanthus*: FEUER and KUJIT (1979). Lythraceae. *Rotala*, monogra.: COOK (1980). Magnoliaceae. *Manglietia*, monogr.: TIÊP (1980). Malpighiaceae. Specialized pollination responsible for floral conservativism: W.R. ANDERSON (1979). Peculiar mode of cryptic self-fertilization: W.R. ANDERSON (1980). Malvaceae. Rev. of *Gossypium* and related genera: FRYXELL (1980a), of *Hibiscus* sect. *Bombicella*: FRYXELL (1980b). Melastomataceae. *Diplectria*: VELDKAMP et al., (1978). *Memecyclon* in Sri Lanka: BREMER (1979), on Malay Peninsula: MAXWELL (1980). *Medinilla*, *Pachycentria*, *Pogonanthera* on Malay Peninsula: MAXWELL (1978). Meliaceae. *Chisocheton*, monogr. and growth-form study:

MABBERLEY (1979a). Monimiaceae. *Hortonia* intermediate between Monimiaceae and Atherospermaceae: ENDRESS (1980c). *Mollinedia* sect. *Exappendiculatae*, rev.: PEIXOTO (1979). Moraceae. *Cecropia* in Amazonia: BERG (1978). *Ficus*, life history, review: JANZEN (1979); co-evolution with pollinators, review: WIEBES (1979); rev. for Ethiopia: AWEKE (1979). Myristicaceae. *Knema*, rev.: DE WILDE (1979). Palynology of *Compsoneura* and *Virola*: WALKER and WALKER (1979). Myrsinaceae: New classification of *Cybianthus*, with key to American genera: AGOSTINI (1980). Myrtaceae. Important morphological study: BRIGGS and JOHNSON (1979). *Psiloxylon*, *Heteropyxis*, and problems of tribal classification: R. SCHMID (1980). Hybridization in *Eucalyptus*: DRAKE (1980). *Eucalyptus Foecundae*-group: BROOKER (1979). *Eugenia*, seed morphology: VAN WYK (1980). *Leptospermum*, leaf anatomy: C.T. JOHNSON (1980). *Melaleuca* in S. Australia: CARRICK and CHORNEY (1979). *Rhodamnia*: SCOTT (1979a). *Xanthomyrtus*: SCOTT (1979b). Olacaceae. Account for Eastern Tropics: SLEUMER (1980). Oleaceae. Flavonoids: HARBORNE and GREEN (1980). Onagraceae. Review of reproductive biology; self-incompatibility superseded by protandry, protogyny, male sterility, etc. as secondary devices to promote outcrossing; repeated evolution of autogamy, especially in herbs, one of the most prominent trends within this family: RAVEN (1979b). Loss of glycosylation of flavonoids correlated with phyletic advancement in Epilobieae: AVERETT et al. (1979). *Circaea* in U.S.S.R.: SKVORTSOV (1979). N-alkane constituents of leaf cuticles in *Epilobium* sect. *Zauschneria* so specific as to allow for the recognition of parents in allopolyploid hybrids: BOWMAN (1980). Cytology of *Ludwigia* stresses isolated position within family: RAVEN and TAI (1979). Genetics and taxonomy in *Oenothera*: STUBBE and RAVEN (1979b), RAVEN et al. (1979). Opiliaceae. Part. rev.: HIEPKO (1979). Ranunculaceae. *Aconitum* subgen. *Lycoctonum*: TAMURA and LAUENER (1979). *Nigella* with unusual ektextine differentiation, *N. integrifolia* to be split off as monotypic genus: SKVARLA and NOWICKE (1979). *Knowltonia*: RASMUSSEN (1979). Resedaceae. Final part of family monograph comprising *Reseda*: ABDALLAH and DE WIT (1978). Rhizophoraceae. Seed morphology of *Anisophyllea* and *Poga*: FLORET (1979). *Rhizophora* in NE. Australia: BUNT and DUKE (1979). Rosaceae. New subgeneric classification of *Crataegus*: EL-GAZZAR (1980). Rubiaceae. *Galium glaucum*-complex in SW Europe: KRENDL (1979). New tribe Hypobathreae: ROBBRECHT (1980). Naucleeae s. str., rev.: RIDSDALE (1978). *Polysphaera*: VERDCOURT (1980a). *Portlandia* complex: AIELLO (1979). *Psyllocarpus*: KIRKBRIDE (1979). *Tricalysia* subgen. *Empogona*: ROBBRECHT (1979). Rutaceae. *Coleonema*: WILLIAMS (1981). *Tetractonia*: HARTLEY (1979). *Raputia* split into 4 genera, rev.: EMMERICH (1978). Sabiaceae. *Sabia*: VAN DE WATER (1980). Sapindaceae. Palynology and evolutionary trends within Nephelieae: R.G. VAN DEN BERG (1978). Saxifragaceae. Flavonoid spectra not in line with sectional classification of N. American *Saxifraga*: MILLER and BOHM (1980). Tendency toward O-methylated flavones with extra hydroxyls in 6- and/or 2'-position: BOHM and MILLER (1979). Wood anatomy of *Deutzia*: STYER and STERN (1979), of *Philadelphus*: STERN and STYER (1979). Scrophulariaceae. New type of haustorial system in parasitic members stresses relationship to Orobanchaceae: OZENDA and CAPDEPON (1979b). *Calceolaria*, part. rev.: MOLAU (1979, 1980). *Campylanthus* with Macaronesian-Arabian disjuction: MILLER (1980). *Euphrasia* in Europe: YEO (1979). *Hebenstetia* and *Dischisma*: ROESSLER (1979). *Linaria*, part. rev.: VIANO (1978). Palynology of *Mimulus*: ARGUE (1980). Biosystematics of *Scrophularia* in Macaronesia: DALGAARD (1979). Solanaceae. Symposium volume on biology, taxonomy, ethnobotany, and phytochemistry: HAWKES et al. (1979). Present state of knowledge of Solanaceae (incl. Nolanaceae) classification, with historic survey, generic index, and bibliography: D'ARCY (1979). Conspectus of S. American subfamilies and tribes: HUNZIKER (1979). Annotated enumeration of Indian genera, with distribution maps: DEB (1979). Apart from *Solanum* and *Nicotiana*, the family is represented in Australia by four rather coherent and otherwise isolated genera probably originated from solanaceous stock ancestral to Cestreae: HAEGI (1979). Biosystematics of N American physaloid genera: AVERETT (1979). *Benthamiella*: S. ARROYO (1980). S American origin of *Brunfelsia* with subsequent differentiation into three distinct groups, one migrating to Antilles: PLOWMAN (1979). Numerical taxonomy of allozyme variation in *Capsicum*: JENSEN et al. (1979). Experimental hybridization of recently discovered African tobacco together with other evidence points to affinity with Southern Pazific *Nicotiana fragrans* which even enhances the puzzle: GERSTEL et al. (1979). *Solanum* in Australia: SYMON (1979a), sect. *Androceras*: WHALEN (1979b), *S. nigrum*-complex: HEISER et al. (1979). Sterculiaceae. *Hermannia* subgen. *Hermannia* in S. Africa: VERDOORN (1980). Stylidiaceae. Natural history of spp. from Arnhem Land, Australia; *Stylidium* flowers mimicking other spp. growing close to them: CARLQUIST (1979). Styracaceae. Leaf anatomy and venation patterns: SCHADEL and DICKISON (1979). Symplocaceae. Palynology of Brazilian *Symplocos*: BARTH (1979); *Symplocos* in New Caledonia: NOOTEBOOM (1980). Theaceae. Part. rev. for Papuasia: BARKER (1980). Thymelaeaceae. Observations from fruit, seed, and seedling denote family as isolated: TAN (1980a). Umbelliferae. *Cicuta* in N. America: MULLIGAN (1980). *Ferulago*, rev. with new method of diagrammatic presentation of diagnostic characters to aid identification: BERNARDI (1979b). *Trachymene* in N. Australia: MACONOCHIE (1980). Valerianaceae. Palynology suggests affinity to Dipsacaceae and, possibly, Caprifoliaceae: PATEL and SKVARLA (1979). *Valeriana* in European Alps: TITZ and TITZ (1979). Verbenaceae. *Glandularia* in N. America: UMBER (1979).

References

ABDALLAH, M.S., DE WIT, H.C.D.: Meded. Landbouwhogeschool Wageningen 78 (14), 99-416 (1978). - ACKERMANN, J.D., WILLIAMS, N.H.: Grana 19, 7-18 (1980). - AESCHIMANN, D., BOCQUET, G.: Candollea 35, 451-495 (1980). - AGOSTINI, G.: Acta Biol. Venez. 10, 129-185 (1980). - AHLSTRAND, L.: Bot Notiser 132, 109-116 (1979). - AIELLO, A.: J. Arnold Arb. 60, 38-126 (1979). - AIRY SHAW, H.K.: Kew Bull 35, 577-700 (1980). - ALBERS, F.: Plant Syst. Evol. 136, 137-167 (1980a); - Phyton (Austria) 20, 95-116 (1980b); - Flora 169, 150-167 (1980c). - ALDRIDGE, A.E.: Evolution within a single genus: Sonchus in Macaronesia, 279-289. In: see ref. BRAMWELL, 1979. - ALEXANDER, J.C.M.: Notes R. Bot. Gard. Edinburgh 37, 387-428 (1979). - ALLORGE, L.: Adansonia 19, 117-124 (1979). - ALMEDA, F., WHIFFIN, T.: Syst. Bot. 5, 294-311 (1980). - ANCIBOR, E.: Bot. J. Linn. Soc. 78, 237-266 (1979). - ANDERSON, G.J.: Systematic and evolutionary consideration of species of Solanum, sect. Basarthrum, 549-562. In: see ref. HAWKES et al., 1979. - ANDERSON, W.R.: Biotropica 11, 219-233 (1979); - Science 207, 892-893 (1980). - ANDERSSON, L.: Nord. J. Bot. 1, 48-56 (1981). - ANGEVINE, M.W., CHABOT, B.F.: Seed germination syndromes in higher plants, 188-206. In: see ref. SOLBRIG et al., 1979. - ANTONOVICS, J.: Science 208, 587-589 (1980). - AREKAL, G.D., RAMASWAMY, S.N.: Bot. Notiser 133, 295-309 (1980). - ARGUE, C.L.: Am. J. Bot. 67, 68-87 (1980). - AROHONKA, T., ROUSI, A.: Ann. Bot. Fennici 17, 258-263 (1980). - AUQUIER, P., STACE, C.A.: Plant Syst. Evol. 136, 47-52 (1980). - ARROYO, M.T.K.: Comments on breeding systems in Neotropical forests, 371-380. In: see ref. LARSEN and HOLM-NIELSEN, 1979. - ARROYO, S.: Bot. Notiser 133, 67-76 (1980). - ASHTON, P.S.: Some geographic trends in morphological variation in the Asian tropics and their possible significance, 35-48. In: see ref. LARSEN and HOLM-NIELSEN, 1979. - ATWOOD, J.T., WILLIAMS, N.H.: Bot. J. Linn. Soc. 78, 141-156 (1979). - AVERETT, J.E.: Biosystematics of the physaloid genera of the Solanaceae in North America, 493-503. In: see ref. HAWKES et al., 1979. - AVERETT, J.E., RAVEN, P.H., BECKER, H.: Am. J. Bot. 66, 1151-1155 (1979). - AWEKE, G.: Meded. Landbouwhogeschool Wageningen 79/3, 1-115 (1979).

BAAS, P., ZWEYPFENNING, R.C.V.J.: Acta Bot. Neerl. 28, 117-155 (1979). - BAAS, P., GEESINK, R., VAN HEEL, W.A., MULLER, J.: Grana 18, 69-89 (1979). - BACHMANN, K., CHAMBERS, K.L., PRICE, H.J.: Plant Syst. Evol. Suppl. 2, 41-66 (1979). - BACHMANN, M.A.: Oekologie und breeding system bei Poa alpina L. 199 pp., Diss. Univ. Zürich (1980). - BAIN, J.F., DENFORD, K.E.: Bot. Notiser 132, 121-129 (1979). - BAKER, H.G., BAKER, I.: Am.

J. Bot. 66, 591-600 (1979). - BANYARD, B.J., JAMES, S.H.: Aust. J. Bot. 27, 27-37 (1979). - BARETTA-KUIPERS, T.: IAWA Bulletin 1979 (2-3), 47-50 (1979). - BARKER, W.R.: Brunonia 3, 1-60 (1980). - BARKWORTH, M.E., McNEILL, J., MAZE, J.: Can. J. Bot. 57, 2539-2553 (1979). - BARTH, O.M.: Grana 18, 99-107 (1979). - BARTHLOTT, W. MARTENS, B.: Trop. Subtrop. Pflanzenw. 30, 1-113 (1979). - BARTHLOTT, W., VOIT, G.: Plant Syst. Evol. 132, 205-229 (1979). - BARTHLOTT, W., ZIEGLER, B.: Ber. Dtsch. Bot. Ges. 93, 391-403 (1980). - BAWA, K.S.: New Zealand J. Bot. 17, 521-524 (1979); - Annu. Rev. Ecol. Syst. 11, 15-39 (1980a); - Evolution 34, 467-474 (1980b). - BEACH, J.H., BAWA, K.S.: Evolution 34, 1138-1142 (1980). - BEACH, J.H., KRESS, W.J.: Syst. Bot. 5, 1-5 (1980). - BECKER, K.M.: Mem. N. Y. Bot. Gard. 31, 1-64 (1979). - BECKMANN, R.L.: Am. J. Bot. 66, 1053-1061 (1979). - BELL, A.D.: Ann. Bot. 43, 209-229 (1979). - BELL, A.D., TOMLINSON, P.B.: Bot. J. Linn. Soc. 80, 125-160 (1980). - BELL, A.J., LACKEY, J.A., POLHILL, R.M.: Biochem. Syst. Ecol. 6, 201-212 (1978). - BELL, E.A.: Toxins in seeds, 143-161. In: Biochemical Aspects of Plant-Animal Co-Evolution, ed. J.B. Harborne. London: Academic Press, 1978. - BENTLEY, B.L.: Biotropica 11, 11-17 (1979). - BERENBAUM, M.: Science 201, 532-533 (1978). - BERG, C.C.: Acta Amazonica 8, 149-182 (1978). - BERNARDI, L.: The New Caledonian genera of Araliaceae and their relationship with those of Oceania and Indonesia, 315-325. In: see ref. LARSEN and HOLM-NIELSEN, 1979a; - Boissera 30, 1-182 (1979b). - BHATTACHARJEE, R.: Notes R. Bot. Gard. Edinburgh 38, 65-96 (1980). - BISBY, F.A., VAUGHAN, J.G., WRIGHT, C.A. (eds.): Chemosystematics: Principles and Practice. The Systematics Association. Spec. Vol. No. 16, 449 pp. London: Academic Press, 1980. - BOCK, C.: Candollea 35, 541-564 (1980). - BOESEWINKEL, F.D., BOUMAN, F.: Acta Bot. Neerl. 29, 103-115 (1980). - BOHM, B.A., COLLINS, F.W.: Biochem. Syst. Ecol. 7, 195-201 (1979). - BONNEMAISON, F., DOMMEE, B., JACQUARD, P.: Oecol. Plant 14, 85-101 (1979). - BORGEN, L.: Karyology of the Canarian flora, 329-346. In: see ref. BRAMWELL, 1979. - BOTHMER, R. VON: Bot. Tidsskr. 74, 117-146 (1979). - BOTHMER, R. VON, JACOBSEN, N.: Bot. Tidsskr. 74, 223-230 (1979). - BOWMAN, R.N.: Am. J. Bot. 67, 671-685 (1980). - BRAMWELL, D. (ed.): Plants and Islands. 459 pp. London-New York: Academic Press, 1979. - BRAYTON, R.D., CAPON, B.: Aliso 9, 581-587 (1980). - BRENAN, J.P.M., ROSS, R., WILLIAMS, J.T. (eds.): Computers in Botanical Collections. 216 pp. London-New York: Plenum Press, 1979. - BREMER, K.: Opera Bot. 50, 1-32 (1979). - BREMER, K., WANNTORP, H.-E.: Nord. J. Bot. 1, 1-3 (1981). - BRIGGS, B.G., JOHNSON, L.A.S.: Proc. Linn. Soc. New South Wales 102, 157-256 (1979). - BRIGHTON, C.A.: Notes R. Bot. Gard. Edinburgh 38, 399-412 (1980). - BROAD, W.J.: Science 211, 1137-1139 (1981). - BROCKWAY, L.H.: Science and Colonial Expansion: The Role of the British Royal Botanic Gardens. 215 pp. New York-London: Academic Press, 1979. - BROOKER, M.I.H.: Brunonia 2, 125-170 (1979). - BRUMMIT, R.K., FERGUSON, I.K., POOLE, M.M.: Pollen Spores 22, 11-16 (1980). - BRUNKEN, J.N.: Bot. J. Linn. Soc. 79, 37-64 (1979). - BRYANT, J.P., KUROPAT, P.J.: Annu. Rev. Ecol. Syst. 11, 261-285 (1980). - BUNT, J.S., DUKE, N.C.: Aust. J. Bot. 27, 657-678 (1979). - BURGER, W.C.: Taxon 28, 385-386 (1979).

CABRERA, A.L., RAGONESE, A.M.: Darwiniana 21, 185-257 (1978). - CABRERA, A.L., ZARDINI, E.M.: Darwiniana 22, 427-492 (1980). - CALDERÓN, C.E., SODERSTROM, T.R.: Smithsonian Contrib. Bot. 44, 1-27 (1980). - CALVIN, M.: Naturwissenschaften 67, 525-533 (1980). - CAMPION-BOURGET, F.: Rev. Cytol. Biol. Veget.-Bot. 3, 199-219 (1980). - CARDONA, M.A., CONTANDRIOPOULOS, J.: Endemism and evolution in the islands of the Western Mediterranean, 133-169. In: see ref. BRAMWELL, 1979. - CARLQUIST, S.: Aliso 9, 411-461 (1979); - Aliso 9, 499-533 (1980). - CAROLIN, R.C.: Brunonia 2, 1-17 (1979); - Aust. J. Bot. 28, 123-137 (1980). - CARR, S.G.M.: Bull. Torrey Bot. Club 107, 456-461 (1981). - CARRICK, J., CHORNEY, K.: J. Adelaide Bot. Gard. 1, 281-319 (1979). - CARTER, C.R.: Telopea 1, 387-393 (1978). - CAUWET-MARC, A.-M.: Candollea 34, 49-86

(1979). - CAVALCANTE, P.: Acta Amazonica 8, 201-215 (1978). - ČEŠKA, A., TRUMPOUR, A.D.: Taxon 28, 329-335 (1979). - CHANDA, S., GHOSH, K., NILSSON, S.: Grana 18, 21-31 (1979). - CHEW, F.S.: Symb. Bot. Upsal. 22, 100-106 (1979). - CHIAPELLA, L.F.: Giorn. Bot. Ital. 113, 237-252 (1979). - CHINNAPPA C.C., MORTON, J.K.: Am. J. Bot. 66, 471-473 (1979). - CLARK, W.D.: Madroño 26, 105-127 (1979). - CLAYTON, W.D., COPE, T.A.: Kew Bull. 35, 135-171 (1980). - CLIFFORD, H.T., WILLIAMS, W.T.: Aust. J. Bot. 28, 261-268 (1980). - COETZEE, J., VAN DER SCHIJFF, H.P.: J. S. Afr. Bot. 45, 93-126 (1979). - COLLINSON, M.E.: Ann. Bot. 46, 603-632 (1980). - CONN, B.J.: Blumea 26, 245-364 (1980). - COOK, C.D.K.: Boissiera 29, 7-156 (1979). - COOPER-DRIVER, G.A., BALICK, M.J.: Bot. Mus. Harvard Univ. 26, 257-265 (1978). - CORADIN, L., GIANNASI, D.E.: Taxon 29, 33-40 (1980). - COSTAS-LIPPMANN, M.: Bot. Gaz. 140, 393-397 (1979). - COUDERC, H.: Bull. Soc. Bot. France 127, Lettr. Bot. 139-149 (1980). - CRAWFORD, D.J.: Bot. Rev. 44, 431-456 (1978). - CREPET, W.L.: Rev. Palaeobot. Palynol 27, 213-328 (1979). - CRISCI, J.V.: Taxon 29, 213-224 (1980). - CRISCI, J.V., STUESSY, T.F.: Syst. Bot. 5, 112-135 (1980). - CRISCI, J.V., HUNZIKER, J.H., PALACIOS, R.A., NARANJO, C.A.: Am. J. Bot. 66, 133-140 (1979). - CROAT, T.B.: Phytologia 42, 319-348 (1979). - CRONQUIST, A.: Chemistry in plant taxonomy: an assessment of where we stand. 1-27. In: see ref. BISBY et al., 1980. - CROSS, R.A.: Kew Bull. 35, 279-289 (1980). - CRUDEN, R.W.: Evolution 31, 32-46 (1977). - CRUDEN, R.W., JENSEN, K.G.: Am. J. Bot. 66, 875-879 (1979). - CRUSIO, W.: Meded. Landbouwhogeschool Wageningen 79/14, 1-48 (1979). - CULLEN, J.: Notes R. Bot. Gard. Edinburgh 39, 1-207 (1980). - CULLEN, J., CHAMBERLAIN, D.F.: Notes R. Bot. Gard. Edinburgh 37, 327-338 (1979). - CUTLER, D.F. Applied Plant Anatomy. 103 pp. London-New York: Longman, 1978. - CUTLER, D.F., HARTMANN, H.: Trop. Subtrop. Pflanzenw. 28, 8-55 (1979). - CUTLER, D.F., BRANDHAM, P.E., CARTER, S., HARRIS, S.J.: Bot. J. Linn. Soc. 80, 293-317 (1980). - CZAJA, A.T.: Stärke und Stärkespeicherung bei Gefäßpflanzen. Versuch einer Amylo-Taxonomie. 264 pp. Stuttgart-New York: Fischer, 1978.

DAHLGREN, R.: Symb. Bot. Ups. 22, 210-221 (1979). - DAHLGREN, R.: Bot. J. Linn. Soc. 80, 91-124 (1980a); - Bot. Notiser 133, 337-341 (1980b). - DAHLGREN, R., NIELSEN, B.J., GOLDBLATT, P. ROURKE, J.P.: Ann. Missouri Bot. Gard. 66, 545-556 (1979). - DALGAARD, V.: Opera Bot. 51, 1-64 (1979). - DANIN, A., BAKER, I., BAKER, H.G.: Israel J. Bot. 27, 177-211 ("1978", 1979). - D'ARCY, W.G.: The classification of the Solanaceae, 3-47. In: see ref. HAWKES et al., 1979. - D'ARCY, W.G., KEATING, R.C.: Ann. Missouri Bot. Gard. 66, 557-571 (1979). - DAVIDSE, G.: Ann. Missouri Bot. Gard. 65, 1133-1254 (1978). - DAVIS, P.H., CULLEN, J.: The Identification of Flowering Plant Families. 2nd ed., 113 pp. Cambridge: Cambridge University Press, 1979. - DEB, D.B.: Solanaceae in India, 87-112. In: see ref. HAWKES et al., 1979. - DEHGAN, B.: Bot. J. Linn. Soc. 80, 257-278 (1980). - DEN NIJS, J.C.M., STERK, A.A.: Bot. Jahrb. Syst. 101, 527-554 (1980). - DENTON, J.F.: Contrib. Univ. Michigan Herb. 11 (4), 197-271 (1978); - Taxon 28, 149-155 (1979). - DE VOS, M.P.: J. S. Afr. Bot. 45, 295-375 (1979). - DEWEY, D.R.: Am. J. Bot. 66, 166-172 (1979). - DE WILDE, W.J.J.O.: Blumea 25, 321-478 (1979). - DICKINSON, T.A.: Bot. Rev. 44, 181-232 (1978). - DICKISON, W.C.: Pollen Spores 21, 31-79 (1979); - Am. J. Bot. 67, 975-981 (1980). - DILCHER, D.L.: Rev. Palaeobot. Palynol. 27, 291-328 (1979). - DOEBLEY, J.F., ILTIS, H.H.: Am. J. Bot. 67, 982-993 (1980). - DRAKE, D.W.: Aust. J. Bot. 28, 167-191 (1980). - DRANSFIELD, J.: Kew Bull. 34, 1-33 (1979). - DUNCAN, T.: Taxon 29, 441-454 (1980a); - Syst. Bot. 5, 136-148 (1980b). - DUNCAN, T., PHILLIPS, R.B., WAGNER, W.H., Jr.: Syst. Bot. 5, 264-293 (1980).

EASTOP, V.: Symb. Bot. Ups. 22, 120-134 (1979). - EDMONDS, J.M.: Biosystematics of Solanum L., section Solanum (Maurella), 529-548. In:

see ref. HAWKES et al., 1979. - EHRENDORFER, F.: Reproductive biology in island plants, 293-306. In: see ref. BRAMWELL, 1979. - EHRENDORFER, F., LUFTENSTEINER, H.W., PETRAK, J.: Plant Syst. Evol. 136, 1-6 (1980). - EHRENDORFER, F., SILBERBAUER-GOTTSBERGER, I., GOTTSBERGER, G.: Plant Syst. Evol. 132, 53-83 (1979). - EL-GAZZAR, A.: Symb. Bot. Upsal. 22 (4), 182-193 (1979); - Bot. Jahrb. Syst. 101, 457-469 (1980). - ELLSTRAND, N.C., LEVIN, D.E.: Syst. Bot. 5, 6-16 (1980). - EMMERICH, M.: Rodriguesia 45, 223-307 (1978). - ENCKE, F., BUCHHEIM, G., SEYBOLD, S.: Zander, Handwörterbuch der Pflanzennamen. 11. Aufl., 844 pp., Stuttgart: Ulmer, 1979. - ENDRESS, P.K.: Bot. Jahrb. Syst. 101, 393-433 (1980a); - Plant Syst. Evol. 134, 79-120 (1980b); - Plant Syst. Evol. 133, 199-221 (1980c). - ENDRESS, P.K., HONEGGER, R.: Grana 19, 177-182 (1980). - ENGELL, K.: Bot. Tidsskr. 73, 155-166 (1978). - ERBEN, M.: Webbia 34, 409-417 (1979). - ESTABROOK, G.F.: Taxon 28, 197-203 (1979); - The compatibility of occurrence patterns of chemicals in plants, 379-397. In: see ref. BISBY et al., 1980.

FAEGRI, K. (K.F.): Taxon 28, 168 (1979); - Taxon 29, 312-314 (1980). - FAHN, A.: Secretory Tissues in Plants. 302 pp. London-New York-San Francisco: Academic Press, 1979. - FAIRBROTHERS, D.E.: Taxon 29, 412-416 (1980). - FAVARGER, C.: Anal. Inst. Bot. Cavaniles 35, 361-393 ("1978", 1980). - FAYED, A.: Mitt. Bot. Staatss. München 15, 425-576 (1979). - FEENY, P.P.: Rec. Adv. Phytochem. 10, 1-40 (1976). - FERGUSON, I.K., SKVARLA, J.J.: Grana 18, 15-20 (1979). - FEUER, S., KUIJT, J.: Bot. Notiser 132, 295-309 (1979); - Am. J. Bot. 67, 34-50 (1980). - FISHER, J.B., HONDA, H.: Am. J. Bot. 66, 633-644 (1979a); - Am. J. Bot. 66, 645-655 (1979b). - FITTER, A.H.: Biol. J. Linn. Soc. 13, 331-339 (1980). - FLORET, J.-J.: Adansonia 19, 109-115 (1979). - FOSBERG, F.R.: Tropical floristic botany – concepts and status – with special attention to tropical islands, 89-105. In: see ref. LARSEN and HOLM-NIELSEN, 1979. - FREY-WYSSLING, A.: Ber. Dtsch. Bot. Ges. 93, 281-287 (1980). - FROEBE, H.A.: Plant Syst. Evol. 133, 223-237 (1980). - FRYXELL, P.A.: The Natural History of the Cotton Tribe (Malvaceae, Gossypieae). 245 pp. College Station (Texas): Texas A and M University Press, 1980a; - USDA (= United States Dept. Agric.) Technical Bull. 1624, 53 pp. (1980b). - FUNK, V.A., RAVEN, P.H.: Taxon 29, 417-419 (1980). - FURLOW, J.J.: Rhodora 81, 1-121, 151-248 (1979).

GARAY, L.A.: Bot. Mus. Leafl. 27, 167-259 (1979). - GARDNER, R.C.: Rhodora 81, 291-343 (1979). - GEERINCK, D.: Bull. Jard. Bot. Nat. Belg. 49, 409-420 (1979). - GENTRY, A.H.: Distribution patterns of neotropical Bignoniaceae: some phytogeographic implications, 339-354. In: see ref. LARSEN and HOLM-NIELSEN, 1979. - GENTRY, A.H., TOMB, A.S.: Ann. Missouri Bot. Gard. 66, 756-777 (1979). - GERSTEL, D.V., BURNS, J.A., BURK, L.G.: J. Heredity 70, 342-344 (1979). - GIANNASI, D.E.: Bot. Rev. 44, 399-429 (1978). - GIBSON, A.C., HORAK, K.E.: Ann. Missouri Gard. 65, 999-1057 ("1978", 1979). - GILLETT, J.B.: Traditional African knowledge of plants: its value today and how to preserve it, 99-101. In: see ref. KUNKEL, 1979; - Kew Bull. 34, 569-587 (1980). - GILMARTIN, A.J.: Taxon 29, 199-212 (1980). - GIVNISH, T.: On the adaptive significance of leaf form, 373-407. In: see ref. SOLBRIG et al., 1979. - GIVNISH, T.J.: Evolution 34, 959-972 (1980). - GOETGHEBEUR, P.: Adansonia 19, 269-305 (1980). - GOLDBLATT, P.: Ann. Missouri Bot. Gard. 66, 851-855 (1979a); - J. S. Afr. Bot. 45, 385-423 (1979b); - Bot. Notiser 133, 239-260 (1980a); - Bot. Notiser 133, 85-95 (1980b). - GOLDBLATT, P., GENTRY, A.H.: Bot. Notiser 132, 475-482 (1979). - GOLLOB, L.: Naturwissenschaften 67, 409-410 (1980). - GOMES, C.M.R., GOTTLIEB, O.R.: Biochem. Syst. Ecol. 8, 81-87 (1980). - GONNET, J.-F.: Biochem. Syst. Ecol. 8, 55-63 (1980). - GORNALL, R.J., BOHM, B.A.: Syst. Bot. 3, 353-368 (1978). - GORNALL, R.J., BOHM, B.A., DAHLGREN, R.: Bot. Notiser 132, 1-30 (1979). - GOTTLIEB, L.D.: The origin of phenotype in a recently evolved species, 264-286. In: see ref. SOLBRIG et al., 1979. - GOTTLIEB, L.D., WEEDEN, N.F.: Evolution 33, 1024-1039 (1979). - GOTTLIEB, O.R.: Micromolecular systematics: principles and practice, 329-352. In: see ref. BISBY et al., 1980. - GOTTLIEB, O.R., KUBITZKI, K.: Biochem. Syst. Ecol. 9, 5-12 (1981). - GOTTSBERGER, G., SILBERBAUER-GOTTSBERGER, I., EHRENDORFER, F.: Plant Syst. Evol. 135, 11-39 (1980). - GOULD, F.W.: Ann. Missouri Bot. Gard. 66, 348-416 (1979). - GOULD, F.W., CLARK, C.A.: Ann. Missouri Bot. Gard. 65, 1088-1132 (1978). - GRAHAM, A., BARKER, G., FREITAS DA SILVA, M.: Grana 19, 79-84 (1980). - GRANBY, R.: Opera Bot. 54, 1-47 (1980). - GRANT, M.C., MITTON, J.B.: Evolution 33, 914-918 (1979). - GRANT, V.: Bot. Gaz. 140, 443-448 (1979). - GRANT, V., GRANT, K.A.: Bot. Gaz. 140, 208-215 (1979a); - Plant Syst. Evol. 133, 29-37 (1979b). - GRAU, J.: Mitt. Bot. Staatss. München 16, 269-332 (1980). - GREEN, S., GREEN, T.L., HESLOP-HARRISON, Y.: Bot. J. Linn. Soc. 78, 99-116 (1979). - GREUTER, W.: The origins and evolution of island floras as exemplified by the Aegean archipelago, 87-106. In: see ref. BRAMWELL, 1979. - GREILHUBER, J.: Plant Syst. Evol. Suppl. 2, 263-280 (1979). - GREY-WILSON, C.: Kew Bull. 34, 689-722 (1980a); - Impatiens of Africa. 235 pp. Rotterdam: Balkema, 1980b. - GROBMAN, A., BONAVIA, D.: Nature 276, 386-387 (1978). - GRUN, P.: Evolution of cultivated potato: a cytoplasmic analysis, 655-665. In: see ref. HAWKES et al., 1979. - GUÉDÈS, M.: Magnolioid island plants and angiosperm evolution, 307-328. In: see ref. BRAMWELL, 1979. - GUINET, P., VASSAL, J., EVANS, C.S., MASLIN, B.R.: Bot. J. Linn. Soc. 80, 53-68 (1980).

HAEGI, L.: Australian genera of the Solanaceae, 121-124. In: see ref. HAWKES et al., 1979. - HAINES, R.W., LYE, K.A.: Bot. J. Linn. Soc. 78, 123-140 (1979). - HAKE, S., WALBOT, V.: Chromosoma (Berl.) 79, 251-270 (1980). - HALL, M.T., MUKHERJEE, A., CROWLEY, W.R.: Bot. Gaz. 140, 364-370 (1979). - HALLÉ, F., OLDEMANN, R.A.A., TOMLINSON, P.B.: Tropical Trees and Forests: an Architectural Analysis. 441 pp. Berlin-Heidelberg-New York: Springer, 1978. - HALLÉ, N.: Adansonia 19, 127-144 (1979). - HAMANN, O.: Bot. J. Linn. Soc. 78, 67-84 (1979). - HAMANN, U., KAPLAN, K., RÜBSAMEN, T.: Bot. Jahrb. Syst. 100, 555-563 (1979). - HAMMER, K., HANELT, P.: Biol. Zentralbl. 99, 325-343 (1980). - HAMRICK, J.L.: Genetic variation and longevity, 84-113. In: see ref. SOLBRIG et al., 1979. - HANCOCK, J.F., BRINGHURST, R.S.: Am. J. Bot. 66, 367-375 (1979); - Evolution 34, 762-768 (1980). - HANSEN, D.H.: Am. J. Bot. 66, 477-484 (1979). - HARBORNE, J.B.: Phytochemistry 18, 1323-1327 (1979); - New experimental approaches to plant chemosystematics, 39-70. In: see ref. BISBY et al., 1980. - HARBORNE, J.B., GREEN, P.S.: Bot. J. Linn. Soc. 81, 155-167 (1980). - HARBORNE, J.B., SMITH, D.M.: Biochem. Syst. Ecol. 6, 127-130 (1978). - HARDIN, J.W.: Am. J. Bot. 66, 576-585 (1979). - HARRIES, H.C.: Bot. Rev. 44, 265-320 ("1978", 1979). - HARRIS, J.A., BISBY, F.A.: Classification from chemical data. 305-327. In: see ref. BISBY et al., 1980. - HARTL, D.L.: Evolution 33, 969-972 (1979). - HARTLEY, T.G.: J. Arnold Arb. 60, 127-153 (1979). - HAWKES, J.G.: Evolution and polyploidy in potato species, 637-645. In: see ref. HAWKES et al., 1979. - HAWKES, J.G., LESTER, R.N., SKELDING, A.D. (eds.): The Biology and Taxonomy of the Solanaceae. 738 pp. Linn. Soc. Symp. Ser. 7. London: Academic Press, 1979. - HEDBERG, I. (ed.): Systematic Botany, Plant Utilization and Biospere Conservation. 157 pp. Stockholm: Almquist and Wiksell, 1979. - HEDBERG, I., HEDBERG, O.: Oikos 33, 297-307 (1979). - HEDGE, I.C., KJAER, A., MALVER, O.: Notes R. Bot. Gard. Edinburgh 38, 247-250 (1980). - HEISER, C.B.: Taxon 28, 71-75 (1979a); - Annu. Rev. Ecol. Syst. 10, 309-326 (1979b). - HEISER, C.B., BURTON, D.L., SCHILLING, E.E.: Biosystematic and taxometric studies of the Solanum nigrum-complex in Eastern N-America, 513-527. In: see ref. HAWKES et al., 1979. - HEITZ, B.: Compt. Rend. Acad. Sci. Paris Sér. D 290, 811-814 (1980). - HESLOP-HARRISON, J.: Am. J. Bot. 66, 737-743 (1979a); - Ann. Missouri Bot. Gard. 66,

813-829 (1979b). - HESSE, M.: Pollen Spores 21, 399-413 (1979a); - Plant Syst. Evol. 131, 277-289 (1979b); - Flora 168, 558-577 (1979c); - Plant Syst. Evol. 132, 107-139 (1979d); - Flora 168, 540-557 (1979e); - Plant Syst. Evol. 133, 135-148 (1979f); - ibid. 134, 229-267 (1980a); - ibid. 135, 253-263 (1980b); - ibid. 136, 41-46 (1980c). - HESSE, M., MORAWETZ, W.: Plant Syst. Evol. 135, 1-10 (1980). - HEYWOOD, V.H. (ed.): Flowering Plants of the World. 336 pp. Oxford: Oxford University Press, 1978; - Veröff. Joachim-Jungius-Ges. Wiss. Hamburg, 43, 97-115 (1980). - HIDEUX, M.: Le pollen, données nouvelles de la microscopie electronique et de l'informatique. 164 pp. Paris: Agence de cooperation culturelle et technique, 1979. - HIEPKO, P.: Willdenowia 9, 13-56 (1979). - HIJWEGEN, T.: Symb. Bot. Ups. 22, 146-165 (1979). - HILL, R.J., PRANCE, G.T., MORI, S.A., STEWARD, W.C., SHIMABUKURU, D., BERNARDI, J.: Acta Amazonica 8, 183-199 (1978). - HILL, R.S.: Bot. Gaz. 141, 213-229 (1980). - HILU, K.W., DE WET, J.M.J., HARLAN, J.R.: Am. J. Bot. 66, 330-333 (1979). - HNATIUK, R.J., MASLIN, B.R.: W. Austr. Herb. Res. Notes 4, 1-144 (1979). - HOLM-NIELSEN, L.B.: Comments on the distribution and evolution of the genus Phyllanthus (Euphorbiaceae), 277-290. In: see ref. LARSEN and HOLM-NIELSEN, 1979. - HOOGLAND, R.D.: Blumea 25, 481-505 (1979). - HOPPER, S.D.: Annu. Rev. Ecol. Syst. 10, 399-422 (1979). - HORN, H.S.: Adaptation from the perspective of optimality, 48-61. In: see ref. SOLBRIG et al., 1979. - HORVAT, F., STAINIER, F.: Pollen Spores 22, 139-172 (1980). - HORVITZ, C.C., BEATTIE, A.J.: Am. J. Bot. 67, 321-326 (1980). - HOWE, H.F.: Ecology 61, 944-959 (1980). - HUBER, K.A.: Diss. Bot. 55, 1-252 (1980). - HUMPHRIES, C.J.: Bull. Brit. Mus. Nat. Hist. (Bot.) 7, 83-142 (1979a); - Endemism and evolution in Macaronesia, 171-199. In: see ref. BRAMWELL, 1979; - Nord. J. Bot. 1, 83-96 (1981). - HUMPHRIES, C.J., RICHARDSON, P.M.: Hennig's method and phytochemistry, 353-378. In: see ref. BISBY et al., 1980. - HUNZIKER, A.T.: South American Solanaceae: a synoptic survey, 49-85. In: see ref. HAWKES et al., 1979. - HURKA, H., BENNEWEG, M.: Biol. Zentralbl. 98, 699-709 (1979). - HUYNH, K.-L.: Bot. Jahrb. Syst. 100, 321-371 (1979a); - Bot. Jahrb. Syst. 100, 473-517 (1979b); - Candollea 35, 385-419 (1980a); - Pollen Spores 22, 173-189 (1980b).

IETSWART, J.H.: A Taxonomic Revision of the Genus Origanum (Labiatae). 153 pp. Leiden Botanical Series, Vol. 4. Leiden: Leiden University Press, 1980. - IHLENFELDT, H.-D.: Mitt. Inst. Allg. Bot. Hamburg 17, 145-163 (1980). - IHLENFELDT, H.-D., GRABOW-SEIDENSTICKER, U.: The genus Sesamum L. and the origin of the cultivated sesame. 53-60. In: see ref. KUNKEL, 1979. - INAMDAR, J.A., ALEYKUTTY, K.M. (SR. AVITA): Plant Syst. Evol. 132, 161-166 (1979). - INGRAM, J.: Gentes Herbarum 11, 427-468 (1980). - ILTIS, H.H., DOEBLEY, J.F.: Am. J. Bot. 67, 994-1004 (1980). - ILTIS, H.H., DOEBLEY, J.F., GUZMAN, M.R., PAZY, B.: Science 203, 186-188 (1979). - IRVING, R.S.: Sida 8, 218-295 (1980).

JAIN, S.: Adaptive strategies: polymorphism, plasticity, and homoeostasis, 160-187. In: see ref. SOLBRIG et al., 1979. - JAIN, S.K., MARTINS, P.S.: Am. J. Bot. 66, 361-366 (1979). - JAIN, S.K., SINGH, R.S.: Bot. Gaz. 140, 356-362 (1979). - JAMES, S.H.: Aust. J. Bot. 27, 17-25 (1979). - JANZEN, D.H.: New horizons in the biology of plant defenses, 331-350. In: see ref. ROSENTHAL and JANZEN, 1979a; - Annu. Rev. Ecol. Syst. 10, 13-51 (1979b). - JEFFREY, C.: Kew Bull. 34, 435-471 (1980a); - Bot. J. Linn. Soc. 81, 233-247 (1980b). - JENSEN, S.R., NIELSEN, B.J.: Taxon 29, 409-411 (1980). - JENSEN, R.J., McLEOD, M.J., ESHBAUGH, W.H., GUTTMANN, S.I.: Taxon 28, 315-327 (1979). - JOHNSON, C.T.: Aust. J. Bot. 28, 77-104 (1980). - JOHNSON, G.B.: Enzyme polymorphism: genetic variation in the physiological phenotype, 62-83. In: see ref. SOLBRIG et al., 1979. - JOHNSTON, B.C.: Taxon 29, 47-51 (1980). - JONES, D.A.: Am. Nat. 113, 445-451 (1979). - JONES, S.B.: Bull. Torrey Bot. Club 106, 79-84 (1979). - JONES, S.B., LUCHSINGER, A.E.: Plant Systematics.

New York: McGraw Hill, 1979. - JONSELL, B.: Symb. Bot. Ups. 22 (4), 2-10 (1979). - JUDD, W.S.: J. Arnold Arb. 60, 477-503 (1979). - JUNG, H.-J., BATZLI, G.O., SEIGLER, D.S.: Biochem. Syst. Ecol. 7, 203-209 (1979).

KÄSTNER, A.: Flora 168, 431-467 (1979). - KAPIL, R.N., BOR, J., BOUMANN, F.: Bot. Jahrb. Syst. 101, 555-573 (1980). - KARATAGLIS, S.S.: Plant Syst. Evol. 134, 215-228 (1980a); - Plant Syst. Evol. 134, 23-31 (1980b). - KARLSTRÖM, P.O.: Bot. Notiser 133, 1-16 (1980). - KEELEY, S.C., JONES, S.B.: Syst. Bot. 4, 195-202 (1979). - KELSEY, R.G., SHAFIZADEH, F.: Phytochemistry 18, 1591-1611 (1979). - KESSELL, S.R.: Am. Nat. 113, 333-350 (1979). - KHAN, R.: Solanum melongena and its ancestral forms, 629-636. In: see ref. HAWKES et al., 1979. - KIRKBRIDE, J.H.: Smithsonian Contrib. Bot. 41, 1-32 (1979). - KLAUS, W.: Plant Syst. Evol. 134, 137-171 (1980). - KLOTZ, G.: Wiss. Z. Friedrich-Schiller-Univ. Jena, Math.-Naturwiss. Reihe 29 (4), 453-479 (1980). - KLOTZ, L.C., KOMAR, N., BLANKEN, R.L., MITCHELL, R.M.: Proc. Nat. Acad. Sci. USA 76, 4516-4520 (1979). - KOEK-NOORMAN, J.: Acta Bot. Neerl. 29, 117-126 (1980). - KOEK-NOORMAN, J., HOGEWEG, P., VAN MAANEN, W.H.M., TER, B.J.H.: Acta Bot. Neerl. 28, 21-43 (1979). - KÖHLER, E.: Feddes Rep. 91, 581-591 (1981). - KONAR, R.N. NAGMANI, R.: Bot. Jahrb. Syst. 101, 267-297 (1980). - KOSTERMANS, A.J.G.H.: Taxon 27, 357-359 (1978). - KREBS, J.R.: Nature 278, 689. (1979). - KRENDL, F.: Ann. Naturhistor. Mus. Wien 82, 291-318 (1979). - KRONER, G.: Mitt. Bot. Staatss. München 16, 1-268 (1980). - KUBITZKI, K., MESQUITA, A.L., GOTTLIEB, O.R.: Biochem. Syst. Ecol. 6, 185-187 (1978). - KUBITZKI, K., KURZ, H., RICHTER, H.G.: J. Arnold Arb. 60, 515-522 (1979). - KUIJT, J.: Symb. Bot. Ups. 22, 194-199 (1979). - KUKKONEN, I., TIMONEN, T.: Symb. Bot. Ups. 22, 166-176 (1979). - KULL, U., KÜHN, B.: Vieraea 8, 187-190 (1979). - KUNKEL, G. (ed.): Taxonomic Aspects of African Economic Botany. 250 pp. Las Palmas da Gran Canaria: AETFAT, 1979. - KWAK, M.M.: Taxon 29, 613-628 (1980).

LACK, H.W., LEUENBERGER, B.: Pollen Spores 21, 415-425 (1979). - LACK, H.W., ERN, H., STRAKA, H.: Willdenowia 10, 37-49 (1980). - LADIZINSKY, G.: Econ. Bot. 33, 281-289 (1979). - LACKEY, J.A.: Bot. Gaz. 139, 436-446 (1978). - LA DUKE, J., CRAWFORD, D.J.: Taxon 28, 307-314 (1979). - LANDOLT, E. (ed.): Biosystematische Untersuchungen in der Familie der Wasserlinsen. Veröff. Geobot. Inst. ETH, Stiftung Rübel, Zürich 70, 1-247 (1980). - LANGENHEIM, J.H., STUBBELBINE, W.H., LINCOLN, D.E., FOSTER, C.E.: Biochem. Syst. Ecol. 6, 299-313 (1978). - LANGENHEIM, J.H., FOSTER, C.E., McGINLEY, R.B.: Biochem. Syst. Ecol. 8, 385-396 (1980). - LARSEN, K., HOLM-NIELSEN, L.B. (eds.): Tropical Botany. 453 pp. London-New York-San Francisco: Academic Press, 1979. - LAVANIA, U.C., SHARMA, A.K.: Bot. Gaz. 141-203 (1980). - LAVIE, P.: Adansonia 19, 175-198 (1979). - LAWRENCE, M.E.: Aust. J. Bot. 28, 151-165 (1980). - LEE, J.A., HARMER, R.: Oikos 35, 254-265 (1980). - LEE, D.W., LOWRY, J.B.: Bot. J. Linn. Soc. 80, 223-242 (1980). - LEEUWENBERG, A.J.M. (ed.): Loganiaceae. In: ENGLER and PRANTL, Nat. Pflanzenfam. ed. 2, 28b I, 1980. - LEINS, P.: Ber. Dtsch. Bot Ges. 92, 717-719 (1979). - LEINS, P., ORTH, C.: Bot. Jahrb. Syst. 100, 372-378 (1979). - LEROY, J.-F.: Compt. Rend. Acad. Sci. Paris, Sér. D 291, 593-596 (1980a); - Compt. Rend. Acad. Sci. Paris, Sér. D 290, 679-682 (1980b). - LEVIN, D.A.: Science 204, 381-384 (1979a); - Pollinator foraging behaviour: genetic implications for plants, 129-153. In: see ref. SOLBRIG et al., 1979 (1979b). - LEVIN, D.A., FUNDERBUG, S.W.: Am. Nat. 114, 784-795 (1979). - LEVIN, D.A. YORK, B.M.: Biochem. Syst. Ecol. 6, 61-76 (1978). - LIEU, S.M.: Can. J. Bot. 57, 2325-2352, 2353-2373 (1979). - LINDER, H.P.: J. S. Afr. Bot. 46, 379-434 (1980); - J. S. Afr. Bot. 47, 13-48 (1981). - LIPPERT, W.: Mitt. Bot. Staatss. München 15, 397-423 (1979). - LIPSCHITZ, S.: Genus *Saussurea* DC. (Asteraceae). 281 pp. Leningrad: Nauka, 1979. - LLOYD, D.G.: Evolution 34,

123-134 (1979a); - Plant Syst. Evol. 131, 71-80 (1979b); - N. Z. J. Bot. 17, 595-606 (1979c). - LOBREAU-CALLEN, D.: Adansonia 20, 29-89 (1980). - LOCK, J.M.: Kew Bull. 35, 299-313 (1980). - LOVETT DOUST, J.: New Phytol. 85, 265-273 (1980). - LOVETT DOUST, J., HARPER, J.L.: New Phytol. 85, 251-264 (1980). - LUCAS, G., SYNGE, H.: The I.U.C.N. Plant Red Data Book. 542 pp. Morges, Switzerland: I.U.C.N., 1978. - LÜÖND, B., LÜÖND, R.: Candollea 35, 87-109 (1980). - LUFTENSTEINER, H.W.: Vegetatio 41, 61-67 (1979).

MAAS, P.J.M.: Neotropical saprophytes, 365-370. In: see ref. LARSEN and HOLM-NIELSEN, 1979. - MABBERLEY, D.J.: Bull. Brit. Mus. Nat. Hist. (Bot.) 6, 301-386 (1979a); - Pachycaul plants and islands, 259-277. In: see ref. BRAMWELL, 1979 (1979b). - MACFARLANE, T.D., WATSON, L.: Taxon 29, 645-666 (1980). - MACONOCHIE, J.R.: J. Adelaide Bot. Gard. 2, 171-185 (1980). - MAGIN, N.: Plant Syst. Evol. 133, 239-259 (1980). - MAGUIRE, B., ASHTON, P.S.: Taxon 29, 225-231 (1980). - MALLOCH, D.W., PIROZYNSKI, K.A., RAVEN, P.H.: Proc. Natl. Acad. Sci. USA 77, 2113-2118 (1980). - MANGELSDORF, P.C., BARGHOORN, E.S., BANERJEE, U.C.: Bot. Mus. Leafl. Harvard Univ. 26, 237-255 (1978). - MANNING, W.C.: Ann. Missouri Bot. Gard. 65, 1058-1087 ("1978", 1979). - MARCHANT, C.J., MACFARLANE, R.M.: Bot. J. Linn. Soc. 81, 135-154 (1980). - MARTINS, P.S., JAIN, S.K.: J. Heredity 71, 29-32 (1980). - MASLIN, B.R.: Nuytsia 2, 266-333 (1978). - MAUGH II, T.H.: Science 206, 436 (1979). - MATTOS, N.F.: Acta Amazonica 9, 241-266 (1979). - MAXWELL, J.F.: Gard. Bull. Singapore 31, 139-216 (1978); - Gard. Bull. Singapore 33, 31-150 (1980). - MAZE, J.: Taxon 29, 667-670 (1980). - McNEILL, J.: Taxon 28, 481-504 (1979a); - Syst. Zool. 28, 465-482 (1979b). - McKEY, D.: Am. Nat. 115, 754-759 (1980). - MEACHAM, C.A.: Syst. Bot. 5, 149-172 (1980). - MEAGHER, T.R.: Evolution 34, 1127-1137 (1980). - MEARS, J.A.: Biochem. Syst. Ecol. 8, 361-370 (1980). - MEEUSE, A.D.J.: Proc. Koninkl. Niederlandse Akad. Wetensch., Ser. C, 82, 343-369 (1979); - The significance of the Gnetatae in connection with the early evolution of the angiosperms, 62-73. In: Glimpses in Plant Research, ed. P.K.K. Nair, New Delhi: Vikas, 1980. - MELZHEIMER, V.: Bot. Jahrb. Syst. 101, 153-190 (1980). - MENEZES, N.L. DE: Evolution in Velloziaceae, with special reference to androecial characters. In: Petaloid Monocotyledons, ed. C.D. Brickell, D.F. Cutler, M. Gregory. Linn. Soc. Symp. Ser. 8. London-New York: Academic Press, 1980. - METCALFE, C.R., CHALK, L.: Anatomy of the Dicotyledons, 2nd ed., Vol. I, Systematic Anatomy of Leaf and Stem, with a Brief History of the Subject. 276 pp. Oxford: Clarendon Press, 1979. - MILLER, A.G.: Notes R. Bot. Gard. Edinburgh 38, 373-385 (1980). - MILLER, J.M., BOHM, B.A.: Biochem. Syst. Ecol. 8, 279-284 (1980). - MOFFETT, R.O.: Bothalia 12, 581-613 (1979). - MOGEA, J.P.: Reinwardtia 9, 461-479 (1980). - MOHANA RAO, P.R.: Israel J. Bot. 28, 44-50 (1979/80). - MOLAU, U.: Bot. Notiser 132, 31-48 (1979); - Bot. Notiser 133, 33-45 (1980). - MONOD, T.: Adansonia 20, 147-164 (1980). - MOORE, D.M.: Origins of temperate island floras, 69-85. In: see ref. BRAMWELL, 1979. - MORI, S.A., ORCHARD, J.E., PRANCE, G.T.: Science 209, 400-403 (1980). - MOUSSEL, B.: Rev. Cyt. Biol. Veget.-Bot. 3, 65-89 (1980). - MULCAHY, D.L.: Science 206, 20-23 (1979). - MULLER, J.: Ann. Missouri Bot. Gard. 66, 593-632 (1979). - MULLIGAN, G.A.: Can. J. Bot. 58, 1755-1767 (1980). - MUNIR, A.A.: Brunonia 1, 407-692 (1978); - J. Adelaide Bot. Gard. 2, 1-138 (1979). - MURRAY, L.E., TAI, W.: Am. J. Bot. 67, 1374-1379 (1980).

NAGL, W.: Plant Syst. Evol. Suppl. 2, 3-25 (1979). - NAGL, W., FUSENIG, H.P.: Plant Syst. Evol., Suppl. 2, 221-234 (1979). - NANNFELDT, J.A.: Symb. Bot. Ups. 22 (3), 1-49 (1979). - NEE, M.: Patterns in biogeography in Solanum, section Acanthophera, 569-580. In: see ref. HAWKES et al., 1979. - NEVO, E., BROWN, A.H.D., ZOHARY, D.: Experientia 35, 1027-1029 (1979a). - NEVO, E., ZOHARY, D., BROWN, A.H.D., HABER, M.:

Evolution 33, 815-833 (1979b). - NIELSEN, I.: Adansonia 19, 199-229 (1979a); - Adansonia 19, 3-37 (1979b). - NIGAUD, M.: Bull. Mus. Nat. Hist. Nat., sect. B 4. ser., 2, 263-277 (1980). - NIKLAS, K.J.: Science 211, 275-277 (1981). - NOOTEBOOM, H.P.: Blumea 26, 411-415 (1980). - NORDAL, I., WAHLSTRØM, R.: Adansonia 20, 179-198 (1980). - NOWICKE, J.W., SKVARLA, J.J.: Ann. Missouri Bot. Gard. 66, 633-700 (1979).

OHASHI, H., MURATA, J.: J. Fac. Sci. Univ. Tokyo, Sect. III Bot. 12, 281-336 (1980). - OHBA, H.: J. Fac. Sci. Univ. Tokyo, Sect. III Bot. 12, 337-405 (1980). - OLIVER, E.G.H.: Bothalia 13, 65-93 (1980). - OLIVER, J.L., RUIZ REJÓN, M.: Taxon 29, 27-32 (1980). - OLSEN, J.S.: Rhodora 81, 449-501 (1979). - OPLER, P.A., BAKER, H.G., FRANKIE, G.W.: Biotropica 12, 40-46 (1980). - ORNDUFF, R.: Plant Syst. Evol. 135, 89-93 (1980). - OZENDA, P., CAPDEPON, M.: Rev. Gén. Bot. 86, 235-298 (1979a); - Bull. Soc. Bot. France 126, Lettr. Bot. 453-460 (1979b).

PAGE, C.N.: Plant Syst. Evol. 135, 71-79 (1980). - PAL, A., SHARMA, A.K.: Cellule 73, 115-134 (1980). - PANDEY, K.K.: The Genus Nicotiana: evolution of incompatibility in flowering plants, 421-434. In: see ref. HAWKES et al., 1979. - PANKHURST, R.J.: Biological Identification. The Principles and Practice of Identification Methods in Biology. 104 pp. London: Arnold, 1978. - PARENTI, L.R.: Biol. J. Linn. Soc. 13, 225-242 (1980). - PARKER, W.H., BOHM, B.A.: Am. J. Bot. 66, 191-197 (1979). - PATEL, V.C., SKVARLA, J.J.: Pollen Spores 21, 81-103 (1979). - PAYNE, W.W.: Taxon 28, 117-132 (1979). - PEDLEY, L.: Austrobaileya 1, 75-234 (1978); - Austrobaileya 1, 235-337 (1979). - PEIXOTO, A.L.: Rodriguesia 50, 135-222 (1979). - PERRY, D.A., LOTAN, J.E.: Evolution 33, 958-968 (1979). - PHENGKAI, C.: Thai For. Bull. 11, 1-103 (1978). - PHILIPP, M.: New Phytol. 85, 557-569 (1980). - PHILIPSON, W.R., BUTTERFIELD, B.G.: Taxon 29, 404-406 (1980). - PHILIPSON, W.R., STONE, B.C.: Taxon 29, 391-403 (1980). - PICKERSGILL, B., HEISER, C.B., McNEILL, J.: Numerical taxonomic studies on variation and domestication in some species of Capsicum, 679-713. In: see ref. HAWKES et al., 1979. - PILBEAM, D.J., POLHILL, R.M., BELL, E.A.: Bot. J. Linn. Soc. 79, 259-266 (1979). - PILS, G.: Plant Syst. Evol. 136, 73-124 (1980). - PLAIZIER, A.C.: Meded. Landbouwhogeschool Wageningen. 80/12, 1-40 (1980). - PLOWMANN, T.: The genus Brunfelsia: a conspectus of the taxonomy and biogeography, 475-491. In: see ref. HAWKES et al., 1979. - PODZORSKI, A.C., CRIBB, P.J.: Kew Bull. 34, 147-186 (1979). - POLATSCHEK, A.: Ann. Naturhistor. Mus. Wien 82, 325-362 (1979). - POOLE, M.M., HUNT, D.R.: Kew Bull. 34, 639-660 (1980). - POPPENDIECK, H.-H.: Bot. Jahrb. Syst. 101, 191-265 (1980). - POWELL, D.A.: J. Arnold Arb. 60, 219-271 (1979). - PRAGLOWSKI, J., GRAFSTRÖM, E.: Bot. Notiser 133, 177-188 (1980). - PRANCE, G.T.: Distribution patterns of lowland neoptropical species with relation to history, dispersal, and ecology, with special reference to Chrysobalanaceae, Caryocaryaceae, and Lecythidaceae, 59-87. In: see ref. LARSEN and HOLM-NIELSEN, 1979. - PRANCE, G.T., MORI, S.A.: Lecythidaceae, part I. Fl. Neotrop. 21, 1-272 (1979). - PRENTICE, H.C.: Bot. J. Linn. Soc. 78, 181-212 (1979). - PRICE, M.V., WASER, N.M.: Nature 277, 294-296 (1979). - PRIMACK, R.B., LLOYD, D.G.: Am. J. Bot. 67, 361-368 (1980). - PULLAIAH, T.: Am. J. Bot. 66, 1119-1127 (1979).

QURAISH, H.B., FARUQI, S.A.: Caryologica 32, 311-327 (1979).

RABINOWITZ, D., RAPP, J.K.: Bull. Torrey Bot. Club 106, 32-36 (1979). - RAMANNA, M.S., HERMSEN, J.G.T.: Genome relationships in tuber-bearing Solanums, 647-653. In: see ref. HAWKES et al., 1979. - RAO, T., DAS, S.: Bot. Notiser 132, 319-324 (1979). - RASMUSSEN, H.: Opera Bot. 53, 1-44 (1979). - RATAJ, K.: Acta Amazonica 8 (1), suppl. 1, 53 pp. (1978). - RAVEN, P.H.: Ann. Missouri Bot. Gard. 61, 166-178 (1974); - RAVEN, P.H.: Acad. Nat. Sci., Spec. Publ. 12, 59-83 (1977); - Future directions in plant population biology, 461-481. In: see ref. SOLBRIG et al., 1979 (1979a); - N. Z. J. Bot. 17, 575-593 (1979b). - RAVEN, P.H., DIETRICH, W., STUBBE, W.: Syst. Bot. 4, 242-252 ("1979", 1980). - RAVEN, P.H., TAI, W.: Ann. Missouri Bot. Gard. 66, 862-879 (1979). - REYNOLDS, J.F., CRAWFORD, D.J.: Am. J. Bot. 67, 1380-1390 (1980). - RHOADES, D.F.: Evolution of plant chemical defense against herbivores, 4-54. In: see ref. ROSENTHAL and JANZEN, 1979. - RHOADES, D.F., CATES, R.G.: Rec. Adv. Phytochem. 10, 168-213 (1976). - RICE, R.L., LINCOLN, D.E., LANGENHEIM, J.H.: Biochem. Syst. Ecol. 6, 45-53 (1978). - RICK, C.M.: Biosystematic studies in Lycopersicon and closely related species of Solanum, 667-677. In: see ref. HAWKES et al., 1979. - RICKSON, F.R.: Am. J. Bot. 66, 80-86 (1979). - RIDSDALE, C.E.: Blumea 24, 307-366 (1978). - RIEPPEL, O.: Biol. Zentralbl. 98, 537-552 (1979). - RINTZ, R.E.: Blumea 26, 81-126 (1980). - ROBBRECHT, E.: Bull. Jard. Bot. Nat. Belg. 49, 239-360 (1979); - Bull. Jard. Bot. Nat. Belg. 50, 69-77 (1980). - ROBESON, D.J., HARBORNE, J.B.: Phytochemistry 19, 2359-2365 (1980). - ROBINSON, H.: Smithsonian Contrib. Bot. 42, 1-20 (1979). - ROBSON, N.K.B.: Kew Bull. 33, 571-584 (1979). - RODMANN, J.E., CHEW, F.S.: Biochem. Syst. Ecol. 8, 43-50 (1980). - ROE, K.E.: Dispersal and speciation in Solanum, section Brevantherum, 563-567. In: see ref. HAWKES et al., 1979. - ROESSLER, H.: Mitt. Bot. Staatss. München 15, 1-160 (1979). - ROMMEL, A.: Mitt. Bot. Staatss. München 15, 243-329 (1979). - ROSENTHAL, G.A., JANZEN, D.H. (eds.): Herbivores. Their Interaction with Secondary Plant Metabolites. 718 pp. New York: Academic Press, 1979. - ROSENTHAL, G.A., DAHLMANN, D.L., JANZEN, D.H.: Science 202, 528-529 (1979). - RUDALL, P.: Bot. J. Linn. Soc. 80, 319-340 (1980). - RUIZ REJÓN, M., POSSE, F., OLIVER, J.L.: Chromosoma 79, 341-348 (1980). - RYE, B.L.: Aust. J. Bot. 27, 547-573 (1979). - RZEDOWSKI, J., KRUSE, H.: Taxon 28, 103-116 (1979).

SALATINO, A., GOTTLIEB, O.R.: Biochem. Syst. Ecol. 8, 133-147 (1980). - SAÑUDO, A.: Webbia 34, 363-408 (1979). - SAUER, W., LEEP, H.J.: Plant Syst. Evol. 131, 81-106 (1979). - SAUER, W., STEGMEIER, R.: Ber. Dtsch. Bot. Ges. 92, 663-687 (1979). - SAVILE, D.B.O.: Symb. Bot. Ups. 22, 135-145 (1979a); - Bot. Rev. 45, 377-503 (1979b). - SCHAAN, M.E., NAGL, W.: Plant Syst. Evol., Suppl. 2, 67-71 (1979). - SCHADEL, W.E., DICKISON, W.C.: J. Arnold Arb. 60, 8-37 (1979). - SCHILLING, E.E., HEISER, C.B.: Am. J. Bot. 66, 709-716 (1979). - SCHLISING, R.A., IKEDA, D.H., MOREY, S.C.: Bot. Gaz. 141, 290-293 (1980). - SCHMID, B.W.: Carex flava L. s. l. im Lichte der r-Selektion. 360 pp. Diss. Univ. Zürich (1980). - SCHMID, R.: Taxon 29, 559-595 (1980). - SCHMIDT, B.: Feddes Rep. 90, 257-361 (1979). - SCHMITT, J.: Evolution 34, 934-943 (1980). - SCHOLZ, H.: The phenomenon of mimetic weeds in the African Pennisetum americanum – a critique, 230-233. In: see ref. KUNKEL, 1979 (1979a); - Willdenowia 9, 209-214 (1979b). - SCHREIBER, A.: Mitt. Bot. Staatss. München 15, 331-361 (1979). - SCHULTES, R.E.: Bot. Mus. Leafl. 26 (9-10), 311-337 (1978); - Solanaceous hallucinogens and their role in the development of New World cultures, 137-160. In: see ref. HAWKES et al., 1979. - SCOGIN, R.: Aliso 9, 555-559 (1980). - SCOTT, A.J.: Kew Bull. 33, 429-459 (1979a); - Kew Bull. 33, 461-484 (1979b). - SEAMAN, F.C., MABRY, T.J.: Biochem. Syst. Ecol. 7, 105-114 (1979). - SEAMAN, F.C., FISCHER, N.H., STUESSY, T.F.: Biochem. Syst. Ecol. 8, 263-271 (1980). - SEBALD, O.: Stuttgarter Beitr. Naturk. Ser. A 341, 1-200 (1980). - SEMPLE, J.C., BROUILLET, L.: Am. J. Bot. 67, 1010-1026 (1980). - SENESSE, S.: Pollen Spores 22, 355-423 (1980). - SENGER, H.A., HAWKES, J.G.: Biol. J. Linn. Soc. 13, 299-313 (1980). - SHULKINA, T.V.: Plant Syst. Evol. 136, 233-246 (1980). - SHUMAKER, K.M., BABBLE, G.M.: Evolution 34, 110-116 (1979). - SILANDER, J.A.: Science 203, 658-660 (1979). - SILANDER, J.A., ANTONOVICS, J.: Evolution 33, 1114-1127 (1979). - SILVERTON, J.W.: Biol. J. Linn. Soc. 14, 235-250 (1980). - SIMMONDS, N.W.: Kew Bull. 35, 235-245 (1980). - SIMPSON, B.B.: Smithsonian Contrib. Bot. 43, 1-62 (1979). - SIMPSON, B.B., SEIGLER, D.S., NEFF, J.L.:

Biochem. Syst. Ecol. 7, 193-194 (1979). - SINGH, H.: Embryology of Gymnosperms. In: Handbuch der Pflanzenanatomie, 2. Aufl., Vol. 10 (2), eds. W. ZIMMERMANN, S. CARLQUIST, P. OZENDA, D. WULF. 302 pp. Berlin-Stuttgart: Bornträger, 1978. - SINGH, P., FENEMORE, P.G., DUGDALE, J.S., RUSSELL, G.B.: Biochem. Syst. Ecol. 6, 103-106 (1978). - SKVARLA, J.J., NOWICKE, J.W.: Am. J. Bot. 66, 162-165 (1979). - SKVORTSOV, A.K.: Ann. Missouri Bot. Gard. 66, 880-892 (1979). - SLEUMER, H.: Blumea 26, 145-168 (1980). - SMALL, E.: The Species Problem in Cannabis: Science and Semantics. Vol. 1, 218 pp., Vol. 2, 156 pp. Toronto: Corpus, 1979. - SODERSTROM, T.R., CALDERÓN, C.E.: Brittonia 31, 433-445 (1980). - SOEJARTO, D.D.: Fieldiana Bot. N.S. 2, I-VII, 1-141 (1980). - SOLBRIG, O.T.: A cost-benefit analysis of recombination in plants, 114-130. In: see ref. SOLBRIG et al., 1979. - SOLBRIG, O.T., JAIN, S., JOHNSON, G.B., RAVEN, P.H. (eds.): Topics in Plant Population Biology. 589 pp. London-Basingstoke: Macmillan Press, 1979. - SPETA, F.: Plant Syst. Evol. 132, 1-26 (1979). - SPIRA, T.P.: Am. J. Bot. 67, 278-284 (1980). - SPORNE, K.R.: New Phytol. 85, 419-449 (1980). - STACE, C.A.: Plant Taxonomy and Biosystematics. 279 pp. London: Arnold, 1979. - STEBBINS, G.L.: Fifty years of plant evolution, 18-41. In: see ref. SOLBRIG et al., 1979. - STEPHENSON, A.G.: Ecology 61, 57-64 (1980). - STERN, W.L., STYER, C.H.: Biol. J. Linn. Soc. 79, 267-289 (1979). - STEVENS, P.F.: Annu. Rev. Ecol. Syst. 11, 333-358 1980). - STEWART, J.: Kew Bull. 34, 239-319 (1979). - STONE, A.R.: Symb. Bot. Ups. 22 (4), 46-61 (1979). - STONE, D.E., SELLERS, S.C., KRESS, W.J.: Ann. Missouri Bot. Gard. 66, 701-730 (1979). - STORK, A.L., SNOGERUP, S., WÜEST, J.: Candollea 35, 421-450 (1980). - STRONG, D.R., LEVIN, D.A.: Am. Nat. 114, 1-22 (1979). - STUBBE, W.: Ber. Dtsch. Bot. Ges. 93, 441-447 (1980). - STUBBE, W., RAVEN, P.H.: Science 204, 327 (1979a); - Plant Syst. Evol. 133, 39-59 (1979b). - STUESSY, T.F.: Taxon 28, 179-195 (1979). - STUTZ, H.C., POPE, C.L., SANDERSON, S.C.: Am. J. Bot. 66, 1181-1193 (1979). - STYER, C.H., STERN, W.L.: Bot. J. Linn. Soc. 79, 291-319 (1979). - SUGIYAMA, M.: J. Fac. Sci. Univ. Tokyo, Sect. III, Bot. 12, 199-279 (1979). - SUMMERFIELD, R.J., BUNTING, A.H. (eds.): Advances in Legume Science. 667 pp. Kew: Royal Botanic Gardens, 1980. - SUTHERLAND, O.W., RUSSELL, G.B., BIGGS, D.R., LANE, G.A.: Biochem. Syst. Evol. 8, 73-75 (1980). - SUZUKI, M.: Bot. Mag. (Tokyo) 92, 333-353 (1979). - SUNDING, P.: Origins of the Macaronesian flora, 13-40. In: see ref. BRAMWELL, 1979. - SWAIN, T.: Phenolics in the environment, pp. 617-640. In: Biochemistry of Plant Phenolics, eds. T. Swain, J.B. Harborne, C. van Sumere (Recent Advances in Phytochemistry, Vol. 12). New York-London: Plenum Press, 1979. - SYLVÉN, E.: Symb. Bot. Ups. 22, 62-69 (1979). - SYMON, D.E.: The genus Solanum in Australia, 125-130. In: see ref. HAWKES et al., 1979 (1979a); - Sex forms in Solanum (Solanaceae) and the role of pollens collecting insects, 385-397. In: see ref. HAWKES et al., 1979 (1979b); - J. Adelaide Bot. Gard. 1, 321-331 (1979c); - J. Adelaide Bot. Gard. 2, 277-292 (1980). - SYNGE, H., TOWNSEND, H.: Survivial or Extinction. 250 pp. Kew: The Bentham-Moxon Trust, Royal Botanic Gardens (1979).

TAKHTAJAN, A.: Bot. Rev. 46, 226-359 (1980). - TAMURA, M., LAUENER, L.A.: Notes Bot. Gard. Edinburgh 37, 431-466 (1979). - TAN, K.: Notes Bot. Gard. Edinburgh 38, 149-164 (1980a); - Notes Bot. Gard. Edinburgh 38, 189-246 (1980b). - TAYLOR, R.J., PATTERSON, T.F.: Taxon 29, 421-469 (1980). - TAYLOR, T.N., MILLAY, M.A.: Rev. Palaebot. Palynol. 27, 329-355 (1979). - TEPPNER, H.: Phyton (Austria) 20, 135-157 (1980). - TER WELLE, B.J.H.: IAWA-Bulletin 1, 43-48 (1980). - T'HART, H., BERENDSEN, W.: Plant Syst. Evol. 135, 107-117 (1980). - THIEN, L.B.: Biotropica 12, 1-13 (1980). - THOMPSON, J.N., WILLSON, M.F.: Evolution 33, 973-982 (1979). - THOMPSON, P.M., MOHLENBROCK, R.H.: J. Arnold Arb. 60, 350-366 (1979). - THORNE, R.F.: Symb. Bot. Ups. 22, 200-209 (1979). - THORPE, J.P.: Biol. J. Linn. Soc. 11, 369-

386 (1979). - TIÊP, N.V.: Feddes Rep. 91, 497-576 (1980). - TITZ, W., TITZ, E.: Ber. Schweiz. Bot. Ges. 89, 251-277 (1979). - TOMLINSON, P.B.: Annu. Rev. Ecol. Syst. 10, 85-107 (1979). - TORNADORE, N., GARBARI, F.: Webbia 33, 379-423 (1979). - TRAN, V.N., CAVANAGH, A.K.: Aust. J. Bot. 28, 39-51 (1980). - TSENG, C.C.: Taxon 29, 407-409 (1980). - TUCKER, J.M., NEILSON, R.P., WULLSTEIN, L.H.: Am. J. Bot. 67, 1265-1267 (1980). - TUCKER, S.C.: Am. J. Bot. 66, 227-236 (1979); - Am. J. Bot. 67, 686-702 (1980).

UHL, C.H.: Rhodora 82, 377-402 (1980). - UHL, N.W., MOORE, H.E.: Ann. Bot. 45, 57-75 (1980). - UMBER, R.E.: Syst. Bot. 4, 72-102 (1979). - URBANSKA-WORYTKIEWICZ, K.: Veröff. Geobot. Inst. ETH, Stiftung Rübel, Zürich, 70, 30-101 (1980a); - Oecol. Plant. 15, 137-150 (1980b).

VALDÉS, B.: Lagascalia 9, 65-107 (1979). - VALDEYRON, G., LLOYD, D.G.: Evolution 33, 673-685 (1979). - VAN DEN BERG, M.E.: Acta Amazonica 9, 43-74 (1979). - VAN DEN BERG, R.G.: Blumea 24, 369-394 (1978). - VAN DE WATER, T.P.M.: Blumea 26, 1-64 (1980). - VAN HEEL, W.A.: Blumea 25, 513-529 (1979). - VAN STEENIS, C.G.G.J.: Bothalia 12, 425-427 (1978); - Bot. J. Linn. Soc. 79, 97-178 (1979); - Rheophytes of the World. 400 pp. Alphen aan den Rijn: Sijthoff and Noordhoff, 1980. - VAN WYK, A.E.: J. S. Afr. Bot. 46, 115-119 (1980). - VAZ, A.M.S. DA FONSECA: Rodriguésia 31, 127-234 (1979). - VEILLON, J.-M.: Candollea 35, 609-640 (1980). - VELDKAMP, J.F., MOERMAN, A.: Blumea 24, 463-477 (1978). - VELDKAMP, J.F., FRANKEN, N.A.P., ROOS, M.C., NAYAR, M.P.: Blumea 24, 405-430 (1978). - VELICZKIN, E.M.: Bot. Zurn. 64, 153-171 (1979). - VENTER, H.J.T.: Meded. Landbouwhoogeschool Wageningen 79/9, 1-128 (1979). - VERDCOURT, B.: Kew Bull. 35, 97-130 (1980a); - The classification of Dolichos L. emend. Verdc., Lablab Adans., Phaseolus L., Vigna Savi, and their allies, 45-48. In: see ref. SUMMERFIELD and BUNTING, 1980 (1980b). - VERDOORN, I.C.: Bothalia 13, 1-63 (1980). - VERLAQUE, R.: Plant Syst. Evol. 134, 33-52 (1980). - VIANO, J.: Candollea 33, 209-267 (1978). - VIGNAL, C.: Adansonia 19, 39-70 (1979); - Adansonia 20, 107-140 (1980). - VOGEL, E.F. DE: Seedlings of Dicotyledons. 465 pp. Wageningen: Tudoc, 1980. -

WAGNER, W.H.: Syst. Bot. 5, 173-193 (1980). - WALKER, J.W., WALKER, A.G.: Ann. Missouri Bot. Gard. 66, 731-755 (1979). - WANNTORP, H.-E.: Taxon 29, 668-670 (1980). - WATERMAN, P.G., MESHAL, J.A., HALL, J.B., SWAINE, M.D.: Biochem. Syst. Ecol. 6, 239-245 (1978). - WEBB, C.J.: Plant Syst. Evol. 132, 299-303 (1979a); - Evolution 33, 662-672 (1979b). - WEBER, H.C.: Naturwissenschaften 66, 367-368 (1979); - Plant Syst. Evol. 136, 217-232 (1980). - WEBER, A.: Plant Syst. Evol. 133, 149-179 (1980a); - Plant Syst. Evol. 134, 183-192 (1980b). - WEBERLING, F., LÖRCHER, H., BÖHMKE, E.: Plant Syst. Evol. 133, 261-283 (1980). - WEBSTER, G.L.: J. Arnold Arb. 60, 403-444 (1979). - WEIDLICH, W.H.: Am. J. Bot. 67, 790-803 (1980). - WELLS, H.: Taxon 29, 53-65 (1980). - WELTER, A.: Compt. Rend. Acad. Sci. Paris, Sér. D 288, 387-390 (1979). - WHALEN, M.D.: Speciation in Solanum, Section Androceras, 581-596. In: see ref. HAWKES et al., 1979 (1979a); - Gentes Herbarum 11, 359-426 (1979b); - Syst. Bot. 4, 203-222 (1979c). - WHITE, F.: Bull. Jard. Bot. Nat. Belg. 48, 245-358 (1978); - Bull. Jard. Bot. Nat. Belg. 50, 445-460 (1980). - WHITE, M.J.D.: Modes of Speciation. 455 pp. San Francisco: Freeman, 1978. - WHITMORE, T.C.: Gard. Bull. Singapore 31, 51-56 (1978); - Plant Syst. Evol. 135, 41-69 (1980). - WHITMORE, T.C., PAGE, C.N.: New Phytol. 84, 407-416 (1980). - WIEBES, J.T.: Annu. Rev. Ecol. Syst. 10, 1-12 (1979). - WIENS, D., BARLOW, B.A.: Bothalia 13, 161-169 (1980). - WILLIAMS, I.: J. S. Afr. Bot. 47, 63-102 (1981). - WILLSON, M.F.: Am. Nat. 113, 777-790 (1979). - WILSON, C.L.: Am. J. Bot. 66, 280-289 (1979). - WILSON, H.D.: Syst. Bot. 5, 253-263 (1980). - WILSON, H.D., HEISER, C.B.: Am. J. Bot. 66, 198-206 (1979). - WILSON,

P.G.: Nuytsia 3, 1-154 (1980). - WOLF, S.J.: Am. J. Bot. 67, 300-308 (1980). - WOON, C., KENG, H.: Gard. Bull. Singapore 32, 1-55 (1979). - WU, L., JAIN, S.: Bot. Gaz. 141, 300-304 (1980).

YAKOVLEV, M.S., ZHUKOVA, G.Y.: Bot. Notiser 133, 323-336 (1980). - YEO, P.F.: Bot. J. Linn. Soc. 77, 223-334 (1979). - YI-ZHEN, X.: Acta Bot. Sinica 22, 323-329 (1980).

ZAVARIN, E., SNAJBERK, K., LEE, C.-J.: Biochem. Syst. Ecol. 6, 177-184 (1978). - ZOHARY, D., PLITMANN, U.: Plant Syst. Evol. 131, 143-156 (1979).

Dr. HANS-HELMUT POPPENDIECK
Professor Dr. KLAUS KUBITZKI
Institut für Allgemeine Botanik und
Botanischer Garten
Postfach 30 27 22
D-2000 Hamburg 36

II. Paläobotanik

Von FRIEDEMANN SCHAARSCHMIDT

1. Allgemeines

Seit dem Auslaufen des "World Report on Palaeobotany" vor über 10 Jahren hat es sich eingebürgert, gegenseitige Information durch regionale Bibliographien zu ermöglichen. Sie geben in der Regel – nach Sachgebieten geordnet – Übersicht nicht nur über die Publikationen, sondern auch über Themen, die in Arbeit sind. Als Vorbild diente die am längsten laufende "Bibliography of American Paleobotany" (WATT 1979, 1980). Für England haben CHALONER und SHEERIN (1980) die Jahre 1978-1979 erfaßt, während LEJAL-NICOL (1979) in noch größerem Abstand die französischsprachigen Arbeiten für die Jahre 1976-1978 referiert. Das deutschsprachige Gebiet betreut der "Arbeitskreis für Paläobotanik und Palynologie" (STURM 1979, BOERSMA und BROEKMEYER 1980). Zu erwähnen wären noch zwei spezielle Bibliographien: Die eine ist von der "International Association of Angiosperm Palaeobotany" für die Jahre 1978 und 1979 herausgegeben worden, die andere ist eine Bibliographie paläobotanischer Bibliographien von WEBER (1979, 1980).

Das Nachschlagewerk "Index of figured plant megafossils" von BOERSMA und BROEKMEYER hat sich inzwischen bestens bewährt. Er enthält für den entsprechenden Zeitabschnitt alle in wissenschaftlichen Publikationen abgebildeten Pflanzenfossilien verzeichnet. Bisher sind folgende Teile erschienen:

"Carboniferous 1971-1975" (1979)
"Triassic 1971-1975" (1980)
"Permian 1971-1975" (1981).

In den folgenden Jahren werden die Teile über Jura- und Devonpflanzen folgen.

2. Thallophyta

Seit seiner Entdeckung im Jahre 1844 durch LOGAN zählt *Prototaxites* zu den am meisten umstrittenen fossilen Pflanzen. Den Namen gab ihr DAWSON 1857, weil er in den verkieselten Stämmen *Taxus*-ähnliche Nadelhölzer zu erkennen glaubte. Nach einem temperamentvollen Streit mit CARRUTHERS (1872) setzt sich dessen auf anatomische Kenntnis beruhende Ansicht durch, daß es sich bei *Prototaxites* um Algenstämme handelt. Freilich blieb weiterhin umstritten, ob sie im Wasser oder auf dem Land lebten, wie z.B. noch LANG (1945) annahm. In der Tat handelt es sich bei *Prototaxites* nach wie vor um ein äußerst rätselhaftes Fossil. In der Regel findet man von ihm die verkieselten Stämme – eine ungewöhnliche Erhaltungsform für eine Alge. Der Vergleich mit den Phaeophyceen geht auf CARRUTHERS zurück, weil nur in dieser Gruppe Thalli vergleichbarer

Größe vorkommen. Aber Vermehrungsorgane, die diese Einordnung belegt hätten, wurden nie eindeutig nachgewiesen. Am häufigsten wurde noch die kugelförmige *Pachytheca* als Samen (der "Konifere" *Prototaxites* von DAWSON) oder vegetative Vermehrungsorgane einer Alge (WO. SCHMIDT 1958) zugeordnet. Als zugehörige Assimilationsorgane galten bald röhrenförmige Gebilde (PENHALLOW 1888), bald flächige Blätter (KRÄUSEL und WEYLAND 1934) oder schmale Bänder (SCHAARSCHMIDT 1974).

Obwohl unausgesprochen Verwundern darüber in der Luft lag, daß in der gesamten Erdgeschichte nur eine Alge im kurzen Zeitraum des Unterdevon in der Lage gewesen sein sollte, echte Versteinerungen in so großer Menge zu liefern, schien KRÄUSEL (1964) mit seiner Deutung als submers lebendem Tang das letzte Wort gesprochen zu haben. Kürzlich hat jedoch NIKLAS (1976) wieder Beweise für die Landpflanzennatur geliefert, indem er auf chemischem Wege Kutin und Suberin sowie andere nur bei Gefäßpflanzen vorkommende organische Verbindungen gefunden hat. Bereits kurz zuvor hatte SCHAARSCHMIDT (1974) sein vermutlich ebenfalls in diesen Formenkreis gehörendes *Mosellophyton hefteri* als eine im Strandbereich lebende temporäre Landpflanze gedeutet und damit auf das "Halophytenproblem" hingewiesen.

An völlig neuartigem Material hat JONKER (1979) nun versucht, das *Prototaxites*-Problem neu aufzurollen. In der Eifel fand er in großer Menge Strukturen im Gestein, die er auf *Prototaxites*-Thalli zurückführt. Sie bestehen aus einem Gewirr von Cauloiden und Phylloiden. Die massiven Stämme anderer Autoren werden durch Zusammenbacken von Cauloiden mit Sediment erklärt. Als Lebensraum wird das Schlickwatt angesehen, in dem die Pflanzen zeitweise trockenfallen konnten.

Auf Grund von morphologischen Ähnlichkeiten und dem Fund von 2 Sorten von Sporen (*Pachytheca* als Cystocarpien sowie Tetrasporen) wird *Prototaxites* als Rhodophycee (Rhodomelaceae) angesehen.

Die "Gyrogonite" von Charophyten lassen sich in ähnlicher Weise wie tierische Mikrofossilien gut für stratigraphische Zwecke verwenden. MÄDLER und STAESCHE (1979) können mit ihrer Hilfe eine grobe Gliederung des türkischen Tertiärs durchführen. SOULIE-MÄRSCHE (1979) rekonstruiert auf Grund detaillierter Kenntnisse des fossilen Materials die phylogenetischen Beziehungen der rezenten Gattungen. Sie lassen sich in acht Linien bis ins jüngere oder z.T. ins ältere Mesozoikum zurückverfolgen und haben offenbar nur eine langsame Evolution durchgemacht.

Ein schon lange bekanntes problematisches pflanzliches Fossil aus dem unteren Devon ist *Parka decipiens*. Es handelt sich um ± runde, ursprünglich offenbar gestielte Plättchen mit runden Warzen auf der Oberfläche, die von NIKLAS (1975/76) mit der Grünalge *Coleochaete* verglichen werden. NEUBER (1979) sieht jedoch mehr Ähnlichkeit mit dem rezenten Lebermoos *Sphaerocarpus*. Dafür würde die einschichtige Wand der "Warzen" sprechen, die dann mit den Involucren von *Sphaerocarpus* gleichzusetzen wäre, die die Archegonien umgeben. Bei den sporenähnlichen Körpern könnte es sich um Keimscheiben handeln, wie sie bei *Sphaerocarpus* aus den Zygoten entstehen.

Bei seiner Bearbeitung der baltischen Bernsteinflora hatte bereits CASPARY (1887, 1907) eine größere Anzahl von Laub- und Lebermoosen beschrieben. Im Gegensatz zu Abdrücken im Sediment sind diese Bernsteininklusen auch in feinen Details ausgezeichnet erhalten und haben GROLLE (1980) zu einer Revision der CASPARYschen Originale angeregt. Die ersten vier vorgestellten Arten gehören zu den Jungermanniales s. str. und lassen sich in die rezenten Gattungen *Cephaloziella*, *Bazzania*, *Jungermannia* und *Radula* als besondere Fossil-Arten einordnen. Alle vier Gattungen haben heute ihre größte Mannigfaltigkeit in den Tropen. In einem Stück ist eine aus 3 Arten bestehende Gesellschaft rindenbewohnender Moose überliefert.

3. Psilophyta

Von allgemeinem Interesse sind zwei Schwerpunkte: Die frühesten Landpflanzen und die Frage nach Psilophyten-Gametophyten. Außerdem setzte SCHWEITZER seine seit 20 Jahren laufende Revision der rheinischen Devonflora in einer Reihe von Mitteilungen fort.

Durch jahrelanges Aufsammeln von Pflanzenhäcksel ("plant debris") hat EDWARDS (1979) aus dem obersten Obersilur (Downtonian) von Dyfed in Süd-Wales eine der frühesten Landpflanzenfloren zusammengetragen. Freilich ist in ihr im wesentlichen nur eine Gattung vertreten, *Cooksonia*, die seit ihrer Erstbeschreibung durch LANG (1937) als älteste sichere Landpflanze gilt. Es handelt sich um ein kleines, dichotom verzweigtes Pflänzchen mit breitovalen Sporangien am Ende der Sprosse und mit Leitbündeln, die einfache Tracheiden enthalten. Nach der Form der Sporangien und ihrem Ansatz werden mehrere Arten unterschieden, in Dyfed ist es neben *C. hemisphaerica* vor allem *C. cambrensis*. Aus einigen Sporangien konnten runde, glatte Sporen gewonnen werden, die jedoch dunkel gefärbt blieben und keine weiteren Einzelheiten zeigten. Nach GENSEL (1980) sind diese Sporen bereits trilet. Bei *Cooksonia* handelt es sich um eine offensichtlich nur kleine Pflanze, von der bisher nur Bruchstücke gefunden wurden. SCHWEITZER (1980b) bildet mit seiner *C. bohemica* das bisher wohl vollständigste Exemplar dieser ersten Tracheophyte der Erdgeschichte ab. Es stammt aus den mit dem Downtonian etwa gleichalten Přidolí-Schichten Böhmens und ist – obwohl eine fast vollständige Pflanze – nur 3,5 cm groß, von der Basis aus reich dichotom verzweigt und von buschigem Habitus mit Sporangien an jedem Zweigende. Zusammen mit *Cooksonia* kommen in Dyfed noch verschiedene Achsen-Fragmente vor, die z.T. glatt sind (*Hostinella*), z.T. aber auch bereits stachel- bzw. blattartige Anhängsel tragen (*Psylophytites*) und die zeigen, daß diese frühe Landflora aus mehreren Pflanzentypen bestanden haben muß. Es treten auch bereits verschiedene Verzweigungstypen auf: Außer der dominanten Dichotomie vor allem noch die sog. K-Verzweigung, wie sie auch von Unterdevonpflanzen, z.B. *Zosterophyllum myretonianum* bekannt ist: An einem Achsenpunkt entspringen 2 oder mehrere büschelförmig nach einer Seite gerichtete kurze Seitenzweige.

Als Thallophyt wird schließlich ein als *Tortilicaulis transwalliensis* bezeichneter Sporangienrest gedeutet. Der tragende Stengel ist spiralig gewunden, wie es auch an der Seta des rezenten Lebermooses *Pellia epiphylla* zu beobachten ist.

Noch älter als die Flora von Dyfed sind Pflanzenreste, die EDWARDS et al. (1979) aus einem Steinbruch bei Cwm Graig Ddu beschreiben; die Einstufung als Bringewoodian Stage (Ludlow) ist durch Tierfossilien gesichert. Tracheophyten sind durch *Hostinella*-Fragmente mit deutlichem Leitbündel und durch *Cooksonia*-Reste nachgewiesen. Bei der Cwm-Graig-Ddu-Flora dürfte es sich somit um die älteste sichere Tracheophyten-Flora überhaupt handeln.

Mit Recht wird von den Autoren das Alter von zwei anderen "Silur"floren angezweifelt: Eine von BOUREAU et al. (1978, 1980) aus Libyen beschriebene und Lycopsiden enthaltende Flora dürfte auf Grund ihrer Entwicklungshöhe ins Mittel- oder Oberdevon gehören. Aufgelebt ist auch erneut die Diskussion um das Alter der australischen *Baragwanathia*-Flora. Diese war ursprünglich von LANG und COOKSON (1935) wegen des gemeinsamen Vorkommens mit Graptolithen als obersilurisch angesehen worden. Da es sich bei *Baragwanathia* um eine relativ hoch organisierte Lycophyte handelt, würde diese zeitliche Einstufung auf ein wesentlich höheres Alter der Landpflanzen schließen lassen, als nach allen anderen Befunden angenommen wird. Es müßten dann einfachere Vorläufer im Ordovizium oder gar im Kambrium gelebt haben. Da jedoch die ältesten einfach gebauten Tracheophyten (*Cooksonia*) nirgends früher als im Obersilur auftreten, liegt hier eine Diskrepanz vor, deren Klärung nicht nur für unsere Kenntnis über die Entwicklung des Pflanzenreiches von grundlegender Bedeutung ist, sondern auch allgemeine geologische Fragen, wie die Entwicklung der Sauerstoff-Atmosphäre, berührt. Seit JÄGER (1970) nachgewiesen hatte, daß die fraglichen Graptolithen im Unterdevon vorkommen, wurde *Baragwanathia* allgemein als unterdevonisch angesehen und der Widerspruch schien bereinigt. Nun hat GARRATT (1977, 1979) auf Grund neuer Graptolithen-Funde erneut behauptet, daß zumindest die unteren *Baragwanathia*-Horizonte ins Ludlow einzustufen seien. Wie bereits CHALONER und SHEERIN (1979), bestreiten auch EDWARDS et al. (1979) dies, weil die fraglichen Graptolithen nicht wirklich mit *Baragwanathia* assoziiert und die ebenfalls angegebene Bivalve und ein Brachiopod bisher nur im Unterdevon nachgewiesen worden seien. Somit kann auch weiterhin davon ausgegangen werden, daß *Baragwanathia* erst im Unterdevon auftritt und die ältesten Landfloren die *Cooksonia*-Floren sind.

Eine gewisse Ähnlichkeit mit *Cooksonia* besitzt die Gattung *Renalia*, die von GENSEL (1976) aus dem Unterdevon von Gaspé (Kanada) beschrieben wurde. Sie besitzt ebenfalls meist endständige, nierenförmige Sporangien, die sich jedoch bereits an einer Dehiszenz öffneten. Auch ist die Pflanze wesentlich größer und verzweigt sich pseudomonopodial. Die Gattung wird nun gleichzeitig von FAIRON-DEMARET (1980a) und SCHWEITZER (1980b) aus dem belgischen bzw. rheinischen Unterdevon angegeben. Sicher ist die Gattung – wie SCHWEITZER meint – auch in Europa weit verbreitet, aber, wie häufig bei Devonpflanzen, ist die Identifizierung schwierig, weil Sporangien äußerst selten zu finden sind.

Die Gattung *Psilophyton* wurde von SCHWEITZER (1980b) für das rheinische Devon neu überarbeitet. Die bereits seit längerer Zeit bekannte Art *P. burnotense* (früher auch als *P. goldschmidtii* bezeichnet) ist eine der häufigsten Arten des rheinischen Unterdevons. Sporangien wurden bisher noch niemals gefunden, so daß nur die bestachelten Achsen bekannt sind, die im unteren Bereich wenig und im oberen pseudomonopodial verzweigt sind. Häufiger sind dagegen sporangientragende Achsen bei *P. arcuatum*. Die Sporangien waren früher nur einzeln gefunden und von KRÄUSEL und WEYLAND (1930) als *Dawsonites arcuatus* bezeichnet worden. Im Gegensatz zu *P. burnotense* sind die Achsen bei *P. arcuatum* nicht bestachelt, aber mit feinen Längsriefen versehen.

GENSEL (1979) konnte bei *P. charientos* aus Nordamerika den anatomischen Bau untersuchen und fand einen komplizierten Tüpfelbau: In den scheinbar großen Tracheiden-Tüpfeln ist ein feines Netzwerk kleiner Tüpfel ausgebildet. Auch die Sporen sind bereits recht hochentwickelt und vom Bau einer *Apiculiretusispora*.

Die im rheinischen Unterdevon massenhaft vorkommende *Taeniocrada* hatten sich KRÄUSEL und WEYLAND (1935) mit flachen Sprossen und untergetaucht lebend vorgestellt. Wie bereits SCHAARSCHMIDT (1974) nimmt nun auch SCHWEITZER für *Taeniocrada* runde Stengel an. Er hält wegen der Sproßlänge (bis 2 m) jedoch an einer submersen Lebensweise fest. STÜRMER und SCHAARSCHMIDT (1980a,b) gelang es, im Röntgenlicht Zellgewebe von *Taeniocrada dubia* sichtbar zu machen.

So groß die Bedeutung der Psilophyten als erste Tracheophyten und erste echte Landpflanzen ist, so wenig sicher ist im Grunde ihre Stellung im natürlichen System, solange man nichts Näheres über ihren Fortpflanzungsmodus weiß. Nachdem LEMOIGNE (1969) archegonienartige Gebilde an *Rhynia*-Achsen gefunden hatte, bis heute (LEMOIGNE 1980) bei dieser Deutung geblieben ist und nun auch D.S. EDWARDS (1980) sie nicht mehr völlig ablehnt, sind Funde von weiteren Psilophyten-Gametophyten von Interesse. So haben REMY und REMY (1980a, 1980b) im unterdevonischen Rhynie-chert gestielte schüsselförmige Organe gesehen (*Lyonophyton rhyniensis*), die sie als Gametangiophore deuten und die im Inneren der Schüssel die Archegonien sowie am Rande die Antheridien getragen haben sollen. Da die beiden als Gametophyten gedeuteten unterdevonischen Pflanzen (*Lyonophyton*, *Sciadophyton*) Gefäßbündel enthalten, leitet REMY (1980) sowohl die Kormophyten als auch die Bryophyten von einem gemeinsamen Vorläufer ab, der einen isomorphen Generationswechsel hatte. Bei den Kormophyten sei der Gametophyt, bei den Moosen der Sporophyt reduziert worden. Ganz ähnlich glauben sie nun auch *Sciadophyton* deuten zu können (REMY et al. 1980): Die Stengel der kleinen Rosettenpflänzchen enden in Anschwellungen, von denen schon KRÄUSEL und WEYLAND (1930) angenommen hatten, daß sie keine Sporangien seien. Sie hatten auch bereits Betrachtungen darüber angestellt, ob es sich bei *Sciadophyton* um *Psilophyten*-Gametophyten handeln könne. Sie waren jedoch zu dem Schluß gekommen, daß eine Deutung als Brutknospen die beste Erklärung sei. Obwohl er keinen schlüssigen Beweis zu geben vermag, neigt SCHWEITZER (1980a) zur Gametophyten-Deutung und nimmt an, daß sich auf den "Anschwellungen" die Archegonien und Antheridien befunden haben, zumal er an einigen Stücken sproßartige Auswüchse beobachtete, in denen er Jungpflanzen von Sporophyten sieht. Weniger vorsichtig sind REMY et al. (1980), die glauben, in einer Ähnlichkeit mit *Lyonophyton* den Beweis für die Gametophytennatur von *Sciadophyton* in Händen zu haben.

Während die Sporangienstände bei *Psilophyton* und *Taeniocrada* rispig sind, haben die Zosterophyllaceae ährenförmige Stände. SCHWEITZER (1979) konnte nun aus dem rheinischen Devon mehrere Gattungen dieser Verwandtschaft nachweisen. Am interessantesten hiervon – weil schon am längsten bekannt – ist *Zosterophyllum rhenanum*. Während KRÄUSEL und WEYLAND (1935) der Meinung waren, daß die Pflanze submers gelebt habe (mit über die Wasserfläche erhobenen Sporangienähren), weist nun SCHWEITZER nach, daß die Stengel ebenfalls "rundlich" gewesen sind. Aus reich verzweigten Rhizomsystemen schließt er auf einen rasigen Wuchs, wie er für *Z. myretonianum* aus Schottland bekannt ist. Diese Ähnlichkeit und die runden Stengel legen den Schluß nahe, daß *Z. rhenanum* in Wirklichkeit wie die anderen Psilophyten auch eine echte Landpflanze war (vgl. *Taeniocrada*!).

Verwandt mit *Zosterophyllum*, jedoch einseitswendige Ähren tragend, ist *Distichophytum mucronatum* (die früher auch *Bucheria ovata* DORF und *Rebuchia ovata* (DORF) HUEBER genannte Pflanze, müßte jedoch richtig *Distichophytum ovatum* heißen!).

Noch stärker abweichend von *Zosterophyllum* ist die Sporangienanordnung bei *Gosslingia*, die bisher nur aus Süd-Wales bekannt ist und nun erstmals mit einer eigenen Art (*G. cordiformis*) aus dem rheinischen Unterdevon angegeben wird: Bei ihr stehen die Sporangien streng einreihig auf der Unterseite von Seitenästen. Schließlich wird noch erstmals aus dem rheinischen Unterdevon – allerdings nur in Form von sterilen Achsen – die Zosterophyllacee *Sawdonia ornata* angegeben, die bereits von DAWSON (1859) zusammen mit nicht bestachelten Resten aus dem Unterdevon von Gaspé (Kanada) als *Psilophyton princeps* und als erste Psilophytale überhaupt beschrieben worden ist. Inzwischen hatte HUEBER (1971) festgestellt, daß die Sporangien an den bestachelten Zweigen (*Sawdonia ornata*) seitenständig sind, während sie an den glatten Sprossen (*Psilophyton princeps*) in endständigen Rispen angeordnet sind. Da solche sporangientragenden Achsen von RAYNOR (1980) im schottischen Unterdevon gefunden wurden, ist es sehr wahrscheinlich, daß diese in Nordamerika weitverbreitete Pflanze auch ein Bestandteil der europäischen Unterdevonflora war. SCHWEITZER nimmt an, daß von *Sawdonia* ausgehend eine phylogenetische Reihe über *Asteroxylon* zu *Drepanophycus* führt.

4. Lycophyta

An reichem neuem Material aus dem rheinischen Unterdevon konnte SCHWEITZER das Lebensbild von *Drepanophycus spinaeformis* in einigen Punkten wesentlich verbessern. So waren die jungen Zweige am Ende nicht wie bei den Farnen und einigen Psilophyten eingerollt, und die Sporangien entsprangen nicht der Oberseite der stachelartigen Blätter, wie KRÄUSEL und WEYLAND (1930) beschrieben haben, sondern saßen zwischen ihnen an separaten Sporangiophoren. Diese Sporangienstellung läßt sich nur schwer mit der Telomtheorie erklären, die für die Stellung der Sporangien auf den Lycopsiden-Blättern die Reduktion lateraler Telomsysteme verantwortlich macht. SCHWEITZER deutet die gestielten Sporangien daher auch folgerichtig wie die sterilen "Stacheln" als Emergenzen. Da z.B. bei *Protolepidodendron* die (ebenfalls gestielten) Sporangien tatsächlich an gegabelten Seitenorganen sitzen, erwägt er eine "diphyletische" Entstehung der Lycopsiden. Vielleicht sollte einmal nachgeprüft werden, ob die Angaben über eine Verbindung zwischen Sporangien und "Stachel" wirklich auf Beobachtungsfehlern beruhen, oder ob bei *Drepanophycus* beide Anheftungsarten nebeneinander vorkommen. Dies würde es erleichtern, auch folgende Reihenfolge anzunehmen:

gestieltes Sporangium
mit "Stachel" verwachsenes Sporangium
gegabeltes Seitenorgan.

Es müßte erlaubt sein, von Zeit zu Zeit die Telomtheorie wenigstens in besonderen Fällen in Frage zu stellen und nach anderen Modellen zu suchen.

Das früher von WEYLAND und BEHRENDT (1968) als Zosterophyllacee gedeutete *Taeniophyton* wird nun von SCHWEITZER (1979) als eine eigene *Drepanophycus*-Art (*D. devonicus*) angesehen, die – wenn es richtig ist, daß das nordamerikanische *Kaulangiophyton* identisch ist – ebenfalls weit verbreitet war. Die Neuuntersuchung des mitteldevonischen *Protolepidodendron scharianum*-Originalmaterials von KRÄUSEL und WEYLAND (1932) durch FAIRON-DEMARET (1980b) zeigt, daß die Blätter wesentlich stärker gespalten sind als am böhmischen Material und daß der Stengel, wie bereits MUSTAFA (1975) und SCHAARSCHMIDT (1976) gezeigt hatten, eine Aktinostele enthält, die dem des nordamerikanischen *P. gilboense* gleicht.

Noch komplizierter gebaut sind die Blätter nach FAIRON-DEMARET (1979) beim ehemaligen *Protolepidodendron* (jetzt *Estinnophyton*) *wahnbachense*: Sie sind zweimal gespalten und tragen auf der Oberseite vier zurückgekrümmte Sporangien.

Damit erweist sich erneut, daß die Lycophyten nicht nur erstaunlich früh im Unterdevon auftreten, sondern auch bereits morphologisch sehr vielgestaltig sind.

Die Lebensgeschichte der baumförmigen Lycopsiden des Karbons ist immer wieder spannend in ihren Eigentümlichkeiten: riesige Bäume, die im wesentlichen aus Rinde bestanden, deren Stämme noch lange über ein "Parichnossystem" Gas austauschen und assimilieren konnten und Bäume, die vielleicht ihr Leben durch eine gewaltige Produktion von Sporen in terminalen Sporenzapfen beendeten. Daß in dieser eigentümlichen Pflanzengruppe sich nicht nur verschiedenartige vegetative Differenzierungen herausgebildet haben, sondern auch in der Fortpflanzung verschiedene Wege der Spezialisierung eingeschlagen wurden, war zu erwarten. Nach PHILLIPS (1979) lassen sich vier Gruppen unterscheiden:

(1) Der *Lepidostrobus*-Typ hat freie bisporangiate Sporophylle. Die Sporangien enthalten zahlreiche (mindestens 4) Megasporen, in denen sich noch im Zapfen Megagametophyten mit mehreren eingesunkenen Archegonien bildeten. Vielleicht haben die Gametophyten teilweise aus der Spore herausgeragt. Die Megasporen wurden durch eine normale Dehiszenz aus dem Sporangium ausgestreut, und die Befruchtung fand vermutlich ähnlich wie bei *Selaginella* und *Isoetes* statt.

(2) Beim *Mazocarpon*-Typ, der bei *Sigillaria* vorkommt, enthalten die Sporangien nur 8-12 Megasporen, die 1-2 mm groß wurden. Im Gegensatz zum *Lepidostrobus*-Typ zerfallen hier die Zapfen in der Weise, daß jede Megaspore umgeben von Parenchym mit einem Stück Megasporangium vereinigt bleibt. Dadurch wird es möglich, die Sutur in die richtige Position zur Bestäubung zu bringen. Vielleicht ist diese Ausbreitungsform eine Anpassung an trockenere Verhältnisse, die den Sigillarien das Überleben der "Westphal/Stephan-Katastrophe" ermöglichte, an der die übrigen Lepidophyten ausstarben!

(3) Beim *Achlamydocarpon*-Typ (vgl. auch LEISMA und PHILLIPS 1979) und dem *Lepidocarpon*-Typ enthalten die Megasporangien nur jeweils eine große Megaspore und fallen gemeinsam mit den Sporophyllen ab. Beide unterscheiden sich im wesentlichen dadurch, daß *Lepidocarpon* ein Integument besitzt, das bei *Achlamydocarpon* noch fehlt. Es wird angenommen, daß die abgefallenen Sporophyll-Sporangien-Einheiten (die mehrere cm groß werden konnten mit bis zu 11 mm großen Megasporen) im Wasser schwimmend direkt durch Mikrosporen bestäubt wurden. Die damit einhergehende Verbreitung durch Wasser hing vielleicht mit regelmäßigen Transgressionen zusammen. Bei *Lepidocarpon* konnten vermutlich noch bereits aufrechte Embryonen schwimmend verbreitet werden.

Während sich PHILLIPS im wesentlichen auf die Megagametophyten beschränkte, hat BRACK-HANES (1978b) Zellen des Mikrogametophyten von *Lepidostrobus schopfii* und Strukturen nachgewiesen, die sie als Zellkerne und Chromosomen deutet.

5. Sphenophyta

Von allgemeinem Interesse ist ein Vergleich von *Sphenophyllum*-Arten aus drei intramontanen (limnischen) Karbonbecken Mitteleuropas durch STORCH (1980). Bei seinen Untersuchungen des Zwickauer Karbons, des Saargebietes und des Innersudetischen Beckens kommt er zu dem Schluß, daß mehr Gemeinsamkeiten zwischen den drei Becken bestehen, als früher z.B. von GOTHAN (zahlr. Arbeiten) und DABER (1963) angenommen worden war. Selbst die Florenprovinzen Euramerika, Gondwana, Cathaysia und Angara sollen mehr gemeinsame Pflanzenarten gehabt haben, als in den Lehrbüchern steht. So soll *Sph. oblongifolium* ein Bestandteil aller Florenprovinzen gewesen sein. Zum Teil werden Unterschiede durch fehlerhafte Benennung vorgetäuscht. Parallelentwicklung in den einzelnen Florenprovinzen lehnt STORCH ab und hält eine Ausbreitung durch Migration für wahrscheinlich. Die übrigbleibenden Unterschiede führt er auf Verschiedenheiten der Verbreitungsgebiete, der Ausbreitungsfähigkeit und der ökologischen Bedingungen zurück. Daß die Unterschiede in der Verbreitung anderer Sippen der Pteridospermen größer sind als bei *Sphenophyllum*, erklärt STORCH mit der Größe der Samen, die Barrieren schwieriger überwinden konnten als die Sporen von *Sphenophyllum*.

Bemerkenswert ist noch, daß Epidermen untersucht werden konnten, die vereinzelt haplocheile Spaltöffnungen trugen und Hinweise auf Sekretzellen zeigten, sowie daß an Blattspitzen die gleichen Kletterhaken beobachtet wurden wie von BATENBURG (1977), die vom Autor jedoch vermutlich zu Unrecht als Welkeerscheinungen gedeutet werden.

Zu den Sphenophyllales wird auch *Trizygia* gerechnet, die in der Gondwana- und Cathaysia-Flora vorkommt. Im Gegensatz zu Sphenophyllum sind bei *Trizygia* die Blätter vorn abgerundet und stehen in den Quirlen paarweise genähert. Während *Sphenophyllum* im euramerischen Gebiet am Ende des Perm ausstirbt, hat *Trizygia* im Gondwanagebiet noch länger, in Japan sogar bis in die obere Trias (Karn) überlebt.

6. Filicophyta

Zu der ausführlichen Darstellung im vorigen Beitrag (Fortschr. Bot. 41) ist aus dem Berichtszeitraum wenig Bemerkenswertes nachzutragen.

GALTIER führte mit seinen Mitarbeitern die Untersuchungen an strukturerhaltenen Coenopteriden fort (GALTIER und SCOTT 1979, HOLMES 1979a,b), und MILLAY und TAYLOR (1980) beschrieben aus dem Pennsylvanian von Nordamerika eine neue *Botroyopteris*-Sporangienform.

Der Schlüssel zum Verständnis der rezenten Marattiaceen liegt in der Kenntnis der karbonischen Arten (MILLAY 1979). In dieser Zeit hatten sie den größten Formenreichtum und existierten die ursprünglicheren Vertreter. So wird die Gattung *Scolecopteris* als der direkte Vorläufer der rezenten *Marattia* angesehen. Unter den nordamerikanischen Vertretern von *Scolecopteris* lassen sich nach dem Verwachsungsgrad der Synangien drei Artengruppen unterscheiden.

Wie sich an Stammabdrücken (*Caulopteris*, *Megaphyton*), versteinerten Psaronien und *Pecopteris*- Blättern ablesen läßt, traten die Marattiaceen sicher bereits im Unterkarbon auf. Über ihre Entwicklungsgeschichte im Mesozoikum ist wenig bekannt, weil hier wenig strukturerhaltenes Material vorkommt.

Zu den Marattiaceen gehört auch der auf der Nordhalbkugel der Alten und der Neuen Welt im Oberkarbon weit verbreitete *Pecopteris unita*, von dem JENNINGS und MILLAY (1979) aus Eisenkonkretionen von Illinois die sternförmigen Synangien aus der Verwandtschaft von *Scolecopteris* und *Cyathotrachus* beschreiben.

Die Zuordnung des in der Unteren Trias weitverbreiteten Farns *Anomopteris mougeoutii* war bisher umstritten. Nun ist es GRAUVOGEL-STAMM und GRAUVOGEL (1980) gelungen, Sporangien mit einem apikalen Anulus sichtbar zu machen. Damit gehört *Anomopteris mougeoutii* eindeutig zu den leptosporangiaten Farnen. Freilich bleibt noch ungewiß, in welche Familie er gestellt werden muß. Weist der Sporangienbau zu den Schizaeaceen, so ähnelt das W-förmige Leitbündel der Rhachis eher den Cyatheaceen und Dicksoniaceen, nicht jedoch den Osmundaceen, bei denen es einem C gleicht.

7. Pteridospermae

Seit ihrer Entdeckung vor 27 Jahren (DELEVORYAS und MORGAN 1954) sind die Callistophytaceen zu einer der bestbekannten fossilen Pflanzengruppen geworden, und inzwischen werden sie als eigenständige Familie der Pteridospermen angesehen. Das reichste Material haben die oberkarbonischen "coal balls" von Illinois, Kansas und Kentucky geliefert, aber auch aus dem französischen Stephan von Autun sind sie bekannt. Bei der einzigen bekannten Gattung *Callistophyton* handelt es sich nach der Rekonstruktion von ROTHWELL (1975) um einen niederliegenden Strauch mit schlanken Zweigen, die zur Ausläuferbildung neigen. Die dünnen, bis 3 cm im Durchmesser messenden Stämme waren reich verzweigt mit Sproßknospen und Adventivwurzeln an den Knoten. Die kleinen farnwedelartigen Blätter sind im Abdruck vermutlich schon lange als *Pecopteris pluckeneti* bekannt (STIDD und BARTHEL 1979).

Besonders gut kennt man den anatomischen Bau der Pflanze: Das parenchymatische Mark der Achse ist von einem Ring mesarcher Leitbündel umgeben, an denen nach außen ein massiver Sekundärholzzylinder anschließt. Die Reproduktionsorgane (ROTHWELL 1980) sitzen – im Gegensatz zu allen übrigen Pteridospermen – auf der abaxialen Seite nicht veränderter, normaler Laubblätter. Die Pollenorgane – ringförmige Synangien – enthalten Zweisackpollen vom Bau *Vesicaspora*. Er gelangte durch den Wind in die Pollenkammer des Samens, wurde in einem Bestäubungstropfen aufgefangen und keimte, wie moderner Koniferenpollen, am distalen Pol zu einem langen, schlanken Pollenschlauch aus. Freie Gameten wurden nicht gebildet.

Alle diese Merkmale lassen die Callistophytaceen als eine sehr eigenständige Familie erscheinen. Nach ihrem farnartigen Habitus sind sie zwar typische Pteridospermen, sie vereinigen jedoch primitive Merkmale in den Blättern mit fortgeschrittenen in der Fortpflanzung, wie wir sie sonst niemals in der Pteridospermen-Cycadophyten-Reihe, sondern nur bei den Cordaiten-Koniferen finden. So nimmt diese "Cordaite ohne Zapfen und mit Farnblättern" einen bemerkenswert eigenständigen Platz ein und wirft ein ganz neues Licht auf die phylogenetischen Vorgänge dieser frühen Gymnospermen.

Die Pollenorgane der Pteridospermen sind erstaunlich mannigfaltig (MILLAY und TAYLOR 1979). Sie dürften sich von Mikrosporangien-Aggregaten der Progymnospermen herleiten, die an wenig modifizierten Blättern gebildet wurden, frei waren und sich durch einen Längsspalt öffneten. Während die unterkarbonischen Vertreter noch weitgehend diesem Schema entsprachen, waren sie am Beginn des Oberkarbon bereits stärker abgewandelt und zu Synangien verwachsen, die aus einem eine Höhlung umschließenden Sporangienkreis bestanden oder blattartig abgeflacht waren. Sie öffneten sich entlang eines innen gelegenen Spaltes. Bei den blattförmigen spreizten sich hierzu die Sporangien bzw. öffneten sich an der Spitze. Vom mittleren Oberkarbon an wurden die Pollenorgane sehr komplex und erzeugten ungeheure Mengen Pollen. Die Callistophytaceen unterscheiden sich von den Lyginopteridaceen vor allem dadurch, daß sich bei ihnen die Synangien nicht an Zweigen, sondern an Blättern bildeten.

Sicher hat bei Pteridospermen die Windbestäubung noch eine große Rolle gespielt (TAYLOR und MILLAY 1979). Da aber bereits mehrere Insektenfamilien entwickelt waren, könnte auch Entomogamie vorgekommen sein, zumal der Pollen z.T. schon wegen seiner Größe zum Windtransport ungeeignet erscheint. Auch spezielle Einrichtungen wie die großen Drüsenhaare an den Kupulen mancher Pteridospermen (*Lyginopteris*, *Callistophyton*) könnten der Bestäubung gedient haben, z.B. als Nektarien.

DABER (1980) macht sich Gedanken über die Bedeutung der "Gabelwedelformen", die vor allem bei Pteridospermen bis ins Mesozoikum verbreitet sind. Er sieht in ihnen ursprüngliche Konstruktionsprinzipien, die sich auf die Psilophyten zurückführen lassen und die bei den Pteridospermen eine Funktion erfüllen. Er deutet sie als echte Dichotomien, nicht als Dichasien, wie durch das Auftreten von Trichotomien belegt wird. Diese Trichotomien führt DABER auf zwei unmittelbar aufeinanderfolgende dichotome Gabelungen zurück und hält sie für ein wichtiges Gestaltungsprinzip des Pteridospermenwedels: Da für die Leitung grösserer Mengen von Assimilaten in den Achsen noch die Voraussetzungen fehlten, werden die Reproduktionsorgane stets an funktionierenden Laubblättern gebildet. Dies bedingt, daß sie zeitlich und räumlich nach den vegetativen Teilen angelegt wurden. So kommt es, daß sie sich stets im oberen Teil eines Blattes und häufig in der mittleren Partie eines trichotomen Blattes befinden. Auf diese Weise lassen sich auch die stets auf den Hauptachsen von *Glossopteris*-Blättern entspringenden Blüten erklären.

Ein Beispiel für eine frühe Pteridosperme ist *Hydrasperma longii* aus dem Devon/Karbon-Übergangsbereich von Schottland und Irland (MATTEN et al. 1980). Jedes Aggregat besteht aus vier Samen, die von einer vierteiligen Kupula umgeben sind und deren jeder Teil sich zweifach gabelt. Die Kupula ist somit ein Achsensystem, dessen Verzweigungsmodus an die Aneurophytales erinnert, von denen die Pteridospermen vermutlich abzuleiten sind.

8. Coniferophyta

Eine im Laufe von Jahrzehnten von LOUIS GRAUVOGEL zusammengetragene reiche Sammlung aus dem Voltziensandstein (untere Trias) der Vogesen ermöglichte in den vergangenen Jahren LÉA GRAUVOGEL-STAMM eine Reihe von gründlichen Untersuchungen insbesondere der Koniferen dieser Zeit. In ihrer neuesten umfangreichen Arbeit (GRAUVOGEL-STAMM 1978) sind die Koniferen wieder der wesentlichste Bestandteil, und besonders die Studie und Rekonstruktion von *Aethophyllum stipulare* zeigt, wie eine vorbildliche Bearbeitung einer fossilen Pflanze aussehen kann. Die Pflanze war von BRONGNIART (1828) und SCHIMPER und MOUGEOT (1844) als Monocotyle angesehen worden. Seit SCHENK (1891) galt sie als eine *Schizoneura* nahestehende Equisetale.

GRAUVOGEL-STAMM kann nun zeigen, daß *Aethophyllum stipulare* eine Konifere aus der Verwandtschaft von *Swedenborgia* ist. Wie bei dieser ist die Samenschuppe fünfzipflig und trägt fünf Samen auf der Oberseite. Der Stamm der bis 2 m hohen Pflanze wird als krautig bezeichnet, da er überwiegend aus Markparenchym besteht. Sekundärholz ist nur in der Wurzel entwickelt. Das Primärholz besteht im Stamm aus einer Eustele und ist in der Wurzel diarch. An den gefundenen Keimpflanzen läßt sich die Abfolge der linealen Blätter erkennen: Die zwei Kotyledonen sind zwei-, die Primordialblätter vier- und die Folgeblätter meist sechsnervig.

Aethophyllum war wenig verzweigt, wuchs aufrecht und trug die schlanken ♀ Ähren terminal und lateral im oberen Teil, die gedrungenen ♂ Kätzchen jedoch weiter unten lateral. Pollenzapfen waren bereits früher von GRAUVOGEL-STAMM isoliert gefunden worden (*Willsiostrobus acuminatus*). Sie sind hypopeltat mit in lange Spitzen ausgezogenen Mikrosporophyllen mit vier bis sechs langen Pollensäcken. Der Pollen hat zwei Luftsäcke mit einer proximalen Laesur.

Auf Grund des Baues der weiblichen Zapfen könnte *Aethophyllum* bei den Swedenborgiaceae und nach der Blattform bei den Podozamitaceae eingereiht werden. Die Aethophyllaceae genannte, beide Organe umfassende neue Familie gehört wie jene in den Übergangsbereich zwischen Voltziales und modernen Koniferen. Anklänge an *Aethophyllum* findet man in rezenten Familien nur bei Araucariaceen im Bau des Mikrosporophylls und bei den Podocarpaceen im Bau des Keimlings, der in der Sektion *Eupodocarpus* ebenfalls zwei zweinervige Kotyledonen hat.

Die isoliert vorkommenden Koniferen-Zapfen des Mesozoikums lassen sich nicht ohne weiteres in das natürliche System einordnen. Sie wurden daher lange Zeit unter den Namen *Masculostrobus* und *Pityanthus* geführt. Es ist einer der Verdienste von GRAUVOGEL-STAMM, daß sie die Vielfalt und Konstruktionsprinzipien dieser Organe klar herausgearbeitet hat. Nachdem SCHAARSCHMIDT (1973) nachgewiesen hatte, daß der Holotypus von *Masculostrobus* viel moderner ist als die frühmesozoischen Zapfen und daß der Holotypus von *Pityanthus* eine Kotspur ist, haben GRAUVOGEL-STAMM und SCHAARSCHMIDT (1978, 1979) ein neues Ordnungsschema der peltaten Mikrosporophyllstände erarbeitet. Neben den beiden jüngeren bisporangiaten Gattungen *Masculostrobus* und *Amydrostrobus* spielen im älteren Mesozoikum vor allem die beiden polysporangiaten Gattungen *Willsiostrobus* und *Darneya* eine Rolle. Davon ist *Willsiostrobus* hypopeltat und polysporangiat mit meist sehr langen Pollensäcken. In der Vielzahl der Pollensäcke erinnert die Gattung an rezente Araucariaceen. Gefunden wurde sie häufig zusammen mit *Yuccites* und ansitzend an *Aethophyllum stipulare* (s.o.). Bei *Darneya* sitzt eine Vielzahl von Pollensäcken mittels verzweigter Stielchen am Stiel des peltaten Mikrosporophylls. Auf Grund dieses Baues ist *Darneya* mit keiner rezenten Blüte zu vergleichen. Wegen der verzweigten Sporangienstielchen sieht GRAUVOGEL-STAMM (1978) eine gewisse Ähnlichkeit mit *Cephalotaxus* und deutet die peltate Schuppe als Metamorphose eines Seitenzweiges und folgerichtig den Zapfen als Infloreszenz. Die Mutterpflanze dürfte in den Gattungen *Voltzia* und *Albertia* zu suchen sein.

Die heutige südhemisphärische Verbreitung der Araucariaceen regt zur Erforschung der Geschichte dieses Areals an. Mit Hilfe anatomischer Details konnte STOCKEY (1980a,b) zwei Zapfen aus dem Jura von England bis zur Sektion bestimmen. Beide gehören zur verhältnismäßig primitiven Sektion *Bunya*, die heute von *Araucaria bidwillii* vertreten wird und die im Mesozoikum ihre Blütezeit gehabt hat.

ROTHWELL und BASINGER (1979) beschreiben aus dem Eozän von Britisch Kolumbien einen versteinerten Pollenzapfen von *Metasequoia*, der sich in der Anatomie und im Pollenbau kaum von der rezenten *Metasequoia glyptostroboides* unterscheidet. Das deutet darauf hin, daß sich diese Gattung, wie auch andere Taxodiaceen, offenbar über lange Zeit hin nicht wesentlich verändert hat.

9. Angiospermae

Bis vor 50 Jahren hat sich die Tertiär-Paläobotanik im wesentlichen darauf beschränkt, mit Hilfe von oberflächlich bestimmten Pflanzenresten – vorwiegend Blättern – einzelne Floren zu beschreiben und daraus weitreichende klimatische Schlüsse zu ziehen. Erst als die Arbeiten KIRCHHEIMERs in den 30er Jahren zeigten, daß Samen und Früchte wesentlich bessere Bestimmungen und Aussagen ermöglichten, ging man daran, die Methodik in der Tertiär-Paläobotanik zu überdenken. Insbesondere galt es, neue Verfahren zu finden, die eine sicherere Identifizierung der Pflanzenfossilien ermöglichen.

So wurde in Europa neben der Karpologie zur besseren Bestimmung der Blätter die Kutikularanalyse ausgebaut, die in der Folgezeit sehr wesentlich zum besseren Verständnis der Vegetationsgeschichte und der pflanzengeographischen Beziehungen beigetragen hat.

Gerade für die Klärung dieser Zusammenhänge ist die europäische Tertiärflora von besonderer Bedeutung, ist doch die in jener Zeit ± gleichmäßig über die Nordhalbkugel verbreitete warm-getönte Flora während der Klimaverschlechterung im Quartär im europäischen Sektor ausgestorben, während sie in Amerika und Asien durch Südwanderung weitgehend erhalten bleiben konnte. So kommt es, daß viele Eigenarten der heutigen Pflanzenareale erst aus der Kenntnis der europäischen Tertiärflora verständlich werden.

Unter diesen Gesichtspunkten sind in jüngster Zeit eine Reihe von Vorkommen mit modernen Methoden vor allem unter paläoklimatischen und pflanzengeographischen Gesichtspunkten neu untersucht worden. Einen wichtigen Impuls hatten die karpologischen Arbeiten von MAI (1964, 1965, 1967) gegeben, in denen er am wechselnden Anteil laurophyller "paläotropischer" und laubwerfender "arktotertiärer" Florenelemente eine rhythmische Klimaentwicklung im Laufe des Tertiärs nachwies. Nach seinen Feststellungen lösen allein im Oligozän und Miozän mehrere wärmere und kühlere Perioden einander ab.

Der Frage, wie weit sich ähnliche Florenentwicklungen auch in anderen Gebieten verfolgen lassen, sind mehrere Bearbeiter in der Miozänen Braunkohle der Oberpfalz nachgegangen (JUNG 1972). Von zentraler Bedeutung sind hierfür zweifellos die karpologischen Untersuchungen durch GREGOR (1975, 1978, 1979, 1980). An Hand einer äußerst reichhaltigen Samen- und Fruchtflora ließen sich sowohl die ökologischen und klimatischen Bedingungen als auch die pflanzengeographischen Zusammenhänge mit anderen Gebieten weitgehend klären. So konnte dieser Autor mehrere Pflanzengesellschaften (überwiegend feuchter Standorte) unterscheiden, von Moor- bis Au- und Galeriewaldgesellschaften. In ihnen kommen mit stark wechselnden Anteilen vor allem Symplocaceen, Mastixiaceen, Vitaceen, Juglandaceen, Fagaceen und Rutaceen vor. Eigentümlich für die Wackersdorfer Braunkohle ist eine starke Beteiligung von *Glyptostrobus* an der Kohlebildung. Aus der Zusammensetzung der Flora läßt sich ablesen, daß die Kohle während eines warmen Abschnittes im Mittelmiozän entstanden ist. Dieser ist in die Klimazone VI nach MAI (1967) einzustufen, in der vorübergehend ein besonders hoher Anteil an laurophyllen paläotropischen Elementen vorhanden war. Das Klima wird als "humides warmgemäßigtes Subtropenklima" bezeichnet. Andere Floren der Oberpfalz haben einen höheren Anteil an laubwerfenden Arten und sind offenbar jünger, vermutlich Obermiozän. Bemerkenswert ist in diesem Zusammenhang, daß auch MÄDLER und STEFFENS (1979) im unteren Obermiozän der Türkei eine kühlere Phase feststellten.

Es scheint, als ob sich einzelne der Klimaphasen, die MAI in Mitteldeutschland gefunden hat, auch in der Früchteflora der Oberpfalz wiederfinden lassen. Weniger deutlich war dies in der Blätterflora erkennbar, die bereits früher von KNOBLOCH und KVAČEK (1976) publiziert worden ist. Auch sie waren zu dem Schluß gekommen, daß sich die Oberpfälzer Miozänwälder im Übergangsbereich zwischen dem subtropischen "Broadleafed Sclerophyllous Evergreen Forest" und dem warm-temperierten "Mixed Mesophytic Forest" befanden. Obwohl einige wichtige Gattungen bzw. Familien als Blattfossilien nicht nachgewiesen werden konnten (vor allem Mastixiaceae, *Symplocos*, *Eurya*), stimmt das Bild im wesentlichen mit den Ergebnissen von GREGOR überein und bestätigt erneut die enge Beziehung der europäischen Tertiärflora zu rezenten Floren Südostasiens und in geringerem Maße zum atlantischen Nordamerika, die noch heute in der Asa-Gray-Disjunktion ihren Ausdruck findet. Einige Anklänge sind auch in Pflanzengesellschaften Mittelamerikas, der Mediterraneis und des Vorderen Orients wiederzuerkennen.

Wenn auch die südostasiatischen Floren am meisten Ähnlichkeiten mit unseren Tertiärfloren haben, so gibt es doch nirgends heute einen Vegetationstyp, der ihnen völlig entspräche. Dies ist auch kein Wunder, handelt es sich doch bei den Veränderungen, die seit dem Miozän stattgefunden haben, nicht nur um einfache Pflanzenwanderungen, sondern um einen sehr komplexen Vorgang, um eine Verzahnung von Arealänderungen mit phylogenetischen Vorgängen, wobei die Arealänderungen im wesentlichen in einer Dezimierung der europäischen Tertiärflora bestanden und bis heute nur die asiatischen bzw. nordamerikanischen Sippen überlebt haben.

Diesen an Makrofossilien gewonnenen Ergebnissen widerspricht die Untersuchung der Wackersdorfer Pollenflora durch THIELE-PFEIFFER (1979, 1980) nicht. Zwar konnte sie die Flora nicht so detailliert erfassen wie GREGOR und die Beschränkung auf einen Tagebau verhinderte eine umfassende Darstellung. Es war aber möglich, zwei kohlebildende Gesellschaften, das baumfreie Riedmoor und den Moorwald sowie den umgebenden mesophilen Mastixioideen-Wald zu erkennen, und auch die Erscheinung, daß das gewöhnlich häufige *Taxodium* (bzw. *Sequoia*) in Wackersdorf durch *Glyptostrobus* als Kohlebildner ersetzt wird, ist in den Pollenbildern erkennbar. Dagegen läßt sich ein dem MAI'schen Zonenschema entsprechender periodischer Wandel in der Sporenflora nicht ablesen. THIELE-PFEIFFER macht hierfür eine länger andauernde subtropische Phase in dem von ihr untersuchten Profil verantwortlich, was auf ein klimatisches Nord-Süd-Gefälle und eine entsprechende Pflanzenzonierung zurückgeführt werden könnte.

Mit einer älteren Klimaschwankung beschäftigen sich MAI und WALTHER (1978). Bei ihren Untersuchungen einer mitteloligozänen Flora im Weißelsterbecken südlich von Leipzig fanden sie Blätter und Früchte einer Flora, die zwar noch überwiegend laurophyllen Charakter hat, aber daneben einen für ihr hohes Alter bereits erstaunlich großen Anteil an laubwerfenden Elementen enthält. Sie gleicht damit eher einer "jüngeren" als einer "älteren" Mastixioideen-Flora und kennzeichnet offenbar eine erste kühlere Phase im ausgehenden Paläogen, zumal dieser "Florenkomplex Haselbach" offenbar im tieferen Mitteloligozän in Europa weiter verbreitet gewesen ist. In seiner Zusammensetzung ähnelt er am ehesten den "Mixed Mesophytic Forests" Mittelchinas und Mitteljapans sowie der Bergwaldgebiete von Südchina, Burma und Laos.

Es zeigt sich immer mehr, daß die vegetationskundlich ausgerichteten Untersuchungen fossiler Floren zu einem wesentlich besseren Verständnis der Veränderungen in der Flora und des Klimaablaufs führen. Das Wesentliche an dieser Methode ist, daß man das Klima einer fossilen Flora dadurch zu ermitteln sucht, daß man von den Klimaansprüchen rezenter Arten die der fossilen Verwandten ableitet. Mit Hilfe dieser in Europa dominierenden "Methode der nächsten lebenden Verwandten" lassen sich recht genaue Aussagen gewinnen. Der Vorteil liegt in der Einheit vegetationskundlicher, ökologischer und paläoklimatischer Aussagen. Die Methode ist jedoch nur ausreichend genau, wenn ein hoher Prozentsatz von Arten bestimmt werden kann bzw. wenn nahe lebende verwandte Arten ermittelt werden können. Sie wird daher umso ungenauer, je älter eine Flora ist. In Nordamerika, wo in grösserem Umfang Alttertiär- und Kreidefloren untersucht werden, gibt man daher häufig der "blattphysiognomischen Methode" den Vorzug. Hierbei wird von den Blattmerkmalen (Blattform, Blattgröße, Ausbildung des Blattrandes, der Blattspitze und -basis sowie Nervaturmuster) auf das Klima geschlossen. So hatte WOLFE (1971) festgestellt, daß im tropischen Regenwald über 75% der Arten ganzrandige Blätter haben. Mit abnehmender Temperatur in den Bergen sinkt dieser Anteil auf 57% und in Richtung auf die Pole gar auf 39%. Allerdings lassen sich diese an rezenten Floren gefundenen Werte nicht ohne weiteres auf fossile Floren übertragen, weil in diesen stets die lokale Vegetation überrepräsentiert ist und das Ergebnis verfälschen kann. Um dies zu umgehen, wäre es erforderlich, durch Feldstudien den

Zusammenhang zwischen rezenten Blatt-Thanatozönosen und der Vegetation zu klären. In einer Modellstudie haben DOLPH und DILCHER (1979) keine einfachen Beziehungen zwischen Blattgröße und Klima in Nord- und Süd-Carolina gefunden. Sie hoffen, künftig durch intensive Gelände-Beobachtungen bessere Kriterien zu erhalten. Die Technik haben ROTH und DILCHER bereits 1978 angekündigt: Eingehende statistische Auswertungen von Blättern aus rezenten Ablagerungen in Seen.

Aus guten Gründen basiert die Systematik des Pflanzenreiches in erster Linie auf den Reproduktionsorganen; in besonderem Maße gilt dies für die Angiospermen, deren Blütenbau Grundlage eines jeden natürlichen Systems ist. Es ist daher ein großer Mangel, daß fossile Angiospermenblüten in der Vergangenheit nur selten beschrieben worden sind. Zwar sind auch früher bereits hin und wieder einzelne Blüten erwähnt worden – erinnert sei an die vorzüglich erhaltenen Blüten aus dem eozänen Bernstein –, aber die Vergänglichkeit angiospermer Blüten ließ sie nur unter besonders günstigen Umständen zu gut untersuchbaren Fossilien werden. Dabei hatte man schon lange auf ein klärendes Wort von Seiten der Paläobotanik gewartet, das den Widerstreit der verschiedenen Theorien über die Entstehung und die Phylogenie der Angiospermen entschieden hätte. Mehrere Umstände haben dazu beigetragen, daß das Problem der frühen Angiospermen-Evolution in jüngster Zeit besonders von amerikanischer Seite verstärkt angegangen werden konnte und bereits zu einigen überraschenden neuen Ergebnissen geführt hat. Die grundlegenden Arbeiten kommen aus der Schule von DELEVORYAS, der sich zusammen mit CREPET bereits während seiner Zeit in New Haven mit Bestäubungsvorgängen bei der Bennettitee *Cycadeoidea* beschäftigt hat (DELEVORYAS 1968, CREPET 1972, 1974). Voraussetzungen für die Arbeiten über frühe Angiospermenblüten waren Fundpunkte mit günstigen Erhaltungsbedingungen in Nordamerika, die man systematisch absammeln konnte, und ständig verfeinerte Untersuchungsmethoden, vor allem der Elektronenmikroskopie. Der Erfolg ist aber wohl im wesentlichsten darauf zurückzuführen, daß man das Problem mit einer klaren biologischen Fragestellung angegangen und daß man von mehreren Seiten gleichzeitig intensiv daran gearbeitet hat. So haben DOYLE und HICKEY (DOYLE et al. 1975, DOYLE und HICKEY 1976, HICKEY 1973, 1978, HICKEY und DOYLE 1977) damit begonnen, die Schichten der Potomac-Gruppe durch stratigraphisch besser fundierte Aufsammlungen neu zu untersuchen. In diesen an der Atlantikküste gelegenen unterkretazischen Ablagerungen ist das erste Auftreten und eine stetige Zunahme von Angiospermen enthalten. HICKEY konnte zeigen, wie sich ausgehend von einfachen elliptischen Blattformen sehr rasch gelappte und zusammengesetzte entwickeln. Parallel hierzu konnte in phylogenetischen Reihen verfolgt werden (DOYLE et al. 1975, DOYLE 1980), wie sich aus einfachem unisulcatem Pollen in den unteren Lagen tricolpater, tricolporater und triporater Pollen entwickelte.

Die ersten modern untersuchten Blüten stammen freilich nicht aus der Kreide, sondern aus der mitteleozänen Clairborne-Formation von Tennessee (CREPET und DILCHER 1974, CREPET et al. 1975), wo über 100 Exemplare geborgen werden konnten. Darunter sind sowohl einfache entomogame radiärsymmetrische Blüten mit meist fünf freien Blütenblättern und zahlreichen Staubblättern als auch Kätzchen mit noch gut entwickeltem bis völlig reduziertem Perianth. Wichtig scheinen unter den Kätzchenblüten vor allem diejenigen zu sein, die wegen ihrer dreilappigen Brakteen und des triporaten Pollenbaues zu den Juglandaceen gestellt werden, und zwar in die Verwandtschaft des *Engelhardia-Oreomunnia-Alfaroa*-Komplexes, der somit bereits im Eozän so vielgestaltig wie heute gewesen sein muß und in gleichem Maße wie die lebenden Arten an Windbestäubung angepaßt war (*Eokachyra* CREPET et al. 1975, *Eoengelhardia* CREPET et al. 1980).

Andere Kätzchen (CREPET 1979) werden zu den Fagaceen (s.a. CREPET und DAGHLIAN 1980) gestellt bzw. mit *Betula* oder *Myrica* in Verbindung gebracht und unterscheiden sich ebenfalls nur wenig von den lebenden Verwandten. Auch einige windblütige Einzelblüten mit kurzer Blütenhülle und lang heraushängenden Staubfäden sind bekannt. So liegt die Schlußfolgerung nahe, daß bereits im Mitteleozän die Anpassung von Blüten und Infloreszenzen an Windblütigkeit im Prinzip so vollkommen war wie bei heutigen Pflanzen.

Die entomogamen Blüten sind überwiegend radiärsymmetrisch und lassen sich auf Grund von bestimmten Merkmalen bereits verschiedenen Bestäubergruppen zuordnen: Große (bis 6 cm Ø!), schüsselförmige Blüten (ähnlich Parietales, Dilleniaceae) wurden vermutlich von Käfern besucht, und kleine offene Blüten sowie Araceen-Spadices deuten auf Bestäubung durch kleine Fliegen hin. Eine bilaterale Blüte mit vermutlich nur zwei Staubgefäßen ist der früheste sichere Nachweis einer melittophilen Blüte. Andere mit glockenförmiger Korolle sowie Mimosoideenähren (CREPET und DILCHER 1977; vgl. auch DAGHLIAN et al. 1980: *Eomimosoidea* aus dem Oligozän von Texas) sind vermutlich ebenfalls von Hymenopteren bestäubt worden. Schmetterlingsblütig war schließlich vermutlich eine unterständige Blüte mit langer Blütenröhre.

Die Übersicht zeigt, daß im Mitteleozän auch die Insektenblütigkeit bereits hoch entwickelt war, allerdings noch nicht ganz den gegenwärtigen Stand erreicht hatte. So fehlen offenbar Nachtfalter-Blüten und Ornithophilie, wie auch Zygomorphie erst in Anfängen erkennbar ist. Besonders hochentwickelte Familien wie Compositen und Orchideen scheinen ebenfalls zu fehlen. Es hat den Anschein, als ob die Windblütigkeit im Mitteleozän vollkommener entwickelt war als die Insektenblütigkeit. Dies wird damit erklärt, daß die Anpassung an vorhandene physikalische Faktoren, wie sie bei Windblütigkeit erforderlich ist, schneller vor sich gehen konnte als die auf Wechselwirkung angewiesene Koevolution von Blüte und tierischem Bestäuber.

Es war nur folgerichtig, nun nach noch älteren Angiospermenblüten zu suchen. DILCHER (1979) gibt einen ersten Überblick über den derzeitigen Stand der Untersuchungen von Blüten aus der Kreide. Als Hinweise auf blütenhöffige Ablagerungen dienten ihm ältere Beschreibungen, wie FONTAINE (1889), der in der Potomac-Gruppe zwei Blüten gefunden hatte, an denen drei bis acht radiär stehende Karpelle zu erkennen waren. Weitere Blütenteile fehlen oder sind nicht erhalten. Neuere Aufsammlungen scheinen in der Potomac-Gruppe nicht gemacht worden zu sein.

Dagegen war DILCHER mit seinen Mitarbeitern (DILCHER et al. 1976, DILCHER 1979) in dem etwas jüngeren Janssen Clay aus der Dakota-Formation von Zentral-Kansas (unteres Cenoman = unterste Oberkreide) und in einigen Fällen dem gleichalten Amboy Clay von New Jersey erfolgreicher. Von hier hatte bereits LESQUEREUX (1874, 1892) Früchte und Fruchtstände beschrieben. Neue Aufsammlungen brachten eine langgestreckte Achse, an der spiralig eine große Anzahl hülsenförmiger Karpelle steht, die vermutlich eine Reihe Samen enthielten. Über Staubblätter und Blütenhülle ist nichts bekannt. Es wurden jedoch mit den Karpellen zusammen Blätter (*Liriophyllum*) gefunden, die Ähnlichkeit mit *Liriodendron* haben und die auf Grund des Vorkommens von gleichen Sekretkörpern zusammengehören könnten. So ist der Schluß berechtigt, daß es sich bei der Pflanze um eine altertümliche Magnoliale oder Hamamelidale handelt. Daneben kommen eine Reihe radiärsymmetische Blüten vor. Besonders zahlreich sind fünfzählige tetrazyklische Blüten: fünf Kelchblätter alternierend mit fünf Blütenblattzipfeln sitzen am Rande eines flach schüsselförmigen Kelches. Die fünf Staubblätter, stehend opponiert vor den Blütenblättern, besitzen vier Pollensäcke und enthalten tricolpaten Pollen. Das Gynäzeum ist ebenfalls fünfzählig. Außer fünfzähligen kommen auch vier- und sechszählige Blüten vor. Neben diesen entomogamen Blüten treten verschiedene Formen von Kätzchen auf, die Früchte oder Pollen enthalten können. Eine Form ähnelt Platanenkätzchen, eine andere besteht aus 100 - 120 vierzipfligen trichterförmigen Blütchen. Wenn man davon ausgeht, daß die Angiospermen in der Unterkreide oder nur kurze Zeit vorher entstanden sind, so haben wir in der Blütenflora von Kansas eine recht frühe Flora vor uns, die durch die Vielfalt der Blütenformen überrascht. Auffällig ist, daß magnoliale Blüten keine bedeutende Rolle spielen und daß windblütige sehr stark vertreten sind. DILCHER (1979) kommt daher zu dem Schluß, daß der Ranales-Komplex nicht die einzige primitive Gruppe ist, sondern höchstens eine von mehreren Linien darstellt und daß die als reduziert angesehenen Blüten der Trochodendrales, Cercidiphyllales, Eupteleales, Hamamelidales und Piperales als ursprünglich und primär als eingeschlechtig anzusehen sind. Er nimmt weiter an, daß sich ein Teil der anemogamen Blüten gemeinsam mit entomogamen aus einer Wurzel herleiten, die eingeschlechtige Blüten besaß. Die Entwicklung und Aufspaltung muß dann während der Unterkreide auch im Blütenbereich sehr rasch gegangen sein, wie wir es bereits von den Blättern und dem Pollen her kennen.

Als primitiv nimmt er folgende Merkmale an: (1) geringe Anzahl von Blütenteilen; (2) freie Blütenteile; (3) Radialsymmetrie; (4) sowohl ein- wie zweigeschlechtige Blüten. Komplizierte Anordnung, große und zahlreiche Blütenteile werden dagegen als abgeleitet angesehen.

Es ist abzusehen, daß durch diese gerade erst begonnenen Untersuchungen die frühe Evolution der Angiospermen in völlig neuem Licht erscheint und daß die Konzepte für "natürliche Systeme" überarbeitet werden müssen. Auch zur Beantwortung der Frage "Woher kommen die Angiospermen?" sind Antworten zu erwarten. Über die hiermit verbundenen Probleme gibt DILCHER (1979) einen Überblick. Als Ursprung der Angiospermen bieten sich die mesozoischen Pflanzengruppen an, die sich von paläozoischen Pteridospermen ableiten lassen, wie die Bennettiteen, Caytoniales und andere mesozoische "Pteridospermen", bei denen wir eine große Vielfalt von Pollen- und Samenorganen finden. Es käme jetzt darauf an, weitere blütenmorphologische Vergleiche zwischen frühen Angiospermen und mesozoischen Gymnospermen anzustellen. Zwei Tendenzen zeichnen sich bereits ab: Die Neigung, den Samen zu bedecken, ist nicht auf Bennettiteen und Caytoniales beschränkt, sondern ist auch bei den mesozoischen Pteridospermen *Leptostrobus*, *Umkomasia* und *Glossopteris* sowie sogar bereits bei manchen paläozoischen Pteridospermen in Form von Kupulen erkennbar und reicht bis ins Unterkarbon und Oberdevon zurück (TAYLOR und MILLAY 1979).

Auch DILCHER geht davon aus, daß die Bestäubung durch Insekten – anfangs sicher durch Pollenfresser – nicht auf Angiospermen beschränkt war, sondern auch bei verschiedenen Gruppen von Gymnospermen bis zurück ins Oberkarbon eine Rolle gespielt hat.

Literatur

ALVAREZ RAMIS C., DOUBINGER, J., GERMER, R.: Palaeontographica B 170, 126-150 (1979). - ASAMA, K., NAITO, G.: Bull. Nation. Sci. Mus. C 4, 89-98 (1979).

BOERSMA, M., BROEKMEYER, L.M.: Index of figured plant megafossils Carboniferous 1971-1975: Spec. Publ. Labor. Palaeobot. Palyn. 3, 1-183 (1979); - Index of figured plant megafossils Triassic 1971-1975: Spec. Publ. Labor. Palaeobot. Palyn. 2, 1-70 (1980a); - Cour. Forsch.-Inst.

Senckenberg 42, 3-26 (1980b); - Index of figured plant megafossils Permian 1971-1975: Spec. Publ. Labor. Palaeobot. Palyn. 3, 1-133 (1981). - BOUREAU, E., LEJAL-NICOL, A., MASSA, D.: C. R. Hebdom. Séances Acad. Sci. 286 D, 1567-1571 (1978). - BRACK-HANES, S.D.: Bot. Gaz. 139, 140-146 (1978a); - Science 200, 1383-1385 (1978b). - BRONGNIART, A.: Ann. Sci. Nat. 15, 435-460 (1928).

CARRUTHERS, W.: Month. Microsc. J. 9, 160-172 (1872). - CASPARY, R.: Schr. königl. physik.-ökol. Ges. Königsberg 27, 1-9 (1887); - Abh. königl. preuß. Landesanst., N.F. 4, 1-182 (1907). - CHALONER, W.G., SHEERIN, A.M.: Report on British Palaeobotany and Palynology 1978-1979 (1980); - Spec. Pap. Palaeont. 23, 145-161 (1979). - CREPET, W.L.: Am. J. Bot. 59, 1048-1056 (1972); - Palaeontographica B 148, 144-169 (1974); - Rev. Palaeobot. Palynol. 27, 213-238 (1979). - CREPET, W.L., DAGHLIAN, C.P.: Am. J. Bot. 67, 739-757 (1980). - CREPET, W.L., DILCHER, D.L.: Am. J. Bot. 64, 714-725 (1977). - CREPET, W.L., DAGHLIAN, C.P., ZAVADA, M.: Rev. Palaeobot. Palynol. 27, 361-370 (1979). - CREPET, W.L., DILCHER, D.L., POTTER, F.W.: Science 185, 781-782 (1974); - Am. J. Bot. 62, 813-823 (1975).

DABER, R.: Geologie 12, 683-699 (1963); - Schriftenr. Geol. Wiss. 16, 7-13 (1980). - DAGHLIAN, C.P., CREPET, W.L., DELEVORYAS, T.: Am. J. Bot. 67, 309-320 (1980). - DAWSON, J.W.: Proc. Am. Ass. Adv. Sci. 10 (II. Nat. Hist.), 174-176 (1857); - Quart. J. Geol. Soc. London 15, 477-488 (1859). - DELEVORYAS, T.: Palaeontographica B 121, 122-133 (1968). - DELEVORYAS, T., MORGAN, J.: Palaeontographica B 96, 12-23 (1954). - DILCHER, D.L.: Rev. Palaeobot. Palynol. 27, 291-328 (1979). - DILCHER, D.L., BASINGER, J.F.: Abstr. Int. Palaeobot. Conf. Reading, 11 (1980). - DILCHER, D.L., CREPET, W.L., BEEKER, C.D.: Science 191, 854-856 (1976). - DOLPH, G.E.: IAAP Misc. Publ. 1, 1-45 (1979); - IAAP Misc. Publ. 2, 1-45 (1980). - DOLPH, G.E., DILCHER, D.L.: Palaeontographica B 170, 151-172 (1979). - DOUBINGER, J., GERMER, R.: Palaeontographica B 175, 1-26 (1980). - DOYLE, J.A.: Abstr. Int. Palaeobot. Conf. Reading, 14, (1980). - DOYLE, J.A., HICKEY, L.J.: 139-206 in: The Origin and Early Evolution of the Angiosperms, 1976. - DOYLE, J.A., VAN CAMPO, M., JUGARDON, B.: Pollen Spores 17, 429-486 (1975).

EDWARDS, D.: Palaeontology 22, 23-52 (1979). - EDWARDS, D., BASSETT, M.G., ROGERSON, E.C.W.: Lethaia 22, 313-324 (1979). - EDWARDS, D.S.: Rev. Palaeobot. Palynol. 29, 177-188 (1980).

FAIRON-DEMARET, M.: Rev. Palaeobot. Palynol. 28, 145-160 (1979); - Abstr. Int. Palaeobot. Conf. Reading, 17 (1980a); - Rev. Palaeobot. Palynol. 29, 201-220 (1980b). - FONTAINE, W.M.: U. S. Geol. Surv. Monogr. 15 (1), 375 S. (1889).

GALTIER, J., SCOTT, A.C.: Palaeontographica B 170, 81-125 (1979). - GARRATT, M.J.: Abstr. 2nd Geol. Conf. Geol. Soc. Austr., Monash, 14 (1977); - Alcheringa 2, 217-224 (1979). - GENSEL, P.G.: Rev. Palaeobot. Palynol. 22, 19-37 (1976); - Palaeontographica B 168, 81-99 (1979); - Rev. Palaeobot. Palynol. 30, 101-132 (1980). - GOCHT, H.: N. Jb. Geol. Paläont., Abh., 157, 344-365 (1979). - GRAUVOGEL-STAMM, L.: Sci. Géol. Mém. 50, 1-225 (1978). - GRAUVOGEL-STAMM, L., GRAUVOGEL, L.: Sci. Géol. Bull. 33, 53-66 (1980). - GRAUVOGEL-STAMM, L., SCHAARSCHMIDT, F.: Sci. Géol. Bull. 31 (2), 105-107 (1978); - Senckenbergiana Lethaea 60, 1-37 (1979). - GREGOR, H.-J.: Diss. Univ. München, 249 S. (1975); - Palaeontographica B 167, 8-103 (1978); - Rev. Palaeobot. Palynol. 28, 311-363 (1979); - Palaeontographica B 174, 7-94 (1980). - GROLLE, R.: Feddes Rep. 91, 183-190, 401-407 (1980).

HICKEY, L.J.: Am. J. Bot. 60, 17-33 (1973); - Cour. Forsch.-Inst. Senckenberg 30, 27-34 (1978). - HICKEY, L.J., DOYLE, J.A.: Bot. Rev. 43, 3-104 (1977). - HOLMES, J.C.: Ann. Bot. 44, 113-119 (1979a); - Can. J. Bot. 57, 1518-1527 (1979b). - HUEBER, F.M.: Taxon 20, 641-642 (1971).

JÄGER, H.: Lethaia 3, 173-182 (1970). - JENNINGS, J.R., MILLAY, M.A.: Palaeontology 22, 913-920 (1979). - JONKER, F.P.: Palaeontographica B 171, 39-56 (1979). - JUNG, W.: Ber. Bayer. Bot. Ges. München 43, 97-108 (1972).

KIMURA, T.: Proc. Japan Acad. 55 (B), 425-430 (1979). - KNOBLOCH, E., KVAČEK, Z.: Rozpr. Ústředn. Úst. Geol. 42, 1-131 (1976). - KRÄUSEL, R.: Fortschr. Geol. Rheinld. Westf. 12, 25-38 (1964). - KRÄUSEL, R., WEYLAND, H.: Abh. preuß. geol. Landesanst., N. F. 131, 1-92 (1930); - Senckenbergiana Lethaea 14, 391-403 (1932); - Palaeontographica B 79, 131-142 (1934); - Palaeontographica B 80, 171-190 (1935).

LANG, W.H.: Philos. Trans. R. Soc. London Ser. B 227, 245-291 (1937); - J. Linn. Soc. Bot. 52, 535-552 (1945). - LANG, W.H., COOKSON, C.: Philos. Transact. R. Soc. London Ser. B 224, 421-449 (1935). - LEISMAN, G.A., PHILLIPS, T.L.: Palaeontographica B 168, 100-128 (1979). - LEJAL-NICOL, A.: Rapport sur la Paléobotanique et la Paléopalynologie, France, Belgique, Suisse 1976-1978, 87 S. (1979) (Labor. de Paléobot.). - LEJAL-NICOL, A., MASSA, D.: Rev. Palaeobot. Palynol. 29, 221-239 (1980). - LEMOIGNE, Y.: Bull. Soc. Linn. Lyon 4, 94-102 (1969); - Int. Palaeobot. Conf. Reading, 32 (1980). - LESQUEREUX, L.: Rep. U. S. Geol. Surv. Terr., 136 S. (1874); - Rep. U. S. Geol. Surv. Monogr. 17, 400 S. (1892).

MÄDLER, K., STAESCHE, U.: Geol. Jahrb. B 33, 81-157 (1979). - MÄDLER, K., STEFFENS, P.: Geol. Jahrb. B 33, 3-33 (1979). - MAI, D.H., WALTHER, H.: Abh. Staatl. Mus. Mineral. Geol. Dresden 28, 1-101 (1978). - MATTEN, L.C., LACEY, W.S., LUCAS, R.C.: Bot. J. Linnean Soc. 81, 249-273 (1980). - MEYEN, S.V.: Rev. Palaeobot. Palynol. 28, 191-201 (1979). - MILLAY, M.A.: Palaeontographica B 169, 1-69 (1979). - MILLAY, M.A., TAYLOR, T.N.: Bot. Rev. 45, 301-375 (1979); - Am. J. Bot. 67, 758-773 (1980). - MUSTAFA, H.: Argumenta Palaeobot. 4, 101-133 (1975).

NEUBER, E.: Neues Jahrb. Geol. Paläontol. Monatsh. 1979 (11), 681-689 (1979). - NIKLAS, K.J.: Trans. R. Soc. Edinburgh 69, 483-499 (1975/76); - Rev. Palaeobot. Palynol. 22, 1-17 (1976).

PENHALLOW, D.P.: Trans. R. Soc. Can., Sect. IV 1888, 27-47 (1888). - PHILLIPS, T.L.: Rev. Palaeobot. Palynol. 27, 239-289 (1979).

RAYNOR, R.: Int. Palaeobot. Conf. Reading 45 (1980). - REMY, W.: Argumenta Palaeobot. 6, 139-155 (1980). - REMY, W., REMY, R.: Argumenta Palaeobot. 6, 37-72 (1980a); - Science 208, 295-296 (1980b). - REMY, W., REMY, R., HASS, H., SCHULTKA, S., FRANZMEYER, F.: Argumenta Palaeobot. 6, 73-94 (1980). - ROTH, J.L., DILCHER, D.L.: Cour. Forsch.-Inst. Senckenberg 30, 165-171 (1978). - ROTHWELL, G.W.: Palaeontographica B 151, 171-196 (1975); - Palaeontographica B 173, 85-106 (1980). - ROTHWELL, G.W., BASINGER, J.F.: Can. J. Bot. 57, 958-970 (1979).

SCHAARSCHMIDT, F.: Cour. Forsch.-Inst. Senckenberg 1, 17-19 (1973); - Paläont. Z 48, 188-204 (1974); - Cour. Forsch.-Inst. Senckenberg 17, 90 (1976). - SCHENK, A. in ZITTEL, K.A.: Traité de Paléontologie. 2. Paléophytologie: 224-896. R. Oldenbourg, München u. Leipzig (1891). - SCHIMPER, W.P., MOUGEOT, A.: Monographie des plantes fossiles du Grès bigarré de la chaine des Vosges: 1-83. G. Engelmann, Leipzig (1844). - SCHMIDT, W.: Palaeontographica B 104, 1-38 (1958). - SCHWEITZER, H.-J.:

Bonner paläobot. Mitt. 3, 1-32 (1979); - ibid. 5, 1-38 (1980a); - ibid. 6, 1-34 (1980b); - ibid. 7, 1-29 (1980c). - SCHWEITZER, H.-J., GIESEN, P.: Palaeontographica B 173, 1-25 (1980). - SOULIÉ-MÄRSCHE, J.: Bull. Centre Rech. Explor.-Prod. Elf-Aquitaine 3, 821-831 (1979). - STIDD, B.M., BARTHEL, M.: Bot. Soc. Am. Miscell. Ser., Publ. 157, 38 (1979). - STOCKEY, R.A.: Palaeontology 23, 657-666 (1980a); - Bot. Gaz. 141, 116-124 (1980b). - STORCH, D.: Schriftenr. Geol. Wiss. 16, 171-273 (1980). - STÜRMER, W., SCHAARSCHMIDT, F.: Kleine Senckenberg-Reihe 11, 19-25 (1980a); - Nat. Mus. 110 (5), 19-25 (1980b). - STURM, M.: Cour. Forsch.-Inst. Senckenberg 37, 3-25 (1979).

TAYLOR, T.N., MILLAY, M.A.: Rev. Palaeobot. Palynol. 27, 329-355 (1979). - TAYLOR, T.N., DILCHER, D.L., DELEVORYAS, T.: Rev. Palaeobot. Palynol. 27, 211-212 (1979). - THIELE-PFEIFFER, H.: Diss. Univ. München, 268 S. (1979); - Palaeontographica B 174, 95-224 (1980).

WATT, A.D.: Bibliography of American Paleobotany for 1978, Bot. Soc. Am., 97 S. (1979); - Bibliography of American Paleobotany for 1979, Bot. Soc. Am., 86 S. (1980). - WEBER, R.: Bibliografias paleobotanicas, Inst. Geol. Univ. nacion. auton. Mexico, 68 S. (1979); - Bibliografias paleobotanicas, Inst. Geol. Univ. nacion. auton. Mexico, Supl. 1, 39 S. (1980). - WEYLAND, H., BERENDT, W.: Palaeontographica B 122, 171-182 (1968). - WOLFE, J.A.: U. S. Geol. Surv. Open-file Rep., 1-114 (1969); - Palaeogeogr., Palaeoclimatol., Palaeoecol. 9, 27-57 (1971).

Dr. FRIEDEMANN SCHAARSCHMIDT
Forschungsinstitut Senckenberg
Senckenberganlage 25
D 6000 Frankfurt 1

E. Geobotany*

I. The History of Flora and Vegetation During the Quaternary

By BURKHARD FRENZEL

The present report deals with recent Sovietic investigations. Only those of general importance could be reviewed.

1. Pleistocene Stratigraphy

One of the prerequisites of each paleobotanical and paleoecological work is the knowledge of the Earth's history during the Quaternary and the recognition of synchronous events all over the world. As has been stressed earlier, our opinions about the division of the Quaternary are changing rapidly, complicating any paleobotanical work. During the last years most interesting observations on the sequence of fossil soils and intercalated loess layers have been reported from Tadzhikistan and the adjacent territories. In contrast to 11 to 12 interglacials hitherto known, DODONOV et al. (1978) report on 20 soil complexes of Pleistocene age within the up to 200 m thick loess layers there, each of them consisting of one to three soils, most of them indicating past interglacial conditions. These observations are corroborated by DODONOV and LOMOV (1980), LAZARENKO et al. (1980), and by DODONOV and PEN'KOV (1977) as well. The oldest loesses have ages of about more than $1.79 \cdot 10^6$ years. They were formed under extremely arid conditions, whereas most of the fossil soils indicate relatively moist subtropical conditions. The tenth fossil soil complex was formed immediately before the Brunhes/Matuyama boundary, i.e., just before 690,000 B.P. The fifth soil complex, consisting of two widely spread and very distinct soils, has an age of between 150,000 ± 21,000 and 125,000 ± 29,000 years. On top of this a layer of reversed magnetization occurs. This is held by DODONOV et al. (1978) to be the Blake event at the end of the last European interglacial. One of the next soil complexes has an age of about 110,000 ± 22,000 years (thermoluminescence dates, LAZARENKO et al., 1980). These dates parallel those of the Karangat-transgression of the Black Sea. Formerly it was held that only one Karangat-transgression had occurred, the sediments of which are characterized by a specific stenohaline marine mollusk fauna. It could be shown now (ARSLANOV et al., 1976) that on the coasts of western Caucasia and the Kerchenskiy proliv three marine terraces with this faunal assemblage are met with, each being connected with river terraces of its own. According to U/Th-datations their ages range from 118,000 ± 3,500 to 139,000 ± 4,000 for the oldest of these Karangat terraces, from 74,000 ± 3,000 to 96,000 ± 5,000 for the second (the classical Karangat) terrace and from middle Würmian time for the youngest Karangat terrace. Possibly only this last-named terrace and its corresponding sediments were investigated by SEMENENKO et al. (1979) in the Sivash area, provided that the [^{14}C]-ages of 42,000 to 27,000 are reliable. ARSLANOV et al. (1976) suggest that the second of the Karangat terraces corresponds to the European last interglacial (Mikulino), the first one being presumably the equivalent of the Odintsovo

*In Memoriam of JOSIAS BRAUN-BLANQUET (1884-1980) and REINHOLD TÜXEN (1899-1980).

interglacial. In view of the serious doubts as to the real stratigraphic position of this last-named interglacial (see below) it may be questioned whether this correlation is the correct one. As to Tadzhikistan and its surroundings, still other absolute dates are available for pronounced fossil soils: soil complex V or VI: about 200,000 years, soil complex VII or VI about 250,000 years (LAZARENKO and RANOV, 1977; DODONOV et al., 1978). PEVZNER and PÉCSI (1980) report on the sequence of fossil soils and loesses in the Hungarian plain, describing various types, faunas, and paleomagnetic setting of several sites. Here the division of the last glaciation by interstadial and stadial phases strongly resembles that of Tadzhikistan, yet it differs somewhat from that of the Central Russian plain. In all these investigations paleomagnetic dates have proved to be a good tool for stratigraphic work. They are also used by PEN'KOV et al. (1976) for their research work on the Pleistocene of the Pamir Mts. It could be shown that sediments of the oldest glaciation there seem to date from about more than $1.8 \cdot 10^6$ or even from $3.0 \cdot 10^6$ years B.P. For observations on the earliest glaciation of Iceland of roughly the same age, see GLADENKOV and GURARIY (1976).

It is in general held that the glacier tongues, traveling down the valleys of the rivers Dnepr and Don, are of the same age, i.e., that they are the equivalents of the Saalian glaciation. Yet it could be shown now by means of paleontological, geological, and paleomagnetic work that this view does not seem to be correct, the Don lobe being older than the Dnepr lobe (VELICHKO et al., 1977: Oka glaciation; KRASNENKOV et al., 1980: older than the Oka glaciation). So moraines and their correlative sediments of the former Dneprovsk glaciation can no longer be used in the Southern and Central Ukraine as synchronous reference layers. The different views as to the ages of the Don-lobe (Oka or Don glaciation, respectively) may have been caused by the fact that only very recently still another moraine could be observed between the classical Likhvin (presumably the equivalent of the Holsteinian) interglacial and the underlying moraine, which up till now was held to be of Oka age (SUDAKOVA and BOL'SHAKOV, 1977; KRASNENKOV et al., 1980), testifying to an up till now unknown strong glacier advance just before the Likhvin interglacial proper. This glaciation is interesting also from another point of view: in Russia the widely accepted view exists that between the Likhvin (= Holstein) and the Mikulino (= Eem) interglacials still another one had existed, the Odintsovo interglacial which is held to have been composed of two warm climate periods and one intercalating cold climate phase. BRESLAV et al. (1979) and PISAREVA et al. (1979) now stress that the youngest of the two warm climate phases in reality had happened before the classical Likhvin interglacial (see also KRASNENKOV and PISAREVA, 1977, and KRASNENKOV et al., 1980). VELICHKEVICH (1979) stresses that a rich macrofossil-flora, found at the very bottom of the interglacial sediments of the Likhvin site, does not indicate a true interglacial, but some colder climate period only. So the view of the Roslavl optimum of the Odintsovo interglacial preceding the Likhvin interglacial could not be corroborated. But considering the fact that only very recently yet another moraine could be found there (see above) this controversial point might be explained quite easily.

The stratigraphy of the quaternary of Eastern Siberia has been investigated by VIRINA and SVITOCH (1976: paleomagnetism), and also by ALESHINSKAYA et al. (1976), using paleobotanical observations including diatom analyses and [^{14}C]-datations. ORESHKINA (1980) dealt with the marine diatom floras of the Kamchatka peninsula, from the middle Miocene onwards. One of the crucial points is the so-called Karginsk interglacial or interstadial, which is held to have occurred, according to several [^{14}C]-datations, between 50,000 and 25,000 B.P. At that time the Arctic Ocean is believed to have experienced a strong transgression on to Northern Siberia, the climate having been at least as warm as today or even warmer (SVITOCH, 1977). Moreover, BARDEEVA and ISAEVA (1980) stress that during the Karginsk interglacial, climate seems to have been even a little more favorable for the forest vegetation of Northern Siberia than during the preceding Kazantsev interglacial, which is held to be the equivalent of the Mikulino (= Eem) interglacial. Future investigations, using other methods of absolute datations, may throw more light on this difficult problem.

2. Vegetation History

a) Caucasia

The Pleistocene history of this most interesting region becomes much clearer now than it has been hitherto. In so-called Old-Euxinian sediments (beginning of the Quaternary) a very rich pollen flora was observed (CHOCHIEVA and MAMATSASHVILI, 1977), comprising some 70 taxa of cryptogams and phanerogams. It seems that already at that time all the recent vegetation belts had developed, though their floristic composition differed from that of today. In the prevailing forests *Pinus*, *Abies*, *Taxodium*, Taxodiaceae, *Sequoia*, *Fagus*, *Castanea*, and *Alnus* dominated at lower elevations, with some *Podocarpus*, *Tsuga*, *Cedrus*, *Cryptomeria*, *Carya*, *Engelhardtia*, and *Liquidambar*. This flora and vegetation changed slowly into that of the Uzunlar period (CHOCHIEVA and MAMATSASHVILI, 1979). At that time the dense forests were most of all governed by *Pinus*, *Fagus*, and *Castanea sativa*, with some admixture of *Abies nordmanniana*, *Taxodium*, *Carpinus caucasica*, *C. orientalis*, *Quercus*, *Ulmus*, *Corylus*, *Zelkova*, *Juglans*, *Pterocarya pterocarpa*, and *Engelhardtia*. The still high amount of exotic plants is remarkable (besides some of those just mentioned: *Podocarpus*, *Tsuga*, *Sequoia*, *Taxodium*, *Glyptostrobus*, *Carya*, *Laurus*, *Magnolia*, *Parrotia*), though their percentage had diminished strongly. This change is explained by changes of climate and by diseases as well. The forest vegetation of the last interglacial or of the beginning of the last glacial period seems to have been governed there most of all by *Carpinus*, *Castanea*, *Fagus*, *Quercus*, and *Tilia*, most of the older exotic plants being already missing (GABUNIYA et al., 1977). Within the contemporaneous sediments a very rich theriofauna was observed. The ensuing geological events are described by TSERETELI and MAYSURADZE (1980), comprising the Karangat-transgression (see above). Since between the first and the second Karangat-transgressions (ARSLANOV et al., 1976) at the mouths of the Caucasian rivers thick clastic sediments were deposited, the vegetation during the cold climate period must have changed seriously. This is corroborated by TSERETELI and MAYSURADZE (1980) for the pleniglacial periods of the last glaciation in Gruziya. Here periglacial phenomena could be observed even at low elevations, and flora and fauna as well point to a lowering of all the vegetation belts at that time of about 1200 to 1400 m against present-day conditions. The synchronous diminution in the amount of human remains is interpreted as indicating the severe cold and dry climate, when even *Rangifer*, *Rhinoceros tichorhinus*, *Alces*, and *Mammuthus primigenius* had lived in Grusinia, Imeretia, and Abkhasia. For the occurrence of loesses on the northern slopes of the Caucasus Mts., see KRIGER and BOTNIKOV (1979); DUBROVO et al. (1979) report on the herb-steppe flora, rich in Compositae, including *Artemisia*, *Ephedra*, Plumbaginaceae, Caryophyllaceae, and Chenopodiaceae, in which *Mammuthus trogontherii* had browsed during the preceding glacial period in the area of Krasnodar. MARUASHVILI (1975, 1977) tries to reconstruct the history of climate in Western Grusiya by means of cave sediments beginning with the end of the last interglacial. The youngest history of the subalpine vegetation at the upper timberline is investigated in Central Balkaria by SEREBRYANNIY et al. (1980). Here during the last 7000 years the composition of the forest vegetation does not seem to have changed seriously. It is thought that man-made changes in the vegetation only date from about 1840 ± 70 B.P., with strongest clearances beginning at about 840 ± 60 B.P., though it seems to me that the anthropogenic factor can be discerned even at 3890 ± 70 B.P., when the upper timberline had shifted downwards, paralleled by an increase in the long-distance transport of *Castanea*.

b) The Black Sea

The transition from Old-Euxinian to Uzunlar sediments just mentioned seems to have happened near the Bosporus according to investigations of Coccolithophoridae, diatoms, and pollen (SHUMENKO and USHAKOVA, 1980) at about 580,000 B.P. It is in general held that during the upper Pleistocene Mediterranean waters could only spread very recently into the Black Sea, thanks to the late Pleistocene and Flandrian eustatic rise of the oceans of the world. Yet SHCHERBAKOV et al. (1979) stress that already during full glacial times, when the level of the Black Sea was lowered by about 80 to 90 m, in the very vicinity of the Bosporus layers of halophilic and halophobic diatoms repeatedly followed each other, thus indicating at that time oscillations in the contributions of Mediterranean waters to the Black Sea and vice versa. The very retarded influx of the bulk of Mediterranean waters into the Black Sea basin during postglacial times is only explained by a strong discharge of freshwater (melt water) via the big Ukrainian rivers into the Black Sea. It is thought that the phase of strongest bioproduction of the Black Sea (7000 to 6000 B.P.) corresponds in time to comparable phases in the Mediterranean and Atlantic Oceans as well. On the history of the shelf to the south of Odessa: SHCHERBAKOV et al. (1979); the geological history of the central part of the Black Sea: KOGAN et al. (1977); the geological history of the Kerchenskiy proliv: BARG et al. (1979); SEMENENKO et al. (1979).

c) The Caspian Sea

Up till now the exact correlation between the transgressions and regressions of the Caspian Sea and glacial and interglacial phases on the continent is to some extent still problematical. So some absolute dates from Pleistocene sediments of the Caspian Sea are welcome. According to SEMENENKO and PEVZNER (1979) the Akchagyl comprises the Kuyal'nik (3,300,000 to 2,420,000 B.P.) and the ensuing Taman (to 1,950,000 B.P.). After a short period of transition the Apsheron follows, beginning at about 1,760,000, and ending at 700,000 B.P. Loess-like sediments seem to date in the Ponto-Caspian area from about 900,000 B.P. This would mean that they are younger than in the vicinity of Vienna (?). The lower Chazar transgression has an age of more than 300,000 years (ARSLANOV et al., 1978), whereas the Upper Chazar transgression could be dated by the same authors to between 76,000 ± 4,000 and 81,000 ± 2,000 (perhaps even 109,000 ± 5,000). Provided that these dates are correct, they would indicate a post-Eemian age of the Upper Chazar transgression! Whereas the Lower Khvalyn-transgression could not be dated, the Upper Khvalyn-transgression embraces the time from 18,000 to 14,000 B.P. It seems to have been terminated by the strong amelioration of climate at about the Alleröd interstadial. The highest shore line of the Caspian Sea during the New-Caspian transgression was formed at about 6,000 to 6,500 B.P. The ages given here could not be corroborated by BAKINOVA et al. (1978), using [^{14}C]-dates only. The geological events at the beginning and during the above-mentioned Akchagyl are discussed by TABOYAKOVA (1981), TAMRAZYAN and OVNATOV (1975), MASLYAEV (1979). SUPRUNOVA and VRONSKIY (1977) report on the coastal vegetation of the Caspian Sea during the Baku, Chazar, and Khvalyn transgressions, respectively. The vegetation of the coastal plains and of the surrounding steppe to desert vegetation had nearly always governed the pollen rain. At least beginning with the Khazar transgressions, *Ephedra* seems to have been a remarkable element. This is important in so far as in general pollen of *Ephedra* had not been indicated much earlier by Soviet botanists (see also below). The climatic conditions during the Khvalyn transgressions must have been of extreme severity, causing there a very open steppe to desert vegetation.

d) The European Part of the USSR

The problem of the stratigraphical position of the Roslavloptimum of the Odintsovo interglacial has already been discussed. According to KRASNENKOV and PISAREVA (1977) this interglacial was preceded in the Volgograd Oblast' by an interstadial, the forest vegetation of which has been characterized even there by *Pinus silvestris*, *P. sibirica*, *Picea* sp., and by *Abies sibirica*, together with Ericaceae and *Sphagnum*. This interstadial was preceded and followed by phases of an astonishingly rich steppe vegetation, provided that the palynological determination of the various chenopodiaceae and *Artemisia* species are reliable. The ensuing interglacial was characterized by a forest-steppe-like vegetation, the forest vegetation of which was dominated by various oak species, *Zelkova*, *Celtis*, *Acer*, various *Ulmus* and *Tilia* species, *Corylus avellana*, and *C. colurna*, etc. The relatively rich steppe vegetation was dominated most of all by several Chenopodiaceae species, together with *Ephedra*. The formerly held view of a twofold climatic optimum could not be corroborated. In the vicinity of Moscow, in the Odintsovo-stratotype area, the composition of the forests strongly resembled that of the site just described. Even here *Zelkova* was of some importance. The interglacial seems to have ended with a pine-spruce-birch-larch phase (BRESLAV et al., 1979). PISAREVA et al. (1979) report on a comparable forest vegetation at Balashicha in the vicinity of Moscow. Here the preceding interstadial was dominated most of all by larch. The interglacial vegetation proper must have been extremely rich in *Quercus*, *Ulmus*, and *Tilia*, with *Carpinus* and *Abies* lacking, whereas even during the interglacial *Pinus haploxylon*, *Larix*, and *Picea* Sect. *Omorica* seem to have lived there. UKRAINTSEVA (1977) reports on pollen finds of *Cistus tauricus* Presl. from the Odintsovo interglacial in the surroundings of Moscow. An astonishingly rich macrofossil flora from beneath the Oka moraine at the Likhvin site has been analyzed by VELICHKEVICH (1979), this flora pointing much more to interstadial than to interglacial conditions. The same author reports on a rich macrofossil-flora from Ruba in the Vitebsk Oblast'. The amount of at least 13 now exotic or even extinct species is worth mention. The new species *Aracites interglacialicus* was described (VELICHKEVICH, 1977). The contemporaneous rodent-fauna is described at some length within the river Don area by KRASNENKOV et al. (1980; see also VELICHKO et al., 1977; for pedogenesis and the clay minerals of the Likhvin stratotype see BOLIKHOVSKIY, 1975).

Thanks to investigations by VELICHKEVICH (1978) of the macrofossil floras of the Mikulino interglacial in the Central part of the USSR, up to 90 taxa could be observed in several sites. At Rjasna on the river Kaslpya even the preceding interstadial was found, in which *Picea obovata*, *Betula* Sect. *Albae*, *B. humilis*, *Juniperus communis*, *Selaginella selagi-*

noides, *S. tetraedra*, *Sparganium* cf. *hyperboreum* and several other plant species seem to have occurred for a relatively long time. *Picea obovata* is lacking there in the interglacial proper, yet *Empetrum nigrum* and *Tilia tomentosa* should be noticed, together with *Azolla interglacialica* (reworked?). A nearly complete sequence of the vegetation history of the Mikulino interglacial is reported from the surroundings of Yaroslavl (VALUEVA and SEREBRYANNIY, 1978). The extremely strong share of oak and elm at the past pollen rain is not paralleled by modern analogs. Even here *Corylus* reaches high percentages, its pollen curve forming two distinct maxima. The documentation of the preceding vegetation history seems to be strongly influenced by reworked pollen, though the authors do not cover of this difficulty. Yet falsifications by reworked pollen are held to explain why the famous "interglacial" site of Mologo-Sheksna has for such a long time been a matter of debate. According to VALUEVA et al. (1979) this "interglacial" in reality has only been an interstadial phase, the modern phytosociological equivalents of which exist in the southern part of Central Siberia and in the adjacent regions of Western Siberia (pollen floras, diatoms). CHEBOTARYOVA et al. (1979) contribute some information on the interglacial vegetation in the northeastern part of Byelorussia. The allegedly contemporaneous theriofauna in the Voronezh-Oblast' was investigated by SHEVYREV et al. (1979), showing the simultaneous occurrence of *Mammuthus primigenius* and of *Palaeoloxodon antiquus*. The problematical position of the Karginsk interglacial or interstadial in Western Siberia has already been discussed. Its position is mainly based on absolute [^{14}C]-datations. In this respect it is worthwhile noticing that CHEBOTARYOVA et al. (1979) describe in Byelorussia equivalents of the Amersfoort and Brørup interstadials from the beginning of the last glaciation, though these interstadials here gave only ages from between 43,800 ± 1,000 and 26,170 ± 165 B.P. (i.e., the same ages as those of the Karginsk interglacial, though in comparison with Central and Western Europe some 30,000 to 40,000 years too young: Upper Volga interstadial). They were characterized most of all by pine forests with a high amount of Ericaeae and *Sphagnum*, together with some *Picea* sp. at the beginning and at the end of the interstadial proper. The ages quoted seem to warn against a chronostratigraphy based only on [^{14}C]-dates. The authors cited, and BURASHNIKOVA et al. (1979), contribute interesting paleoclimatic calculations of the last glaciation in the European and Asiatic parts of the USSR. The chronological background of the younger part of the last glaciation is given for the European part of the USSR by GVOZDOVER and SULERZHITSKIY (1979), and by VELICHKO et al. (1976). For contemporaneous changes in the steppe-like vegetation of the lower Don area, see PRASLOV et al. (1980).

In about the same region ZOZULIN (1976) discusses the Holocene history of the forest vegetation according to recent distribution patterns of several plant species.

Up till now paleobotany has been widely used in the USSR as a tool for geological and archaeological investigations, botanical problems being only of secondary importance. Now the situation has changed. An example for this is the investigation of the history of a peat bog in Lithuania (BITVINSKAS et al., 1978), in which primarily the changes in the peat-forming plant species are analyzed, together with spontaneous and man-made changes in the vegetation of the surrounding areas. Here a marked influence of man seems to date from the second part of the subboreal, whereas the present-day microrelief of the ombrogenous bog was only formed during the last 2,500 years. Another example for the new situation of pollen analysis is the investigation by SEREBRYANNAYA (1980) on the Holocene forest history of the western part of the Middle-Russian hills. This strongly deforested region has, at least since the beginning of the Atlantic zone, been covered by oak forests, with some *Tilia*, *Ulmus*, and *Picea*. The sporadic occurrence of *Carpinus* already at that time points to possible refuge areas of this tree species in the eastern part of Europe. The preceding phase of a steppe or forest-steppe vegetation is remarkable in that *Ephedra* seems repeatedly to have been of some importance. Taking all the papers cited here together might indicate a pleni- and late glacial center of distribution of *Ephedra* within the Middle-Russian hills and in the surroundings of the Caucasus Mts., the vast loess areas of the Ukraine not being so rich in *Ephedra* finds (see also Central Asia). SEREBRYANNAYA points to the relatively late impact of man on the vegetation, this only dating from sub-Atlantic times. Yet the opening of the forest vegetation at the Atlantic/Subboreal transition there and the increasing amount of minerogenic material in the subboreal peat seem to point to much earlier clearances. This is corroborated by GUMAN's (1978) analysis of the influence of man on the vegetation history in the Pskov Oblast'. This paper seems to be one of the earliest to distinguish various cerealia genera. It is palynologically deduced from the observations made, that some islands of a late glacial steppe vegetation occurred there even just before the onset of human impact at the beginning of Neolithic times. Some remarks on the terrestrialization of sink holes in limestones near Ulyanovsk on the river Volga: BLAGOVESHCHENSKIY and BLAGOVESHCHENSKAYA (1978). TUGANAEV and EFIMOVA (1981) and TUGANAEV and BARANOV (1978) report on cultivated plants and weeds in the river Kama region during the tenth to fourteenth centuries. The cultivation of *Secale* seems to have begun there only during or even after the tenth century. This is said to hold true for the whole of the forest-steppe area. Several interesting "southern" plants like *Elisanthe noctiflora*, *Stachys neglecta*, and *Hordeum lacunculiforme* are reported from the sites investigated.

e) Northeastern Europe

SMIRNOVA investigated the diatom floras of the Mikulino interglacial in then existing estuaries of the Severnaya Dvina and the Vychegda rivers (1976, 1979). The maximal transgression is said to have happened here at the transition from the preceding glacial period to the interglacial proper. The diatom floras investigated exhibit the transition from oligotrophic freshwater, via brackish conditions to marine environments and back again to oligotrophic freshwater. At the beginning of this sequence, *Larix sibirica*, *Picea abies*, *Scirpus lacustris*, and *Potamogeton heterophyllus* thrived in the vicinity (SMIRNOVA, 1976). In the western part of the Bol'shezemelskaya Tundra marine and lake sediments are widely distributed, preceding the accumulation of moraines of the last glaciation. The diatom floras of these aquatic sediments show the transition from marine to freshwater environments with a high amount of reworked marine frustules (LOSEVA, 1978). At the same time the surrounding terrestrial vegetation changed at least twice from some type of a tundra vegetation (together with *Artemisia* and Chenopodiaceae) to birch- or even birch-pine-spruce-forests (ARSLANOV et al., 1977). This is held to have occurred between about 45,000 and 38,000 B.P. (^{14}C). It may be questioned whether the dates, which seem to corroborate the age of the Karginsk interglacial are correct, thus ruling out the possibility of a misinterpretation of interglacial sediments of perhaps Mikulino age. Interesting calculations about the Holocene climate of Karelia on the background of former and present-day distribution patterns of the most important plant communities: KLIMANOV and ELINA (1980). These paleoclimatic studies are paralleled by investigations on the Holocene history of the North-Eurasian area of permafrost (BOBOV, 1977), on the disturbances of the heat budget of the soil within the same region during the late Pleistocene (NESTEROV et al., 1980), and on relics of Pleistocene glacier ice in the northern part of Western Siberia (KAPLYANSKAYA, TARNOGRADSKIY, 1976).

f) Northeastern Siberia

It could be seen from the foregoing that Russian scientists pay much attention to the evolution and the ecological significance of diatom floras. This also holds true for Siberia, as will be seen later. ORESHKINA (1980) deals at some length with marine diatoms off the eastern coast of the Kamchatka peninsula, from the middle Miocene onwards. Earlier observations on a very slow evolution can now be corroborated, rendering the distinction of only three diatom zones in the whole of the Pleistocene marine record possible. The chronological background of the evolution of the terrestrial macro- and microfloras of Northeastern Siberia during younger Cenozoic times is carefully analyzed by DANILOV (1980; containing lists of the rich marine and terrestrial faunas as well), and by SHER et al. (1977). According to these investigations the famous Omoloy flora dates from middle to upper Oligocene times, whereas the lignite-layer there was formed during the Miocene. On the other hand, SHER et al. (1977) point to the existence of a tundra and forest tundra-like vegetation on perennially frozen ground already before $2.0 \cdot 10^6$ years ago at the lower course of the river Kolyma (in this paper correlation of Eastern Siberian rodent faunas with those of Europe). The vegetation history of the upper Cenozoic in the lowlands of Western Kamchatka is briefly summarized by ALESHINSKAYA et al. (1976; macro- and microfossils, diatoms). Here the strong northerly distribution of *Myrica* during one of the younger Pleistocene interglacial phases is remarkable, together with observations favoring the view of present-day (or even better) vegetational and climatical conditions at about 33,000 to 39,400 years ago. This again raises the question as to the reality of the Karginsk interglacial (see DANILOV, 1980; DANILOV et al., 1975; BESPALIY et al., 1979; SVITOCH et al., 1977, too). ALESHINSKAYA's et al. (1976) work, as well as that of the authors to be mentioned later in this chapter, show unequivocally that the pleniglacial open vegetation in Northeastern Siberia differed strongly from its contemporaneous counterparts in Europe and Western Siberia in the important share of different herbs (most of all Caryophyllaceae, Rosaceae, Ranunculaceae, Onagraceae, *Armeria*, Valerianaceae, various Compositae), the strongly reduced amount of Chenopodiaceae, the lack of *Ephedra*, and the extremely strong contribution of various *Selaginella* and *Lycopodium* species (most of all in the younger part of the Pleistocene), together with repeatedly high amounts of *Artemisia*, grasses, and Cyperaceae, whereas during interstadial or even interglacial times the significance of the ericaceous plants has been much more pronounced than in other parts of Northern Eurasia. This points to a somewhat isolated evolution of the tundra vegetation in Northeastern Siberia and to a much more reduced influence of the edificators of steppe-like plant communities there than is observed in Europe and Western Siberia. New taxa from middle or young-Pleistocene sediments: *Larix gracilis*, *Pinus longipelta* (GRICHUK, 1979); *Amphorochara grambastii*, resembling the South American *Lamprothamnium longifolium* var. *bulbillifera*; *Chara angarica*, f. *angarica* and *tenera* (KRASAVINA, 1978). A wealth of information about the middle and young Pleistocene vegetation history is found in the contributions by GRICHUK (1979), KAPLINA and LOZHKIN (1979), KAPLINA et al. (1980b), dealing with the lower courses of the rivers Indigirka and Kolyma, and SHOFMAN et al. (1977) for the river Vilyuiy and its tributaries. Beside this KAPLINA's et al. (1980b) paper is worth mention in view of the criticism of $[^{14}C]$-datations as the only basis for young Pleistocene stratigraphy:

paleontologically dated sediments from the middle Pleistocene gave only radiocarbon ages of about 31,000 to 45,000 years B.P. (cf. the Karginsk problem mentioned).

In the lowlands of Northeastern Siberia some 30- to 50-m-thick sediments, mostly composed of silt, with some admixture of organic material and sand or gravel, are widely spread. They are called the sediments of the Vorontsov- or Yedom- or ice-complex series (thanks to their strong amount of permafrost ice). The age of these fluviatile and lacustrine sediments is a matter of debate. It can be shown now that its lowermost part in general was formed at least during the last interglacial, yet most of the sediments were only accumulated during the last glaciation (LOZHKIN, 1977, New Siberian Islands; BOLIKHOVSKAYA and BOLIKHOVSKIY, 1979, KAPLINA et al., 1980a, river Indigirka; KAPLINA et al., 1978, lower course of the river Kolyma; SVITOCH et al., 1979, Chaunskaya Guba). In all the papers cited a wealth of botanical information is given, KAPLINA et al. (1980b) contributing to the synchroneous beetle faunas as well. When the bulk of the Yedom sediments was formed, climate has been of extreme severity. According to the then existing permafrost features, the annual mean air temperature must have been less than -20°C. The sediments of this series are rich in remains of the mammoth fauna, though most of these fossils originate in the fossil soils of the Yedom series, becoming very rare in the pleniglacial sediments themselves, indicating the difficult situation of animal life during full glacial times there (BOLIKHOVSKAYA and BOLIKHOVSKIY, 1979). This can also be seen from absolute ages of remains of *Mammuthus primigenius* on the island Severnaya Zemlya (MAKEEV et al., 1979). Here two clusters of ages for these remains exist, one concentrating on the period between 25,000 and 19,000 B.P., the other on about 10,000 to 9,000 B.P., when the vegetation cover was richer and denser there than at present. This raises the question as to the nourishment of these big herbivores. Horse, bison, and mammoth here seem to have preferred grasses, herbs, and mosses (UKRAINTSEVA, 1979; UKRAINTSEVA et al., 1978; DUBROVO et al., 1980), bison preferring *Polemonium* and *Phlox sibirica* (UKRAINTSEVA et al., 1978). According to the plant remains found in the stomachs of these animals, the vegetation must have been a very open one, nearly completely lacking in trees and shrubs. The absolute ages in general concentrate on the period between 29,500 ± 1,000 and about 44,000 B.P. The age of the mammoth baby from Kirgilyach (44,000 B.P. for its soft parts) is questioned by DUBROVO et al. (1980), in view of the geological and paleobotanical situation of the cadaver, rendering an age of about 12,000 to 15,000 B.P. much more reasonable (for the present-day vegetation in the immediate vicinity: UKRAINTSEVA and KOZHEVNIKOV, 1979). It is held by UKRAINTSEVA (1979) and by UKRAINTSEVA (1979) and by UKRAINTSEVA et al. (1978) that an important reason for the death of several of these frozen big herbivores during the [^{14}C]-dated "Karginsk interglacial" could have been the increase in the thickness of the active layer above the upper boundary of permafrost during phases of warmer climate, causing the spreading of bogs and dangerous solifluction layers, this enabling on the other hand the rapid conservation of the carcasses in oxygen-free sediments. Yet at least as regards the Holocene warming, this had happened only from about 12,000 to 11,000 onward, when most of the animals cited had already become extinct. The Holocene warming caused the formation of widely distributed thermokarst sediments, the so-called Alas series, when immediately at the beginning of the Holocene vegetation was much richer in the present-day tundra zone of Northeastern Siberia than it is today (KAPLINA and LOZHKIN, 1979; LOZHKIN et al., 1975). A correlation of the Alaskan and Yakutian young and middle Pleistocene and Holocene vegetation and climate history: SVITOCH, 1977. Chronology of man's activity in northeasternmost Yakutia: SHILO et al. (1976). The oldest finds up till now point to a domestication of the ren: 710 ± 40 B.P. The Holocene history of vegetation, including the coastal marine flora and faunas, on the western coasts of the Bering Sea: GOL'DFARB et al., 1980; DANILOV et al., 1979).

g) The Soviet Far East

The Pleistocene history of this most important refuge area for plant and animal taxa of the Tertiary is even today not well understood. Thus investigations by MAKHOVA (1978) on the middle and old Pleistocene vegetation history of the Zeya basin, and by ALEKSEEV and GOLUBEVA (1980) on the young Pleistocene evolution of vegetation to the southwest of Vladivostok are welcome. Both these papers report on the repeated shift of the boundary between steppe- and exacting forest taxa during approximately the last 500,000 years. These observations are paralleled by investigations by KOSTYAEV (1979) and by KOSTYAEV et al. (1979) on the history of the permafrost region within the same area. Here it can be shown by means of macrofossil analysis that some plant species, like *Carex paradoxa* and *C. elata*, advanced during warmer periods about 1000 km farther to the east and some 250 km more to the north than nowadays (KOSTYAEV et al., 1979). The complicated Holocene paleoecological changes at the mouth of river Amur are investigated by TROITSKAYA et al. (1978), taking into consideration foraminifera, diatoms, silicoflagellates, and sporomorphs. It is concluded that the first strong advance of broad-leaved thermophilous trees after the last glaciation had already occurred at about the Alleröd interstadial, though this phase could not be dated exactly. *Pinus koraiensis* seems to have reached the area only at the beginning of sub-Atlantic times. Most of the changes in the paleoecological conditions are explained by changes in climate, yet observations also exist favoring the view of a strong influence on the paleoecology of a changing depth of the sea and of sandbars being built up. This seems to be a most promising attempt to reavaluate the observations made by TROITSKAYA et al. (1978), since pollen analytical work on the subboreal and sub-Atlantic periods in Northern Sakhalin (ALEKSANDROVA and BROVKO, 1979) do not agree with the scheme drawn by the authors just cited. Here equivalents of a "Grenzhorizont" could be found, dating from about 2,600 B.P. It is concluded that between 4,300 and 3,400 B.P. *Myrica*, *Quercus mongolica*, *Ulmus*, *Osmunda*, and treelike *Betula* spread far to the north after a short cold spell, in which *Picea*, *Pinus pumila*, and *Alnaster* dominated the pollen rain. It is thought that a deterioration of climate after about 3,400 B.P. had caused the readvance of spruce and the retreat of *Myrica* and *Pinus pumila*. In view of the simultaneous increase in the NAP-content and the spreading of *Artemisia*, one wonders whether these changes might also have been influenced to some extent by the impact of man. The influence of the rising sea level during late glacial and Holocene times on the cutting off of several islands from the continents is investigated by VELIZHANIN (1976). Since this had happened in general at only about 11,000 to 7,000 years ago, it is concluded that flora and fauna of these islands should be the last descendents of full glacial biocoenoses. For the changes of sea level and their timing, see DENISOV (1979), STAROV (1977), VNUCHKOV et al. (1976).

h) Central Asia

As has been pointed out previously, the early Pleistocene flora of the Pamir Mts. was very rich in various tree species. According to PEN'KOV et al. (1976) the Kokbaysk series, dating from the earliest Pleistocene, contains among 80% - 40% tree pollen (AP) 33 to 35 tree and shrub species, most of all *Pinus*, *Cedrus*, *Picea*, *Alnus*, *Betula*, *Quercus*, *Corylus*, *Zelkova*, and *Ostrya*. The site is situated at an elevation of 4000 m a.s.l. Since these interglacial sediments are covered and underlain by moraines, a strong uplift of the region cannot have happened meanwhile. This holds even more true if it is noticed that the old and middle Pleistocene Akdzhar and Alichur series have only 50% to 15% AP or 15% to 0% AP, respectively, with 15 to 17 or 6 tree species, respectively. In the foothill area of the Pamir Mts. and the Tian Shan, thick loess layers cover watersheds and slopes below 2000 to 2300 m. The numerous fossil soils of these loesses contain pollen floras, the AP content of which in general was higher than that of the NAP. *Pinus* Sect. *Diploxylon*, *Betula* Sect. *Albae*, and *Alnus* nearly always dominate, with some admixture of *Picea* Sect. *Eupicea*, *Corylus*, *Carpinus* cf. *orientalis*, *Ostrya*, *Quercus* cf. *robur*, and *Ulmus* cf. *pumila*. Only during the formation of the four youngest soil complexes had the prevailing vegetation changed, grasses, herbs, *Artemisia*, and Chenopodiaceae now dominating, with an increasing amount of *Ephedra* (up to more than 20%!) (LAZARENKO et al., 1980). GUDOSHNIKOV (1978) studies the history of the *Abies sibirica* forests of the Kuznetskiy Alatau, in view of the recent distribituon patterns of its bryoflora. The mollusk fauna of peat layers and silts in the Minusinsk basin (Northeastern Altai Mts.) characterizes the end Pliocene age of the sediments of the ninth river terrace and of its cover sands. Here already at that time a semidesert had developed, which was governed primarily by various Chenopodiaceae, *Artemisia*, and up to 20% to 50% *Ephedra*. These high percentages of *Ephedra* are unknown today. Higher up in the profile the amount of this plant diminishes. The later shift to a steppe to forest steppe vegetation is remarkable. Indications for the existence of thermophilous broad-leaved forests are lacking (LAUKHIN, 1979). The Holocene history of the Barguzin National Park (on the east coast of lake Baikal) is studied by KOL'TSOVA et al. (1979). The formation of wood peat began at 7,940 ± 420 B.P., the pollen flora of which shows clearly the long history of the steppe and forest communities. It is striking to note that besides a small share of *Ephedra* and other steppe plants the immigration of *Betula* Sect. *Nanae* and of *Pinus sibirica* only happened between 7,940 and 3,600 B.P., after a pronounced period of spruce dominance. From a paleoclimatological point of view the Holocene history of Issyk-Kul' (TROFIMOV, 1978), and of lake Aral (KHRUSTALEV and REZNIKOV, 1976) is worth mention. Here changes of climate and of the discharge of rivers, being caused to some extent by changes in their beds, contribute to the rising and sinking lake levels. Thus the changes in the lake levels cannot always be interpreted only in terms of temperature and precipitation.

References*

ALEKSANDROVA, A.N., BROVKO, P.F.: DAN 246, 1478-1481 (1979). - ALEKSEEV, M.N., GOLUBEVA, L.V.: Bjull. Kom. 50, 96-107 (1980). - ALESHINSKAYA, Z.V., BOYARSKAYA, T.D., VOSKRESENSKAYA, T.N., SVITOCH, A.A.: DAN 226, 155-158 (1976). - ARSLANOV, K.A., IZMAILOV, Y.A., OSTORVSKIY, A.B., TERTYCHNIY, N.I., SHCHEGLOV, A.P.: DAN 226, 159-162 (1976). - ARSLANOV, K.A., BERDOVSKAYA, G.N., ZAITSEVA, G.Y., LAVROV, A.S., NIKOFOROVA, L.D.: DAN 233, 188-191 (1977). - ARSLANOV, K.A., GERASIMOVA, S.A.,

* Abbreviations: DAN = Doklady Akademii Nauk SSSR; Bjull. Kom. = Bjulleten' Komissi po izučeniju četvertičnogo perioda; Izv. AN = Izvestija Akademii Nauk SSSR

LEONT'EV, O.K., LOKSHIN, N.V., MAMEDOV, A.V., RYCHAGOV, G.I., TERTYCHNIY, N.I., SIRINOV, N.S.: Bjull. Kom. 48, 39-48 (1978).

BAKHINOVA, V.P., ZUBAKOV, V.A., ITSIKZON, E.M., CHERPANOV, E.V., LIPIN, V.M.: Bjull. Kom. 48, 208-212 (1978). - BARDEEVA, M.A., ISAEVA, L.L.: DAN 251, 169-172 (1980). - BARG, I.M., YALOVENKO, I.P., YATSNEKO, Y.G.: DAN 246, 164-167 (1979). - BESPALIY, V.G., IVANOV, V.F., POLOVOVA, T.P.: DAN 249, 932-934 (1979). - BITVINSKAS, T., GRIGYALITE, M., SAVUKINENE, N.: Stratigraphy and development of the Uzhpyakyu Tiryalis peat bog, 56-61 (russ.). In: Usloviya sredy i radial'niy prirost derev' ev, ed. T.T. BITVINSKAS. Inst. Bot. Akad. Nauk Litovskoy SSR, Kaunas 1978. - BLAGOVESHCHENSKIY, I.V., BLAGOVESHCHENSKAYA, N.V.: Bot. Zhurn. 63, 1778-1788 (1978). - BOBOV, N.G.: DAN 232, 1410-1413 (1977). - BOLIKHOVSKAYA, N.S., BOLIKHOVSKIY, V.F.: DAN 247, 409-412 (1979). - BOLIKHOVSKIY, V.F.: DAN 224, 1149-1151 (1975). - BRESLAV, S.L., VALUEVA, M.N., MAUDINA, M.I.: DAN 248, 161-166 (1979). - BURASHNIKOVA, T.A., MURATOVA, M.V., SUETOVA, I.A.: DAN 244, 723-727 (1979).

CHEBOTARYOVA, N.S., DERYUGINA, V.S., IVANOVA, N.V., KUER'EROVA, L.V., TYURINA, L.S.: Izv. AN, ser. geol., No. 1, 102-117 (1979). - CHOCHIEVA, K.I., MAMATSASHVILI, N.S.: DAN 235, 1148-1151 (1977); - DAN 245, 692-696 (1979).

DANILOV, I.D.: Izv. AN, ser. geol., No. 6, 53-62 (1980). - DANILOV, I.D., NEDESHEVA, G.N., RYABOVA, E.I.: DAN 225, 393-395 (1975). - DANILOV, I.D., NEDESHEVA, G.N., POLYAKOVA, E.I., SMIRNOVA, T.I.: DAN 249, 1416-1420 (1979). - DENISOV, S.V.: DAN 247, 650-652 (1979). - DODONOV, A.E., LOMOV, S.P.: Izv. AN, ser. geol., No. 8, 75-85 (1980). - DODONOV, A.E., PEN'KOV, A.V.: Bjull. Kom. 47, 67-76 (1977). - DODONOV, A.E., RANOV, V.A., PEN'KOV, A.V.: Bjull. Kom. 48, 86-102 (1978). - DUBROVO, I.A., RENGARTEN, N.V., ZELIKZON, E.M.: Bjull. Kom. 49, 63-77 (1979). - DUBROVO, I.A., GITERMAN, R.E., GORLOVA, R.N., RENGARTEN, N.V.: Izv. AN, ser. geol., No. 10, 84-100 (1980).

GABUNIYA, L.K., TUSHABRAMISHVILI, D.M., VEKUA, A.K.: Bjull. Kom. 47, 107-111 (1977). - GLADENKOV, YU.B., GURARIY, G.Z.: DAN 230, 1173-1175 (1976). - GOL'DFARB, YU.I., LOZHKIN, A.V., TEREKHOVA, V.E.: DAN 250, 410-412 (1980). - GRICHUK, M.P.: Bjull. Kom. 49, 49-62 (1979). - GUDOSHNIKOV, S.V.: Bot. Zhurn. 63, 11-19 (1978). - GUMAN, M.A.: Bot. Zhurn. 63, 1415-1429 (1978). - GVOZDOVER, M.D., SULERZHITSKIY, L.D.: Bjull. Kom. 49, 144-146 (1979).

KAPLINA, T.N., LOZHKIN, A.V.: DAN 233, 1162-1164 (1977); - Izv. AN, ser. geol., No. 2, 69-76 (1979). - KAPLINA, T.N., GITERMAN, R.E., LAKHTINA, O.V., ABRASHOV, B.A., KISELEV, S.V., SHER, A.V.: Bjull. Kom. 48, 49-65 (1978). - KAPLINA, T.N., SHILOVA, G.N., PIRUMOVA, L.G.: Izv. AN, ser. geol., No. 9, 74-81 (1980a). - KAPLINA, T.N., SHER, A.V., GITERMAN, R.E., ZAZHIGIN, V.S., KISELEV, S.V., LOZHKIN, A.V., NIKITIN, V.P.: Bjull. Kom. 50, 73-95 (1980b). - KAPLYANSKAYA, F.A., TARNOGRADSKIY, V.D.: DAN 231, 1185-1187 (1976). - KHRUSTALEV, YU.P., REZNIKOV, S.A.: DAN 231, 981-984 (1976). - KLIMANOV, V.A., ELINA, G.A.: DAN 252, 419-423 (1980). - KOGAN, L.I., MALOVITSKIY, YA.P., MOSKALENKO, V.N., SHINKUS, K.M.: DAN 233, 450-452 (1977). - KOL'TSOVA, V.G., STARIKOV, E.V., ZHIDOVLENKO, V.A.: Bjull. Kom. 49, 121-124 (1979). - KOSTYAEV, A.G.: DAN 244, 983-986 (1979). - KOSTYAEV, A.G., ZAIKINA, N.G., KOLOSOVA, G.N., SMIRNOVA, T.I.: DAN 244, 1227-1231 (1979). - KRASAVINA, L.K.: Bot. Zhurn. 63, 226-233 (1978). - KRASNENKOV, R.V., PISAREVA, V.V.: DAN 232, 1154-1157 (1977). - KRASNENKOV, R.V., IOSIFOVA, YU.I., SHULESHKINA, E.A., LIBERMAN, YU.N.: DAN 252, 677-680 (1980). - KRIGER, N.I., BOTNIKOV, V.I.: DAN 245, 902-905 (1979).

LAUKHIN, S.A.: Izv. AN, ser. geol., No. 5, 121-131 (1979). - LAZARENKO, A.A., RANOV, V.A.: Bjull. Kom. 47, 45-57 (1977). - LAZARENKO, A.A., BOLIKHOVSKAYA, N.S., SEMENOV, V.V.: Izv. AN, No. 5, 53-66 (1980). - LOSEVA, E.I.: Bjull. Kom. 48, 103-112 (1978). - LOZHKIN, A.V.: DAN 235, 435-437 (1977). - LOZHKIN, A.V., PROKHOROVA, T.P., PARIY, V.P.: DAN 224, 1395-1398 (1975).

MAKEEV, V.M., ARSLANOV, KH.A., GARUTT, V.E.: DAN 245, 421-424 (1979). - MAKHOVA, YU.V.: Bjull. Kom. 48, 170-179 (1978). - MARUASHVILI, L.I.: DAN 225, 651-654 (1975); - Bjull. Kom. 47, 3-16 (1977). - MASLYAEV, G.A.: DAN 247, 661-663 (1979).

NESTEROV, I.I., STAVITSKIY, B.P., KURCHIKOV, A.R.: DAN 250, 418-421 (1980).

ORESHKINA, T.V.: Izv. AN, ser. geol., No. 11, 57-66 (1980).

PEN'KOV, A.V., NIKONOV, A.A., PAKHOMOV, M.M.: DAN 229, 691-694 (1976). - PEVZNER, M.A., PECSI, M.: Bjull. Kom. 50, 24-34 (1980). - PISAREVA, V.V., VELICHKEVICH, F.YU., SHIK, S.M.: DAN 248, 185-190 (1979). - PRASLOV, N.D., IVANOVA, M.A., MALYASOVA, E.S.: Bjull. Kom. 50, 168-175 (1980).

SEMENENKO, V.N., PEVZNER, M.A.: Izv. AN, ser. geol., No. 1, 5-15 (1979). - SEMENENKO, V.N., LUPARENKO, A.V., KOVALYUKH, N.N.: Bjull. Kom. 49, 17-25 (1979). - SEREBRYANNAYA, T.A.: Bjull. Kom. 50, 178-184 (1980). - SEREBRYANNIY, L.R., GEY, N.A., DZHINORIDZE, R.N., IL'VES, E.O., MALYASOVA, E.S., SKOBEEVA, E.I.: Bjull. Kom. 50, 123-137 (1980). - SHCHERBAKOV, F.A., KUPRIN, P.N., MOGRUNOV, YU.G.: Bjull. Kom. 49, 3-16 (1979). - SHER, A.V., VIRINA, E.I., ZAZHIGIN, V.S.: DAN 234, 1171-1174 (1977). - SHEVYREV, A.T., RASKATOV, G.I., ALEKSEEVA, L.I.: Bjull. Kom. 49, 39-48 (1979). - SHILO, N.A., DIKOV, N.N., LOZHKIN, A.V.: DAN 231, 1204-1205 (1976). - SHOFMAN, I.L., KIND, N.V., PAKHOMOV, M.M., PROKOPCHUK, B.I., VINOGRADOVA, S.N., SULERZHITSKIY, L.D., FOROVA, V.S.: Bjull. Kom. 47, 100-107 (1977). - SHUMENKO, S.I., USHAKOVA, M.G.: DAN 251, 213-215 (1980). - SMIRNOVA, V.M.: DAN 231, 1200-1203 (1976); - DAN 246, 171-174 (1979). - STAVROV, V.N.: DAN 235, 902-905 (1977). - SUDAKOVA, N.G., BOL'SHAKOV, V.A.: DAN 233, 1176-1179 (1977). - SUPRUNOVA, N.I., VRONSKIY, V.A.: Bjull. Kom. 47, 87-94 (1977). - SVITOCH, A.A.: DAN 232, 1161-1164 (1977). - SVITOCH, A.A., PARUNIN, O.B., SARAZHINSKAYA, N.I., TIMASHKOVA, T.A., SHLYUKOV, A.I.: DAN 233, 447-450 (1977). - SVITOCH, A.A., BAZILEVSKAYA, L.I., BOYARSKAYA, T.D.: DAN 245, 1462-1466 (1979).

TABOYAKOVA, V.YA.: Izv. AN, ser. geol., No. 2, 143-149 (1981). - TAMRAZYAN, G.P., OVNATONOV, S.T.: DAN 225, 168-171 (1975). - TROFIMOV, A.K.: Bjull. Kom. 48, 79-85 (1978). - TROITSKAYA, T.S., KARAULOVA, L.P., TSAR'KO, E.I.: Bjull. Kom. 48, 66-78 (1978). - TSERETELI, D.V., MAYSURADZE, LG.M.: Bjull. Kom. 50, 116-122 (1980). - TUGANAEV, V.V., BARANOV, E.I.: Bot. Zhurn. 63, 1035-1037 (1978). - TUGANAEV, V.V., EFIMOVA, T.P.: Bot. Zhurn. 66, 561-564 (1981).

UKRAINTSEVA, V.V.: Bot. Zhurn. 62, 553-554 (1977); - Bot. Zhurn. 64, 318-330 (1979). - UKRAINTSEVA, V.V., KOZHEVNIKOV, YU.P.: Bot. Zhurn. 64, 1091-1098 (1979). - UKRAINTSEVA, V.V., FLEROV, K.K., SOLONEVICH, N.G.: Bot. Zhurn. 63, 1001-1004 (1978).

VALUEVA, M.N., SEREBRYANNIY, L.R.: Bjull. Kom. 48, 113-122 (1978). - VALUEVA, M.N., GUZMAN, A.A., ZAIKINA, N.G., SEMENENKO, L.T., TSUKUROVA, A.M.: Bjull. Kom. 49, 26-38 (1979). - VELICHKEVICH, F.YU.: DAN 233, 1158-1161 (1977); - Bjull. Kom. 48, 123-136 (1978); - DAN 245, 682-684

(1979). - VELICHKO, A.A., ARSLANOV, KH.A., KURENKOVA, E.I.: DAN 228, 713-716 (1976). - VELICHKO, A.A., UDARTSEV, V.P., MOROZOVA, T.D., MARKOVA, A.K., PEVZNER, M.A., GRIBCHENKO, YU.N., SYCHEVA, S.A.: DAN 232, 1142-1145 (1977). - VELIZHANIN, A.G.: DAN 231, 205-207 (1976). - VIRINA, E.I., SVITOCH, A.A.: DAN 230, 399-401 (1976). - VNUCHKOV, V.A., KAPLIN, P.A., SHLYUKOV, A.I.: DAN 228, 914-916 (1976).

ZOZULIN, G.M.: Bot. Zhurn. 61, 39-47 (1976).

Professor Dr. BURKHARD FRENZEL
Botanisches Institut der Universität
D 7000 Stuttgart 70 (Hohenheim)

II. Vegetation Science (Sociological Geobotany)

By RÜDIGER KNAPP

1. Fundamental Perspectives, Comprehensive Volumes, Bibliographies

Vegetation science continues to integrate an increasing number of theoretical and practical perspectives. Thus, their conceptual structure becomes more and more implemented. This diversification has evolved partly in connection with the accomplishments of certain studies emphasizing special features of the local vegetation in particular regions. Simultaneously, the resulting amplified cognition of diversified problems contributes to a more comprehensive understanding of interrelations between regional and conceptual branches of vegetation science. On the other hand, progress has been achieved by exchange of ideas between groups and "schools" of scientists from various countries. The methodological principles of vegetation science have thus been broadened fundamentally in recent time.

This expansion of scope is paralleled by rapidly expanding numbers of publications in vegetation science. Thus, the selection of the publications quotable in our reviews in *Progress in Botany* has to be more extensive than in some other fields. Aiming to mitigate this condition, these annual reviews both contain a selection of topics in vegetation science and avoid quoting a particular publication repeatedly in successive volumes. Therefore, a quotation, missing perhaps in the present review, may be included in a later or earlier volume. The reviews in the succeeding volumes of *Progress in Botany* should be regarded as an entity covering the whole field of vegetation science.

New comprehensive volumes contain articles and discussions (mainly in connection with congresses and symposia) dealing with special fields, such as evolution of plant communities (WILMANNS and TÜXEN, 1980), epharmony in vegetation (WILMANNS and TÜXEN, 1980), data processing in phytosociology (VAN DER MAAREL et al., 1980), vegetation on peat soils (GÉHU, 1980). A summary of the indicator values (light, temperature, continentality, soil reaction, water and nitrogen supply) of Central European species by ELLENBERG (1979) is available in a revised edition of this work.

Volumes on tropical vegetation emphasize their special structures and their plant life forms (VARESCHI, 1980) and certain relations to anthropogenic changes (HARRIS, 1980; several parts in: HALL et al., 1979; SINGH and GOPAL, 1978; WALKER, 1979).

The vegetation of nature protection areas is documented in several monographs, embracing various fields and often containing vegetation maps. Examples are areas in the following regions: Venezuela (STEYERMARK and HUBER, 1978), Chile (VILLAGRAN, 1980), Florida (WARD, 1979), Camargue, France (SAUVAGE, 1979, vegetation map: MOLINIER and DEVAUX, 1978), French Alps (AMANDIER and GASQUEZ, 1978), northern Finland (SÖYRINKI and SAARI, 1980), NW German bog areas (SCHNEEKLOTH and J. TÜXEN, 1978;

WITTIG, 1980), forest regions in southern Germany (PETERMANN and SEIBERT, 1979; SCHWABE-BRAUN, 1979a,b; WITSCHEL, 1980), endangered German riverine landscapes (ZAHLHEIMER, 1979; also DIERSCHKE, 1978, 1979; DIERSCHKE and JECKEL, 1980; R. TÜXEN, 1979a). Another approach to vegetational protection is the detailed investigation of endangered edificators and of ways for their conservation (e.g., *Taxus baccata*: LEUTHOLD, 1980; *Buxus sempervirens*: OZENDA, 1980b). On the other hand, the anthropogenic vegetation of urbanized areas (HÜLBUSCH et al., 1979; HÜLBUSCH, 1980) and around rural settlements (LORENZONI, 1979; SCHUSTER, 1980) has been treated in extensive monographs.

New bibliographies concern the vegetation of Kansas (KÜCHLER, 1980a), of Chile (RAMÍREZ, 1980), of the Bahamas (KNAPP, 1980a), of India (MEHER-HOMJI and GUPTA, 1980), of France (GÉHU and BIGOT, 1980), of Czechoslovakia (NEUHÄUSLOVÁ-NOVOTNÁ and NEUHÄUSL, 1980), of Hungary (SOÓ, 1980), population biology in the USSR vegetation (RABOTNOV, 1980), fire and vegetation (ECKELS et al., 1980), vegetation maps in South America (KÜCHLER, 1980b), application of remote sensing techniques in vegetation mapping, bird species and vegetation (KNAPP, 1980a). New issues of the bibliography series on vegetational classes contain the *Scheuchzerietea palustris* (R. TÜXEN et al., 1979) and the *Cetrario-Loiseleurietea* (R. TÜXEN and INSELBERG, 1979).

Additional books and monographic publications will be quoted in connection with the special fields concerned, treated in the next chapters.

2. General Results and Methods

a) Niche Concepts in Vegetation

One of the reasons for the coexistence of several taxa in a plant community is the phenomenon that various species can use different parts of the resources of the same site. This partitioning in utilization of the local environmental resources has been frequently defined and numerically modeled, mainly in North America, under the auspices of the niche concepts (e.g., CODY, 1978; HUTCHINSON, 1978; KEMP and WILLIAMS, 1980; PARRISH and BAZZAZ, 1979; WATSON, 1980) contributing to theoretical phytosociology. These discussions extend the knowledge of the internal structure and of the interactive systems within plant communities (e.g., KNAPP, 1980b,c). A "niche" is a manifestation of the place of a species in the functional whole of a biotic community, in its definition by biotic and abiotic environmental conditions and in its spatial or seasonal position. Thus, a niche seems to be, at least under favorable conditions, less implied by the primary site attributes than by the influences of other organisms of the communities. These organisms can create modified or even new (e.g., for epiphytes) growth conditions, but they can contract also the actual niche space by effects of competition ("realized niche", "post-interactive niche"). Comparing three fundamental techniques of biomathematic treatment of niche overlap (SLOBODCHIKOFF and SCHULZ, 1980; also HANSKI, 1978; HURLBERT, 1978; LAWLOR, 1980; MAJOR, 1980; PETRAITIS, 1979), the curve-averaging method has the advantage of convenient geometric interpretations and of a simple transition from discrete to continuous data; however, there are difficulties in applying it to niche overlaps along several resource dimensions. The symmetrical Alpha method has the same advantages; but it is also convenient for computing niche overlaps along several resource dimensions and beyond that for tests of the independence in the use of these different resource dimensions. The actual competition pressure resulting from niche overlaps can be more adequately assessed numerically with the asymmetrical Alpha method.

b) Biomathematic Approaches in Vegetation Dynamics

The growing interest in inductive studies on vegetation dynamics (examples of comprehensive reviews: BEEFTINK, 1979; JAHN, 1980; JOHNSON, 1979; VAN DER MAAREL and WERGER, 1979; MALMGREN, 1979; NUMATA, 1979) is combined with augmented approaches of numerical modeling on phenomena in vegetational succession. Indeed, this application of such biomathematic methods seems to be in all ways appropriate, for example, in predicting future vegetational conditions, one of the paramount objectives of successional studies. These manifold approaches have been classified according to some main problems in the succession theory (VAN DER MAAREL and WERGER, 1979): (1) The character of the directionality (e.g., uni-directionality or multi-directionality) of successions can be assessed with techniques used also for numerical classification and ordination (e.g., AUSTIN, 1980; VAN DER MAAREL, 1980; MATTHEWS, 1979; PERSSON, 1980; REGNÉLL, 1980; SJÖRS, 1980; SWAINE and GREIG-SMITH, 1980). The importance of such analyses has become obvious, since inductive data suggest that straightforward directionality, once generally accepted, seems to be by no means always prevalent (e.g., GLENN-LEWIN, 1980; VAN DER HULST, 1979a, 1980; JENSÉN, 1979; LONDO, 1978; SJÖRS, 1980; VARTIAINEN, 1980). (2) The historical impact, the influence of the former development on future successional trends, can be estimated by means of Markovian chains or similar models (e.g., VAN DER HULST, 1979b; NOBLE and SLATYER, 1980). (3) The quantity in the coincidence of certain plant population developments and of the successional course, respectively (e.g., DEBUSSCHE et al., 1980; FALIŃSKI, 1980; LIETH, 1980b; PEET and CHRISTENSEN, 1980a; PETERSON and BAZZAZ, 1980; TURKINGTON and HARPER, 1979) can also be elucidated by Markovian chain models or systems of related numerical significance. (4) Numerical evaluations in changes of diversity and of stability during successional series (e.g., CAMPBELL and VAN DER MEULEN, 1980; DANIN, 1978; HOUSSARD et al., 1980; MALMER et al., 1978; TRABAUD and LEPART, 1980) seem to be accessible on the basis of multivariate analysis.

Another classification of successional models focusing on North American forest vegetation suggests three organizational categories contributing to the primary basic parameters of the numerical formulations: (1) Attributes of single trees. (2) Relatively small-sized spatial units, either gaps in the forest canopy or sample quadrats. (3) Collective attributes of forest stands (SHUGART and WEST, 1980).

c) Actual Aspects of Syntaxonomy

The classification of plant communities and their allocation to vegetational classes, orders, alliances, associations, and their subdivisions (syntaxonomy) continues to be a central goal of several studies (examples of syntaxonomic synopses: BÖTTCHER, 1980; BALÁTOVÁ-TULÁČKOVÁ, 1978; BOURNÉRIAS, 1979; ELLENBERG, 1979; GÉHU, 1979; JENÍK et al., 1980; KRAUSE, 1978; MATUSZKIEWICZ, 1980a; MIYAWAKI, 1979; NEUHÄUSL and NEUHÄUSLOVÁ-NOVOTNÁ, 1979; RIVAS-MARTÍNEZ, 1979). More detailed syntaxonomic differentiations serve to increase knowledge of the parallelism between the distribution of certain vegetational units and of environmental conditions. Another trend is the augmented differentiation of vegetational units on the basis of dominance and of optimal behavior by certain species in connection with limited spatial and syndynamical structures, e.g., at the ecotone between forests and grasslands or in particular successional stages. As a by-product in this connection, the percentage of syntaxa with pronounced dominance of certain life-forms (therophytes, geophytes, etc.) is increasing. Thus, the S.I.G.M.A. syntaxonomy is approaching synusial concepts (on synusia: IPATOV and KIRIKOVA, 1980; NORIN, 1978). Coincidences of classification on the basis of vegetational formations and of syntaxonomical methods can thus be observed more frequently. Generally, this development seems to promote a unification of various geobotanic and ecologic approaches.

Some problems in the congruence of syntaxa and of vegetational formations are indicated by discussions on the relations of shrub communities with *Salix caprea* and *Sambucus* to those with *Prunus spinosa*, *Crataegus*, *Rosa*, etc. (*Prunetalia*; DE BOLOS, 1979; LEPŠ and PRACH, 1981; MATUSZKIEWICZ, 1980a; OBERDORFER, 1980; RIVAS-MARTÍNEZ and GÉHU, 1978; also WILMANNS, 1980).

The problems also continue to be discussed, whether several herbs and grasses can be evaluated as characteristic species of forests with trees of limited light interception (e.g., *Quercus*, *Tilia*, *Sorbus*, etc.) or of skirt (seam) associations (*Trifolio-Geranietea*) (e.g., ALTHUIS et al., 1979; HÜLBUSCH, 1979; OBERDORFER, 1980; PASSARGE, 1980; WILMANNS et al., 1979).

The indigenous vegetational classes mostly convene ultimately in the vegetational circles (holosigmeta), spatially coincident with phytogeographic flora regions (e.g., Mediterranean region, etc.). Exceptions are syntaxa founded on species presumed to occur spontaneously in several flora regions. But recent monographic revisions have revealed rather frequently that such species are not homogeneous taxonomically. For instance, the nearly cosmopolitan distribution of *Lemna minor*, assumed formerly, has been found to be nonexistent. Taxa, well defined recently, occurring in East Asia (*Lemna japonica*), in the interior of North America (*L. turionifera*), in south-eastern North America (*L. obscura*), and in tropical South America (*L. ecuadoriensis*), greatly restrict the actual distributional area of *Lemna minor* s.str. (LANDOLT, 1980; URBANSKA-WORYTKIEWICZ, 1980).

A syntaxonomic synopsis of the *Lemna* communities (Prodromus: SCHWABE-BRAUN and R. TÜXEN, 1981; also MÉRIAUX, 1978; PASSARGE, 1978; PEDROTTI, 1979; POTT, 1980; WIEGLEB, 1978; ZAHLHEIMER, 1979) includes instructive considerations of the syntaxon definition with their necessarily specialized methodology (on problems in syntaxa very poor in species see also R. TÜXEN, 1979b). For instance, the basis of a typification seems to be inevitably a table with a series of homogeneous relevés; a single relevé can apparently not document all indispensable characteristic attributes of a syntaxon.

d) Integrated Multi-Regional Vegetation Maps

The project of a modern vegetation map of Europe, integrating the enormous progress in geobotanic survey and phytosociological methodology of the last decades, has been initiated in the year 1975 on the basis of broad-scale international team work. Meanwhile, a series of three maps has been published, representing the potentially natural vegetation of the western part of Europe (including Scandinavia, Italy, etc.) in 104 mapping units (signatures) (scale 1:3 millions) (OZENDA et al., 1979). These maps are regarded by OZENDA (1980a) and TRAUTMANN (1980) as a first draft, which has to be continuously improved and substantiated. In the remaining European regions, in addition to discussions on principles and on general mapping problems (DONIŢA, 1980; GRIBOVA and ISACHENKO, 1980; MATUSZKIEWICZ, 1980b; MIŠIĆ, 1980;, NEUHÄUSL, 1980; OZENDA, 1980; TRAUTMANN and BOHN, 1980), special methods and achievements in various countries have been reported (details: BONDEV, 1980; FEKETE, 1980; FUKAREK, 1980; HOFMANN, 1980; LAVRENKO et al., 1980; MORAVEC and MICHALKO, 1980; SCAMONI, 1980; SCHLÜTER, 1980). It is intended to consider on these maps the zonal, etagal, edaphic, and geographically regional structure of the potentially natural vegetation, including the azonal and extrazonal phenomena and important manifestations of vegetational history. The mapping units are elaborated on phytocoenological basis, realizing that their contents are mainly vegetational mosaics and complexes in view of the scale. This project is appropriate to promote further international collaboration and methodical harmonizing in vegetation mapping. Additionally, it is hoped to produce more detailed modern vegetation maps. Examples are vegetation maps in scale 1:200,000 covering large parts of France (BOURNÉRIAS and LAVERGNE, 1979; DUPIAS and VILKS, 1979; JOVET, 1979; TIMBAL, 1979; 1:250,000 : GÉHU, 1979b). The new vegetation map of the whole area of China (1:4 millions, WU, 1979b) integrates the research of several contributors.

3. Boreal and Subarctic Vegetation

a) Boreal and Subarctic Vegetation in Highly Oceanic Climates

Woodlands with dominant *Betula* species are zonally characteristic in certain boreal and subarctic regions with most oceanic climates, frequently constituting the timber-line in these regions. *Betula ermani* predominates in this vegetation in north-eastern Asia from Japan to Kamchatka (WATANABE, 1979). The much better-known birch woodlands in Fennoscandia, Iceland, and southern Greenland, with dominant taxa of the *Betula pubescens* group (EUROLA and VORREN, 1980; LÖVE and LÖVE, 1979; THANNHEISER, 1977) have recently been studied in their timber-line dynamics (KULLMAN, 1979). The *Betula* tree-limit has been rising at 75% of 213 localities in Sweden during the period 1915-1975, apparently owing to diminished pressure of snow cover by climatic amelioration. Little or no change of such tree-limits has been stated on sites with shallow snow cover. Tall herb and tall grass communities occur in the oceanic Subarctic (Kamchatka: MOROSOV, 1978, 1980; Norway: BLAŽKOVÁ, 1981), representing terminal successional stages on certain sites not suitable for woodland or intermediate stages and anthropogenic substitute vegetation.

b) Mire Vegetation in Northern Europe

The mire vegetation of northern Europe (e.g., DIERSSEN, 1980; KIVINEN et al., 1979; SONESSON, 1980; R. TÜXEN, 1980) attracts new interest for several reasons, among them as an indicator basis for physio-geographic regional subdivisions (BLAGOVESCHENSKY and BLAGOVESCHENSKAYA, 1978; BOTCH and VASILEVICH, 1980; ELINA and YURKOVSKAYA, 1980; EUROLA and VORREN, 1980). These mires seem to be to some extent more appropriate for delimitations of vegetational zones and their sections than woody upland vegetation. Such woody natural vegetation is much more influenced anthropogenically. It has been totally destroyed, for instance, in large parts of Norway (e.g., on the Lofoten islands) by the action of man, over at least 4000 years (VORREN, 1979a). On the basis of characters of mire vegetation, among others by special sociological groups with *Sphagnum*, Cyperaceae, Ericaceae, and *Salix*, six vegetational zones (including subzones), each of them divided in four sections, are recognizable in northern Fennoscandia (EUROLA and KAAKINEN, 1978; EUROLA and VORREN, 1980; VORREN, 1979b; 1980). Comparisons with early vegetation studies, detailed maps, and photographs, but also analyses of peat deposits impart knowledge of long-term alterations in the mosaics and complexes of this boreal mire vegetation (HANSEN et al., 1978; JENSÉN, 1979; KUZNETZOV, 1978; PETERSEN, 1980; RUUHIJÄRVI et al., 1979; SJÖRS, 1980; in north-western Germany: GROSSE-BRAUCKMANN, 1979; SCHWAAR, 1979b; J. TÜXEN, 1979, 1980a,b).

4. Temperate, Meridional, and Subtropical Vegetation

a) Forest and Shrub Vegetation in East Asia

The spatial basis framework of the East Asian forest vegetation, unique among the temperate woody formations in their phanerophytic species diversity, is the distribution of the potentially natural, zonally dominating plant communities, recently characterized for Japan (MIYAWAKI, 1979) and for China (WU, 1979a; parallels to western Europe: ELLENBERG, 1980). However, with the exception of rather inaccessible slopes or climatically unfavorable regions (e.g., in Japan: FRANKLIN et al., 1979; KASHIMURA et al., 1979; MASUZAWA and KONTA, 1979; SAITO, 1979; SHIDEI, 1979), most of the now existing East Asian forests are anthropogenic successional stages or afforestations. Thus, the evergreen broad-leaved forests, approaching the potentially natural vegetation in large warm temperate and subtropical lowland regions (with species of *Castanopsis*, *Camellia*, *Elaeocarpus*, *Eurya*, *Machilus*, *Osmanthus*, *Photinia*, etc.), are actually often restricted to groves surrounding sanctuaries or temples (on their structure and regeneration: TAGAWA, 1979; TAODA, 1979). Anthropogenic deciduous or coniferous woody vegetation dominates now mostly in the forest and shrub lands of these regions (KISHIMOTO et al., 1978; MIYAWAKI and SUZUKI, 1980; NAKANISHI and HATTORI, 1979; OKUTOMI, 1979). The vast destruction of the natural evergreen broad-leaved forests started with the advance of rice agriculture, in southern Japan about 4000 years ago (MAEDA, 1976; northern Japan: YAMANAKA, 1979). A very great number of the East Asian woody species are components of successional stages, now mostly anthropogenic. Seral vegetation in Japan is promoted by earthquakes, volcanism, and catastrophic storms (ELLENBERG, 1980). An obvious example is the natural pioneer shrub vegetation at high, climatically per-humid elevations of Japan, with dominant species of *Weigela*, *Salix*, and *Alnus* (additionally *Hydrangea*, *Deutzia*, *Coriaria*, etc.) on recent volcanic deposits, screes, and landslides (OHBA and SUGAWARA, 1979).

Another important part of the East Asian woody species comprises plant communities on rocky places, in valley bottoms, or in swamps, constituting partly terminal stages on special sites, recently studied in Japan. Among the alluvial and swamp forests (ISHIKAWA, 1979; MAKITA et al., 1979; MIYAWAKI, 1979; OHNO, 1979; OKUDA, 1979; SASAKI, 1979), rich *Ulmus-Fraxinus* communities are structurally similar to vicariant European and North American riverine forests, but comprise a great number of special East Asian genera. Forests with dominant *Pterocarya rhoifolia* and *Fraxinus spaethiana*, including numerous ferns, grow on the bottom of narrow mountain valleys on continuously moist or wet soils.

The woody vegetation on rocky sites in southern Japan at low altitudes (below 700 m) includes natural occurrences of interesting evergreen species (among the several conifers *Sciadopitys verticillata*; now planted world-wide in subtropical gardens: *Pittosporum tobira*, *Euonymus japonicus*, etc.; HUKUSIMA, 1979; NAKANISHI, 1979; TOYOHARA, 1979; on related *Pinus* communities in southern China: CHAU and LO, 1980).

b) Vegetation of South-Eastern North America

The high diversification of the south-eastern North American vegetation is promoted by the intensive entanglement of boreal, temperate, subtropical, and even tropical phytogeographic irradiations, implied by the fundamental repeated vegetational changes in connection with the Pleistocene alterations of warm and cold climatic periods (e.g., DELCOURT and DELCOURT, 1979). New results confirm that *Pinus* species (*P. caribaea*, *P. taeda*, *P. australis*, etc.) are a high-percentage constituent of the original vegetation of the south-eastern plains, in spite of the fact that they are theoretically (disregarding natural fires) mostly not components of terminal successional stages (BARRY, 1980; CHRISTENSEN, 1979; GOLDEN, 1979; LANE, 1980; McKINLEY and DAY, 1979; PEET and CHRISTENSEN, 1980a; SNYDER, 1980; STEINBERG, 1980; SWITZER et al., 1979). In addition to natural fires, catastrophic tropical storms (hurricanes), affecting nearly the whole region and occurring more than 20 times in 55 years (1901-1955) in parts of Florida and Alabama (LIETH, 1979), are an agent of spontaneous vegetation destruction, simultaneously releasing successions (additionally "freezing rain" storms, mainly in the south-eastern Appalachian Piedmont). The deciduous forests, most rich in woody species in North America, are being constantly studied, among them the southern Appalachian mixed forests (BARDEN, 1980; LORIMER, 1980; PITTILLO and SMATHERS, 1979; WENTWORTH, 1980) and certain floodplain forests (ROBERTSON, 1978; ROBERTSON et al., 1978) (on upland hardwood forests of the Piedmont and other less diversified deciduous forests: LIETH, 1980a; McEVOY et al., 1980; PEET and CHRISTENSEN, 1980b; SKEEN et al., 1980; DE WITT and WARE, 1979).

An important part of the high species numbers of these regions are components of plant communities on special sites. Important groups among them are the cedar glades (BASKIN and BASKIN, 1978; 1979a; 1979b), the grassy balds of the southern Appalachians (BARDEN, 1978; LINDSAY and BRATTON, 1979; PITTILLO and SMATHERS, 1979), the swamps of the coastal plains (GADDY and SMATHERS, 1980; HICKMAN and NEUHAUSER, 1978; KIRK, 1979; McKINLEY and DAY, 1979; MONTAGUE and DAY, 1980; PARRISH and RYKIEL, 1979; SCHLESINGER, 1978a; 1978b; WOOD and BRENNEMAN, 1980) comprising open vegetation (partly with *Sarracenia* and *Dionaea*) and forests with *Taxodium distichum* and *Nyssa* (BRINSON et al., 1980; EWEL and MITSCH, 1978; MITSCH and EWEL, 1979; MITSCH et al., 1979; WHIPPLE, 1980), and the manifold plant communities near the marine coast line (BARRY, 1980; BELLIS, 1980; FORE and PETERSON, 1980; HOPKINSON et al., 1980; HOSIER and EATON, 1980; TYNDALL and LEVY, 1979; STEINBERG with 4 maps, 1980).

5. Tropical Vegetation

a) Vegetation of Monsoonal South East Asia

A broad-scale survey on the present vegetation of Vietnam (THOMASIUS, 1980; THAI VAN TRUNG, 1980) differentiates 21 mapping units on a mainly physiognomic basis, adding data on standing biomass and annual primary production. The signature of evergreen broad-leaved lowland (below 700-1000 m) rain forests with 3-4 tree layers and emergent trees up to more than 50 m high covers less than 10% on this map of Vietnam only. In addition to the prevailing woody vegetation dependent on the monsoonal summer rains (with several Dipterocarpaceae, Leguminosae, in less extreme hot regions also Fagaceae), evergreen sclerophyllous woodlands (up to 15 m high, only 1-2 tree layers, with several Myrtaceae), regarded as connected with winter rains, have been found in large planar, colline, and sub-montane areas. The highly diversified, mainly evergreen mountain forests above 1000 m (above sea level; also UHLIG, 1977) with many Lauraceae, Fagaceae, and interesting conifers (e.g., *Fokienia*, *Cunninghamia*, *Libocedrus*, *Cryptomeria*, *Podocarpus*, etc.) are treated rather summarily (compiled in 4 mapping units). This vegetation has been differentiated in much more detail in tropical southern China (WU, 1979a,b).

b) Central Polynesian and Micronesian Vegetation

The prevalently small islands of central Polynesia and of Micronesia are covered by a highly diversified vegetation, up to some hundred meters above sea level. On Eastern Samoa with an area of only 195 km^2 (highest elevation 632 m) with abundant rains (up to more than 5000 mm annual mean), most of the humid vegetational formation groups have been stated (WHISTLER, 1980), discerned generally for the whole hot tropical Pacific regions (for comparison in Western Samoa: OLLIER et al., 1979; WHISTLER, 1978). The terrestrial vegetation of atolls with relatively flat land surfaces is much less manifold, all the more so as anthropogenic influences are frequently intensive (e.g., SABATH, 1978; STODDART and COE, 1979; Polynesian land use influences: KIRCH, 1978).

c) Vegetation of the Antilles, of the Bahamas, and of the Bermudas

The neotropical vegetation of Guadeloupe, Martinique, and adjacent islands has been studied in new projects (CURRAT and MATHURIN, 1979; DOREAU and VIVIER, 1979; DE FOUCAULT, 1978; JEREMIE and RAYNAL-ROQUES, 1978; STEHLÉ, 1980), including vegetation mapping (PORTECOP, 1979) and an analysis of the first stages on deposits of the volcano Soufrière (eruptions 1976, 1977; STEHLÉ, 1979). Cuba (BORHIDI, 1979; SCHUBERT, 1979, 1980), the Bahamas (CORRELL, 1979; KNAPP, 1980a), and the Bermudas (KNAPP, 1980d), also predominantly neotropical, have some physiognomical traits transitional to the vegetation of the south-eastern North American coastal plains, among them woodlands with species of *Pinus* and *Juniperus*.

6. Vegetation of the Extra-Tropical Southern Hemisphere

a) Vegetation of Extra-Tropical South America

In studies on extra-tropical South America, a certain emphasis existed formerly on orientating vegetation mapping of large regions and on the characterization of the main vegetational zonation (e.g., zonal synopsis: LASSALLE, 1980; vegetation maps: NW Argentine, VERVOOST, 1979; southern Brasil, KLEIN, 1978; Andine altitudinal zonation: DONOSO, 1978; GAJARDO, 1980; GARLEFF, 1977; QUINTANILLA, 1979; RODRIGUEZ-GARCIA et al., 1978; SCHULMEYER, 1978). The concentration has now shifted more to special studies on the structure, function, dynamics, and problems of environmental conservation in groups of plant communities, for instance, in the *Nothofagus* forests and related woody vegetation (PISANO, 1977; ROMERO, 1980; STEUBING et al., 1980; VEBLEN et al., 1978, 1979, 1980; VILLAGRAN, 1980; VILLAGRAN et al., 1979; WEINBERGER, 1978; Juan Fernandez: SCHWAAR, 1979a). The conservational function is even more important in the Chilean matorral and related shrub lands or open woods (ARMESTO and GUTIERREZ, 1978; ARMESTO and MARTÍNEZ, 1978; COWLING and CAMPBELL, 1980; GILIBERTO and ESTAY, 1978; GUTIERREZ and FUENTES, 1979; JAKSIĆ and FUENTES, 1980; MONTENEGRO et al., 1979), in the arid areas of north-western Argentine (ANCIBOR, 1980; RUTHSATZ, 1978), or in the steppes and semi-deserts of Patagonia (MONTENEGRO et al., 1978; ROIG et al., 1980; SEKI, 1979; SORIANO et al., 1980). Special sites considered are the sand dunes (in coastal southern Brasilia: PFADENHAUER, 1978, 1980; in the interior of Argentine: ESKUCHE, 1979) and various aquatic or otherwise wet or moist biotopes (PALMA et al., 1978; RAMÍREZ et al., 1979; SAN MARTIN, 1979; SANTELICAS, 1980; SCHWAAR, 1979c). The vegetation above the timber-line (CABRERA et al., 1977; VEBLEN and ASHTON, 1979) approximates in the southern-most part of the continent the sea level, merging with the Antarctic (see below) tree-less zones (BLISS, 1979). (Irradiations of its special attributes beyond the equator in the Paramo vegetation: CLEEF, 1978).

b) Antarctic Vegetation

The Antarctic vegetation south of 50° is now studied in numerous areas sufficient to substantiate general traits of the zonation, structure, and classification (ALEKSANDROVA, 1980; BLISS, 1979; HOLDGATE, 1977; LLANO, 1977; LONGTON, 1979). On the island between 50° and 60° (latitude; Falkland, South Georgia, Macquarie, Gough, Edward, Marion islands, etc.), plant communities with grasses, forbs, herbs, and dwarf shrubs are frequent, either forming partly rather closed vegetation (in the "Low Antarctic") or only very scattered, open vegetation (in the "High Antarctic", BLISS, 1979). The vegetation south of 60° can only be compared with the climatically most severe and most marginal geobotanical zones of the Arctic (polar semi-desert or desert). Islands south of 60° (S Orkney, S Shetland, etc.) and a few parts of the Antarctic Peninsula (Grahamland) are inhabitated by not more than two angiosperms (*Deschampsia antarctica, Colobanthus quitensis*), but by several lichen and moss species, forming nine subformations (in the sense of LONGTON, 1979). Four of these subformations with only few associations are represented in the extremely poor vegetation of the Antarctic continent (except of the above-mentioned parts of the Peninsula).

As compared with the Arctic plant cover of the same latitude on the northern hemisphere, the Antarctic vegetation is much less diversified. The poor representation of angiosperms is especially conspicuous. Therefore, mosses and also lichens are relatively much more important components of the Antarctic terrestrial vegetation (DODGE, 1979; KASHIWADANI, 1979; LINDSAY, 1978; NAKANISHI, 1977; SAVICH-LYUBITSKAYA, 1978; SMITH and GIMINGHAM, 1976). The primary productivity of the Antarctic vegetation does not always seem to be excessively scanty (BLISS, 1979; FENTON, 1980; PRIDDLE, 1980). Annual net productions of 500-1500 g/m^2 are reported in certain Antarctic island grass-forb and herb meadow communities. As possible reasons, the following attributes of this Antarctic vegetation are considered: (1) The relatively long growing season due to the extreme climatic oceanity. (2) High nutrient level owing to the extreme numbers of marine birds and sea mammals concentrating on the small land surfaces of the Antarctic islands.

References

ALEKSANDROVA, V.D.: The Arctic and Antarctic. 247 pp. New York: Cambridge Univ. Press 1980. - ALTHUIS, H. VAN, KEYSERS, G.: Bull. Soc. R. Bot. Belgique 112, 96-127 (1979). - AMANDIER, L., GASQUEZ, J.: Boissiera 27, 1-153 (1978). - ANCIBOR, E.: Bol. Soc. Argent. 19, 157-202 (1980). - ARMESTO, J., MARTÍNEZ, J.A.: J. Ecol. 66, 881-890 (1978). - AUSTIN, M.P.: Vegetatio 43, 87-94 (1980).

BALÁTOVÁ-TULÁČKOVÁ: Rozpr. Československ. Akad. Věd. M. P. V. 88 (3), 1-113 (1978). - BARDEN, L.S.: Castanea 43, 238-246 (1978); - Oikos 35, 16-19 (1980). - BARRY, J.M.: Natural Vegetation of South Carolina. 238 pp. Columbia: South Carolina University Press 1980. - BASKIN, J.M., BASKIN, C.C.: Rhodora 80, 545-557 (1978); - Bull. Torrey Bot. Club 106, 176-181 (1979a); - Castanea 44, 193-202 (1979b). - BEEFTINK, W.G.: Vegetatio 40, 101-105 (1979). - BELLIS, V.: Veröff. Geobot. Inst. Eidg. Tech. Hochsch. Stift. Rübel Zürich 69, 121-144 (1980). - BLAGOVESCHENSKY, I.V., BLAGOVESCHENSKAYA, N.V.: Bot. Zh. 63, 1178-1189 (1978). - BLAŽKOVÁ, D.: Folia Geobot. Phytotax. 16, 45-59 (1981). - BLISS, L.C.: Can. J. Bot. 57, 2167-2178 (1979). - BÖTTCHER, H.: Phytocoenologia 7, 8-20 (1980). - BOLOS, O. DE: Doc. Phytosociol. N. Sér. 4, 69-74 (1979). - BONDEV, I.A.: Folia Geobot. Phytotax. 15, 189-191 (1980). - BORHIDI, A. (ed.): Acta Bot. Hungar. 25, 263-301 (1979). - BOTCH, M.S., VASILEVICH, V.I.: Bot. Zh. 65, 27-38 (1980). - BOURNÉRIAS, M.: Guide des groupements végétaux de la région parisienne. 509 pp. Paris: SEDES 1979. - BOURNÉRIAS, M., LAVERGNE, D.: Carte de végétation: Châlons sur Marne. Toulouse: C.N.R.S. 1979. - BRADSHAW, A.D., CHADWICK, M.J.: Studies Ecol. (Oxford) 6, 1-314 (1980). - BRINSON, M.N., BRADSHAW, H.D., HOLMES, R.N., ELKINS, J.R.: Ecology 61, 827-835 (1980).

CABRERA, A.L., KIESLING, R., ZARDINI, E.: Obra Cent. Museo La Plata 3, 213-227 (1977). - CAMPBELL, B.M., VAN DER MEULEN, F.: Vegetatio 43, 43-47 (1980). - CHAU, K.C., LO, W.K.: Plant Soil 56, 243-254 (1980). - CHRISTENSEN, N.L.: Veröff. Geobot. Inst. Eidg. Techn. Hochsch. Stift. Rübel Zürich 68, 246-262 (1979). - CLEEF, A.M.: Erdwiss. Forschungen 11, 365-390 (1978). - CODY, M.L.: Am. J. Bot. 65, 1107-1116 (1978). - CORRELL, D.S.: Taxon 28, 35-40 (1979). - COWLING, R.M., CAMPBELL, B.M.: Vegetatio 42, 191-197 (1980). - CURRAT, G., MATHURIN, G.: Bull. A.P.B.G. (Paris) 3, 345-350 (1979).

DANIN, A.: Flora 167, 409-422 (1978). - DEBUSSCHE, M., ESCARRÉ, J., LEPART, J.: Vegetatio 42, 73-82 (1980). - DELCOURT, P.A., DELCOURT, H.R.: Veröff. Geobot. Inst. Eidg. Tech. Hochsch. Stift. Rübel Zürich 68, 79-107 (1979). - DIERSCHKE, H.: Naturopa 31, 29-32 (1978); - Doc. Phytosociol. N. Ser. 4, 235-252 (1979); - Verhandl. Ges. Ökol. 8, 205-212 (1980). - DIERSCHKE, H., JECKEL, G.: Mitt. Florist. Soziol. Arbeitsgem. 22, 77-81 (1980). - DIERSSEN, K.: Colloq. Phytosociol. 7, 391 (1980). - DODGE, C.W.: Bull. Torrey Bot. Club 106, 262-263 (1979). - DONIȚA, N.: Folia Geobot. Phytotax. 15, 178-180 (1980). - DONOSO, C.: Bol. Téc. Fac. Ci. Forest. (Santiago) 54, 1-27 (1978). - DOREAU, M., VIVIER, M.: Agron. Tropic. 34, 356-371 (1979). - DUPIAS, G., VILKS, A.: Carte de la végétation: Poitiers. Toulouse: C.N.R.S. 1979.

ECKELS, K., TÜXEN, R., RIESS, W., KNAPP, R.: Excerpta Bot. B 20, 1-16 (1980). - ELINA, G.A., YURKOVSKAYA, T.K.: Bot. Zh. 65, 958-970 (1980). - ELLENBERG, H.: Scripta Geobot. 9, 1-122 (1979); - Phytocoenologia 7, 507-511 (1980). - ELLENBERG, H., FRÄNZLE, O., MÜLLER, P.: Forschungsber. Umweltbundesamt 78-101, 1-144 (1978). - ESKUCHE, U.: Doc. Phytosociol. N. Sér. 4, 253-266 (1979). - EUROLA, S., KAAKINEN, E.: Suotyyppiopas. 87 pp. Helsinki: Poorvo, Juva 1978. - EUROLA, S., VORREN, K.-D.: Aquilo Ser. Bot. 17, 39-56 (1980). - EWEL, K.C., MITSCH, W.J.: Florida Sci. 41, 25-31 (1978).

FALIŃSKI, J.: Vegetatio 43, 23-28 (1980). - FEKETE, G.: Folia Geobot. Phytotax. 15, 193-196 (1980). - FENTON, J.H.C.: J. Ecol. 68, 211-228 (1980). - FITTER, A., SMITH, C. (eds.): A Wood in Ascam. 164 pp. York: Ebor 1979. - FORE, P.L.: PETERSON, R.D.: Gulf of Mexico Coastal Ecosystems Workshop. 214 pp. Washington, D.C.: U.S.D.I. 1980. - FOUCAULT, B. DE: Doc. Phytosociol. N. Sér. 2, 181-189 (1978). - FRANKLIN, J.F., MAEDA, T., OHSUMI, Y., MATSUI, M., YAGI, H., HAWK, G.M.: Ecol. Monogr. 49, 311-334 (1979). - FUKAREK, P.: Folia Geobot. Phytotax. 15, 183-189 (1980).

GADDY, L.L., SMATHERS, G.A.: Veröff. Geobot. Inst. Eidg. Tech. Hochsch. Stift. Rübel Zürich 69, 171-182 (1980). - GAJARDO, R.: Bol. Téc. Fac. Ci. Forest. (Santiago) 57, 1-27 (1980). - GARLEFF, K.: Göttinger Geogr. Abh. 68, 1-150 (1977). - GÉHU, J.M.: Etude phytocoenotique des vases et prés salés. 514 pp. Lille: Minist. Environm. 1979a; - Carte phytosociologique de la végétation naturelle potentielle du N France. Lille: OREAM 1979b; - (ed.) Colloq. Phytosociol. 7, 1-556 (1980). - GÉHU, J.M., BIGOT, M.: Excerpta Bot. B. 20, 207-239 (1980). - GILIBERTO, J., ESTAY, H.: Bot. Gaz. 139, 236-240 (1978). - GLENN-LEWIN, D.C.: Vegetatio 42, 141-148 (1980). - GOLDEN, M.S.: Ecology 60, 770-782 (1979). - GRIBOVA, S.A., ISACHENKO, T.I.: Bot. Zh. 65, 609-612 (1980). - GROSSE-BRAUCKMANN, G.: Ber. Symp. Int. Ver. Vegetationskd. 1978, 111-132 (1979). - GUTIÉRREZ, J.R., FUENTES, E.R.: Oecol. Plant. 14, 151-158 (1979).

HABER, W. (ed.): Verhandl. Ges. Ökol. 8, 9-267 (1980). - HALL, A.E., CANNELL, G.H., LAWTON, H.W. (eds.): Agriculture in Semi-Arid Environments. 340 pp. Berlin, Heidelberg, New York: Springer 1979. - HANSEN, B., JENSEN, J., MOGENSEN, G.S.: Bot. Tidsskr. 73, 1-19 (1978). - HANSKI, I.: Ecology 59, 168-174 (1978). - HARRIS, D.R. (ed.): Human Ecology in Savanna Environments. 522 pp. London: Academic Press 1980. - HICKMAN, J.L., NEUHAUSER, J.A.: Bartonia 45, 30-36 (1978). - HOFMANN, G.: Folia Geobot. Phytotax. 15, 201-205 (1980). - HOLDGATE, M.W.: Trans. R. Soc. London B 279, 5-25 (1977). - HOPKINSON, C.S., GOSSELINK, J.G., PARRONDO, R.T.: Ecology 61, 1091-1098 (1980). - HOSIER, P.E., EATON, T.E.: J. Appl. Ecol. 17, 173-182 (1980). - HOUSSARD, C., ESCARRÉ, J., ROMANE, F.: Vegetatio 43, 59-72 (1980). - HÜLBUSCH, K.H.: Doc. Phytosociol. N. Sér. 4, 451-462 (1979); - Mitt. Florist. Soziol. Arbeitsgem. 22, 51-75 (1980). - HÜLBUSCH, K.H., BÄUERLE, H., HESSE, F., KIENAST, D.: Urbs et Regio 11, 1-215 (1979). - HUKUSIMA, T.: Bull. Yokohama Phytosociol. Soc. 16, 251-266 (1979). - HULST, R. VAN: Vegetatio 39, 85-96 (1979a); - ibid. 40, 3-14 (1979b); - ibid. 43, 147-151 (1980). - HURLBERT, S.H.: Ecology 59, 67-77 (1978). - HUTCHINSON, G.E.:

An Introduction to Population Ecology. 260 pp. New Haven: Yale University Press 1978.

IPATOV, V.S., KIRIKOVA, L.A.: Bot. Zh. 65, 470-477 (1980). - ISHIKAWA, S.: Ecol. Rev. (Japan) 19, 67-74 (1979).

JAHN, G.: Forstwiss. Cbl. 99, 297-323 (1980). - JAKSIĆ, F.M., FUENTES, E.R.: J. Ecol. 68, 665-670 (1980). - JENÍK, J., BUREŠ, L., BUREŠOVÁ, Z.: Folia Geobot. Phytotax. 15, 1-28 (1980). - JENSÉN, E.W.: Acta Phytogeogr. Suecica 66, 1-120 (1979). - JEREMIE, J., RAYNAL-ROQUES, A.: Adansonia 18, 279-290 (1978). - JOHNSON, E.A.: Ecology 60, 238-240 (1979). - JOVET, P.: Carte de la végêtation: Melun. Toulouse: C.N.R.S. 1979.

KASHIMURA, T., OHMINATO, T., KOJIMA, M.: Bull. Yokohama Phytosociol. Soc. 16, 189-196 (1979). - KASHIWADANI, H.: Mem. Nat. Inst. Polar Res. (Tokyo) Spec. Iss. 11, 42-46 (1979). - KEMP, P.R., WILLIAMS, G.J.: Ecology 61, 846-858 (1980). - KIRCH, P.V.: Econ. Bot. 32, 157-181 (1978). - KIRK, P.W. (ed.): The Great Dismal Swamp. 464 pp. Charlottesville: Virginia University Press 1979. - KIVINEN, E., HEIKURAINEN, L., PAKARINEN, P.: Classification of Peat and Peatlands. 367 pp. Helsinki: Internat. Peat Society 1979. - KLEIN, R.M.: Mapa Fitogeográfica de Estado de Santa Catarina. Itajaí: Herbário Ed. 1978. - KNAPP, R.: Excerpta Bot. B 20, 97-124, 137-144, 241-263 (1980a); - Angew. Bot. 54, 125-138 (1980b); - Ber. Symp. Int. Ver. Vegetationskd. 1979, 37-52 (1980c); - Phytocoenologia 7, 475-491 (1980d). - KRAUSE, A.: Decheniana 131, 52-60 (1978). - KÜCHLER, A.W.: Excerpta Bot. B 20, 53-60 (1980a); - Univ. Kansas Publ. Library Ser. 45, 1-324 (1980b). - KULLMAN, L.: Acta Phytogeogr. Suecica 65, 1-121 (1979). - KUZNETZOV, O.L.: Bot. Zh. 63, 933-935 (1978).

LANDOLT, E.: Veröff. Geobot. Inst. Eidg. Tech. Hochsch. Stift. Rübel Zürich 70, 13-29 (1980). - LANE, D.M.: Veröff. Geobot. Inst. Eidg. Tech. Hochsch. Stift. Rübel Zürich 69, 114-120 (1980). - LASSALLE, J.C.: Ber. Geobot. Inst. Eidg. Tech. Hochsch. Stift. Rübel Zürich 47, 87-131 (1980). - LAVRENKO, E.M., ISAČENKO, T.I., GRIBOVA, S.A.: Folia Geobot. Phytotax. 15, 165-171 (1980). - LAWLOR, L.R.: Ecology 61, 245-251 (1980). - LEPŠ, J., PRACH, K.: Folia Geobot. Phytotax. 16, 61-72 (1981). - LEUTHOLD, C.: Veröff. Geobot. Inst. Eidg. Tech. Hochsch. Stift. Rübel Zürich 67, 1-217 (1980). - LIETH, H.: Veröff. Geobot. Inst. Eidg. Tech. Hochsch. Stift. Rübel Zürich 68, 25-50 (1979). - ibid. 69, 162-170 (1980a); - MAB-Mitt. (Bonn) 6, 91-100 (1980b). - LINDSAY, D.C.: Bryologist 81, 268-276 (1978). - LINDSAY, M.M., BRATTON, S.P.: Bull. Torrey Bot. Club 106, 264-275 (1979). - LLANO, G.A. (ed.): Adaptations within Antarctic Ecosystems. 1110 pp. Washington, D.C.: Smithsonian Inst. 1977. - LÖVE, Á., LÖVE, D.: Phytocoenologia 6, 94-105 (1979). - LONDO, G.: Phytocoenosis 7, 165-175 (1978). - LONGTON, R.E.: Can. J. Bot. 57, 2264-2278 (1979). - LORENZONI, G.G. (ed.): Notiz. Soc. Ital. Fitosociol. 15, 1-208 (1979). - LORIMER, C.G.: Ecology 61, 1169-1184 (1980).

MAAREL, E. VAN DER: Vegetatio 42, 95-102 (1980). - MAAREL, E. VAN DER, WERGER, M.J.A.: Phytocoenosis 7, 257-278 (1978). - MAAREL, E. VAN DER, ORLOCI, L., PIGNATTI, S. (eds.): Advances Veget. Sci. 1, 1-228 (1980). - MAJOR, J.: Phytocoenologia 7, 26-34 (1980). - MAKITA, H., MIYAGI, T., MIURA, O., KIKUCHI, T.: Bull, Yokohama Phytosociol. Soc. 16, 237-244 (1979). - MALMER, N., LINDGREN, L., PERSSON, S.: Vegetatio 36, 17-36 (1978). - MALMGREN, V.: Sven. Bot. Tidskr. 73, 177-188 (1979). - MASUZAWA, T., KONTA, F.: Bull. Yokohama Phytosociol. Soc. 16, 197-202 (1979). - MATTHEWS, J.A.: J. Ecol. 67, 255-272 (1979). - MATUSZKIEWICZ, W.: Mitt. Florist. Soziol. Arbeitsgem. 22, 19-50 (1980a); - Folia Geobot. Phytotax. 15, 171-173 (1980b). - McEVOY, T.J., SHARIK, T.L., SMITH, D.W.: Am. Midland Nat. 103, 96-105 (1980). - McKINLEY, C., DAY, F.P.: Bull. Torrey Bot. Club 106, 20-28 (1979). - MEHER-HOMJI, V.M., GUPTA, R.K.: Excerpta Bot. B 20, 17-38 (1980). - MÉRIAUX, J.-L.: Doc. Phytosociol. N. Sér. 3, 1-244 (1978). - MIŠIĆ, V.: Folia Geobot. Phytotax. 15, 181-182 (1980). - MITSCH, W.J., EWEL, K.C.: Am. Midland Nat. 101, 417-426 (1979). - MITSCH, W.J., DORGE, C.L., WIEMHOFF, J.R.: Ecology 60, 1116-1124 (1979). - MIYAWAKI, A.: Bull. Yokohama Phytosociol. Soc. 16, 49-70 (1979). - MIYAWAKI, A., SUZUKI, K.: Phytocoenologia 7, 492-506 (1980). - MOLINIER, R., DEVAUX, J.P.: Biol. Écol. Médit. 5, 159-195 (1978). - MONTAGUE, K.A., DAY, F.P.: Am. Midland Nat. 103, 83-87 (1980). - MONTENEGRO, G., HOFFMANN, A.J., ALJARO, M.E., HOFFMANN, A.E.: Bot. J. Linn. Soc. 77, 177-202 (1978). - MONTENEGRO, G., ALJARO, M.E., KUMMEROW, J.: Bot. Gaz. 140, 114-119 (1979). - MORAVEC, J., MICHALKO, J.: Folia Geobot. Phytotax. 15, 196-199 (1980). - MOROSOV, V.L.: Bot. Zh. 63, 682-689 (1978); - ibid. 65, 538-543 (1980).

NAKANISHI, H.: Bull. Yokohama Phytosociol. Soc. 16, 289-302 (1979). - NAKANISHI, S.: Antarctic Res. 59, 68-96 (1977). - NAKANISHI, S., HATTORI, T.: Bull. Yokohama Phytosociol. Soc. 16, 113-140 (1979). - NEUHÄUSL, R.: Folia Geobot. Phytotax. 15, 174-176 (1980). - NEUHÄUSL, R., NEUHÄUSLOVÁ-NOVOTNÁ, Z.: Stud. ČSAV (Praha) 2, 1-203 (1979). - NEUHÄUSLOVÁ-NOVOTNÁ, Z., NEUHÄUSL, R.: Excerpta Bot. B 20, 145-170 (1980). - NIKLFELD, H.: Stapfia 4, 1-229 (1979). - NOBLE, I.R., SLATYER, R.O.: Vegetatio 43, 5-22 (1980). - NORIN, B.N.: Bot. Zh. 62, 544-547 (1978). - NUMATA, M.: Bull. Yokohama Phytosociol. Soc. 16, 71-92 (1979).

OBERDORFER, E.: Mitt. Florist. Soziol. Arbeitsgem. 22, 11-18 (1980). - OHBA, T., SUGAWARA, H.: Bull. Yokohama Phytosociol. Soc. 16, 267-280 (1979). - OHNO, K.: Bull. Yokohama Phytosociol. Soc. 16, 227-236 (1979). - OKUDA, S.: Bull. Yokohama Phytosociol. Soc. 16, 203-212 (1979). - OKUTOMI, K.: Bull. Yokohama Phytosociol. Soc. 16, 145-160 (1979). - OLLIER, C., WHISTLER, W.A., AMERSON, A.B.: O le Pupu-Pue National Park. 79+83 pp. Suva: U.N. Develop. Adv. Pacif. 1979. - OZENDA, P.: Folia Geobot. Phytotax. 15, 156-160 (182 (1980). - OZENDA, P., NOIRFALISE, A., TOMASELLI, R., TRAUTMANN, W.: Vegetation Map of the Council of Europe Member States. Strasbourg: Publication Section, Council of Europe, 1979.

PALMA, B., SAIZ, F., PIZARRO, C.: Anales Mus. Hist. Nat. Valparaiso 11, 61-67 (1978). - PARENT, G.H.: Mém. Soc. R. Bot. Belgique 8, 1-72 (1980). - PARRISH, F.K., RYKIEL, E.J.: J. Elisha Mitchell Soc. 95, 17-31 (1979). - PARRISH, J.A.D., BAZZAZ, F.A.: Ecology 60, 597-610 (1979). - PASSARGE, H.: Folia Geobot. Phytotax. 13, 1-16 (1978); - Mitt. Florist. Soziol. Arbeitsgem. 22, 111-124 (1980). - PEDROTTI, F.: Doc. Phytosociol. N. Sér. 4, 795-802 (1979). - PEET, R.K., CHRISTENSEN, N.L.: Vegetatio 42, 131-140 (1980a); - Veröff. Geobot. Inst. Eidg. Tech. Hochsch. Stift. Rübel Zürich 69, 14-39 (1980b). - PERSSON, S.: Vegetatio 43, 103-122 (1980). - PETERMANN, R., SEIBERT, P.: Nationalpark Bayerischer Wald 4, 1-142 (1979). - PETERSEN, P.M.: Bot. Tidskr. 75, 77-88 (1980). - PETERSON, D.L., BAZZAZ, F.A.: Ecology 59, 1005-1013 (1978). - PETRAITIS, P.S.: Ecology 60, 703-710 (1979). - PFADENHAUER, J.: Rev. Brasil. Biol. 38, 827-836 (1978); - Phytocoenologia 8, 321-364 (1980). - PISANO, E.: Anales Inst. Patagon 8, 121-250 (1977). - PITTILLO, J.D., SMATHERS, G.A.: Veröff. Geobot. Inst. Eidg. Tech. Hochsch. Stift. Rübel Zürich 68, 206-245 (1979). - PORTECOP, J.: Doc. Cart. Ecol. 21, 1-78 (1979). - POTT, R.: Abh. Landesmus. Naturkd. Münster Westf. 42 (2), 1-156 (1980). - PRIDDLE, J.: J. Ecol. 68, 141-166 (1980).

QUINTANILLA, V.G.: Biogeogr. 16, 49-68 (1979).

RABOTNOV, T.A.: Excerpta Bot. B 20, 171-190 (1980). - RAMÍREZ, C.: Excerpta Bot. B 20, 61-64 (1980). - RAMÍREZ, C., ROMERO, M., RIVEROS, M.: Aquatic Bot. 7, 241-253 (1979). - REGNÉLL, G.: Vegetatio 43, 123-130 (1980). - RIVAS-MARTÍNEZ, S.: Lazaroa (Madrid)) 1, 1-128 (1979). - RIVAS-MARTÍNEZ, S., GÉHU, J.-M.: Doc. Phytosociol. N. Sér. 3, 371-423 (1978). - ROBERTSON, P.A.: Vegetatio 37, 45-54 (1978). - ROBERTSON, P.A., WEAVER, G.T., CAVANAUGH, J.A.: Ecol. Monogr. 48, 249-267 (1978). - RODRIGUEZ-GARCIA, D., SOUROUILLE, A., GALLOPIN, G., MONTANA, C.: An. Parq. Nacion. (Buenos Aires) 14, 161-248 (1978). - ROIG, F.A., MARCO, G. DE, WILLOUD, C.: Bol. Soc. Argent. 19, 331-338 (1980). - ROMERO, E.J.: Bol. Soc. Argent. 19, 289-308 (1980). - RUTHSATZ, B.: Darwiniana 21, 491-539 (1978). - RUUHIJÄRVI, R., REINIKAINEN, A., LINDHOLM, T.: Lammi Notes 2, 14-19 (1979).

SABATH, M.D.: Pacif. Sci. 31, 321-333 (1978). - SAITO, K.: Bull. Yokohama Phytosociol. Soc. 16, 177-188 (1979). - SAN MARTIN, J.: Bol. Inst. Ecol. Chile (Talca) 4, 36-38 (1979). - SANTELICAS, B.: Phycologia 19, 1-12 (1980). - SASAKI, Y.: Bull. Yokohama Phytosociol. Soc. 16, 213-226 (1979). - SAUVAGE, C. (ed).: Terre et Vie (Paris) Suppl. 2, 3-282 (1979). - SAVICH-LYUBITSKAYA, L.I.: Bot. Zh. 62, 272-290 (1978). - SCAMONI, A.: Folia Geobot. Phytotax. 15, 177-178, 191-193 (1980). - SCHLESINGER, W.H.: Am. Nat. 112, 949-954 (1978a); - Ecol. Monogr. 48, 43-65 (1978b). - SCHLÜTER, H.: Folia Geobot. Phytotax. 15, 180-181 (1980). - SCHNEEKLOTH, H., TÜXEN, J.: Veröff. Niedersächs. Inst. Landeskd. Landesentw. Univ. Göttingen A 96 (5), 1-220 (1978). - SCHUBERT, R. (ed.): Feddes Repert. 90, 19-50, 313-326 (1979, 1980). - SCHULMEYER, D.: Bol. Inform. Inst. Geogr. (Santiago) 2, 11-27 (1978). - SCHUSTER, H.J.: Dissert. Bot. 53, 1-478 (1980). - SCHWAAR, J.: Phytocoenologia 6, 514-523 (1979a); - Telma (Hannover) 9, 53-61 (1979b); - Amazoniana 6, 601-609 (1979c). - SCHWABE-BRAUN, A.: Waldschutzgebiete (Freiburg i.B.) 1, 1-68 (1979a); - Urbs et Regio (Kassel) 18, 1-212 (1979b). - SCHWABE-BRAUN, A., TÜXEN, R.: Prodromus Europ. Pflanzenges. 4, 1-140 (1981). - SEKI, T.: Bull. Yokohama Phytosociol. Soc. 16, 349-356 (1979). - SHIDEI, T.: Bull. Yokohama Phytosociol. Soc. 16, 93-100 (1979). - SHUGART, H.H., WEST, D.C.: Bioscience 30, 308-313 (1980). - SINGH, J.S., GOPAL, B. (eds.): Glimpses of Ecology. 592 pp. Jaipur: International Scientific Publications 1978. - SJÖRS, H.: Vegetatio 42, 1-4 (1980). - SKEEN, J.N., CARTER, M.E., RAGSDALE, H.L.: Bull. Torrey Bot. Club 107, 1-6 (1980). - SLOBODCHIKOFF, C.N., SCHULZ, W.C.: Ecology 61, 1051-1055 (1980). - SMITH, C.J.: Ecology of the English Chalk. 574 pp. London: Academic Press 1980. - SMITH, R.I.L., GIMINGHAM, C.H.: Brit. Antarct. Surv. Bull. 43, 25-47 (1976). - SNYDER, J.R.: Veröff. Geobot. Inst. Eidg. Tech. Hochsch. Stift. Rübel Zürich 69, 40-113 (1980). - SÖYRINKI, N., SAARI, V.: Acta Bot. Fenn. 114, 1-148 (1980). - SONESSON, M. (ed.): Ecol. Bull. (Stockholm) 30, 1-313 (1980). - SOÓ, R.V.: Synopsis Syst.-Geobot. Flor. Veget. Hungariae 6, 1-557 (1980). - SORIANO, A., SALA, O.E., LEÓN, R.J.C.: Bol. Soc. Argent. 19, 309-330 (1980). - STEHLÉ, H.: Bull. Soc. Bot. France 126, 349-359, 127, 159-168 (1979); - ibid. 127, 275-287, 373-386 (1980). - STEINBERG, B.: Florida Sci. 43, 7-12 (1980). - STEUBING, L., RAMÍREZ, C., ALBERDI, M.: Vegetatio 39, 25-34 (1979). - STEYERMARK, J.A., HUBER, O.: Flora y vegetación des las Montañas del Avila. 971 pp. Caracas: INCAFO 1978. - STODDART, D.R., COE, M.J.: Atoll. Res. Bull. 223, 3-18, 27-42 (1979). - SWAINE, M.D., GREIG-SMITH, P.: J. Ecol. 68, 33-42 (1980). - SWITZER, G.L., SHELTON, M.G., NELSON, L.E.: Ecology 60, 1162-1171 (1979).

TAGAWA, H.: Bull. Yokohama Phytosociol. Soc. 16, 379-392 (1979). - TAODA, H.: Bull. Yokohama Phytosociol. Soc. 16, 161-166 (1979). - THAI VAN TRUNG: Zit. in THOMASIUS, H.: Petermanns Geogr. Mitt. 124, 23-46

(1980). - THANNHEISER, D.: Nat. Canad. 104, 151-156 (1977). - THOMASIUS, H.: Petermanns Geogr. Mitt. 124, 23-46 (1980). - TIMBAL, J.: Carte de la végétation: Metz. Toulouse: C.N.R.S. 1979. - TOYOHARA, G.: Bull. Yokohama Phytosociol. Soc. 16, 167-176 (1979). - TRABAUD, D., LEPART, J.: Vegetatio 42, 49-58 (1980). - TRAUTMANN, W.: Folia Geobot. Phytotax. 15, 182 (1980). - TRAUTMANN, W., BOHN, U.: Folia Geobot. Phytotax. 15, 160-165 (1980). - TÜXEN, J.: Ber. Symp. Int. Ver. Vegetationskd. 1978, 133-151 (1979); - Phytocoenologia 7, 142-165 (1980a); - Colloq. Phytosociol. 7, 371-381 (1980b). - TÜXEN, R.: Ber. Symp. Int. Ver. Vegetationskd. 1977, 219-224 (1979a); - ibid. 1978, 363-365 (1979b); - Colloq. Phytosociol. 7, 383-391 (1980). - TÜXEN, R., INSELBERG, A.: Bibliographia Phytosociol. Syntax. 34, 1-44 (1979). - TÜXEN, R., LOIDI, X., NAKAMURA, Y.: Bibliographia Phytosociol. Syntax. 33, 1-111 (1979). - TURKINGTON, R., HARPER, J.L.: J. Ecol. 67, 219-230 (1979). - TYNDALL, R.W., LEVY, G.F.: J. Elisha Mitchell Soc. 94, 1-15 (1979).

UHLIG, S.K.: Petermanns Geogr. Mitt. 121, 190-191 (1977). - URBANSKA-WORYTKIEWICZ, K.: Veröff. Geobot. Inst. Eidg. Tech. Hochsch. Stift. Rübel Zürich 70, 30-101 (1980).

VARESCHI, V.: Vegetationsökologie der Tropen. 294 pp. Stuttgart: Ulmer 1980. - VARTIAINEN, T.: Acta Bot. Fenn. 115, 1-105 (1980). - VEBLEN, T.T., ASHTON, D.H.: Vegetatio 40, 39-48 (1979). - VEBLEN, T.T., SCHLEGEL, F.M., ESCOBAR, R.B.: J. Ecol. 68, 1-32, 397-404 (1980). - VERVOOST, F.B.: Conservación Natur. (Tucumán) 1, 7-50 (1979). - VILLAGRAN, C.: Dissert. Bot. 54, 1-165 (1980). - VILLAGRAN, C., TRONCOSO, A., MUÑOZ, M.: Arch. Biol. Med. Exp. (Santiago) 12, 485-486 (1979). - VORREN, K.-D.: Norw. Arch. Rev. 12, 1-21 (1979a); - Tromura Naturwit. (Tromsö) 3, 1-118, 5, 1-181 (1979b); - ibid. 8, 1-103 (1980).

WALKER, B.H. (ed.): Management of Semiarid Ecosystems. 398 pp. Amsterdam: Elsevier 1979. - WARD, D.B.: Rare Endangered Biota Florida 5, 1-175 (1979). - WATANABE, S.: Bull. Yokohama Phytosociol. Soc. 16, 101-112 (1979). - WATSON, M.A.: Bull. Torrey Bot. Club 107, 346-372 (1980). - WEINBERGER, P.: Anales Parques Nat. (Buenos Aires) 14, 133-160 (1978). - WENTWORTH, T.R.: Veröff. Geobot. Inst. Eidg. Tech. Hochsch. Stift. Rübel Zürich 69, 145-161 (1980). - WHIPPLE, S.A.: Bull. Torrey Bot. Club 107, 71-76 (1980). - WHISTLER, W.A.: Pacif. Sci. 32, 79-94 (1978); - Allertonia 2 (2), 1-190 (1980). - WIEGLEB, G.: Vegetatio 38, 165-174 (1978). - WILMANNS, O.: Mitt. Florist. Soziol. Arbeitsgem. 22, 125-134 (1980). - WILMANNS, O., TÜXEN, R. (eds.): Ber. Symp. Int. Ver. Vegetationskd. 1978, 1-636 (1979); - ibid. 1979, 1-466 (1980). - WILMANNS, O., SCHWABE-BRAUN, A., EMTER, M.: Doc. Phytosociol. N. Sér. 4, 983-1024 (1979). - WITSCHEL, M.: Beih. Veröff. Naturschutz Landschaftspfl. Baden-Württ. 17, 1-212 (1980). - WITT, R. DE, WARE, S.: Castanea 44, 163-174 (1979). - WITTIG, R.: Schriftenr. Landesanst. Ökol. Landsch. Forstpl. Nordrhein-Westf. 5, 1-230 (1980). - WOOD, G.W., BRENNEMAN, R.E.: J. Wildlife Manag. 44, 420-427 (1980). - WU ZHENGYI (ed.): Vegetation of China. 1375 pp. Peking: Acad. Sin. 1979a; - Vegetation Map of China. Peking: Acad. Sin. 1979b.

YOSHINO, M.: Erdkunde 34, 81-87 (1980).

ZAHLHEIMER, W.A.: Hoppea 38, 3-398 (1979).

Professor Dr. RÜDIGER KNAPP
Botanisches Institut I
der Universität
Senckenbergstr. 17-25
D 6300 Giessen

III. Experimental Ecology (Ecological Geobotany)

By WOLFGANG SCHMIDT

1. General

Ecology has traditionally been about plants and animals: microbial ecology has been considered as something best left to microbiologists. At some time in their lives, many "macro-ecologists" may wish to find out about microbial ecology, and the book by LYNCH and POOLE (1979) might well be a good starting point. Aerobiology has usually been regarded as the study of airborne micro-organisms, primarily viruses, bacteria, and fungi, but also including spores of bryophytes and pteridophytes and pollen grains of higher plants. The aim of a synthesis by EDMONDS (1979) was to place aerobiology within a cohesive framework by means of ecological systems analysis. HORN et al. (1979) published nine papers on the analysis of ecological systems covering a very wide range of subjects and strategies. The intention of a book edited by FREYSEN and WOLDENDORP (1978) was to present a generalized picture of current knowledge on the demography, population ecology, and population genetics of plants on the one hand, and on ecophysiology and interactions among and between plants, soil, and other components of the ecosystem on the other.

In a broad, but immediately useful generalization GRIME (1979) delineates three types of plant strategies: competitive, ruderal, and stress-tolerating. The volume of THORP and GIBBONS (1978) consists of forty-nine selected papers from a symposium of *Energy and Environmental Stress in Aquatic Systems*, while KOZLOWSKI (1979) has gathered together a good deal of information and given many useful North American references on *Tree Growth and Environmental Stresses*.

Again several books summarize the results obtained from ecosystem study which often formed part of the IBP. On the basis of sheet analysis the results of IBP grassland ecosystem research groups were compared by KALINOWSKA and MOCHNACKA (1976). An ecological perspective on amenity grassland was edited by RORISON and HUNT (1980), while NUMATA (1979) summarized the *Ecology of Grasslands and Bamboolands in the World*. The book of SINCLAIR and NORTON-GRIFFITHS (1980) is devoted to the dynamics of the well-known Serengeti Ecosystem. GOODALL et al. (1979) describe the structure and functioning of arid-land ecosystems all over the world. WERGER (1978) gives a descriptive account of the ecology and biogeography of southern Africa. The informative book of BRADSHAW and CHADWICK (1980) considers the problems of derelict and degraded land on a world-wide scale.

Two books summarize the Swedish contribution to the IBP: the book edited by PERSSON (1980) deals with the coniferous forest ecosystem, and within that ecosystem, the specific plant, soil, and abiotic processes. The book further treats the use of simulation models in forest ecosystem appraisal, especially with regard to energy exchange.

SONESSON (1980) describes the structure and functioning of a tundra ecosystem in northern Sweden. The plant cover lives under nutrient stress, cut off from a nutrient supply from the substrate by permafrost. Further studies on arctic and alpine ecosystems were published by BLISS (1977), RADFORTH and BRAWNER (1977), MANI (1979), IVES and ZIMINA (1978), and FRANZ (1979).

Thoughtful considerations of factors controlling stability and breakdown of ecosystems, as well as the restoration of degraded ecosystems, were edited by HOLDGATE and WOODMAN (1978). Historical aspects of ecological instability of a Roman Iron Age human community, the so-called Alvar settlement on Öland (Sweden), were discussed by ENCKELL et al. (1979) in ecological terms of ecosystem complexity and flow of energy, matter, and information. The exploitation of the less organized Alvar system by the more complex systems outside the area entailed an output of energy from the marginal land of the Alvar, and input in the form of, e.g., work-saving implements (energy liberators) which in their turn increased the output.

The search for a functional relationship between diversity and stability has thus far been futile. Recent advances in cybernetics suggest that progress may be achieved if diversity, stability, and redundancy are considered to be cofactors in determining the key-dependent variable – the capacity for self-organization (ULANOWICZ, 1979, 1980). Species diversity in different ecosystems was measured by DANIN (1978), NILSSON and NILSSON (1978), and REED (1978), while ROUTLEDGE (1980) developed methods for estimating the diversity of large, uncensused communities. A general mathematical model is presented by CROWLEY (1978) to show one reason why ecosystems of larger effective size should persist longer than smaller ones: oscillations of population densities tend to be displaced farther from extinction thresholds – even in spatially homogeneous systems.

2. Productivity and Ecosystem Research

a) Photosynthesis, Respiration, and Carbon Cycle

α) Modelling. A description of photosynthesis in a single leaf is developed by TENHUNEN et al. (1976a, 1977) that separates physiological sub-processes and is practical to apply as an ecological tool. In further publications TENHUNEN et al. (1976b), TENHUNEN and WESTRIN (1979), and TENHUNEN et al. (1980a, 1980b) presented a physiologically based steady-state model of whole leaf photosynthesis (WHOLEPHOT) and used it to describe net photosynthesis daily time courses in *Prunus armeniaca*, *Phaseolus vulgaris*, and *Glycine max*. Net photosynthesis rates are calculated in response to incident light intensity, leaf temperature, air CO_2 concentration, and leaf diffusion resistance measured at 5-min intervals. The steady-state calculations closely approximate the observed net photosynthesis rates for a broad range of weather conditions and leaf stomatal behavior. Other models of leaf photosynthesis and respiration were developed by HALL (1979) and LONG and INCOLL (1979) which adequately predicted CO_2 assimilation responses by a C_3 species, *Atriplex patula*, and of the temperate C_4 species, *Spartina townsendii*, to light, CO_2, O_2, and temperature in controlled environments. Functional relationships between the controlling environmental factors and the physiological processes of photosynthesis and diffusion resistance are used by KINERSON (1979) in a computer simulation model to examine production and water use aspects of *Betula papyrifera*. A mathematical compartments model based on differential equations and describing the distribution of CO_2 uptake in grassland is presented by WAREMBOURG et al. (1979). A method for approximating the photosynthetic production of stand members inside the canopy is described by KELLOMÄKI et al. (1979). KERNER et al. (1977) give a description of variable geometric needle model by which the structure of the assimilation system of a dominating *Picea abies* tree of closed stand is computed according to its needle surface.

β) Carbon Fixation Pathways. Research during recent years has demonstrated the ecological importance of the three main pathways of photosynthetic carbon fixation: C_3 photosynthesis, C_4 photosynthesis, and crassulacean acid metabolism (CAM). As one result, it is generally predicted that C_4 and CAM plants will become more abundant with increasing temperature and decreasing precipitation. A considerable body of literature has appeared establishing the use of the stable isotope ratio, $^{13}C/^{12}C$ ($\delta^{13}C$ value) as a reliable means by which the photosynthetic mode of higher plants can be determined. Mediterranean species of Poaceae in the northern Sahara predominantly had $\delta^{13}C$ values indicating the C_3 pathway. By contrast, nearly all species belonging to the Saharo-Arabian and/or Sudanian group showed a C_4-like carbon isotope composition (WINTER et al., 1976). $\delta^{13}C$ values in succulents from Madagascar were indicative of CAM in 17 species. Only three species of the Euphorbiaceae with succulent stems and non-succulent leaves showed C_3-like $\delta^{13}C$ values for both leaves and stems (WINTER, 1979). EDER et al. (1979) proved that all Central European *Euphorbia* species belong to the C_3-type, except for *E. humifusa* and *E. maculata* which are C_4-plants. These two species are not native to Central Europe, but introduced from Asia and N. America, respectively. A floristic analysis of low altitude grasslands in Kenya suggests that nearly all species at these low altitudes are of the C_4 type (TIESZEN et al., 1979a). At high altitudes, however, nearly all grasses are of the C_3 type. The variation in $\delta^{13}C$-values of the live grass vegetation with altitude confirms these distributional patterns. Nearly two-thirds of both the native and exotic grasses of the Hawaiian Islands are C_4 species. C_4 taxa dominate the species composition and mainly the coverage of grasses up to 1000 m, while C_3 taxa predominate above 1400 m. This elevation of floristic balance between C_4 and C_3 grasses corresponds to a low monthly mean minimum temperature of approximately 9°C and a mean maximum temperature for the warmest month of about 21°C (RUNDEL, 1980). The distribution of biomass of species differing in photosynthetic pathway along an altitudinal transect in south eastern Wyoming grassland also showed decreasing C_4 biomass with increasing elevation, while the percentage of C_3 biomass increased. All elevations had a significantly higher percentage of C_4 biomass in August than in June, reflecting the warm season growth characteristic of C_4 plants. CAM species were present at all elevations, but showed no trends in biomass distribution (BOUTTON et al., 1980a). Measurements of $\delta^{13}C$ values of North American and European bryophytes from arid and humid regions show consistent values of C_3 metabolism (RUNDEL et al., 1979).

Ecological data from separate studies on plant species in California and Central Europe were analyzed by DOLINER and JOLLIFFE (1979). The occurrence of C_4 species was greater where summer or winter temperatures were relatively high and moisture availability relatively low. No significant differences between the C_3 and C_4 groups were found for other environmental factors, including light, soil nitrogen, soil salinity, and continentality of climate. These results support the suggestion from previous physiological studies.

The implications of a reduced quantum yield (initial slope of the photosynthetic light response curve) in C_4 plants and temperature dependence of quantum yield in C_3 plants on total canopy primary production were investigated by EHLERINGER (1978) using computer simulations. The biomass of C_3 species of coastal and inland habitats of Israel and the Sinai was more dominant than that of C_4 species on all sites studied by WINTER and THROUGHTON (1978), and this includes the arido-active species at the desert habitats. Especially CAM was of secondary importance in arido-active species.

The $\delta^{13}C$ values of two C_3 halophytes, *Salicornia europaea* ssp. *rubra* and *Puccinellia nuttalliana*, native to inland saline areas of Canada, seemed to reflect the water potential of the soil, but may depend also on other factors (GUY et al., 1980). Comparative laboratory gas exchange and relative growth rate experiments were conducted by DE JONG (1978a, 1978b) on native California coastal strand species. *Atriplex leucophylla*, a C_4 species, had a photosynthetic temperature optimum substantially higher than leaf temperatures normally experienced on the beach during the growing season. It had higher mesophyll conductances and water use efficiencies at all salinity levels than the C_3 species *A. californica*. The $\delta^{13}C$ values of *Phleum pratense* were found to vary both with temperature and with nutrient level; the highest values were found under optimum growth conditions (BENDER and BERGE, 1979).

In general, the abundance ratio of the carbon isotopes in terrestrial plants is about 19‰ lower than that in atmospheric CO_2. The same applies to the biogenic CO_2 released from the soil. This difference in isotopic composition makes it possible to evaluate the amount of biogenic CO_2 that is re-assimilated by low-lying leaves in a thick forest environment. It is estimated by VOGEL (1978) that 15% of the carbon incorporated into leaves at 2 m above ground level in a Bavarian forest is derived from the soil. The average magnitude of the $\delta^{13}C$ value reduction in the lower forest layers in the Amazon forest was about -5‰. If the only cause of this reduction is the lower $\delta^{13}C$ value of the soil CO_2, MEDINA and MINCHIN (1980) estimated that the contribution of soil CO_2 to the photosynthesis of the shade flora is 20%.

BOUTTON et al. (1978) tested the hypothesis that grasses with the C_4 photosynthetic pathway are avoided as a food source by insect herbivores in natural communities. Results showed no statistically significant differences in utilization of C_3-C_4 species. According to BOUTTON et al. (1980b), the preference for C_3 species of Arphia (Orthoptera) along an altitudinal transect is presumably related to their higher nutritional value and digestibility, in spite of the fact that more time and energy must be spent to locate these food plants in some of the habitats studied. The isotopic analysis of rumen contents allowed TIESZEN et al. (1979b) to differentiate East African herbivores between grazers and browsers and to determine the quantitative dependence of each animal on C_3 and C_4 photosynthetic types.

The role of *Spartina alterniflora* in the production of detritus found in North American marsh soils and waters has been questioned because the carbon isotope ratio of marsh detritus is different from that of living *Spartina*. PETERSON et al. (1980) present a hypothesis to explain this paradox. Anaerobic decomposition of *Spartina* and organic matter from other sources results in the production of energy-rich reduced inorganic sulfur compounds. Bacteria use the energy of these inorganic compounds to fix CO_2 from marsh waters. Thus the bacteria obtain their energy but not their carbon from *Spartina*. Nearly the same results were obtained by THAYER et al. (1978) who compared the $\delta^{13}C$ values of producer and consumer organisms of a *Zostera marina* ecosystem. The majority of organisms analyzed appeared to be linked more directly to a plankton-carbon food chain than to the seagrass-carbon system in this relatively young *Zostera marina* bed.

KLUGE and TING (1978) and KLUGE (1979) review most of the results obtained during the past ten years investigations on the CAM phenomenon. Two main aspects of CAM are considered, i.e., the flow of carbon during the diurnal CAM cycle, including its regulation, and the verification of the hypothesis that CAM is an ecological adaptation of plants which facilitates survival on arid habitats. Measurements of $\delta^{13}C$ values of central Mexican Crassulaceae collected over a broad habitat range show also consistent patterns of CAM activity with no indications of substantial flexible photosynthetic pathways between C_3 and CAM (RUNDEL et al., 1979). In the stem succulent *Opuntia basilaris* C_3 fixation comprised 30% of the total carbon fixed by the potted, well-watered plants (GULMON and BLOOM, 1979). On many occasions following summer rainfall stomata of *O. inermis*, a CAM plant introduced in Australia opened for 4 h in the late afternoon permitting net C_3 fixation, which may contribute about 25% of the total carbon assimilated. Further investigations on CAM and C_3 photosynthesis of *Opuntia* species were published by GERWICK and WILLIAMS (1978) and KOCH and KENNEDY (1980). MEDINA et al. (1977), WILLERT et al. (1977), and SCHUBER and KLUGE (1979) investigated the ecophysiology of CAM plants in the family of the Bromeliaceae and Mesembryanthemaceae and the genus *Sempervivum*, respectively. Seasonal as well as environmental shift from C_3 to CAM photosynthesis is described in detail by SCHULZE et al. (1976) and SCHULZE and SCHULZE (1976) in *Welwitschia mirabilis*, by LANGE and ZUBER (1977) in *Frerea indica* and by WINTER et al. (1978) in *Mesembryanthemum crystallinum*.

In the European *Sedum acre* $\delta^{13}C$ values suggest that in situ at the natural stand CO_2 was fixed mainly via the C_3 mode of photosynthesis rather than via the CAM mode. However, experimental water stress shifts the CO_2 exchange pattern from the C_3 type to CAM type (KLUGE, 1977). Special aspects of water relations, mineral ion composition, and CAM photosynthesis were discussed by WINTER et al. (1976), NOBEL (1977), and LÜTTGE and BALL (1977).

γ) Relations to Environmental Factors. The temperature- and water-related photosynthetic capacity of higher plants under desert conditions in the Negev (Israel) is described in detail by EVENARI et al. (1977), LANGE et al. (1978), and SCHULZE et al. (1980a, 1980b). SZAREK and WOODHOUSE (1977, 1978a, 1978b) and MONSON and SZAREK (1979) measured gas exchange and water relations of several Sonoran Desert plants throughout the seasons. *Ambrosia deltoidea*, a drought-deciduous shrub, is mainly winter-spring active since maximum photosynthesis rates of 38 mg CO_2 dm^{-2} h^{-1} were measured at this time. *Olneya tesota*, a nonriparian, but deep-rooted microphyllous tree, is evergreen and photosynthetically active throughout the entire year, although demonstrating maximum photosynthesis rates of 12 mg CO_2 dm^{-2} h^{-1} in spring and summer. The photosynthetic characteristics of the C_3 desert winter annual *Camissonia claviformis* are presented by EHLERINGER et al. (1979). This species has one of the highest photosynthetic capacities of any higher C_3 plant, with rates exceeding 95 mg CO_2 dm^{-2} h^{-1} at midday irradiances and optimum leaf temperatures. The photosynthetic temperature optimum is at 20°C, in concert with the prevailing cool winter temperatures of the Death Valley desert habitat. Under optimal field conditions, the CO_2 uptake rate of *Hilaria rigida*, a C_4 grass of the Colorado Desert, can be 106 mg CO_2 dm^{-2} h^{-1}, higher than has been reported for any other species. The accompanying high water use efficiency may help explain the success of this hardy grass in the deserts (NOBEL, 1980).

The maximum photosynthetic rate at light saturation for the desert fern *Notholaena parryi* was only about 8 mg CO_2 dm^{-2} h^{-1}, a rather typical rate for ferns living in rock outcroppings protected from direct sunlight and moderated diurnal variations in air temperature (NOBEL, 1978). Patterns and mechanisms of recovery of photosynthetic activity of resurrection *Selaginella* species from desert habitats were compared by EICKMEIER (1979, 1980) in the laboratory. Results support the hypothesis that poikilohydric plants from increasingly xeric environments are capable of more rapid photosynthetic recovery. *Barbula aurea*, a desert moss of North America, is unable to absorb sufficient water from a saturated atmosphere. Unlike lichens, which are much more successful ecologically in deserts, it requires liquid water for growth (RUNDEL and LANGE, 1980). However, *Ramalina maciformis*, one of the largest macrolichens in the Negev desert, according to KAPPEN et al. (1979, 1980) and LANGE (1980), uses a minimum of supply of water for activation of photosynthesis. All experimental data formed the basis of a correlation model for net photosynthesis of *R. maciformis*.

CO_2-exchange in *Cladina* and *Cladonia* lichens from subarctic and temperate habitats was compared by LECHOWICZ (1978) and LECHOWICZ and ADAMS (1979). With the exception of rainy periods, the arctic lichen *Cetraria cucullata* exhibited a strong diurnal pattern with peak photosynthetic activity occurring between 3.00 and 7.00 h over the summer period (MOSER and NASH, 1978). *Sticta filix* and *Pseudocyphallaria delisei*, two of the most widespread Tasmanian lichens, reach greatest dominance in areas of precipitation of approximately 1600 mm yr^{-1}. Habitats with mean annual precipitation of 500-1000 mm are apparently too arid to allow *S. filix* to maintain necessary levels of CO_2 fixation for survival (RUNDEL et al., 1979).

Frond-like segments of the shoots of *Hylocomium splendens* photosynthesize for only two years. Underground tissues and translocation are absent in *H. splendens* but occur in *Polytrichum commune*. In hot dry conditions during summer in Swedish Lapland, *P. commune* continues to photosynthesize whereas *H. splendens* loses water, stops photosynthesizing, and loses mass (CALLAGHAN et al., 1978). The CO_2 exchange of the *Empetrum nigrum-Sphagnum fuscum* community of a raised bog was studied by SILVOLA and KEIKKINGEN (1979) and SILVOLA and HANSKI (1979). At the optimum temperature (10° to 15^oC) the maximum net CO_2 exchange of *Empetrum nigrum* was ca. 2 and that of *Sphagnum fuscum* ca. 2.5 mg CO_2 dm^{-2} h^{-1}. The total respiration in peat increased exponentially from 0.5 to 3.5 mg CO_2 dm^{-2} h^{-1} with increasing temperature from 5° to 30^oC.

Agropyron smithii (a C_3 species) and *Bouteloua gracilis* (a C_4 species) occur in the same habitat of the shortgrass prairie ecosystem in Colorado. Growth and gas exchange were measured by BROWN and TRLICA (1977a, 1977b), CHUNG and TRLICA (1980), and KEMP and WILLIAMS (1980) under natural conditions and on plants grown in controlled-environment chambers. Results suggest that the temporal displacement of growth between *A. smithii* and *B. gracilis* in their native is partly a function of the differential temperature responses of photosynthesis and related metabolic processes. The similarities between the species in their responses to water stress suggest the seasonal moisture gradients are not as important a parameter along which niche separation has occurred as are seasonal temperature gradients. An empirical model for predicting net photosynthesis and dark respiration in the field was developed and tested by DETLING et al. (1978) for *B. gracilis*.

Results on photosynthetic capacities of 13 tundra plant species support the generalization that species with an evergreen growth form have lower photosynthetic capacities than species with other growth forms. However, these low photosynthetic capacities in evergreen shrubs are associated with relatively extended leaf longevities (JOHNSON and TIESZEN, 1976).

The relationships between photosynthesis, energy utilization, carbon metabolism, and cover in Mediterranean evergreen shrub vegetation of Chile and California is described by OECHEL and LAWRENCE (1979) and OECHEL and MUSTAFA (1979). In general, the species with the greatest carbon uptake were best represented in the vegetation. Photosynthesis of tropical and subtropical trees and shrubs were measured by SQUIRE (1977), SAMSUDDIN and IMPENS (1979), and FERRAR (1980). Photosynthetic rates of *Eucalyptus* species originating in Australian habitats were highly correlated with leaf nitrogen content (MOONEY et al., 1978).

Conifers are widely cultivated in forestry because of their rapid growth rates. This is despite generally low rates of net photosynthesis especially under field conditions. Conifers, therefore, probably have special strategies that make them productive. According to SCHULZE et al. (1977a, 1977b) and FUCHS et al. (1977), the annual CO_2 uptake by evergreen *P. abies* was higher than deciduous *F. sylvatica* not because of a long growing season, but because of the longevity of its needles which during their total life time (an average of 5 years) have a two or three times greater CO_2 uptake than a deciduous leaf in one summer season. The relatively small investment in current year needles produces an annually low, but long-lasting assimilation of CO_2. Studies on photosynthesis in *Picea sitchensis* were continued by WATTS and NEILSON (1978) and LEVERENZ and JARVIS (1979, 1980), in *Pinus sylvestris* by PELKONEN (1980) and PELKONEN and HARI (1980). In New Zealand, *Pinus radiata* had a high photosynthetic capacity with a measured maximum of 10.2 mg CO_2 dm^{-2} h^{-1}. However, at temperatures greater than 25°C, even at the highest irradiance level, total respiration was greater than photosynthesis (ROOK and CORSON, 1978; BENECKE, 1980). Field response of photosynthesis to CO_2 enhancement in *Pinus ponderosa* suggests that net photosynthesis will increase in linear relation to rising atmospheric CO_2 during the next few decades. Severe damage from oxidant air pollution reduced, but did not eliminate the response to CO_2 enhancement (GREEN and WRIGHT, 1977).

Significant differences in the CO_2 exchange and the transpiration of light and shade crowns of several trees species were confirmed by KÜNSTLE and MITSCHERLICH (1978). Photosynthesis of a dominant, 18.9 m tall *Quercus alba* tree averaged 7.0, 5.6, and 4.6 mg CO_2 dm^{-2} h^{-1} for the top, middle, and bottom crown position. High leaf temperatures and mid-day stomatal closure appeared to limit photosynthesis in the upper crown, while light appeared limiting in the lower (AUBUCHON et al., 1978). All three tree species studied by WALLACE and DUNN (1980) were able to acclimate to a high light environment, as evidenced by their higher photosynthetic rates. However, species-specific physiological responses to high light intensity may account for each species known growth response following canopy opening.

δ) Chlorophyll Content. Leaf chloroplasts and chlorophyll content were studied by GORYSHINA (1980a, 1980b) in the Russian forest-steppe Tilieto-Quercetum aegopodiosum, in a vertical profile. Chloroplast size and chlorophyll content per chloroplast increase from top of the tree crown to the herbaceous understorey; plastid number per unit of cell volume, and the total plastid surface per cm^2 of leaf decrease. Chloroplasts of leaves of summerherbs are larger and contain more chlorphyll than chloroplasts of spring herbs. It was found by BARCIKOWSKI (1978) that the shade-tolerant *Majanthemum bifolium* produced less chlorophyll in the habitat exposed to sunlight, but in spite of this showed a higher increase in dry matter per unit leaf area. The amount of total chlorphylls per leaf dry weight of the seasonal dimorphic Mediterranean plants *Sarcopoterium spinosum* and *Phlomis fruticosa* is higher in winter leaves compared with summer leaves (MARGARIS, 1977).

ε) Methods. Investigations by KOCH and ELLER (1977) on global radiation losses in cuvettes, in dependence upon their exposure, showed losses of 20% of the daily radiation sum due to absorption by the cuvette material. This effect must primarily be considered when the light saturation of the experimental plants in the chamber is not surpassed. Young plants of *Lactuca sativa* and *Rumex alpinus* grown in assimilation cuvettes show different development of photosynthetic capacity depending on whether they are made from material allowing (quartz) or not allowing (Plexiglas) transmission of UV light. According to BOGENRIEDER and KLEIN (1980), normal glass and UV-transparent Plexiglas are for various reasons not suitable as substitutes for quartz. A description is given by STEFFAN and

ULLEMEYER (1977) for the self-construction of a Plexiglas chamber which is suitable for assimilation and respiration experiments with $^{14}CO_2$.

Primary production of phytoplankton in a eutrophic lake and of laboratory cultures of *Selenastrum cypricornutum* was investigated by ANDERSEN and SAND-JENSEN (1980) using the O_2 and ^{14}C light and dark bottle techniques. The quotient Q (gross oxygen production divided by carbon fixation) ranged between 1.14 and 1.35 at light saturation, but at low light intensities it increased to values ranging from 2 to 8. This pattern is ascribed to underestimation of gross photosynthesis by ^{14}C fixation rates due to reassimilation of respired CO_2 which is most important at low light levels.

ζ) Respiration. CO_2 effluxes from different parts of *Agropyron dasystachyum* plants and from soil samples of grassland sites were measured by REDMANN and ABOUGUENDIA (1978). Respiration of green leaves (1.46 mg CO_2 $g^{-1}h^{-1}$) was significantly higher than microbial respiration in moist, dead leaf samples (0.79 mg CO_2 $g^{-1}h^{-1}$) or litter (0.75 mg CO_2 $g^{-1}h^{-1}$). Average respiration rates of roots from soil cores (0.24 mg CO_2 $g^{-1}h^{-1}$) were lower than many values reported in the literature, probably because the root population sampled included inactive, suberized, and senescent roots. Root respiration rates of both *Lolium multiflorum* and *Hordeum distichum* were greatest one or to weeks before heading. The dark respiration rate integrated over the growing period was about half of the gross photosynthesis (MOGENSEN, 1977). Dial cycles of bole and root respiration were observed by EDWARDS and McLAUGHLIN (1978) in *Quercus alba* and *Liriodendron tulipifera* trees with highest rates occurring between 9.00 h and midnight, and lowest rates between noon and 15.00 h. Average respiration of current year twigs (315 mg CO_2 $m^{-2}h^{-1}$) was considerably higher during the growing season than that of older branches (175 mg CO_2 $m^{-2}h^{-1}$) of *L. tulipifera*. This high respiratory activity and large surface area of small branches and twigs indicates that they should be considered important in any attempt to quantify gas exchange and energy flow through forested ecosystems (McLAUGHLIN et al., 1978). A previous method of fitting a linear relationship between specific respiration and growth rates, using time-course data and based on equating coefficients in polynomials, is extended by CHANTER (1977).

η) Carbon Cycle. The book of BOLIN et al. (1979) covers most aspects of the carbon cycle, with emphasis on model calculations, sources, and sinks of carbon, the kinetics of exchange and net transfer between the various compartments of the carbon cycle, the increasing level of CO_2 in the atmosphere of the earth, the assessment of primary production on land and at sea, and the impact of man on the natural environment. The carbon cycle of a *Pinus taeda* plantation was examined by KINERSON et al. (1977) during its 12th through 16th years from planting. Net primary production during the study period averaged 2056 g C $m^{-2}y^{-1}$. With autotrophic respiration equal to 2068 g C, the calculated gross production was 4124 g C $m^{-2}y^{-1}$. Heterotrophic respiration of 694 g C $m^{-2}y^{-1}$resulted a net ecosytem production of 1362 g C $m^{-2}y^{-1}$.

b) Biomass and Productivity

α) Algae and Aquatic Ecosystems. DUNBAR (1979) summarized many of the results obtained world wide in the International Biological Program, Marine Productivity Section. The reported studies were involved in arctic, temperate, and tropical regions and include the Atlantic, Pacific, Indian, and Arctic Oceans. The results have clarified, quantified, and to an important extent altered our concepts of marine production and productivity. However, marine ecosystems are very complex, and many environmental factors that have a potential for widespread influence on production rates have hardly been adequately investigated or even in some cases clearly recognized as important. General aspects of the annual course of marine plankton communities were described by SMETACEK (1978), while COLLINS (1980) developed a mathematical model of phytoplankton growth. During the eutrophication phase of Lake Constance, the density of phytoplankton rose in the deep Obersee to more than double and in the shallow Gnadensee to quadruple; the "sigmoid" increase was strongest in the 1960's. Especially the small phytoplankton forms (Cryptomonads and centric Diatomeae) increased enormously and displaced the larger forms (Pennales) to a new summer peak. Since 1975, the decrease of phosphate, caused by waste water purification, allows us to expect an equivalent decrease of the phytoplankton density and an increase of the oxygen content in the deep water (GELLER, 1980; LEHN, 1980).

Phytoplankton dynamics and growth of eutrophic lakes of temperate regions were studied by HICKEL (1978), ANDERSEN (1976), LASTEIN (1976), COVENEY et al. (1977), JONASSON and ADALSTEINSSON (1979), and HEANEY and TALLING (1980), of tropical lakes by MELACK (1979). In general, availability of P seems to limit the phytoplankton production. Annual production of three oligotrophic antarctic lakes varies between 3.3 and 9 g C m^{-2} (PRIDDLE, 1980). In a small, oligotrophic subalpine lake lying in a heavily forested coniferous watershed, terrestrial inputs to the sediments, and the subsequent cycling of nutrients in the sediments, have a major influence on phytoplankton productivity during most of the growing season (RICHEY and WISSMAR, 1979).

The prokaryotic and eukaryotic algal components of soil were quantified and identified generically by HUNT et al. (1979) in two successional fields and a climax forest. The greatest number of algal genera as well as the highest counts of algal cells were found in the 1-yr old field, fewer were observed in the 11-yr old field and the forest.

β) Lichens and Mosses and Their Communities. Epiphytic lichen biomass of *Abies balsamea* forests ranged from 120 to 1603 kg/ha, with about 80% of the biomass located on branches (LANG et al., 1980).

Production and growth ecology of *Sphagnum* species in British and Finnish raised bogs were described by BOATMAN (1977) and PAKARINEN (1978). The rate of upward growth of three antarctic moss banks, 0.9-1.3 mm yr^{-1}, is approximately half the annual shoot growth. The rate of peat accumulation, 89-158 g $m^{-2}yr^{-1}$, is about half the rate of production, 162-350 g $m^{-2}yr^{-1}$. The former is similar to accumulation rates of peat elsewhere in the world (FENTON, 1980). The pattern of growth of *Polytrichum alpestre*, a common moss in the maritime Antarctic, is described by COLLINS (1976). The relationship between weight of the annual increment and the spatial density approximates to the density effect law applied to higher plants in which yield per unit area is little affected by variation in density. Growth of *Tomenthypnum nitens*, a tall boreal forest moss of Canada, was measured using small wire stakes; its net primary production (NPP) averaged 190 g $m^{-2}yr^{-1}$. Growth of *Hylocomium splendens*, another boreal forest moss, was estimated by measuring dry weight changes of segments. NPP averaged 79 g $m^{-2}yr^{-1}$. In both cases, the most important environmental factor appears to be total precipitation, followed by evaporation stress (BUSBY and al., 1978). Bryophytes formed ca. 90% of the standing crop of the ground vegetation in a moist Quercetum petraeae in Wales and contained a correspondingly high proportion of the mineral-element resources (RIELEY et al., 1979). The influence of mineral fertilization on the species composition of bryoflora and moss biomass was found to be unfavorable (MICKIEWICZ, 1976).

γ) Herbs and Grasses in Different Ecosystems, Mainly Grasslands. Effects of thermal effluent on standing crop and NPP of *Elodea canadensis* and other submerged macrophytes were described by HAAG and GORHAM (1977). After the closure of the Grevelingen estuary (The Netherlands) in 1971, the intertidal *Zostera marina* population extended downwards to 5 m below lake level, probably owing to the increased transparency of the water. The annual above-ground net primary production (ANPP) of *Z. marina*, based on doubled maximum standing crop values in July-August, was estimated by NIENHUIS and DE BREE (1977) at 50 g C m^{-2} in 1968, 121 g C m^{-2} in 1973, and 91 g C m^{-2} in 1975.

Five harvest methods were used by LINTHURST and REIMOLD (1978) to estimate the ANPP of angiosperm species in North American coastal salt marshes. Differences as great as tenfold were found between these methods. It was concluded that the morphology of the species, the location, and the general environmental conditions affect the results of any single method. Production and mortality of six marsh macrophytes were measured by HOPKINSON et al. (1980) using phenometric methods. Annual production estimates averaged 1.8 times higher than annual mortality estimates, although in a steady-state system the rates should be about equal. Estimates of mean annual NPP in a Georgia salt marsh, computed from changes in biomass and disappearance of dead plant material, were: creekbank *Spartina alterniflora* stands 3700 g m^{-2}, high marsh *S. alterniflora* stands 1300 g m^{-2}, and *Juncus roemerianus* stands 2200 g m^{-2} (GALLAGHER et al., 1980). Calculated annual ANPP for 7 plant species commonly found in Louisiana's coastal marshes varied between 6043 g m^{-2} (*Spartina patens*) and 1355 g m^{-2} (*S. cynosuroides*) (HOPKINSON et al., 1978). This is a generally higher level of production compared to studies of WHITE et al. (1978), ZEDLER et al. (1980), and PARSONS and DE LA CRUZ (1980). On the other hand, estimates of ANPP by LINTHURST and REIMOLD (1978) ranged from 616 g $m^{-2}yr^{-1}$ for highmarsh *Juncus gerardii* in Maine to 6040 g $m^{-2}yr^{-1}$ for *Spartina cynosuroides* in Georgia. Below-ground biomass (BGB) of a *Juncus roemerianus* marsh in Mississippi ranged from 9.7 to 12.4 kg m^{-2} with the peak biomass occurring in April. Annual below-ground net primary production (BNPP) estimated from expected maximum minus expected minimum biomass fitted into a periodic regression model was 1.36 kg m^{-2} (DE LA CRUZ and HACKNEY, 1977). The decomposition rate of roots and rhizomes was greatest in the top 10 cm (20% mass loss yr^{-1}) of the marsh soil (HACKNEY and DE LA CRUZ, 1980).

Production of lake reed communities is analyzed by SZCZPANSKI (1978), KRISCH (1978), RAGHI-ATRI and BORNKAMM (1978), and ERNST (1979) against the background of environmental conditions. ANPP in a *Carex rostrata* wetland was 540 g $m^{-2}yr^{-1}$ based on harvest data, but 820 g $m^{-2}yr^{-1}$ based on weight of shoots presented in permanent plots. When shoot mortality was added, ANPP was 1080 g $m^{-2}yr^{-1}$. Maximum BNPP was estimated at 260 g $m^{-2}yr^{-1}$; NPP thus was 1340 g $m^{-2}yr^{-1}$ (BERNARD and HANKINSON, 1979). Seasonal production of a *Carex*-dominated ecosystem showed a strong bimodal pattern with peak productivities in mid-June (15.3 g $m^{-2}d^{-1}$) and mid-July (4.3 g $m^{-2}d^{-1}$). Terminal standing crop after a 130-day growth period was 807 g m^{-2} (AUCLAIR et al., 1976).

Strategies of growth and population dynamics of tundra plants, especially of *Carex bigelowii*, were described by CALLAGHAN and COLLINS (1976) and CALLAGHAN (1976). The quantity of growing root tips was analyzed by KUMMEROW and RUSSELL (1980) in an Alaskan tundra site. By June 10, the aboveground fraction of the vegetation had initiated the flush of spring growth and flowering, while less than 5 active root tips cm^{-3} were found. By June 25, this value had increased to 30 root tips cm^{-3}. Low root temperatures were responsible for the time lag between shoot growth initiation and the beginning of root growth. In early August, root growth stopped completely in spite of adequate soil temperatures and accumulated carbohydrate for root growth.

The entire phytomass of an alpine grass heath community (Caricetum curvulae) amounts to 1900-2700 g m^{-2}, the entire biomass 1450-2010 g m^{-2}. The main part of dead material remains of standing dead which gets decomposed by fungi. The lack of litter-fall may explain the low litter-mass of 50-160 g m^{-2}. NPP amounts to 100-160 g $m^{-2}y^{-1}$ in a production period of 105-128 days (GRABHERR et al., 1978). The book of HEAL and PERKINS (1978) represents a major contribution to the study of production ecology of upland peat and grassland communities and summarizes much of the British IBP effort, especially of the well-known Moor House Programme, the British contribution to the tundra biome studies.

FRENCH (1979) summarizes and synthesizes data collected in the North American Grassland Biome Program with an emphasis on grassland primary production and the response of grassland to ecosystem stress. The book of INNIS (1978) and the paper by PARTON and RISSER (1980) describe the major simulation modelling exercise of the Grassland Biome. The structure and function of ten western North American grasslands including biomass and productivity studies were investigated by SIMS et al. (1978), SIMS and SINGH (1978a, 1978b), LAUENROTH and WHITMAN (1977), and LAUENROTH et al. (1978). Peak live biomass ranged from 84 to 336 g m^{-2} and showed a linear increase with increasing amounts of growing-season precipitation up to 450 mm; at higher values of precipitation increases in live biomass tended to level out. BNPP was significantly higher on grazed treatments than ungrazed grasslands. In general, BNPP increased with decreasing levels of long-term mean annual temperature, while there was an apparent linear increase in ANPP with increasing precipitation up to approximately 500 mm yr^{-1}. NPP ranged from 225 to 1425 g $m^{-2}yr^{-1}$. Approximately 46% and 58% of its variability were explained by annual precipitation in ungrazed and grazed grasslands, respectively.

Classically evaluating water-use efficiency as annual ANPP divided by annual evapotranspiration, forests are the most efficient (0.9 to 1.8 g/1000 g water), followed by shortgrass prairie (0.2 to 0.7), then hot deserts (0.1 to 0.3) (WEBB et al., 1978). A grassland primary producer model for simulating intra-seasonal biomass dynamics as a function of temperature, moisture, light, and nitrogen was developed by BROWN and TRLICA (1977c), PARTON et al. (1978), and DETLING et al. (1979) for *Bouteloua gracilis*, the dominant C_4 grass of the North American shortgrass prairie.

Seasonal changes in biomass allocation were determined by BELL et al. (1979) and CLARK and BURK (1980) for desert annuals. In all species, roots comprised 12%-22% of total biomass until late in flowering, when root allocation declined. Maximum investment in reproductive structures varied among species from 16% to 50% of total biomass. The energy allocation patterns of three *Lupinus* species, an annual, a perennial herb, and a shrub were investigated by PITELKA (1977) and related to life cycle differences as well as ecological differences among the species.

The influence of mineral fertilization on the structure, yield and floristic composition of meadows and their grasses is described by TRACZYK et al. (1976), KOWALCZYK and LEKAWSKA (1977), RUTKOWSKA et al. (1976), PLEWCZYNSKA-KURAS (1976), SPEIDEL (1976), WILLEMS (1980), ELIAS and CHADWICK (1979), and LAUENROTH et al. (1978). The effects of water supply on dry matter production were analyzed by GALES (1979) in *Lolium perenne*, by FRANCOIS and RENARD (1979) in *Festuca arundinacea*, while TRIPATHI and GUPTA (1980) compared the growth response of the two closely related sympatric fodder grasses, *Bothriochloa pertusa* and *Dichanthium annulatum* to increasing density and herbage removal.

Several studies were designed to investigate the pattern of yields of old-field communities during succession (REED, 1977; PEROZZI and BAZZAZ, 1978; TSUCHIDA and NUMATA, 1979; KROH and STEPHENSON, 1980; SCHREIBER, 1980; CAMPINO, 1980) or of the growth and reproductive strategies of special successional species (FAILLE and FARDJAH, 1977; BOSTOCK and BENTON, 1979; BOSTOCK, 1980; LEE and BAZZAZ, 1980; WEAVER and CAVES,

1980). Quantitative changes in the ageing process of *Ammophila arenaria* during primary succession on dunes were studied by WALLEN (1980). The decline of *Ammophila* grass vegetation during a ten year period was mainly quantified as a decrease in the yearly formation of above-ground biomass (AGB) from 400 g m^{-2} to 125 g m^{-2}, a decrease of the maximum leaf area index (LAI) from 0.75 to 0.25 and a decrease of the yearly formation of below-ground organs from 200 g m^{-2} to zero.

δ) Woody Plants and Forest Ecosystems. Annual ANPP in ungrazed plots of the Southern Guinea savanna was 273 g m^{-2}, litter production was 162 g m^{-2}, and decomposition was 179 g m^{-2}. In grazed plots the corresponding figures were 316 g m^{-2}, 144 g m^{-2}, and 148 g m^{-2}, respectively; cattle consumed 141 g m^{-2}. Litter disappearance was greatest in the dry season, largely due to consumption by fungus-growing termites (OHIAGU and WOOD, 1979). ANPP of the herb layer of a South African tree savanna was studied by GRUNOW et al.(1980). NPP of the herbaceous and woody species in seven facies of the humid Lamto savannas varies between 2150 and 3580 g $m^{-2}y^{-1}$ (MENAUT and CESAR, 1979).

Herb layer ANPP was estimated by FORD and NEWBOULD (1977) during a *Castanea sativa* coppice cycle. At 1, 2, 5, 9, and 15 years after coppicing it was 13.5, 216, 95.6, 27.9, and 1.6 g $m^{-2}yr^{-1}$. Standing crop in pure stands of *Mercurialis perennis* shoots in mixed deciduous woodland was measured by HUTCHINGS (1978) by dimension analysis, a technique not previously used in herb production studies. It is accurate and allows several measurements to be made on the same material. Maximum standing crop (97 g m^{-2}) occurred in May, after maximum shoot density, and thereafter fell linearly until September. Biomass, productivity, and vegetative reproduction of *Rubus fruticosus* in mixed deciduous woodland were studied by TAYLOR (1980), HUTCHINGS (1978), and KIRBY (1980).

Estimates of ANPP for a variety of deciduous forest ecosystems within the lake states (U.S.A.) had a narrow range from 7.1 to 10.4 t $ha^{-1}yr^{-1}$. The contribution to stand production by shrubs and ground vegetation ranged from 9% to 17% of the ANPP (CROW, 1978). Leaf production in mature North American hardwood forests was estimated by ABER (1979) and COVINGTON and ABER (1980) to reach about 3 t $ha^{-1}yr^{-1}$. Living biomass, organic matter distribution, and organic matter production budgets were determined by GRIER and LOGAN (1977) for plant communities dominated by large 450-yr-old *Pseudotsuga menziesii*, up to 175 cm diam and 80 m tall. ANPP (trees only) ranged from 492 to 976 t ha^{-1}, total leaf biomass 10.4 -16.3 t ha^{-1}, and total organic matter accumulations 1008 - 1514 t ha^{-1}. AGB increment was negative, ranging from -2.9 to -6.2 t ha^{-1}. ANPP ranged from 6.3 to 10.1 t $ha^{-1}yr^{-1}$ and consisted entirely of detritus.

The annual production and the dynamic changes of root systems were investigated by PERSSON (1978, 1979, 1980a, 1980b) in a 15-20-yr-old *Pinus sylvestris* stand in Sweden. BNPP in the fine roots (<2 mm in diameter) of *P. sylvestris*, *Calluna vulgaris*, and *Vaccinium vitis-idea* was estimated as 183, 38, and 135 g $m^{-2}y^{-1}$, respectively, in the total stand. Using data obtained by sequential core sampling the annual turnover of the fine roots was estimated to 35, 34, and 135 g m^{-2}, respectively, in the F/H layer and to 183, 42, and 57 g m^{-2}, respectively, in the mineral soil. The largest supply of dead *P. sylvestris* fine roots went to the mineral soil, whereas that of the superficially distributed dwarf shrubs ended up in the F/H layer.

Age structure, biomass, productivity, and respiration in heath plants, especially *Calluna vulgaris* were studied by WALLEN (1980), BANNISTER (1978), CHAPMAN (1979), and MILLER (1979). AGB of *Calluna* heathland in

Northwest Germany ranged from 600 to 1100 g m^{-2} in old heathlands and from 80 to 500 g m^{-2} on mowed or fire-regenerated areas (MUHLE et al., 1979). KELLOMÄKI et al. (1977) studied the daily production of structural biomass in *Vaccinium myrtillus* and *V. vitis-idaea* by means of the height growth of plants. The phytomass of alpine dwarf shrub communities diminishes with increasing altitude from 5150 g m^{-2} (*Vaccinium* heath, LAI 5.3) to 1680 g m^{-2} (Loiseleurietum, LAI 2.6). According to SCHMIDT (1977), the ANPP decreases in the same way (*Vaccinium* heath 485 g $m^{-2}yr^{-1}$; Loiseleurietum 108 g $m^{-2}yr^{-1}$). These alpine dwarf communities represent mature, non-expanding systems: no remarkable variation in the above-ground biomass could be observed during the period of investigation; most of the dry matter produced within a year by the above-ground parts is transferred to the litter layer.

Root systems as well as seasonal changes of fine root density of chaparral shrubs of California and Chile were studied by KUMMEROW et al. (1977, 1978) and HOFFMANN and KUMMEROW (1978). The root area always exceeded the shoot area, with the corresponding ratios ranging from 6 for *Arctostaphylos pungens* to 40 for *Haplopappus pinifolius*. The fine root density of 64 g m^{-2} under the canopy was significantly higher than in the unshaded area. However, the corresponding value of 45 g m^{-2} for the open ground is still high enough to make the establishment of other shrubs difficult. Detailed studies on productivity, carbon and reproductive allocation on the semi-desert shrubs and trees *Anogeissus pendula*, *Larrea tridentata*, *Artemisia tridentata*, *Simmondsia chinensis*, *Zygophyllum eurypterum*, and *Atriplex confertifolia* were published by VYAS et al. (1978), CUNNINGHAM et al. (1979), CUNNINGHAM and REYNOLDS (1978), STURGES and TRLICA (1978), WALLACE and RUNDEL (1979), MOORE and BHADRESA (1978), and HODGKINSON et al. (1978).

AGB of a *Chusquea culeou* bamboo thicket in Chile was approximately 156-162 t ha^{-1}, annual ANPP was approximately 10-11 t ha^{-1}. In the understorey of a mixed *Nothofagus* forest near the timberline in the Andes, the bamboo *Chusquea tenuiflora* AGB was only 13 t ha^{-1} (VEBLEN et al., 1980).

The biomass and productivity in five montane rain forest-types in Jamaica was estimated by TANNER (1980). AGB varied from 229 t ha^{-1} in Mor Ridge forest to 407 t ha^{-1} in well-developed Mull Ridge forest. BGB was determined only in Mor Ridge forest (54 t ha^{-1}). The total LAI was 5.5 in the Mor Ridge forest and 5.7 in the well-developed Mull Ridge forest. First estimates by dendrometers of increase in live biomass of the standing crop ranged from 0.5 t $ha^{-1}yr^{-1}$ (Mor Ridge forest) to 3.5 t $ha^{-1}yr^{-1}$ (Gap forest); estimates of loss of live AGB of the standing crop ranged from 0.6 t $ha^{-1}yr^{-1}$ (Gap forest) to 2.7 t $ha^{-1}yr^{-1}$ (Wet slope forest). Average rate of BGB accumulation in the surface root mat of an Amazonian rain forest was 1.2 t $^{-2}ha^{-1}yr^{-1}$, and total BGB increment was 2.0 t $ha^{-1}yr^{-1}$. Root growth rates relative to shoot growth rates were similar to ratios for a temperate forest as compared by JORDAN and ESCALANTE (1980). However, turnover rates of roots were higher at the Amazonian site. High turnover rates result in a relatively large proportion of the roots in smaller size classes which have a large surface area in relation to their volume and thus are efficient nutrient traps which is important in the nutrient-poor rain forest.

A fundamental question in current ecosystem research is the extent to which net organic matter increment controls nutrient input-output behavior. The large uncertainties associated with NPP and heterotrophic respiration make it very difficult to verify statistically an ecosystem as steady-state. The *Populus tremuloides* forest ecosystem reported by GOSZ (1980) appears to be degrading (i.e., negative net ecosystem production), a condition which would allow an important test of the influence of net ecosystem production on the nutrient regulation of an ecosystem. NPP was estimated at 11.74 t $ha^{-1}yr^{-1}$, respiration was 13.8 t $ha^{-1}yr^{-1}$, and net ecosystem production was -2.06 t $ha^{-1}yr^{-1}$. The standard error of the estimate for the sum of woody biomass increment and litter fall was only 0.12 t $ha^{-1}yr^{-1}$.

ε) Methods. HUNT (1978) summarized the techniques of growth analysis. A quantitative method for monitoring the effect of the inherent growth rhythm on growth throughout the whole growth period is presented by VUOKKO et al. (1977). An automatic device has been designed by MILNE et al. (1977) to measure the shoot length of *Picea sitchensis* and other plant species. Using the properties of the reflected solar radiation, a radiometric device working simultaneously in two wavelengths (677 and 795 nm) has been developed by METHY (1977) for a non-destructive estimate of the aerial biomass of grasses.

ζ) Energy Content. The calorific values of woody and herbaceous plants dominating Greek Mediterranean ecosystems appear reduced when compared with corresponding ones measured in ecosystems of colder environments (ADAMANDIADOU et al., 1978). Leaves of *Quercus ilex* growing at the slopes of Mt. Etna (Sicily) reach their highest energy content in December, twigs in May. At both periods, the energy content of leaves from the upper parts of the crown was higher than that of the lower parts (LEONARDI and LINSER-BOURDELLON, 1978). Differences in energy content due to daily course, season, seasonal climate, and specific organic characteristics were also observed by BRZOSKA (1979) and BRZOSKA and ENGEL (1980) in *Alnus viridis* and by CASPERS (1977) in several herbaceous plants from old-field and meadow communities. The energy content of an alpine *Caricetum curvulae* increases in the sequence: lichens, mosses, herbs, graminoides. During the vegetation period, a total amount of energy of 2.5 to 3.0 MJ $m^{-2}yr^{-1}$ is absorbed; about one half is bound by the AGB, the other half by the BGB (GRABHERR et al., 1980).

The calorific contents of macrophytes and algaes were determined by LITTLER and MURRAY (1978) in an unpolluted rocky intertidal habitat and in a nearby polluted habitat. The algae that characterized the unpolluted community usually contained comparatively fewer joules per unit weight. These forms had relatively more structural tissues and hence allocated relatively less energy to rapid growth and reproduction.

c) Litter Fall, Decomposition, and Nutrient Cycle

α) General. The renewed vigor among ecologists in quantifying the circulation of organic matter and mineral elements is becoming obvious by the increasing number of books and papers. One reason for this explosion in publication is the use of modern tools (AAS, X-ray diffraction, X-ray fluorescence, plasma techniques, etc.) by which analytical time has been cut down by several orders of magnitude compared to the techniques that were available just 30 to 40 years ago. However, every new or resurrected field of science needs a contemporary framework, and the book of FORTESCUE (1980) offers one in and for environmental geochemistry. Also the work of HARBORNE (1977) is intended to provide a comprehensive view of ecological biochemistry in all its aspects. The second volume on metals transfer and ecological mass balances edited by NRIAGU (1976) represents a compendium of papers presented at the Second International Symposium on Environmental Biogeochemistry. Special aspects of mineral nutrient cycling were summarized by CLARK et al. (1978), FRISSEL (1978), and FENCHEL and BLACKBURN (1979). Papers from a symposia of the British Ecological Society about the role of terrestrial and aquatic organisms in decomposition processes were edited by ANDERSON and MACFADYEN (1977). Aspects of nutrient cycling and ecosystem stability were discussed by HARWELL et al. (1977) and DEANGELIS (1980), both using a mathematical approach. A compartmental model of phosphorus in soil-plant-animal ecosystem is described by KATZNELSON (1977).

β) Aquatic Ecosystems. Transport and transformation of phosphorus and organic debris in forest stream ecosystems were studied by MEYER and LIKENS (1979) and BILBY and LIKENS (1980) in the well-known Hubbard Brook Experimental Forest (U.S.A.). Incident precipitation in this area averages about pH 4 as a result of increased pollution from sulfuric and nitric acids. To determine the effect of this increased acidity on the ecology of aquatic ecosystems, dilute concentrations of sulfuric acid were added by HALL et al. (1980) to a stream in the Hubbard Brook Experimental Forest. The stream was maintained at pH 4 from April to September, 1977. With increased acidity stream water concentrations of Al, Ca, Mg, K, and probably Mn, Fe, and Cd were elevated. Stream acidification decreased species diversity, increased representation of community dominants, and decreased the complexity of the food web. Changes in the output of ions from a watershed in Colorado as a result of the acidification of precipitation were also noticed by LEWIS and GRANT (1979). Abrupt acidity fluctuations from near-neutrality to just above pH 3 at a broadland site induced a spectacular invasion by the acidophilous alga *Tribonema minus*, while most aquatic macrophytes perished (GOSLING and BAKER, 1980). The major factors controlling streamwater concentrations of elements in North American watersheds are factors related to supply (precipitation chemistry, relative evapotranspiration, rock weathering) and to plant accumulation of nutrients. Rapidly growing successional ecosystems can accumulate a large fraction of inputs of nutrient elements, particularly during the growing season. Steady-state ecosystems, those with no net biomass accumulation, have nutrient outputs equal to nutrient inputs. The increased losses of nutrient elements from clearcut watersheds can be explained in these terms. Clearcut or otherwise disturbed ecosystems may have mineralization in excess of plant uptake, leading to element outputs in excess of inputs. Field experiments by VITOUSEK (1977) demonstrated that the prevention of plant uptake by root trenching led to soil NO_3^- concentrations comparable to those observed in streamwater in the clearcut watershed. Dynamics of mineral nutrients and organic matter during leaf decomposition in forest streams were investigated by DAWSON (1976) and MEYER (1980), recycling of nutrients in an enclosed aquatic community of decomposing macrophytes (*Myriophyllum spicatum*) by KISTRITZ (1978). Terrestrial plant litter is believed to play a role in the detrital energy and nutrient system within oligotrophic, subalpine lakes (RAU, 1976), too.

Decomposition processes in the profundal region of lakes, including O_2 and nitrate uptake, CO_2, CH_4, ammonia and sulphide accumulation, and loss of methane and N_2 as bubbles, were studied by the use of experimental enclosures for the duration of thermal stratification (JONES and SIMON, 1980). P fractions of rain, dust, or snow were analyzed by PETERS (1977) to shed light on the role of aeolian P in the nutrient budgets of lakes. Experiments by PETERS and MACINTYRE (1976) strengthen their conclusion that a large quantity of orthophosphate is available for algal use in East African lakes. The seasonal range and variation of P, N, Si, and Mn were studied by PHILIPS (1977) in three shallow lakes in Great Britain.

γ) Grasslands and Herb Communities. The chemical composition in some maritime antarctic cryptogams and phanerogams was analyzed by SMITH (1978) and WALTON and SMITH (1979). The concentrations within the non-vascular plants tend to be proportional to those occurring in the substratum, although some degree of accumulation, despite the absence of specialized storage tissues, is suspected. This simple mechanism of uptake and storage of mineral ions is considered to be a response to the moisture regime of the habitat and to the concentration of the elements in the substratum or in precipitation. Higher plants from soils associated with seal and bird colonies had higher levels of P and N, while the burning of sub-antarctic tussock grassland caused increases in these elements which lasted for more than a year.

Nutrient status of different plant organs of tundra plants was examined by CALLAGHAN (1980) and CHAPIN et al. (1978, 1979, 1980). In the graminoid *Eriophorum vaginatum*, spring production of roots as well as leaves depleted N reserves in the rhizome. Similar dependence of root production upon sorted nutrient reserves is suggested in deciduous shrubs like *Salix pulchra* and *Betula nana*, but not in the evergreen shrub *Ledum palustre*. CHAPIN et al. (1979) suggest that the *E. vaginatum* tussock growth form, with its favourable thermal and nutritional micro-environment, may be largely responsible for the high productivity and widespread distribution of *E. vaginatum* at tundra and bog sites: within-tussock soils thaw faster, reach maximum summer temperature sooner, are 6°-8°C warmer, and are thermally more stable than soils at comparable depths between tussocks. Soil organic carbon and, presumably, organically bound soil nutrients, such as N, P, and Ca cycle 3-10 times more rapidly within the tussock than beneath it. Nutrient turnover studies in treeless alpine ecosystems showed an obvious increase of N in the underground phytomass toward the end of the growing season (GÖKCEOGLU and REHDER, 1977; REHDER and SCHÄFER, 1978).

Mg and Na levels decreased in a *Carex*-dominated ecosystem from upland to open water. Soil P increased markedly at water's edge related to mineral input by sedimentation. The pattern of N, P, K, and Ca coincided closely with total shoot production and litter mass levels suggesting closed biotic cycles of these elements (AUCLAIR et al., 1976). N, P, and K levels in *Carex rostrata* were high in overwintering tissues in the spring and in tissues of newly emerged shoots. They then declined as the shoots aged. The yearly nutrient budget in *C. rostrata*, based on the differences between maximum and minimum nutrient biomass, are 10.4 g m^{-2}N, 1.4 g m^{-2}P, 12.6 g m^{-2}K, 2.0 g m^{-2}Ca, and 1.2 g m^{-2}Mg (BERNARD and HANKINSON, 1979). Metal inputs to an ombrotrophic *Sphagnum* bog are 88, 132, and 54 mg $m^{-2}yr^{-1}$ for K, Mg, and Pb, respectively. K and Mg are vertically distributed in the bog profile in accord with the inhomogeneous ion exchange chemistry of peat, while Pb is distributed in accord with historical trends in atmospheric Pb fallout (HEMOND, 1980). The influence of community and edaphic variables on tissue concentration and nutrient cycle was assessed by AUCLAIR (1979) for six species of a *Scirpus-Equisetum* wetland, by GARTEN et al. (1977) for woody and herbaceous species from a coastal foodplain community, by GOLDBERG et al. (1980) for *Ammophila breviligulata*, by SPRENT et al. (1978) for *Myrica gale*, by MORTON (1977) for *Molinia caerulea*, by GAUDET (1977), BRINSON (1977), MITSCH et al. (1979), DEGHI et al. (1980), and BRINSON et al. (1980) for different types of alluvial swamp forests. WOODWELL et al. (1979) discussed the exchanges of inorganic N between tidal water and an estuarine marsh community.

A N budget is presented by WOODMANSEE et al. (1978) for a shortgrass prairie ecosystem. Total N uptake by plants from soil solution was 2.9 g $m^{-2}yr^{-1}$. Aerial portions of plants were allocated 1.9 g N $m^{-2}yr^{-1}$, although apparently 26% of this amount came from internally recycled sources. Mineralization of N from plant residues was sufficient to account for all of the N taken up by plants from soil solution. Study was made by CLARK (1977) of the rates at which and the routes by which N once taken up by the shortgrass *Bouteloua gracilis* moves into litter and soil compartments and from them to new plant growth. WOODMANSEE and DUNCAN (1980) estimated that 57-125 kg N ha^{-1} and 13.1-14.7 kg P ha^{-1} were mineralized during growing seasons in an Mediterranean annual grassland. However, N uptake by plants was 68-119 kg N ha^{-1} and P uptake 10.4-14.5 kg P ha^{-1}. While the data showed that P was stable during the 3 years of observation, indicating its cycle was relatively closed, the data of N showed that its cycle was open. Large inputs of N from symbiotic fixation in the 1st year of the study and large losses of N from the system were estimated. A simulation model of the P cycle in semiarid grasslands was developed and tested by COLE et al. (1977). The most critical informational needs revealed by model development and operation were in the areas of activity and morphology of roots and the rates of mineralization of organic P as affected by soil depth.

A 5-year effort by a US-IBP Desert Biome group has produced a comprehensive understanding of the N cycle in salt desert shrub dominated ecosystems (WEST and SKUJINS, 1977). Net physical transfers are minor in these systems. Biological processes are more important and unique. Blue-green algal components of soil cryptogamic crusts have been found to be able to fix comparatively high (25 kg N $ha^{-1}yr^{-1}$) amounts, however, over 70% of this is lost through volatilization and denitrification. Furthermore, *Atriplex* and *Artemisia* litter inhibits both fixation and denitrification. Only about 12 kg N $ha^{-1}yr^{-1}$ is taken up by new above-ground vascular plant growth. Amounts of litter from six shrub species in the northern Mojave Desert ranged from 7% to 83% of their respective live above-ground biomass (STROJAN et al., 1979).

Annual leaf-fall of the woody vegetation in Southern Guinea savanna was 2.4 t ha^{-1}, annual wood-fall 1.4 t ha^{-1}, both with a maximum in January after the annual bush fires (COLLINS, 1977). Annual grass production in ungrazed plots of the same location was 2.7 t ha^{-1}, litter production was 1.6 t ha^{-1}, and decomposition was 1.8 t ha^{-1}. Litter disappearance was greatest in the dry season, largely due to consumption by fungus-growing termites (OHIAGU and WOOD, 1979). Just before the annual savanna burning, nutrient contents in the above-ground plant material are very low (0.28% N; 0.06% P; 0.28% K), so that the losses by fire could be limited (VILLECOURT et al., 1979). The N input by rainfall in savanna areas seemed to be very high (19 kg N $ha^{-1}yr^{-1}$ for an Ivory Coast station), perhaps according to annual burning. Mineral N losses through leaching are negligible (VILLECOURT and ROOSE, 1978).

Special aspects of decomposition and disappearance of dead plant material in grassland were studied by ABOUGUENDIA and WHITMAN (1979), MOSZYNSKA and TATUR (1978), MOCHNACKA-LAWACZ (1978), JAKUBCZYK (1978), CZERWINSKI et al. (1978), KLEIN (1977), VOSSBRINCK et al. (1979), and NEWBERRY (1979). A model has been developed by HUNT (1977) to simulate the dynamics of decomposers and substrates in grasslands. SCHWARTZKOPF (1978) describes a new open-chamber technique for measuring CO_2 evolution from soil in situ which minimizes measurement errors due to soil disturbance and chamber effects. WIEDER and CARREL (1979) used ^{14}C as an indicator of energy flow in litter decomposition studies with *Festuca arundinacea*.

N uptake from applied nutrient solutions was evaluated by HAINES (1977) in two old fields to test the idea that plant communities become more efficient trappers and retainers of plant nutrients during succession. Results suggest that NO_3^- uptake decreased, while NH_4^+ uptake increased with succession. This apparent shift from a nitrate to an ammonia nitrogen economy during succession has also been reported by ANDEL et al. (1979) and is evolutionarily significant as an energy, N, and cation-saving mechanism. The relation of the internal N supply to the total N supply, its dependence on site conditions, and on the age of old-field communities is discussed by HAHN et al. (1979), while BAKELAAR and ODUM (1978) studied the effects of nutrient enrichment on the plant community of an 8-year-old field successional ecosystem. WOLF (1979) investigated the decomposition of organic material in abandoned meadows.

ε) Forest Ecosystems. Ecologists will be pleased to know that the extensive research program from the Hubbard Brook Experimental Forest has been synthesized by LIKENS et al. (1977) as a book entitled *Biogeochemistry of a Forested Ecosystem*, dealing with the hydrology, chemistry, input-output budgets, weathering, and nutrient cycles for this watershed system. Special aspects of forest nutrient cycling and element behavior based on Hubbard Brook datas were presented by WHITTAKER et al. (1979) and FINN (1980). Nutrient leaching from fertilized and undisturbed forest watershed was studied by HARRIMAN (1978), GOSZ (1980), and MARTIN (1979). The effects of stem girdling on biogeochemical cycles within a mixed deciduous forest was investigated by EDWARDS and ROSS-TODD (1979) and JOHNSON and EDWARDS (1979). WEARY and MERRIAM (1978) described litter decomposition in an *Acer rubrum* woodland under natural conditions and under insecticide treatment. Nutrient and trace element influx from gross rainfall and snowfall, and levels in throughfall, stemflow and litterfall of a North Dakota gallery forest exhibited appreciable seasonal and yearly variation (KILLINGBECK and WALI, 1978). *Pinus* stands cycled significantly more of the cations in throughfall during the dormant season than did other deciduous forest types studied by HENDERSON et al. (1977).

Further investigations on nutrient movement through coniferous forests were presented by RAPP and CABANETTES (1980) of a *Pinus pinea* woodland, by FELLER (1977) and SOLLINS et al. (1980) of *Pseudotsuga menziesii* forest, by BARES and WALI (1979) of *Picea mariana* and *Larix laricina* peatlands, and by FAHEY (1979) and CRONAN (1980) of subalpine coniferous forest ecosystems. The annual nutrient uptake in a *Taxodium distichum* swamp-forest is rather small compared to that in upland forests. Foliage abscission and foliar leaching by rainfall return 73% to 91% of the annual nutrient uptakes each year: therefore, there are only small permanent additions of nutrients to the above-water pools each year. Except for K which is apparently reabsorbed from foliage before abscission, the *Taxodium* trees do not appear to conserve foliar nutrients. However, nutrient conservation may be effected by minimizing the amount of foliage (SCHLESINGER, 1978).

Total annual nutrient transfer by stem flow in an Amazonian lowland tropical rain forest was approximately equal to total growing season nutrient transfer in temperate zone studies, but lower than total annual nutrient transfer by stem flow in a montane rain forest (JORDAN, 1978). Analysis of root mat samples supports the ability of the root mat on the surface of poor, depleted tropical soils to efficiently take up dissolved nutrients before they percolate down to mineral soil. Other studies by STARK and JORDAN (1978) have shown that one of the uptake mechanisms is mycorrhizal fungi. Furthermore, it is hypothesized by JORDAN et al. (1979) that the suppression of nitrifying bacteria results in N conservation in the tropical rain forest. Monocultures of the successional tropical tree *Cecropia obtusifolia* had higher nutrient standing crop one year after clear-cutting of the tropical rain forest than plots allowed to regenerate naturally by forbs (HARCOMBE, 1977).

The quantity, temporal distribution, and mineral-element content of litterfall in two tropical forest types of Australia was studied by BRASELL et al. (1980), of four sites in the montane rain forest of Jamaica by TANNER (1980b), in two tropical deciduous forest communities of India by GAUR and PANDEY (1978), in a montane rain forest in New Guinea by EDWARDS (1977), and EDWARDS and GRUBB (1977) and in Australian *Eucalyptus* forests by BIRK and SIMPSON (1980). The total annual litter production of an undisturbed rain forest of the Guatemalan highlands was between 9.1 and 11.5 $t\cdot ha^{-1}yr^{-1}$. Compared with other tropical forests, this value is unusually high. Leaf litter comprised 67% to 77% of the total (KUNKEL-WESTPHAL and KUNKEL, 1979). Further investigations on energy flow, nutrient changes, and respiration during the decomposition of litter in tropical rain forest environment were made by HAINES and FOSTER (1977), MEDINA et al. (1980), and IRMLER and FURCH (1980).

To estimate the rate of consumption of leaf litter by termites on the forest floor in an equatorial rain forest ecosystem of West Malaysia, newly fallen leaves were marked and distributed on the ground.

22% to 32% of the daily leaf-litter fall was transported by termites to their mounds, so that it was considered by MATSUMOTO and ABE (1979) that the termites had an important role in the detritus food chain of the ecosystem.

The forest floor is an important component of forest ecosystems since it represents a stage in the nutrient transfer of biogeochemical cycles. The successional development of the forest floor and its nutritional role was studied by SWITZER et al. (1979), while BLANK et al. (1980) estimated the nutrient uptake by a diverse spring ephemeral forest floor cummunity. The influence of chemical composition, addition of *Rubus idaeus* leaves, and N supply on decomposition rate and dynamics of N and P in *Fagus sylvatica* litter was studied by STAAF (1980). The different responses suggested that the N availability to decomposer microbes in the litter, in combination with a low lignin content, resulted in increased decomposition rate. The initial N concentration seemed to influence weight loss only during the initial period of decomposition. P was also transferred into litter, but was retained less efficiently than N, suggesting a non-limiting availability for decomposer activity.

The influence of animals and microorganisms on temperate forest decomposition of different species of leaf litter and dead wood was investigated by NILSSON (1978), KHEIRALLAH (1979), HOWARD and HOWARD (1979, 1980), CHRISTENSEN (1977), LANG and FORMAN (1978), MEENTEMEYER (1978), STRZELCZYK et al. (1978), VOGT et al. (1980), KOTOWSKI (1979), BELL et al. (1978), and GRIGAL and McCOLL (1977). In general, the decomposition effected by microorganisms alone was less than that by microorganisms and mesofauna; this, however, was less than the one by microorganisms, meso- and macrofauna. According to HERLITZIUS and HERLITZIUS (1977), in acid soil phytophagous-excrements effected a very high decompositon, while in lime soil this was not to be found. Litter bag studies by McBRAYER and CROMACK (1980) in a *Quercus* forest showed that both microfloral and invertebrate decomposers were active under snow cover, but activity was thought to cease periodically when air temperatures fell below the insulation value of snow.

References

ABER, J.D.: Ecology 60, 18-23 (1979). - ABOUGUENDIA, Z.M., WHITMAN, W.C.: Oecologia 42, 23-29 (1979). - ADAMANDIADOU, S., SIAFACA, L., MARGARIS, N.S.: Flora 167, 574-584 (1978). - ANDEL, J. VAN, ERNST, W., NELISSEN, H.: Verh. Ges. Ökologie 7, 361-368 (1979). - ANDERSEN, F.O.: Oikos 27, 243-250 (1976). - ANDERSEN, J.M., SAND-JENSEN, K.: Oikos 35, 359-364 (1980). - ANDERSEN, J.M., MACFADYEN, A. (eds.): The Role of Terrestrial and Aquatic Organisms in Decomposition Processes. 17th Symposium of the British Ecological Society. 474 pp. Oxford: Blackwell 1977. - AUBUCHON, R.R., THOMPSON, D.R., HINCKLEY, T.M.: Oecologia 35, 295-306 (1978). - AUCLAIR, A.N.D.: Ecology 60, 337-348 (1979). - AUCLAIR, A.N.D., BOUCHARD, A., PAJACZKOWSKI, J.: Oecologia 26, 9-31 (1976).

BAKELAAR, R.G., ODUM, E.P.: Ecology 59, 660-665 (1978). - BANNISTER, P.: J. Ecol. 66, 117-131 (1978). - BARCIKOWSKI, A.: Acta Univ. N. C. Biologia 21, 51-59 (1978). - BARES, R.H., WALI, M.K.: Vegetatio 40, 79-94 (1979). - BELL, D.T., JOHNSON, F.L., GILMORE, A.R.: Oikos 30, 76-82 (1978). - BELL, K.L., HIATT, H.D., NILES, W.E.: J. Ecol. 67, 781-787 (1979). - BENDER, M.M., BERGE, A.J.: Oecologia 44, 117-118

(1979). - BENECKE, U.: Oecologia 44, 192-198 (1980). - BERNARD, J.M., HANKINSON, G.: Oikos 32, 328-336 (1979). - BILBY, R.E., LIKENS, G.E.: Ecology 61, 1107-1113 (1980). - BIRK, E.M., SIMPSON, R.W.: Ecology 61, 481-485 (1980). - BLANK, J.L., OLSON, R.K., VITOUSEK, P.M.: Oecologia 47, 96-98 (1980). - BLISS, L.C. (ed.): Truelove Lowland, Devon Island, Canada. 714 pp. Edmonton: Univ. Alberta Press 1977. - BOATMAN, D.J.: J. Ecol. 65, 119-126 (1977). - BOGENRIEDER, A., KLEIN, R.: Flora 169, 510-523 (1980). - BOLIN, B., DEGENS, E.T., KEMPE, S., KETNER, P.: The Global Carbon Cycle. 491 pp. Chichester: Wiley 1979. - BOSTOCK, S.J.: Oikos 34, 359-363 (1980). - BOSTOCK, S.J., BENTON, R.A.: J. Ecol. 67, 91-107 (1979). - BOUTTON, T.W., CAMERON, A.T., SMITH, B.N.: Oecologia 36, 21-32 (1978). - BOUTTON, T.W., HARRISON, A.T., SMITH, B.N.: Oecologia 45, 287-298 (1980a). - BOUTTON, T.W., SMITH, B.N., HARRISON, A.T.: Oecologia 45, 299-306 (1980b). - BRADSHAW, A.D., CHADWICK, M.J.: The Restoration of Land; the Ecology and Reclamation of Derelict and Degraded Land. 314 pp. Oxford, London, Edinburgh, Boston, Melbourne: Blackwell 1980. - BRASELL, H.M., UNWIN, G.L., STOCKER, G.C.: J. Ecol. 68, 123-139 (1980). - BRINSON, M.M.: Ecology 58, 601-609 (1977). - BRINSON, M.M., BRADSHAW, H.D., HOLMES, R.N., ELKINS, J.B. Jr.: Ecology 61, 827-835 (1980). - BROWN, L.F., TRLICA, M.J.: J. Appl. Ecol. 14, 197-204 (1977a); - J. Appl. Ecol. 14, 205-213 (1977b); - J. Appl. Ecol. 14, 215-224 (1977c). - BRZOSKA, W.: Verh. Ges. Ökologie 7, 421-427 (1979). - BRZOSKA, W., ENGEL, L.: Verh. Ges. Ökologie 8, 463-468 (1980). - BUSBY, J.R., BLISS, L.C., HAMILTON, C.D.: Ecol. Monographs 48, 95-110 (1978).

CALLAGHAN, T.V.: Oikos 27, 402-413 (1976); - Oikos 35, 373-386 (1980). - CALLAGHAN, T.V., COLLINS, N.J.: Oikos 27, 383-388 (1976). - CALLAGHAN, T.V., COLLINS, N.J., CALLAGHAN, C.H.: Oikos 31, 73-88 (1978). - CAMPINO, I.: Verh. Ges. Ökologie 8, 221-226 (1980). - CASPERS, N.: Oecologia 26, 379-383 (1977). - CHANTER, D.O.: J. Appl. Ecol. 14, 269-278 (1977). - CHAPIN, F.S. III, BARSDATE, R.J., BAREL, D.: Oikos 31, 189-199 (1978). - CHAPIN, F.S. III, CLEVE, K. VAN, CHAPIN, M.C.: J. Ecol. 67, 169-189 (1979). - CHAPIN, F.S. III, JOHNSON, D.A., McKENDRICK, J.D.: J. Ecol. 68, 189-209 (1980). - CHAPMAN, S.B.: J. Ecol. 67, 1-20 (1979). - CHRISTENSEN, O.: Oikos 28, 177-186 (1977). - CHUNG, H.-H., TRLICA, M.J.: Oecologia 47, 190-195 (1980). - CLARK, D.D., BURK, J.H.: Oecologia 46, 86-91 (1980). - CLARK, F.E.: Ecology 58, 1322-1333 (1977). - CLARK, F.E., COLE, C.V., BOWMAN, R.A.: Nutrient cycling, 659-712. In: Grassland, Systems Analysis and Man, eds. A.J. BREYMEYER, G.M. VAN DYNE. Cambridge: Cambridge University Press 1978. - COLE, C.V., INNIS, G.S., STEWART, J.W.B.: Ecology 58, 1-15 (1977). - COLLINS, C.D.: Ecology 61, 639-649 (1980). - COLLINS, N.J.: Oikos 27, 389-401 (1976). - COLLINS, N.M.: Oecologia 28, 163-175 (1977). - COVENEY, M.F., CRONBERG, G., ENELL, M., LARSSON, K., OLOFSSON, L.: Oikos 29, 5-21 (1977). - COVINGTON, W.W., ABER, J.D.: Ecology 61, 200-204 (1980). - CRONAN, C.S.: Oikos 34, 272-281 (1980). - CROW, T.R.: Ecology 59, 265-273 (1978). - CROWLEY, P.H.: Oecologia 35, 185-195 (1978). - CUNNINGHAM, G.L., REYNOLDS, J.F.: Ecology 59, 37-52 (1978). - CUNNINGHAM, G.L., SYVERTSEN, J.P., REYNOLDS, J.F., WILLSON, J.M.: Oecologia 40, 113-123 (1979). - CZERWINSKI, Z., TRACZYK, T., PRACZ, J.: Polish Ecol. Stud. 4, 267-281 (1978).

DANIN, A.: Flora 167, 409-422 (1978). - DAWSON, F.H.: Oikos 27, 13-18 (1976). - DE ANGELIS, D.L.: Ecology 61, 764-771 (1980). - DEGHI, G.S., CARTER EWEL, K., MITSCH, W.J.: J. Appl. Ecol. 17, 397-408 (1980). - DE JONG, T.M.: Oecologia 34, 343-351 (1978a); - Oecologia 36, 59-68 (1978b). - DE LA CRUZ, A.A., HACKNEY, C.T.: Ecology 58, 1165-1170 (1977). - DETLING, J.K., PARTON, W.J., HUNT, H.W.: Oecologia 33, 137-147 (1978). - DETLING, J.K., DYER, M.I., WINN, D.T.: Oecologia 41, 127-134 (1979). - DOLINER, L.H., JOLLIFFE, P.A.: Oecologia 38, 23-34

(1979). - DUNBAR, M.J. (ed.) Marine Production Mechanisms. 338 pp. Cambridge, London, New York: Cambridge University Press 1979.

EDER, A., STICHLER, W., ZIEGLER, H.: Flora 168, 227-240 (1979). - EDMONDS, R.L. (ed.): Aerobiology: the Ecological Systems Approach. 386 pp. Stroudsburg/PA: Dowden, Hutchinson and Ross 1979. - EDWARDS, N.T., McLAUGHLIN, S.B.: Oikos 31, 200-206 (1978). - EDWARDS, N.T., ROSS-TODD, B.M.: Oecologia 40, 247-257 (1979). - EDWARDS, P.J.: J. Ecol. 65, 971-992 (1977). - EDWARDS, P.J., GRUBB, P.J.: J. Ecol. 65, 943-969 (1977). - EHLERINGER, J.R.: Oecologia 31, 255-267 (1978). - EHLERINGER, J., MOONEY, H.A., BERRY, J.A.: Ecology 60, 280-286 (1979). - EICKMEIER, W.G.: Oecologia 39, 93-106 (1979); - Oecologia 46, 380-385 (1980). - ELIAS, C.O., CHADWICK, M.J.: J. Appl. Ecol. 16, 537-544 (1979). - ENCKELL, P.H., KÖNIGSSON, E.S., KÖNIGSSON, L.-K.: Oikos 33, 328-349 (1979). - ERNST, W.: Phytocoenologia 6, 74-84 (1979). - EVENARI, M., LANGE, O.L., SCHULZE, E.-D., KAPPEN, L., BUSCHBOM, U.: Flora 166, 383-414 (1977).

FAHEY, T.J.: Oikos 32, 422-428 (1979). - FAILLE, A., FARDJAH, M.: Oecol. Plant. 12, 323-341 (1977). - FELLER, M.C.: Ecology 58, 1269-1283 (1977). - FENCHEL, T., BLACKBURN, T.H.: Bacteria and Mineral Cycling. 226 pp. London, New York, San Francisco: Academic Press 1979. - FENTON, J.H.C.: J. Ecol. 68, 211-228 (1980). - FERRAR, P.J.: Oecologia 47, 204-212 (1980). - FINN, J.T.: Ecology 61, 562-571 (1980). - FORD, E.D., NEWBOULD, P.J.: J. Ecol. 65, 201-212 (1977). - FORTESCUE, J.A.C.: Environmental Geochemistry: A Holistic Approach. Ecological Series, Vol. 35. 346 pp. Berlin, Heidelberg, New York: Springer 1980. - FRANCOIS, J., RENARD, C.: Oecol. Plant 14, 417-433 (1979). - FRANZ, H.: Ökologie der Hochgebirge. 495 pp. Stuttgart: Ulmer 1979. - FRENCH, N. (ed.): Perspectives in Grassland Ecology. Ecological Studies, Vol. 32. 204 pp. Berlin, Heidelberg, New York: Springer 1979. - FREYSEN, A.H.J., WOLDENDORP, J.W. (eds.): Structure and Functioning of Plant Populations. 324 pp. Amsterdam: North-Holland 1978. - FRISSEL, M.J. (ed.): Cycling of Mineral Nutrients in Agricultural Ecosystems. Developments in Agricultural and Managed Forest Ecology, Vol. 3. 356 pp. Amsterdam: Elsevier 1978. - FUCHS, M., SCHULZE, E.-D., FUCHS, M.I.: Oecologia 29, 329-340 (1977.

GALES, K.: J. Appl. Ecol. 16, 863-877 (1979). - GALLAGHER, J.L., REIMOLD, R.J., LINTHURST, R.A., PFEIFFER, W.J.: Ecology 61, 303-312 (1980). - GARTEN, C.T., GENTRY, J.B., SHARITZ, R.R.: Ecology 58, 979-992 (1977). - GAUDET, J.J.: Ecology 58, 415-422 (1977). - GAUR, J.P., PANDEY, H.N.: Oikos 30, 570-575 (1978). - GELLER, W.: Verh. Ges. Ökologie 8, 373-382 (1980). - GERWICK, C., WILLIAMS, G.J. III: Oecologia 35, 149-159 (1978). - GÖKCEOGLU, M., REHDER, H.: Oecologia 28, 317-331 (1977). - GOLDBERG, M., TABROFF, N.R., TAMARIN, R.H.: Ecology 61, 1029-1033 (1980). - GOODALL, D.W., PERRY, R.A., HOWES, K.M.W. (eds.): Arid-land Ecosystems. 881 pp. Cambridge: Cambridge University Press 1979. - GORYSHINA, T.K.: Oecol. Plant. 15, 47-54 (1980a); - Oecol. Plant. 15, 201-208 (1980b). - GOSLING, L.M., BAKER, S.J.: J. Appl. Ecol. 17, 479-490 (1980). - GOSZ, J.R.: Ecology 61, 507-514 (1980a); - Ecology 61, 515-521 (1980b). - GRABHERR, G., MÄHR, E., REISIGL, H.: Oecol. Plant. 13, 227-251 (1978). - GRABHERR, G., BRZOSKA, W., HOFER, H., REISIGL, H.: Oecol. Plant. 15, 307-316 (1980). - GREEN, K., WRIGHT, R.: Ecology 58, 687-692 (1977). - GRIER, C.C., LOGAN, R.S.: Ecol. Monographs 47, 373-400 (1977). - GRIGAL, D.F., McCOLL, J.G.: J. Appl. Ecol. 14, 531-538 (1977). - GRIME, J.P.: Plant Strategies and Vegetation Processes. 222 pp. Chichester: Wiley 1979. - GRUNOW, J.O., GROENEVELD, H.T., DU TOIT, S.H.C.: J. Ecol. 68, 877-889 (1980). - GULMON, S.L., BLOOM, A.J.: Oecologia 38, 217-222 (1979). - GUY, R.D., REID, D.M., KROUSE, H.R.: Oecologia 44, 241-247 (1980).

HAAG, R.W., GORHAM, P.R.: J. Appl. Ecol. 14, 835-852 (1977). - HACKNEY, C.T., DE LA CRUZ, A.A.: Ecology 61, 226-231 (1980). - HAHN, W., WOLF, A., SCHMIDT, W.: Verh. Ges. Ökologie 7, 369-380 (1979). - HAINES, B., FOSTER, R.B.: J. Ecol. 65, 147-155 (1977). - HAINES, B.L.: Oecologia 26, 295-303 (1977). - HALL, A.E.: Oecologia 143, 299-316 (1979). - HALL, R.J., LIKENS, G.E., FIANCE, S.B., HENDREY, G.R.: Ecology 61, 976-989 (1980). - HARBONE, J.B.: Introduction to Ecological Biochemistry. 244 pp. London: Academic Press 1977. - HARCOMBE, P.A.: Ecology 58, 1375-1383 (1977). - HARRIMAN, R.: J. Appl. Ecol. 15, 933-942 (1978). - HARWELL, M.A., CROPPER, W.P. Jr., RAGSDALE, H.L.: Ecology 58, 660-666 (1977). - HEAL, O.W., PERKINS, D.F. (eds.): Production Ecology of British Moors and Montane Grasslands. Ecological Studies, Vol. 27. 426 pp. Berlin, Heidelberg, New York: Springer 1978. - HEANEY, S.I., TALLING, J.F.: J. Ecol. 68, 75-94 (1980). - HEMOND, H.F.: Ecol. Monographs 50, 507-526 (1980). - HENDERSON, G.S., HARRIS, W.F., TODD, D.E. Jr., GRIZZARD, T.: J. Ecol. 65, 365-374 (1977). - HERLITZIUS, R., HERLITZIUS, H.: Oecologia 30, 147-171 (1977). - HICKEL, B.: Verh. Ges. Ökologie 6, 119-126 (1978). - HODGKINSON, K.C., JOHNSON, P.S., NORTON, B.E.: Oecologia 34, 353-362 (1978). - HOFFMANN, A., KUMMEROW, J.: Oecologia 32, 57-69 (1978). - HOLDGATE, M.W., WOODMAN, M.J. (eds.): The Breakdown and Restoration of Ecosystems, Series 1, Ecology 3. 496 pp. New York: Plenum 1978. - HOPKINSON, C.S., GOSSELINK, J.G., PARRONDO, R.T.: Ecology 59, 760-769 (1978); - Ecology 61, 1091-1098 (1980). - HORN, D.J., MITCHELL, R.D., STAIRS, G.R.: Analysis of Ecological Systems. 312 pp. Ohio: Ohio State University Press 1979. - HOWARD, D.M., HOWARD, P.J.A.: Oikos 34, 115-124 (1980). - HOWARD, P.J.A., HOWARD, D.M.: Oikos 33, 457-465 (1979). - HUNT, H.W.: Ecology 58, 469-484 (1977). - HUNT, M.E., FLOYD, G.L., STOUT, B.B.: Ecology 60, 362-375 (1979). - HUNT, R.: Plant Growth Analysis. 67 pp. London: Arnold 1978. - HUTCHINGS, M.J.: Oikos 31, 351-357 (1978).

INNIS, D.S. (ed.): Grassland Simulation Model. Ecological Studies, Vol. 26. 298 pp. Berlin, Heidelberg, New York: Springer 1978. - IRMLER, U., FURCH, K.: Pedobiologia 20, 118-130 (1980). - IVES, J.D., ZIMINA, R.P. (eds.): Mountain geoecology and land-use implications. Proc. Symp. Intern. Geogr. Union, Comm. on High Altitude Geoecology, Caucasus Mts., USSR, July 1976. Arctic and Alp. Res. 10 (2), 159-548 (1978).

JAKUBCZYK, H.: Polish Ecol. Stud. 4, 179-208 (1978). - JOHNSON, D.A., TIESZEN, L.L.: Oecologia 24, 159-173 (1976). - JOHNSON, D.W., EDWARDS, N.T.: Oecologia 40, 259-271 (1979). - JONASSON, P.M., ADALSTEINSSON, H.: Oikos 32, 113-138 (1979). - JONES, J.G., SIMON, B.M.: J. Ecol. 68, 493-512 (1980). - JORDAN, C.F.: Oikos 31, 257-263 (1978). - JORDAN, C.F., ESCALANTE, G.: Ecology 61, 14-18 (1980). - JORDAN, C.F., TODD, R.L., ESCALANTE, G.: Oecologia 39, 123-128 (1979).

KALINOWSKA, A., MOCHNACKA, H.: Polish Ecol. Stud. 2, 163-169 (1976). - KAPPEN, L., LANGE, O.L., SCHULZE, E.-D., EVENARI, M., BUSCHBOM, U.: Flora 168, 85-108 (1979). - KAPPEN, L., LANGE, O.L., SCHULZE, E.-D., BUSCHBOM, U., EVENARI, M.: Flora 169, 216-220 (1980). - KATZNELSON, J.: Oecologia 26, 325-334 (1977). - KELLOMÄKI, S., HARI, P., VUOKKO, R., VÄISÄNEN, E., KANNINEN, M.: Oikos 29, 143-149 (1977). - KELLOMÄKI, S., SALMINEN, R., HARI, P., VENTILÄ, M., KANNINEN, M., KAUPPI, P., SMOLANDER, H.: J. Appl. Ecol. 16, 243-252 (1979). - KEMP, P.R., WILLIAMS, G.J. III: Ecology 61, 846-858 (1980). - KERNER, H., GROSS, E., KOCH, W.: Flora 166, 449-459 (1977). - KHEIRALLAH, A.M.: Oikos 33, 466-471 (1979). - KILLINGBECK, K.T., WALI, M.K.: Oikos 30, 29-60 (1978). - KINERSON, R.S.: Oecologia 39, 37-49 (1979). - KINERSON, R.S., RALSTON, C.W., WELLS, C.G.: Oecologia 29, 1-10 (1977). - KIRBY, K.J.: J. Ecol. 68, 513-520 (1980). - KISTRITZ, R.U.: Oikos 30, 561-569 (1978). - KLEIN, D.A.: Ecology 58, 184-190 (1977). - KLUGE, M.: Oecologia 29,

77-83 (1977); - Ber. Dtsch. Bot. Ges. 92, 95-107 (1979). - KLUGE, M., TING, I.P.: Crassulacean Acid Metabolism. Analysis of an Ecological Adaptation. 220 pp. Berlin, Heidelberg, New York: Springer 1978. - KOCH, K.E., KENNEDY, R.A.: Oecologia 45, 390-395 (1980). - KOCH, W., ELLER, B.M.: Flora 166, 279-288 (1977). - KOTOWSKI, W.: Ekol. Polska. 27, 427-436 (1979). - KOWALCZYK, J., LEKAWSKA, I.: Polish Ecol. Stud. 3, 55-65 (1977). - KOZLOWSKI, T.T.: Tree Growth and Environmental Stresses. 192 pp. Seattle: University of Washington Press 1979. - KRISCH, H.: Arch. Naturschutz Landschaftsforsch. 18, 121-140 (1978). - KROH, G.C., STEPHENSON, S.N.: Oecologia 45, 366-371 (1980). - KÜNSTLE, E., MITSCHERLICH, G.: Angew. Bot. 52, 233-252 (1978). - KUNKEL-WESTPHAL, I., KUNKEL, P.: J. Ecol. 67, 665-686 (1979). - KUMMEROW, J., RUSSELL, M.: Oecologia 47, 196-199 (1980). - KUMMEROW, J., KRAUSE, D., JOW, W.: Oecologia 29, 163-177 (1977); - Oecologia 37, 201-212 (1978).

LANG, G.E., FORMAN, R.T.T.: Ecology 59, 580-595 (1978). - LANG, G.E., REINERS, W.A., PIKE, L.H.: Ecology 61, 541-550 (1980). - LANGE, O.L.: Oecologia 45, 82-87 (1980). - LANGE, O.L., ZUBER, M.: Oecologia 31, 67-72 (1977). - LANGE, O.L., SCHULZE, E.-D., EVENARI, M., KAPPEN, L., BUSCHBOM, U.: Oecologia 34, 89-100 (1978). - LASTEIN, E.: Oikos 27, 44-49 (1976). - LAUENROTH, W.K., WHITMAN, W.C.: Oecologia 27, 339-351 (1977). - LAUENROTH, W.K., DODD, J.L., SIMS, P.L.: Oecologia 36, 211-222 (1978). - LECHOWICZ, M.J.: Oecologia 32, 225-237 (1978). - LECHOWICZ, M.J., ADAMS, M.S.: Photosynthetica 13, 155-162 (1979). - LEE, T.D., BAZZAZ, F.A.: J. Ecol. 68, 813-821 (1980). - LEHN, H.: Verh. Ges. Ökologie 8, 363-372 (1980). - LEONARDI, S., LINSER-BOURDELLON, A.: Flora 167, 35-39 (1978). - LEVERENZ, J.W., JARVIS, P.G.: J. Appl. Ecol. 16, 919-932 (1979); - J. Appl. Ecol. 17, 59-68 (1980). - LEWIS, W.M. Jr., GRANT, M.C.: Ecology 60, 1093-1097 (1979). - LIKENS, G.E., BORMANN, F.H., PIERCE, R.S., EATON, J.S., JOHNSON, N.M.: Biogeochemistry of a Forested Ecosystem. 146 pp. Berlin, Heidelberg, New York: Springer 1977. - LINTHURST, R.A., REIMOLD, R.J.: Ecology 59, 945-955 (1978a); - J. Appl. Ecol. 15, 919-931 (1978b). - LITTLER, M.M., MURRAY, S.N.: J. Appl. Ecol. 15, 583-595 (1978). - LONG, S.P., INCOLL, L.D.: J. Appl. Ecol. 16, 879-891 (1979). - LÜTTGE, U., BALL, E.: Oecologia 31, 85-94 (1977). - LYNCH, J.M., POOLE, M.J. (eds.): Microbial Ecology: a Conceptual Approach. 226 pp. Oxford: Blackwell 1979.

MAITRE, J.P.: Oecol. Plant. 15, 179-184 (1980). - MANI, M.S.: Ecology and Phytogeography of High-altitude Plants of the Northwest Himalaya. 205 pp. London: Chapman & Hall 1979. - MARGARIS, N.S.: Oecol. Plant. 12, 343-350 (1977). - MARTIN, C.W.: Ecology 60, 36-42 (1979). - MATSUMOTO, T., ABE, T.: Oecologia 38, 261-274 (1979). - McBRAYER, J.F., CROMACK, K. Jr.: Pedobiologia 20, 47-54 (1980). - McLAUGHLIN, S.B., McCONATHY, R.K., DINGER, B.E.: J. Appl. Ecol. 15, 327-334 (1978). - MEDINA, E., MINCHIN, P.: Oecologia 45, 377-378 (1980). - MEDINA, E., DELGADO, M., TROUGHTON, J.H., MEDINA, J.D.: Flora 166, 137-152 (1977). - MEDINA, E., KLINGE, H., JORDAN, C., HERRERA, R.: Flora 170, 240-250 (1980). - MEENTEMEYER, V.: Ecology 59, 465-472 (1978). - MELACK, J.M.: Oecologia 44, 1-7 (1979). - MENAUT, J.C., CESAR, J.: Ecology 60, 1197-1210 (1979). - METHY, M.: Oecol. Plant. 12, 395-401 (1977). - MEYER, J.L.: Oikos 34, 44-53 (1980). - MEYER, J.L., LIKENS, G.E.: Ecology 60, 1255-1269 (1979). - MICKIEWICZ, J.: Polish Ecol. Stud. 2, 57-62 (1976). - MILLER, G.R.: J. Ecol. 67, 109-129 (1979). - MILNE, R., SMITH, S.K., FORD, E.D.: J. Appl. Ecol. 14, 523-529 (1977). - MITSCH, W.J., DORGE, C.L., WIEMHOFF, J.R.: Ecology 60, 1116-1124 (1979). - MOCHNACKA-LAWACZ, H.: Polish Ecol. Stud. 4, 151-155 (1978). - MOGENSEN, V.O.: J. Appl. Ecol. 14, 243-252 (1977). - MONSON, R.K., SZAREK, S.R.: Oecologia 41, 317-327 (1979). - MOONEY, H.A., FERRAR, P.J., SLATYER, R.O.: Oecologia 36, 103-111 (1978). - MOORE, P.D., BHADRESA, R.: J. Appl. Ecol. 15, 837-845 (1978). - MORTON, A.J.: J. Ecol. 65, 993-999 (1977). - MOSER, T.J., NASH, T.H.

III: Oecologia 34, 37-43 (1978). - MOSZYNSKA, B., TATUR, A.: Polish Ecol. Stud. 4, 143-150 (1978). - MUHLE, O., BONNEMANN, I., RIECKMANN, P.: Verh. Ges. Ökologie 7, 287-294 (1979).

NEWBERY, D. McC.: J. Appl. Ecol. 16, 613-622 (1979). - NIENHUIS, P.H., DE BREE, B.H.H.: Hydrobiologia 52, 55-66 (1977). - NILSSON, I.: Oikos 30, 133-148 (1978). - NILSSON, S.G., NILSSON, I.N.: Ecology 59, 473-480 (1978). - NOBEL, P.S.: Oecologia 27, 117-133 (1977); - Oecologia 31, 293-309 (1978); - Ecology 61, 252-258 (1980). - NRIAGU, J.O. (ed.): Environmental Biogeochemistry, Volume 2: Metals Transfer and Ecological Mass Balances. Second International Symposium on Environmental Biogeochemistry. 370 pp. Michigan: Ann Arbor Science Publishers 1976. - NUMATA, M. (ed.): Ecology of Grasslands and Bamboolands in the World. 299 pp. Jena: Fischer 1979.

OECHEL, W.C., LAWRENCE, W.T.: Oecologia 39, 321-335 (1979). - OECHEL, W.C., MUSTAFA, J.: Oecologia 41, 305-315 (1979). - OHIAGU, C.E., WOOD, T.G.: Oecologia 40, 155-165 (1979). - OSMOND, C.B., NOTT, D.L., FIRTH, P.M.: Oecologia 40, 331-350 (1979).

PAKARINEN, P.: Ann. Bot. Fennici 15, 15-26 (1978). - PARSONS, K.A., DE LA CRUZ, A.A.: Ecology 61, 1045-1050 (1980). - PARTON, W.J., RISSER, P.G.: Oecologia 46, 223-234 (1980). - PARTON, W.J., SINGH, J.S., COLEMAN, D.C.: J. Appl. Ecol. 15, 515-542 (1978). - PELKONEN, P.: Flora 169, 386-397 (1980). - PELKONEN, P., HARI, P.: Flora 169, 398-404 (1980). - PEROZZI, R.E., BAZZAZ, F.A.: Oikos 31, 89-93 (1978). - PERSSON, H.: Oikos 30, 508-519 (1978); - Vegetatio 41, 101-109 (1979); - Oikos 34, 77-87 (1980a); - Acta Phytogeogr. Suec. 68, 101-110 (1980b). - PERSSON, T. (ed.): Structure and Function of Northern Coniferous Forests. An Ecosystem Study. 611 pp. Stockholm: The Editorial Service 1980. - PETERS, R.H.: J. Fish. Res. Board Can. 34, 918-924 (1977). - PETERS, R.H., MacINTYRE, S.: Oecologia 25, 313-319 (1976). - PETERSON, B.J., HOWARTH, R.W., LIPSCHULTZ, F., ASHENDORF, D.: Oikos 34, 173-177 (1980). - PHILLIPS, G.L.: J. Ecol. 65, 447-474 (1977). - PITELKA, L.F.: Ecology 58, 1055-1065 (1977). - PLEWCZYNSKA-KURAS, U.: Polish Ecol. Stud. 2, 63-74 (1976). - PRIDDLE, J.: J. Ecol. 68, 141-153 (1980).

RADFORTH, N.W., BRAWNER, C.O. (eds.): Muskeg and the Northern Environment in Canada. 399 pp. Toronto: Toronto University Press 1977. - RAGHI-ATRI, F., BORNKAMM, R.: Verh. Ges. Ökologie 6, 357-360 (1978). - RAPP, M., CABANETTES, A.: Oecol. Plant. 15, 151-164 (1980). - RAU, G.H.: Oikos 27, 153-160 (1976). - REDMANN, R.E., ABOUGUENDIA, Z.M.: Oecologia 36, 69-79 (1978). - REED, C.: Ecology 59, 481-488 (1978). - REED, F.C.P. III: Oecologia 30, 43-53 (1977). - REHDER, H., SCHÄFER, A.: Oecologia 34, 309-327 (1978). - RICHEY, J.E., WISSMAR, R.C.: Ecology 60, 318-328 (1979). - RIELEY, J.O., RICHARDS, P.W., BEBBINGTON, A.D.L.: J. Appl. Ecol. 67, 497-527 (1979). - ROOK, D.A., CORSON, M.J.: Oecologia 36, 371-382 (1978). - RORISON, I.H., HUNT, R. (eds.): Amenity Grassland. 274 pp. Chichester, New York, Brisbane, Toronto: Wiley 1980. - ROUTLEDGE, R.D.: Ecology 61, 276-281 (1980). - RUNDEL, P.W.: Oecologia 45, 354-359 (1980). - RUNDEL, P.W., LANGE, O.L.: Flora 169, 329-335 (1980). - RUNDEL, P.W., BRATT, G.C., LANGE, O.L.: Bryologist 82, 171-180 (1979a). - RUNDEL, P.W., RUNDEL, J.A., ZIEGLER, H., STICHLER, W.: Oecologia 38, 45-50 (1979b). - RUNDEL, P.W., STICHLER, W., ZANDER, R.H., ZIEGLER, H.: Oecologia 44, 91-94 (1979c). - RUTKOWSKA, B., STANKO-BRODKOWA, B., LEWICKA, E., DEBSKA, Z.: Polish Ecol. Stud. 2, 91-101 (1976).

SAMSUDDIN, Z., IMPENS, I.: Oecologia 37, 361-363 (1979). - SCHLESINGER, W.H.: Ecol. Monographs 48, 43-65 (1978). - SCHMIDT, L.: Oecol. Plant. 12, 195-213 (1977). - SCHREIBER, K.-F.: Veh. Ges. Ökologie 8, 185-203

(1980). - SCHUBER, M., KLUGE, M.: Flora 168, 205-216 (1979). - SCHULZE, E.-D., SCHULZE, I.: Madoqua 9, 5-13 (1976). - SCHULZE, E.-D., ZIEGLER, H., STICHLER, W.: Oecologia 24, 323-334 (1976). - SCHULZE, E.-D., FUCHS, M.I., FUCHS, M.: Oecologia 29, 43-61 (1977a). - SCHULZE, E.-D., FUCHS, M., FUCHS, M.I.: Oecologia 30, 239-248 (1977b). - SCHULZE, E.-D., HALL, A.E., LANGE, O.L., EVENARI, M., KAPPEN, L., BUSCHBOM, U.: Oecologia 45, 11-18 (1980a). - SCHULZE, E.-D., LANGE, O.L., EVENARI, M., KAPPEN, L., BUSCHBOM, U.: Oecologia 45, 19-25 (1980b). - SCHWARTZKOPF, S.H.: Ecology 59, 1062-1068 (1978). - SILVOLA, J., HANSKI, I.: Oecologia 37, 285-295 (1979). - SILVOLA, J., HEIKKINEN, S.: Oecologia 37, 273-283 (1979). - SIMS, P.L., SINGH, J.S.: J. Ecol. 66, 547-572 (1978a); - J. Ecol. 66, 573-597 (1978b). - SIMS, P.L., SINGH, J.S., LAUENROTH, W.K.: J. Ecol. 66, 251-285 (1978). - SINCLAIR, A.R.E., NORTON-GRIFFITHS, M. (eds.): Serengeti: Dynamics of an Ecosystem. 389 pp. Chicago: Chicago University Press 1980. - SMETACEK, V.: Verh. Ges. Ökologie 6, 23-28 (1978). - SMITH, R.I.L.: J. Appl. Ecol. 66, 891-909 (1978). - SOLLINS, P., GRIER, C.C., McCORISON, F.M., CROMACK, K. Jr., FOGEL, R., FREDRIKSEN, R.L.: Ecol. Monographs 50, 261-285 (1980). - SONESSON, M. (ed.): Ecology of a Subarctic Mire. 315 pp. Stockholm: The Editorial Service/NFR 1980. - SPEIDEL, B.: Polish Ecol. Stud. 2, 77-89 (1976). - SPRENT, J.I., SCOTT, R., PERRY, K.M.: J. Ecol. 66, 657-668 (1978). - SQUIRE, G.R.: J. Appl. Ecol. 14, 303-316 (1977). - STAAF, H.: Oikos 35, 55-62 (1980). - STARK, N.M., JORDAN, C.F.: Ecology 59, 434-437 (1978). - STEFFAN, H., ULLEMEYER, H.: Angew. Bot. 51, 287-292 (1977). - STROJAN, C.L., TURNER, F.B., CASTETTER, R.: Ecology 60, 891-900 (1979). - STRZELCZYK, E., STOPINSKI, M., DZIADOWIEC, H.: Acta Univ. N. C. Biologia 21, 115-128 (1978). - STURGES, D.L., TRLICA, M.J.: Ecology 59, 1282-1285 (1978). - SWITZER, G.L., SHELTON, M.G., NELSON, L.E.: Ecology 60, 1162-1171 (1979). - SZAREK, S.R., WOODHOUSE, R.M.: Oecologia 28, 365-375 (1977); - Oecologia 35, 285-294 (1978a); - Oecologia 37, 221-229 (1978b). - SZCZEPANSKI, A.: Polish Ecol. Stud. 4, 45-94 (1978).

TANNER, E.V.J.: J. Ecol. 68, 573-588 (1980a); - J. Ecol. 68, 833-848 (1980b). - TAYLOR, K.: J. Ecol. 68, 51-62 (1980). - TENHUNEN, J.D., WESTRIN, S.S.: Oecologia 41, 145-162 (1979). - TENHUNEN, J.D., YOCUM, C.S., GATES, D.M.: Oecologia 26, 89-100 (1976a). - TENHUNEN, J.D., WEBER, J.A., YOCUM, C.S., GATES, D.M.: Oecologia 26, 101-119 (1976b). - TENHUNEN, J.D., WEBER, J.A., FILIPEK, L.H., GATES, D.M.: Oecologia 30, 189-207 (1977). - TENHUNEN, J.D., LANGE, O.L., HARLEY, P.C., MEYER, A., GATES, D.M.: Oecologia 46, 314-321 (1980a). - TENHUNEN, J.D., MEYER, A., LANGE, O.L., GATES, D.M.: Oecologia 45, 147-155 (1980b). - THAYER, G.W., PARKER, P.L., LA CROIX, M.W., FRY, B.: Oecologia 35, 1-12 (1978). - THORP, J.H., GIBBONS, J.W., WHITFIELD, J.: Energy and Environmental Stress in Aquatic Systems. 876 pp. Washington, U.S.A.: Technical Information Center 1978. - TIESZEN, L.L., SENYIMBA, M.M., IMBAMBA, S.K., TROUGHTON, J.H.: Oecologia 37, 337-350 (1979a). - TIESZEN, L.L., HEIN, D., QVORTRUP, S.A., TROUGHTON, J.H., IMBAMBA, S.K.: Oecologia 37, 351-359 (1979b). - TRACZYK, T., TRACZYK, H., PASTERNAK, D.: Polish Ecol. Stud. 2, 39-47 (1976). - TRIPATHI, R.S., GUPTA, G.P.: Oikos 34, 219-226 (1980). - TSUCHIDA, K., NUMATA, M.: Vegetatio 39, 15-23 (1979).

ULANOWICZ, R.E.: Oecologia 43, 295-298 (1979); - J. Theor. Biol. 85, 223-245 (1980).

VEBLEN, T.T., SCHLEGEL, F.M., ESCOBAR, B.R.: J. Ecol. 68, 397-404 (1980). - VILLECOURT, P., ROOSE, E.: Rev. Ecol. Biol. Sol. 15, 1-20 (1978). - VILLECOURT, P., SCHMIDT, W., CESAR, J.: Rev. Ecol. Biol. Sol. 16, 9-15 (1979). - VITOUSEK, P.M.: Ecol. Monographs 47, 65-87 (1977). - VOGEL, J.C.: Oecol. Plant 13, 89-94 (1978). - VOGT, K.A., EDMONDS, R.L., ANTOS, G.C., VOGT, D.J.: Oikos 35, 72-79 (1980). - VOSSBRINCK, C.R., COLEMAN, D.C., WOOLLEY, T.A.: Ecology 60, 265-271 (1979). -

VUOKKO, R., KELLOMÄKI, S., HARI, P.: Oikos 29, 137-142 (1977). - VYAS, L.N., SHRIMAL, R.L., JINDAL, K.: Flora 167, 457-465 (1978).

WALLACE, C.S., RUNDEL, P.W.: Oecologia 44, 34-39 (1979). - WALLACE, L.L., DUNN, E.L.: Oecologia 45, 331-340 (1980). - WALLEN, B.: Oikos 34, 227-238 (1980a); - Oikos 35, 20-30 (1980b). - WALTON, D.W.H., SMITH, R.I.L.: Br. Antarct. Surv. Bull. 49, 117-135 (1979). - WAREMBOURG, F.R., PAUL, E.A., RANDELL, R.L., MORE, R.B.: Oecol. Plant. 14, 1-12 (1979). - WATTS, W.R., NEILSON, R.E.: J. Appl. Ecol. 15, 245-255 (1978). - WEARY, G.C., MERRIAM, H.G.: Ecology 59, 180-184 (1978). - WEAVER, S.E., CAVERS, P.B.: J. Appl. Ecol. 17, 505-513 (1980). - WEBB, W., SZAREK, S., LAUENROTH, W., KINERSON, R., SMITH, M.: Ecology 59, 1239-1247 (1978). - WEST, N.E., SKUJINS, J.: Oecol. Plant. 12, 45-53 (1977). - WERGER, M.J.A. (ed.): Biogeography and Ecology of Southern Africa. 1439 pp. The Hague: Junk 1978. - WHITE, D.A., WEISS, T.E., TRAPANI, J.M., THIEN, L.B.: Ecology 59, 751-759 (1978). - WHITTAKER, R.H., LIKENS, G.E., BORMANN, F.H., EATON, J.S., SICCAMA, T.G.: Ecology 60, 203-220 (1979). - WIEDER, R.K., CARREL, J.E.: Oecologia 44, 31-33 (1979). - WILLEMS, J.H.: Proceedings C 83, 279-306 (1980). - WILLERT, D.J. VON, THOMAS, D.A., LOBIN, W., CURDTS, E.: Oecologia 29, 67-76 (1977). - WINTER, K.: Oecologia 40, 103-112 (1979). - WINTER, K., TROUGHTON, J.H.: Flora 167, 1-34 (1978). - WINTER, K., TROUGHTON, J.H., CARD, K.A.: Oecologia 25, 115-123 (1976a). - WINTER, K., TROUGHTON, J.H., EVENARI, M., LÄUCHLI, A., LÜTTGE, U.: Oecologia 25, 125-143 (1976b). - WINTER, K., LÜTTGE, U., WINTER, E., TROUGHTON, J.H.: Oecologia 34, 225-237 (1978). - WOLF, G.: Schriftenr. Vegetationskd. 13, 5-118 (1979). - WOODMANSEE, R.G., DUNCAN, D.A.: Ecology 61, 893-904 (1980). - WOODMANSEE, R.G., DODD, J.L., BOWMAN, R.A., CLARK, F.E., DICKINSON, C.E.: Oecologia 34, 363-376 (1978). - WOODWELL, G.M., HALL, C.A.S., WHITNEY, D.E., HOUGHTON, R.A.: Ecology 60, 695-702 (1979).

ZEDLER, J.B., WINFIELD, T., WILLIAMS, P.: Oecologia 44, 236-240 (1980).

Priv.-Doz. Dr. WOLFGANG SCHMIDT
Lehrstuhl für Geobotanik
Untere Karspüle 2
D 3400 Göttingen

F. Special Topics

I. Floral Ecology

Report on the Years 1978(76) to 1981

By Stefan Vogel

1. General

Textbook, new revised edition: FAEGRI and v.d. PIJL (1979); popular introduction: HOLM (1979). History and current tendencies: BAKER (1979), FAEGRI (1978), MEEUSE (1980); field crops: BRANTJES (1979a).

2. Metabolism and Movements

The functionality of a flower will be misinterpreted if considered exclusively in terms of pollination. As a living system, it involves other manifold activities during and after anthesis: self-maintenance, final abscission of floral parts, seed and fruit formation, dispersal, etc. All of these interact and need coordination in the developmental sequence, a problem of internal engineering (v.d. PIJL, 1978). Some flowers contribute considerably to their own carbohydrate requirements and that of the fruits by photosynthesis: measurements of BAZZAZ et al. (1979) show that this may be important in flowers which precede leaf development in deciduous trees, those of *Acer platanoides*, for example, produce more than half of the total C amount required for the growth of mature fruits.

Physiological changes during defloration were studied by NICHOLS (1977) in carnation and by STEAD and MOORE (1977, 1979) in foxglove. Ethylene production was found to increase with age; it again appears responsible for the final fading of the corolla, an event independent as such but accelerated by pollination. In *Dianthus*, pollination (probably via some other hormone) stimulates a pronounced surge of ethylene in the style which causes the petals to roll up 2 - 3 days later. Flax (investigated by ADDICOTT, 1977) and *Digitalis* shed their corolla when still turgid and unwilted, the first within the single day of anthesis and unaffected by pollination. Pollen loads (but not glass powder) on the stigma provoke earlier shedding in *Digitalis*. The signal, whose speed was estimated at 0/4 mm/h, must continue to be effective because it would fail if the style was cut soon after pollination. Rupture of the corolla is preformed by an abscission layer similar to that in foliage leaves, its weakening by pectinase and cellulase activity is a vital process. Abscisic acid is probably involved; when artificially applied on young *Linum* petals it accelerated their shedding.

Experiments and observations were made on stylar movements which promote outcrossing or effect selfpollination. In *Hibiscus trionum* five stylar branches, when unpollinated, curve down during anthesis and contact anthers for selfing. The movement of the whole set was arrested, or its straight position resumed, by application of at least 30 pollen grains (BUTTROSE et al., 1977). All five stylar branches react in unison even if only one is pollinated. Grains lose this property if washed, heated, or aged. The projecting style of the nodding flowers of *Lilium martagon* was found by BRANTJES and BOS (1980) to be phototropic, curving toward unilateral light sources (e.g., forest edges in the wild), from where hawkmoths usually make their initial approach to flowers. This stylar movement effects herkogamy: the stigma is contacted first and may take up cross pollen before the visitor touches the anthers when circling round the flower to feed. If grown in diffuse light, the style bends nastically and keeps to a median position. According to ACKERMAN and MESLER (1979), the rostellum of *Listera cordata*, embracing the loose pollinia when untouched, refolds upon irritation of three sensitive hairs of the rostellar tip. Simultaneously, the glue is released which cements the freed pollinia to the visitor. Through a lifting movement the rostellum exposes the stigma from the second day onward, effecting dichogamy which is unusual in orchids. After HOLM (1978b) the (haptonastic?) style of the smoke bush *Conostylis dorrienii* (Proteac.) smashes forward with a noise when an insect enters the small gullet flower. Two staminodia act as triggers; simultaneously the anthers explode, dusting the visitor's forehead. Explosive pollen release evolved in ocimoid genera of Lamiaceae at least twice. BRANTJES and DE VOS (1981) describe such a mechanism in *Hyptis pauliana* (a birdflower) whose anthers, at first kept under tension by the lower lip, flip upward when the latter is touched, ejecting powdery pollen.

3. Means of Attraction

a) Production of Scents and Resin

Decomposition of hydrogen peroxide, the end-product of the peculiar "uncoupled" respiration process, is thought to be the principal cause of thermogenicity in Arum lilies. The fact that heat so often co-occurs with (and never arises without) smell in aroid osmophores is not fully explained by its ecological role as a volatilizer. According to B. MEEUSE (1978), evidence is accumulating for close, though not completely understood, metabolic connections with the synthesis of nauseous amines which is, in a broader sense, respiratory, too.

Scents of various European orchids were gas-chromatographed by NILSSON (1978a). Scent composition in *Platanthera chlorantha* was found to vary in that release of low molecular compounds (methylbenzoate, monoterpenes) increased at night, while sesquiterpenes (e.g., caryophyllene) showed no obvious periodism. The former, accompanied by visual cues, seem alone responsible for moth attraction. No genuine musk was detected in the odor of *Adoxa* by BRUNKE and HAMMERSCHMIDT (1980). Instead, a mixture of five rather ubiquitous C_6-C_8 alcools and aldehydes accounts for the musky touch, as confirmed by re-synthesis.

Dalechampia spathulata (section *Crematophyllum*, Euphorb.) has "perfume blossoms" adapted to scent-collecting Euglossine males (bibliography of these bees and their relations to orchids: WILLIAMS, 1978). This discovery made by ARMBRUSTER and WEBSTER (1979) in the plant's native Mexican haunts reveals another unusual strategy within the genus; species of section *Scandentes* are known (as also was confirmed by WEBSTER and ARMBRUSTER, 1979, in Mexico) to produce sticky resin as a reward for nest-building females (workers) of various bee genera which pollinate these flowers. Remarkably, perfume and resin are secreted by the same kind of glands amid the pseudanthia. A Madagascan species, the writer may add, lacks any gland and has typical pollen flowers. Resinous substances collected by pollinators in *Monstera* and *Clusia* spp. may also be used as nesting material (RAMIREZ and GOMEZ, 1978).

b) Visual Patterns and Pigmentation

General: PROCTOR and PROCTOR (1978). Certain insects are potentially able to categorize inflorescences as to size. According to field research of SCHAFFER and SCHAFFER (1979), selectivity in favor of bigger and higher flower stalks of a bee-pollinated *Agave* became apparent as long as total number of available stalks was high. With rising competition clusters were more evenly visited. Analogous behavior was observed in the moth *Tegeticula* in relation to *Yucca*, its host. The authors believe that pollinator selectivity may have played its part in the evolution of increased floral display, culminating in the hapaxanthy of many of these Agavaceae. In Australian pyrophytic *Banksias*, a short-lived, "mature-die" species, flowering once and for all, has more showy heads with higher attraction investment (reflected in seed set) than a fire-resistant "resprouter" blooming repeatedly (CARPENTER and RECHER, 1979).

The correlation of flower cluster size and attractivity, as expressed in pollinia transfer and seed set rates, was also found to be positive by SCHEMSKE (1980b) in a white-flowered *Brassavola*, pollinated by hawkmoths; and similar results were obtained in North American *Geranium* and *Phlox* spp. (WILLSON et al., 1979a), and *Asclepias* (WILLSON et al., 1979b; WILLSON and BERTIN, 1979). Larger inflorescences were visited more often, and as the number of flowers per stem decreased, so did the probability of receiving a visit. However, size preference could not be ascribed to some definite natural pollinator in these cases and results were less clear, when (introduced) *Apis* was removed from countings. Trials of BELL and LINDSEY (1978) with manipulated *Daucus* umbels had negative results in that neither umbel shape, nor size, nor pattern were primary factors in attraction.

Honeybees favored a yellow form of *Portulaca grandiflora* over a predominant purple one (which was otherwise identical). MOGFORD (1980) concluded that this is a primary preference (but some pre-conditioning cannot be ruled out). It was also the yellow form in mixed populations of white and yellow-flowered *Raphanus raphanistrum*, observed by KAY (1978), that received more attention by most visitor types. Others, including *Bombus terrestris*, were chiefly inconsistent. Assortative pollination may thus determine the maintenance and relative frequency of these color forms but it varies in strength according to local faunas.

Auxiliary calyx-borne semaphylls in Rubiaceae, consisting of one single enlarged and colored sepal, were surveyed by LEPPIK (1977). The peculiar feature evolved ± independently in nine, partly ornithophilous genera. Others instead developed a fruit wing with the same structural concept. Exposition of the floral wings (alae) of *Erythrina fusca* (Fabac.) signals to birds the flower's previous visitation and depletion. In the area where the chief pollinators are dark orange-chested male oriols (*Icterus spurius*), a variety with identically colored floral wings occurs. MORTON (1979) maintains this has developed to attract the birds' attention and to promote out-crossing by advertising trees already visited. Color changes have been similarly explained and various observers have provided new evidence.

Young, rewarding corollas of *Pachyptera hymenaea* (Bignon., S. America) are dark lavender and turn white when old. Pollinating Euglossines etc. confine themselves to the former, according to BARROWS (1977). Nectar-probing Apids and Lepidoptera quickly learn to seek out the white (and scented) flowers of *Lantana camara*, planted in Crete (KUGLER, 1980b). The pink phase, signaling defloration, was neglected. The same was stated for the yellow variety (turning scarlet) by MATHUR and MOHAN RAM (1978) in India, who claim, following experiments, that pollination is the trigger of color shift. Even thrips, only found inside yellow flowers, reportedly find these by color. In *Diervillea lonicera* (SCHOEN, 1977b) and two lupins (WAINWRIGHT, 1978), the nectar guides change from yellow to purple or red. Again, this "empty" phase was no longer visited by bees and a diurnal sphingid, *Hemaris*.

HARBORNE and SMITH (1978a), in a comparative analysis of anthocyanin chemistry of Polemoniaceae, find little correlation between floral ecology and glycosidic pattern or mode of acylation (often specific to one genus), whereas types of anthocyanidin (aglyca) often show that relation strongly, viz. the significance of pelargonidin for the scarlet of birdflowers. Chemistry of ultraviolet-absorbant pigments, visible to insects but not to man, was further elucidated by HARBORNE and SMITH (1978b) who state that in asteroid Composites four distinct flavonoid classes may account for UV absorptive guide marks (range 350-400 nm) of ray florets: (1) Ordinary flavonol glycosides (quercetin 3- and 7-glucosides). (2, 3) 6- or 8-substituted flavonol glycosides (gossypetin; patuletin, quercetagetin) or (4) Aurone and chalcone glycosides ("anthochlors"). A complement of five anthochlors: marein, sulphurein, coreopsin, butein, and sulphuretin, forms the UV-negative "bull's eye" of *Coreopsis gigantea* (SCOGIN et al., 1977). Similar patterns are species-characteristic in *Rudbeckia* (ABRAMSON and McCREA, 1977) and eschscholzioid Papaveraceae (CLARK, 1979).

Di-, tri-, and tetrachromatic sensory capacities have now been found responsible for color vision of flower insects (review: BRANTJES, 1979b). Such a variety, depending on classes, renders attempts to create a unified nomenclature of "insect colors" futile. Even the rare "insect white" (KAY, 1979: *Bryonia*, *Verbascum lychnitis*, white *Raphanus*) might only hold true for bees. KEVAN (1978, 1979) stresses that results from equal-energy experiments should not be considered absolute: sunlight to which the sensory apparatus is adapted has a very unequal energy distribution in different wavelengths, of which UV is the weakest energetically. Hence the bee's high sensitivity to this waveband would appear a compensative adjustment: UV as a color is no more stimulating than any others. But it is also important for bee orientation in the absence of sun, when skylight polarization which is strongest in UV serves as a cue (SILBERGLIED, 1979; BRINES and GOULD, 1979). Training experiments of GOLDSMITH (1980) with UV-lighted feeders which corroborate UV sensitivity of hummingbirds around 370 nm, merit special attention.

c) Nectaries; Nectar and Oil Secretion

Based on new ultrastructural (SCHNEPF, 1977; FAHN, 1979) and physiological evidence (LÜTTGE, 1977), workers are now inclined to accept that nectar extrusion is granulocrine, a kind of exocytosis. Vesicles carrying (pre-) nectar to the plasmalemma stem mainly from the Golgi apparatus (e.g., in Bromeliads, SCHNEPF), the ER or both (*Musa*, FAHN and BENOUAICHE, 1979) and are likely to be short-lived (2-3 min). The symplast is the site of intercellular pre-nectar flux (GUNNING and HUGHES, 1976). At least at one site of the system, beginning with unloading of phloem, active transport of sugars across biomembranes through phosphorylation is involved (LÜTTGE; SHUEL, 1978; SHUEL and TSAO, 1978), but also electrochemical gradients are discussed as the driving force. External and internal factors determine the temporal pattern of a flower's nectaring. Removal by visitors stimulates secretion; remaining liquid may be re-absorbed at the end of anthesis (CORBET, 1978).

In C_3 plants, i.e., those with normal type of CO_2 fixation (most bee food plants), the isotope ^{13}C is underrepresented including their nectar sugars. Detecting falsifications of honey with cereal starch hydrolysates or cane sugar is possible: C_4 plant products (malate type of metabolism) have a relatively high ^{13}C content. The test is not infallible, however, because the ^{13}C level in nectar of CAM plants (crassulacean acid metabolism) is as high as in the former (ZIEGLER et al., 1979).

Our knowledge about the occurrence of nonsugar nectar components has been much amplified by the work of BAKER (1977). Amino acids seem omnipresent in nectar: alanine, arginine, serine, threonine, and proline predominate. Generally amounting to 0.2-12.5 μmol/ml nectar, amino acids appear too scanty to be of nutritive value to most pollinators, but may play a part in taste. However, statistics show that butterfly-, noctuid-, wasp-, and dungfly-pollinated flowers have higher levels of amino acids than other types, indicating a nutritional demand of their (nonpollen-feeding!) visitors. Extranuptial nectars differ from floral ones (including intraspecific) by a relatively better representation of cysteine-related and nonproteine amino acids (BAKER et al., 1978). Evidence is given by FAHN and BENOUAICHE that proteinic matter in *Musa* nectar is co-secreted and not (as formerly suspected) microbial in origin. It is also recorded in *Fremontia* by SCOGIN (1979). Two iso-flavonoids cause fluorescence in this nectar.

The connective glands of *Mouriri* (Melastom.) produce fatty oil after BUCHMANN (1979); this is collected by Exomalopsine bees which (along with Euglossids) "buzz-pollinate" the flower. In malpighiaceous flower oils, BAKER (1978) traced small amounts of the main nectar sugars and amino acids, further indications (also supported anatomically) that elaiophors may be derived from (extranuptial) nectaries extant in some Old World genera of the family. Nevertheless, W. ANDERSON (1979) hypothesizes that elaiophors are ancestral and secondarily lost in the Paleotropic members due to the absence of appropriate oil bees.

d) Pollen and Its Imitation; Pollen Attachment

Whether starch or fat is the chief assimilate stored in a pollen grain depends on systematic position and, to some degree, on floral ecology. Broad scale tests performed by BAKER and BAKER (1979) show that one of the two kinds usually prevails. Occurrence of starch pollen (which is, on the average, larger in size) was found to be weakly correlated with anemo- and autogamy, while a surmised relationship to primitive or beetle-pollinated flowers was only partly supported. Pollen whose tubes must traverse long styles also have copious starch reserves. Typical entomophilous pollen is starchless and oil-containing: hence its digestibility for bees and flies. It is especially prevalent in pollen flowers. Sternotribic fodder pollen of heterantherous, nectarless flowers of *Lecythis pisonis* (collected by *Xylocopa*) differs from the nototribic fertilization pollen in being inviable and soon turning black (MORI et al., 1980). Different handling of heterantherous *Commelina* by various insects: BRANTJES (1980).

VOGEL (1978), discussing pollen flower syndromes, delineates tendencies toward deceit in oligandric and heterantherous pollen flowers: true edible pollen is gradually replaced by yellow dummies of great attractive power, detracting the foragers' activity from the scarce fertilization pollen which often has a cryptic color. Yellow guide marks, frequent in gullet flowers, are deduced by OSCHE (1979) from superoptimal signal copies of pollen-rich anthers, now generalized so as to include also nectar advertisement.

Exine patterns of Polemoniaceae, though manifold, exhibit no apparent association with the various types of zoophilous pollination occurring in the family (TAYLOR and LEVIN, 1975). Their adaptivity, if any, may relate to stigmatic contact and germination rather than to mode of transfer. Viscine threads connecting pollen grains or tetrads are not tapetal in origin but possibly develop from tetrad wall material during meiosis. They are of exinal (sporopolleninic) nature, root in the pollen exine, and are neither sticky nor viscous per se. They bear additional pollenkitt in Onagraceae but not in *Rhododendron* (HESSE, 1977, 1980a,c). Threads not only entangle grains in the visitor's body hairs but join the pollen in clumps and thus help to cover small stigmata sufficiently to fertilize multiovulate ovaries (CRUDEN and JENSEN, 1979; CRUDEN and HERMANN-PARKER, 1979; first record of threads in Caesalpiniaceae). Scanning microscopy has revealed a surprising diversity of viscine threads in Onagraceae (SKVARLA et al., 1978); smooth threads (prevailing in autogamous and bee-provisioning pollen) or twisted, moniliform, annulate, and "beaded" types occur, with much taxonomic specifity.

Both clip and arms of the asclepioid translator consist of a solidified emulsion, product of a highly coordinated secretion process, as shown by EM studies of SCHNEPF et al. (1979). How translators join the maturing pollinia in anthers is as yet unknown. The pollen of all advanced Apocynaceae is stuck to insects' mouth parts by an adhesive secreted by the stylar head, whose complex function has been analysed by SCHICK (1980). Often mistaken for the receptive slime, but separate from the proper stigma, the fluid resembles the asclepiad translator in origin and composition (emulsion of a mucous polysaccharide and terpenes).

4. Foraging Strategies of Pollinators

Only about the 3 first seconds during approach and nectar uptake are decisive for honeybees to learn the color of a new source. Further stay is irrelevant for coding. A learning process links certain scent, color, and pattern cues which subsequently will not be noticed unless combined. As also daytime of availability enters memory in association with signals, a bee is able to establish a repeatable foraging schedule with various food plants (LINDAUER, 1976).

By offering to honeybees artificial "patches" of two flower types different in quality and reward, "majoring" of better source could be quantified by WADDINGTON and HOLDEN (1979). Speed of decision increased with degree of difference; but distinct individualism was found. Only concentrated (and rapidly refilling) nectar sources induce a bee to fill her crop completely before returning to hive, while dilute loads do so upon 3/4 filling (PFLUMM, 1977). The pollen collected by a large honeybee colony would amount to 30 kg/year, as calculated by McLELLAN (1976). In Scotland, three to six pollen types (including *Ranunculus*) made up over 85% of total collection at a given time; preferences varied among colonies.

Information about bionomics of oligo- or monolectic solitary bees has been accumulated in surveys of LINSLEY (1978) and EICKWORT and GINSBERG (1980). Prenatal conditioning through larval provision is the hypothetical basis of pollen specifity and -recognition in oligoleges (which usually are more catholic with regard to nectar sources). But temporal adjustment to alternative, unrelated pollen hosts by learning occurs if the blooming time of main host is delayed. Rising temperature or (in deserts) rain may trigger synchronism of bees and hosts, whereas the time of day the flowers open is learned. Nocturnal bees depend on moonlight (*Sphegodogastra*) or can forage in dark (*Caupolicana, Ptiloglossa*). Oligoleges more often depend on their food plants than vice versa. Observations of STRICKLER (1979) on *Hoplites anthocopoides*, monoleges on *Echium vulgare*, demonstrated that they handle their flower (including pollen and nectar foraging) much faster than generalist bees and thus are likely to produce more offspring per unit time. When caged, they started nest-founding only if *Echium* was available. According to TORCHIO and TEPEDINO (1980), *Osmia* spp. adapt sex ratio of progeny to blooming peak of (spring) food plants: females outnumber males as long as pollen harvest is most promising; subsequently ♂♂ prevail.

Comparative examinations have been made of harvest methods and transport of pollen and oil (MICHENER et al., 1978; THORP, 1979) as well as of scopal hair microstructure (ROBERTS and VALLESPIR, 1978). Pilosity of baskets closely reflects pollen grain sizes in oligoleges, ranging from fluffy for fine pollen (e.g., Melastomatac.) to loose and stout for coarse pollen (e.g., Cucurbitac.). Parallel adaptations to loading of threated *Oenothera* pollen evolved three times independently. Megachilids use sternal scopa for directly scraping off the pollen presenter of Cynareae, their favorite hosts but, because they limit themselves to the male phase, they are inferior pollinators (WESTERKAMP, 1978). Poricidal anthers occur in 400 genera of 65 families and many bee genera (except, e.g., *Apis*, Megachilids, Andrenids) are known to shake out the powdery pollen by "buzzing" (wing muscle vibration of 0 - 300 Hz), for which BUCHMANN and HURLEY (1978) constructed a physical model. BERNHARDT and MONTALVO (1979) describe an example of buzz pollination by bumblebees in *Echeandia*.

There are now experimental analyses of how bumblebees forage and how they defray their caloric needs with nectar (general: HEINRICH, 1979b). Gross preferences are governed by tongue lengths of species or caste (HEINRICH, 1976a; MACIOR, 1978d). Flower constancy developed during the first one or two flights in freshly emerged *Bombus vagans* workers which probed 6.3 kinds of flowers (average) in their first 50 visits and then restricted themselves to 1.2 in the interval of 200 - 250 flower visits. Besides a majored, most rewarding source, interest in other flower kinds continues which the forager will major when the main source is exhausted. Handling accuracy in complex flowers increases in consecutive flights (HEINRICH, 1979a). The common attitude of *Bombus* when exploiting a raceme (to start with the lowermost flower richest in nectar and then ascend to the younger, less rewarding ones before departing) must be especially meaningful for cross-pollination of proterandric inflorescences (PYKE, 1978). Thoroughness in exploiting a flower depends on the amount of nectar available in the patch. As long as a basal pool of surplus secretion was accumulated in the corollas of *Chilopsis*, *Bombus sonorus* specialized in it, leaving rest nectar dispensed in five capillary grooves unexplored. After depletion of pool nectar bees began to extract grooves, paying shorter visits to (40%) more flowers per unit time. WITHAM (1977) demonstrated that this flexibility is energetically more rewarding than constancy of behavior and enhances cross pollination. By a tipping mechanism, the bat-pollinated *Oroxylum* flower benefits similarly by parcelling its total nectar volume in small bits accessible per visit, thus ±36 visits being necessary for exhaustion of one flower (GOULD, 1978).

The unusually high incidence of nectar thieving observed in an *Asclepias* by WILLSON et al. (1979b) and in two ornithophilous Acanthaceae (pierced by short-billed hummingbirds and *Coereba*; McDADE and KINSMAN, 1980) may be a consequence of habitat disturbance. BARROWS (1980) lists records of nectar robbing by alien *Xylocopa* and *Apis* in Hawaii; also a nectar-foraging wasp (*Epipona*) bites holes (YOUNG, 1980). Alcaloids in nectar and other potential deterrents to unwanted guests were found by BAKER (1977) to be more frequent in tropical lowland plants (with high parasite impact) than in mountain biotopes (see also FEINSINGER and SWARM, 1978). *Apis mellifera* exerted an odd type of theft on thistle heads, as was reported by LAROCA and WINSTON (1978): bees preferred to comb pollen from the fur of co-visiting bumblebee males (which did not mind the treatment).

5. Flower Classes and Their Evolution

a) Zoophily

Symposium: RICHARDS (1978). CREPET (1979) proposes that zoophily has preceded the angiosperm flower since the Carboniferous. In medullosan Pteridosperms, both pollen (too heavy for wind transport) and ovules may have attracted pollinators with chewing mouth parts. Such interactions, once established, were further improved through development of monocliny in Bennettitaleans (which protected ovules by scales). A. MEEUSE (1978b) using Gnetatae as a model, claims the pollination drops of gymnospermous ancestors (replaced by nectaries when angiovuly evolved) as the initial attractant to insects but not necessarily beetles. In PROCTOR's view (1978), strategies of angiospermous floral evolution provided complementary alternatives: secondary anemophily (appropriate to gregarious growth), euphilic specialization (reproductive isolation established by diversity and, hence, ethological barriers) and allophily (mainly physiological barriers). Allophily, for which *Euphorbia* subgen. *Chamaesyce* (EHRENFELD, 1979) and *Helianthus annuus* (HURD et al., 1980; 300 bee visitor spp. alone) present examples as well as euphilic styles, are products of "stochastic" interactions rather than coevolution (PAULUS, 1978, and JANZEN, 1980, criticize a too liberal use of this term which should only connote cases of certain common origin of ecologically connected species). Oldest preserved euphilic nectar flowers are from the Middle Eocene (CREPET, 1979). Evolution of euphily interfered with continental drift. The intrageneric origin of similar bat or bird flowers, nowadays visited by nonrelated pollinators in separate continents, remains enigmatic (VOGEL, 1980). Bird flowers of the southern Eurasian borderline, now obsolete (like Madeiran *Muschia*, visited by lizards today, ELVERS, 1978), testify to the northward extension of ornithophily in the Tertiary.

Onagraceae form an impressive example of euphilic radiation (37% bee-, 31% bird-, 17% hawkmoth-, 6% noctuid-, 7% syrphid-, and tachinid-pollinated). In his analysis, RAVEN (1979) regards ornithophily as basic in some genera (*Fuchsia, Lopezia*), but derived from insect flowers in others. Distribution of floral biotypes was surveyed for the floras of Australia (ARMSTRONG, 1979) and of New Zealand (GODLEY, 1979).

α) Cantharophily and Myiophily. The idea that beetles sponsored the origin of the angiospermous flower still needs support by critical field work, particularly on the surviving most primitive members of Magnoliales. First direct observations made on Winteraceae, *Degeneria*, and *Austrobaileya* therefore excite attention. THIEN's reports (1980) from the South Pacific area suggest presence of beetle syndrome (fruity odor, proterogyny, petal movements) in *Zygogynum* and *Belliolum*, but hardly so in *Drymis* and the famous *Degeneria* (shown in color pictures), though Nitidulids were seen on the latter. The large, pendant, proterogynic blossom of *Degeneria* has a peculiar and ecologically not fully comprehended set-up and movement of floral parts. A slimy secretion of staminodes, foul odor, and type of coloration indicate myiophily. Adaptation to carrion flies is even more evident (and could be confirmed in Queensland) in brown-spotted, fish-smelling *Austrobaileya*, of which ENDRESS

(1980b) gives a detailed description. These results fit the framework suggested by *Magnolia* (THIEN, 1974) and Annonaceae (GOTTSBERGER, 1970): primitive morphologies seem superimposed by almost euphilic specializations. Neither fly nor beetle pollination is so simple as to reflect an archaic condition. GOTTSBERGER (1977), though aware of this, maintains that evolutionary age of Coleoptera and their flowers alone provide a stringent argument for their primacy. In his view secondary polyandry in Rosidae and Dilleniidae was also a response to beetles (but see VOGEL, 1978). GOTTSBERGER et al. (1980) devoted another study to *Drymis* which fits best the classical concept of an unspecialized, primitive flower.

In *D. brasiliensis* they found, besides other insects, various beetles, particularly weevils, taking pollen and sweet stigmatic exudate. But THIEN saw flies and bees in South-Eastern *D. piperita* (and the writer bumblebees on Patagonian *D. winteri*). Thus the flowers, all similar (white, open, long-lasting, pleasantly scented), cannot be called cantharophilous. They apparently remained "schematic" and unaffected by a changing pollinator fauna during a long history.

Nocturnal *Nymphaea amazonica* is (as are congeners and *Victoria*) pollinated by *Cyclocephala* which probably feed on the staminodia (PRANCE, 1980b), while the expected cantharophily of *Nelumbo pentapetala* could not be confirmed by SOHMER and SHEFTON (1978). Its pollination mode remains problematic, as none of the many guests really "fitted", all leaving food-bodies intact. Based on collectors' records and pollen analysis, WILLEMSTEIN (1978) listed all plants known to be visited by Cetonids and (Central European) Cerambycids.

Small milichiid flies (♀♀) are attracted by myiophilous *Bulbophyllum warmingianum* in Brazil (SAZIMA, 1978). Passive movements of the lip through breezes are probably involved in attraction and pollinia deposition. The effectiveness of fungus gnats for pollination of forest herbs in the cool spring, such as *Listera cordata* and *Scoliopus bigelovii* in California, is again underlined by MESLER et al. (1980). That Phoridae also feed on nectar is learned from a list of food plants of this dipteran family, given by BAUMANN (1978b).

β) Melittophily. General conclusions of MICHENER (1979) in his useful "biogeography of bees": Apoidea are best represented in xeric, warm temperate climates (Mediterranean, California, Argentina) but comparatively poorly developed in tropics. Moist heat, detrimental to larval provision and immatures, may be a reason. The eastward-declining number of genera known from South America (315) through Africa (175) to South Asia (89) is remarkable: influence by *Apis* dominance in the East? Apparently there is little correlation between the numbers of (regional) bee and angiosperm species. Scoliids, putative ancestors of Apoidea, effectively transfer pollen but do not collect it (KUGLER, 1980a). Large *Scolia* species, observed in an *Echinops*, have tongues up to 6-8 mm long and exhibit certain flower constancy. Carpenter bees are the main pollinators of *Canavalia* species (Transvaal: STIRTON, 1977b; Java: SASTRAPADJA et al., 1975) and a *Plectranthus* (STIRTON, 1977c). KWAK (1979c) states all *Rhinanthus*, *Melampyrum*, and *Pedicularis* of the Netherlands are bumblebee-adapted, but not specialized to any one species or caste, while *P. furbishiae*, the only nectariferous summer-blooming lousewort of temperate North America, is exclusively visited by *B. vagans* workers (MACIOR, 1978c).

γ) Sphingophily. Hawkmoths fail to warm up for flight in ambient temperature below 15°C, possibly a limiting factor in altitudinal distribution of sphingophiles: CRUDEN et al. (1976) counted lower seed set in montane compared to lowland Mexican populations of *Calliandra*, *Oenothera*,

Mirabilis, (pollination of *M. jalapa* by *Phlegetontius* in Argentina: VALLA and ANCIBOR, 1978) etc. Exceptional is *Aquilegia pubescens* from above 2800 m (Rocky Mts.): pollinating moths are suspected to switch over to daytime activity at these heights.

δ) Ornithophily. Its importance in Australia (and the extent of work still to be done here) becomes obvious from data given by FORD et al. (1979) and (for the Southwest) KEIGHERY (1980): more than 100 species of birds visit flowers. Among those truly adapted, Trichoglossids probably antedate Meliphagids. Bird pollination of considerable antiquity on the continent is suggested by occurrence of various entirely ornithophilous, endemic genera. Around 1000 species are involved, 560 of 16 families alone in the SW. About half of the two hundred eucalypts receive bird visits, though proper ornithophiles are a minority. Like other Australian (and Capensic) bird flowers, *Anigozanthus humilis* blooms close to ground; wattlebirds drink standing and prefer hopping between plants (HOPPER and BURBIDGE, 1978).

The ornithophily of pan-tropic *Erythrina* (110 spp.), subject of a symposium volume, is probably very old. Only two African and 52 American species are specialized to sunbirds and hummingbirds, respectively, while all remaining Old World and (12) New World coral trees attract casual nectar-feeders among passerine birds (incl. parrots and woodpeckers). FEINSINGER et al. (1979), TOLEDO and HERNANDEZ (1979), and STEINER (1979) enumerate not less than 18 genera of 10 families seen on four Neotropical species, GUILLARMOND (1979) records similar numbers for African *E. caffra*. Typical features: nontubular corollas facing a perch, with copious (up to 280 µl), sucrose-dominated nectar. A parallel among Bombacaceae: *Bernoullia* (TOLEDO, 1977).

STILES (1979) who monographed Costa Rican *Heliconia*, believes this genus has governed the evolution of hermit hummingbirds (Phaetorninae) whose bills fit the corollas best. Pollen taken from bills and feathers of African (IWARSSON, 1979) and American flower birds (BOEKE and ORTIZ-CRESPO, 1978), including museum skins, allowed determination of their food plants. Sunbirds pollinate cauliflorous *Halleria lucida* in the Cape: STIRTON (1977).

ε) Mammal Pollination. Of the three W. Malaysian flower bats, one *Macroglossus* species mainly feeds from *Sonneratia*, another from wild bananas: both sleep close to food plants. *Eonycteris spelaea* has a wider range of sources and may travel 38 km (or more) from roost to feeding site (START and MARSHALL, 1976). When a flock of leaf-nose bats is around a flowering tree (of *Lafoensia*), a given flower is visited only once or twice, possibly due to recognition of previous landings by scent marks. But group foraging is not a constant pattern in *Phyllostomus* which passes over to single visiting when flowers become scarce (SAZIMA and SAZIMA, 1977). Chiropterophilous *Passiflora mucronata* adjusts to unilateral approach by zygomorphism (SAZIMA and SAZIMA, 1978). Confirming VOGEL's (1969) deductions, first observations were also published on *Marcgravia* (SAZIMA and SAZIMA, 1980) and another *Markea* (VOSS et al., 1980), both typical "trap-liners" and epiphytes with a pendant raceme which in *Markea* develops for months up to 200 flowers, one to two of which open each night; corollas are extremely short-lasting (4 h) but abound in nectar (390 µl). Bat (Pteropine) pollination in West African *Meranthes polyandra* (Chrysobalanc.), paralleling Brazilian *Couepia* spp., is described by LACK (1978), and also an aberrant species of *Lecythis* (*L. poiteaui*) in French Guiana has the syndrome (MORI et al., 1978). Another notable discovery is that of DANIEL (1979) quoted by GODLEY (1979): *Mystacina*, a Microchiropteran endemic to New Zealand, exhibits nectar-feeding habit, as deduced from pollen carried in the fur, in addition to cage experiments. *Collospermum* (Liliac.) is one of the food plants.

A promising new chapter seems to be opened by growing evidence that nonflying mammals partake in pollination. Observations made in southern continents substantiated old predictions by PORSCH: several species of *Banksia* from various parts of Australia appear to be regularly visited for nectar by mouse-like nocturnal marsupials (of which *Tarsipes* is most specialized (HOLM, 1978a; WIENS et al., 1979), and native rats (*Rattus fuscipes*; CARPENTER, 1978), *Proteas* of the Cape were found exploited in a convergent manner by feral species of mice (*Acomys*, *Aethomys*, *Praomys*, *Rhabdomys*) and shrews (*Crocidura*) by ROURKE and WIENS (1977), WIENS and ROURKE (1978), ROURKE (1980). Most of these animals must be facultative nectar-feeders, since the source is not available year-round. The mammal-pollinated species of both proteaceous genera share dwarf growth form, dense flower heads, dull reddish flowers with a yeasty odor at night, and abundant nectar which collects in longitudinal, interfloral troughs (*Banksia*) or the head center (*Protea*). ARMSTRONG (1979) who enumerates 21 Australian marsupials likely to feed on flowers, coins the word "therophily" to cover mammal pollination including bats. This agrees with the general conformity of syndromes (except geophily in nonchiropterophiles), while his terminological subdividing of placental, marsupial, and rodent pollination is unjustified as visitors' taxonomy is not reflected in the syndrome. Rats (*Oryzomys*, *Peromyscus*) were also observed lapping nectar from nocturnal flowers of a new epiphyte, *Blakea* n.sp. (Melastom.) in Costa Rica (LUMER, 1980), while Cebus monkeys were mainly destructive on a Brazilian *Combretum* (PRANCE, 1980a). Similar behavior of Madagascan lemurs is quoted by SUSSMAN and RAVEN (1978).

In a conjectural paper, the latter authors suggest that flower visits by modern nonflying mammals are probably not a secondary phenomenon but may have locally survived from Upper Cretaceous or Paleogene times, when small climbing marsupials and, later, prosimians may generally have fed from flowers (dental features of fossils give some indications), before bats entered the scene. It is tempting to hypothesize that the syndrome we call chiropterophilous was initiated this way before it was disjunctively taken over by competitive bats in most parts of the tropics.

b) Anemophily and Hydrogamy

NIKLAS (1981) concludes from simulation experiments that, by channeling airflow, pre-integumentary lobes bordering naked Pteridosperm ovules aided wind-borne pollen to settle on the nucellar apex. While gymnosperms totally lack pollenkitt, at least traces of it are omnipresent in angiospermous wind pollen (possible exception: Ericaceae), a fact hinting at its entomophilous ancestry. HESSE (1978, 1979a,b,d; 1980b,d), in comparative TEM-analyses studied the origin and efficacy of pollenkitt with emphasis on families and genera including either type of pollination.

The pollenkitt was found to be inactivated in different ways: (1) Retention of bulk at the loculus wall (*Fraxinus*, *Populus*), (2) its withdrawal into exine cavities (Polygonaceae, *Acer*, *Mercurialis*), (3) loss of stickiness by chemical alteration (*Quercus*) or volatilization (*Castanea*, *Plantago*). But as pollenkitt features do not always fit the pollination status of the flower, conclusions should be drawn with caution. Moreover, the apparent great diversity of kitt substances seems incongruent with the monophyletic origin HESSE attributes them. LEROY (1980) again challenges wind pollination (and unisexuality) as being always secondary in angiosperms; *Myrica*, *Cercidiphyllum*, and some others belong, in his opinion, to the very rare instances where it could be primary.

Flowers of other groups reveal contradictory or intermediate features rendering the question difficult as to whether their way of pollen transfer is "undecided" or transient. According to UHL and MOORE (1977), the majority of palms tend toward a primitive entomophily on the basis of surplus pollen (polyandry in about 70 genera), nectar (*Ptychosperma*), and heat (scent) development (*Bactris*, *Nypa*). Although liberating much air-borne pollen, oil palms require assisted pollination in Malaysian plantations, in contrast to Cameroon, home country of *Elaeis*. This points to lack of appropriate vectors. In fact, SYED (1979) found pollen-eating beetles (*Elaeidobius*) do the job in Africa; the attractant of ♀ flowers is not clear, however. In *Arceuthobium* (Loranth.) visitors neglect pistillate plants which are devoid of nectaries present in staminate flowers; hence, wind pollination is presumed by PLAYER (1979) to occur.

In *Plantago* (HAMMER, 1978) and *Thalictrum* (MELAMPY and HAYWORTH, 1980) the perfection of secondary anemophily still shows a ± linear series of intergrades, reflected (among other traits) in an increasing pollen/ovule ratio from one species to another. In plantains this line culminates in *P. lanceolata* (but Syrphids still play a role here: STELLEMAN, 1978), while another trend toward autogamy declines p/o ratio (*P. coronopus*, *P. maior*). In *Thalictrum*, occurrence of anemophily (followed by dioecy) is explained by the authors as a response to insufficient insect visitation due to lack of nectar, as still observable in related *Isopyrum*. The existence of partly hermaphroditic anemophiles (*Th. polygamum*) argues against earlier suggestion of dioecy having preceded wind pollination in *Thalictrum*. Anemophilous features of *Poterium minus* compared to *Sanguisorba maior* were analyzed by SAID (1979) (pollen: HESSE, 1979c).

DUCKER et al. (1978) and PETTITT (1980) continued research on hydrogamy in marine Cymodoceaceae and Hydrocharitaceae. Stands of *Thalassia* flower synchronously with spring tides, when water fluctuation is strongest across the undersea platforms. Embedded in thecal slime, the (round) pollen grains drift in masses, while *Halophila* releases moniliforme tubes containing several grains. *Enhalus* is ephydrogamous: abscised male flowers float to buoyant pistillate corollas.

6. Pollination of Particular Groups

a) Various Angiosperms

Apiaceae. The common view that umbellifers are typically allophilic requires revision, since recent work has revealed that insect visitation is less promiscuous than expected, and effective pollen vectors are a minor part in the spectrum. Although beetles outnumbered bees as guests of North American *Thaspium* and *Zizia*, up to 96% of the pollen was transferred by the latter, among which, moreover, an oligolege (*Andrena ziziae*) predominated (BELL and LINDSEY, 1977). GRACE and NELSON (1981) counted 59 taxa visiting *Heracleum sphondylium* and 48 visiting *H. mantegazzianum* (introduced), growing side by side in a British meadow; but only 23 species were common to both. Differences in umbel height are one of the factors held responsible for selective foraging which forms a notable hybridization barrier. Apiaceae in general are preferential nectar hosts of phorid flies (especially in montaneous biotopes) according to BAUMANN (1978a) who identified 28 species (16 alone in *Heracleum sphond.*). KERNER's classical concept of geitonogamy in umbellifers also needs reexamination. In species with proterandry (which prevails in the family) CRUDEN and HERMANN-PARKER (1977) found minimal overlap

of ♂ and ♀ phases within individuals. In a single umbel, the florets are receptive when they all have passed the ♂ stage and, moreover, the lateral, secondary umbels start (male) flowering only when the ♀ stage of the primary umbel has ceased. Proterogyny also occurs in the family: *Sanicula*, *Hydrocotyle*, *Erigenia*, *Thaspium*, *Zizia* spp. Another way to ensure outbreeding is through development of ± complete dioecy, investigated in five genera in New Zealand by WEBB (1979). The gynodioecious species (*Gingidia*, *Scandia*, *Lignocarpa*) have proportions of ♀ plants which vary between populations and regulate outbreeding, whereas seed set in hermaphroditic individuals was preferentially effected by selfing.
Araceae. CROAT (1980) compared the flowering behavior of 30 *Anthurium* species whose open spadices exhibit either the "male Euglossine syndrome" with fragrant odors or are myiophilous. The latter produce a sweet, watery stigmatic fluid and a yeasty smell. The single florets are principally proterogynous, though many spadices never developed a stigmatic phase; the anthers emerge sequentially.
Cactaceae. GRANT and GRANT (1979b) presume that the melittophilous Cactaceae and the single birdflower (*Echinocereus triglochidiatus*) of the N. American Southwest are products of autochthonous radiation, while moth- and bat-pollinated members developed their syndrome in the tropics. The melittophiles comprise over 90% of species (Jamaica: 50%), the majority belonging to *Opuntia* (GRANT and HURD, 1979), *Echinocereus* (GRANT and GRANT, 1979a), and *Mammillaria* whose pollen flowers are exploited by over a hundred mostly polylectic bee taxa (GRANT et al., 1979). Although *Carpophilus* beetles are regular visitors, they are destructive and ineffective because they rarely make contact with the stigma (GRANT and CONNELL, 1979). *Opuntia fulgida* exhibits nocturnal anthesis and is probably foraged by vespertine carpenter bees.
Marantaceae. Fourteen Central American species forming an advanced group of *Calathea* are peculiar in that their flowers do not unfold spontaneously. Their exclusive pollinators, Euglossine bees when foraging at dawn (±0500 h) force the flower open by gnawing the corolla tip, a distinct behavioral act (KENNEDY, 1978). The subsequent tongue insertion trigger the typical spring mechanism which smashes the stigma onto the bee's chin and smears it with pollen. In KENNEDY's opinion the failure of autonomous spreading evolved through consistent break-up of flower buds, still observed in "normal" species.
Valerianaceae. The pollination mechanisms of *Valerianella* are presumed to have developed in two directions starting from a medium-sized funnel-form bee-flower: autogamy (minute-flowered weeds) and adaptation to Lepidoptera (white, long-tubed corollas of N. American species). *Fedia* belongs to the latter type (ERNET, 1977). Pollination in *Valeriana officinalis*: KONON (1978).

b) Orchids

Because the pollination process is rarely seen, anthecological concepts, even for many European terrestrial orchids, are mostly conjectural. New observations by professionals and high-level amateurs show that it is simply a matter of patience to see the right visitor at the right time. As to nectariferous species, well-documented accounts from Swedish populations of *Epipactis palustris*, *Platanthera chlorantha*, and *Herminium* are dealt with by NILSSON (1978a,b, 1979b). *E. palustris*, though scentless and poor in nectar, attracts a broad array of insects but, nevertheless, is no generalist; solitary wasps only (in particular *Eumenes* ♂♂) match the lip structure well and thus were chief vectors. Scent-conducted, tiny Chalcidid wasps frequent small-flowered *Herminium* for nectar, getting pollinia fixed on forelegs, while most pollinia of *Platanthera* were attached on the eyes of Noctuids which drink while fluttering, with forelegs rested on sepals. Short-tongued species can forage from a nectar-filled capillary furrow ending near spur mouth. N. American *Tipularia discolor* is also phalenophilous (WIGHAM and McWETHY, 1980); pollinia uptake and deposit was most efficient when nectar supplies in the spur declined, forcing the moths to probe deeper.

As regards the deceitful attraction mechanism in Ophrydeae, putative transient stages in, for example, *Barlia* (TESCHNER, 1976) and *Himantoglossum* (TESCHNER,

1980) deserve special attention, as they suggest how complete deceit may have developed from mutualism. *Himantoglossum* still produces some nectar in a furrow of the spur. In Germany, *H. hircinum* was seen to be probed by four *Andrena* spp. (both sexes) and bumblebees during occasional side-forages from nearby main food-plants. *Orchis* s.l. which completely lacks nectar (except *O. coriophora*, to which DAFNI and IVRY [1979] add *O. sancta*), also seems to display a nutritional mimicry, but the deception situations exploited by the various specialists are heterogeneous. When growing jointly with *O. coriophora* in Israel, similar but "dry" *O. collina* profits from rare "mistake" pollination by bees foraging on the former (DAFNI and IVRY). Early-flowering *Dactylorrhiza sambucina*, observed in Sweden by NILSSON (1980) who witnessed, on the average one visit every 6.4 h during 134 h of observation, appears specialized in deceiving unconditioned, scouring *Bombus* queens newly emerged from hibernation. Ninety-one percent of queens with pollinia had no pollen in their corbiculae and hence were on their initial nectar-foraging trips. The flowers do not mimic a rewarding model but act spontaneously by showiness and scent; attractivity soon ceases with growing experience of queens. VÖTH's observation (1980) of pollinized *Ceratina* males busy in *Serapias* at high noon, suggests that "night-shelter theory" is still not fully settled. The author surmises that ♂ bees expect to find virgin females in the nest-hole-like corollas for mating.

In a survey of pseudocopulation in *Ophrys*, PAULUS and GACK (1980) report that *Colletes cunicularis* visited a form of *O. fusca* near Malaga, in the same (reverse) manner as do *Andrena* spp. elsewhere. *Colletes* was hitherto known as the pollinator of *O. specodes*. It is inferred that there are geographic races with vicarious partners. *O. fusca* even attracted North American *Andrena* spp. (TENGÖ, 1979). As the regular partners of *O. bertolinii*, two species of *Chalicodoma* have been established in Italy by BÜEL (1978). With *Chalicodoma* (head-pollinated), a new bee genus and another family (Megachilids) enter the list of (8) aculeate genera, known as *Ophrys* pollinators. PAULUS and GACK demonstrated the importance of inflorescence height by exposing *O. speculum* flowers in different levels above ground to *Campsoscolia* ♂♂ which, after a matinal foraging period, swarmed gregariously 10-20 cm above the soil surface: frequency and duration of visits strongly decreased from flowers lying on the ground to those at 15 cm level. That readiness of *Campsoscolia* ♂♂ to "pseudocopulate" continues after eclosion of females, and visits to *Ophrys* occur along with genuine copulations was confirmed by GRAF (1977). Mandibular and head extracts of *Andrena* ♀♀ are attractive to ♂♂ (TENGÖ and BERGSTRÖM, 1977). Among presented single synthetic compounds (identical to those known as natural components), straight chain alcools, especially 1-octanol, proved most effective, but mixtures were always superior. BERGSTRÖM (1978), in a biochemical synopsis of the *Ophrys* syndrome, admits that attempts to associate *Ophrys* volatiles with ♀ pheromones of their bees are still unsatisfactory. Since also certain manipulations of ♂♂ on the flowers are not fully explained as part of a copulation behavior, it would pay to look for new aspects of the phenomenon.

Of the bees (deceitfully) allured by *Cypripedium calceolus*, some merely alighted, others are trapped but escape via entrance. Medium-sized *Andrena* spp. only (preferably *A. haemorrhoea* ♀♀) free themselves through the exit holes, as pollination requires (NILSSON, 1979a).

c) Ficus

Various papers have contributed to a better understanding of the life cycle of the common fig. The old belief that "caprificus" which is used to fertilize the fruit-bearing main crop, represents a wild variety of the latter, is not correct. In fact, caprifigs and females occur in both the feral and cultivated state. The reproductive system is principally gynodioecious, and the caprificus, though bisexual, functions as the pollen mate (VALDEYRON and LLOYD, 1979). Its short-styled "gall-flowers" set normal seed when pollinated. Given some overlap of syconium maturation in neighboring caprifigs, these may reproduce themselves sexually if wasps spare some of the florets they pollinate an egg. Usually seed set fails due to the lack of pollen from the antecedent syconia (NEEMAN and GALIL, 1978). While the summer syconia ("profichi") are mainly pollen and wasp donors for the long-styled morph (edible figs), the other generations ("mammoni", "mamme") are poor in ♂ florets and specialized to maintain the wasps' life cycle. Although the female morph is useless for the animal which cannot oviposit in long-styled flowers, many wasps are, nevertheless, attracted (deceived by the external identity of syconia with those of the caprificus), pollinate them and die without setting progeny. In northern countries, wasps either fail to reproduce at all or appear in two generations only. For optimal yield, edible figs require three syconia and three wasp generations in the caprifig. Their autumnal syconia, left unfertilized by lack of wasps, ripe as edible figs in some varieties which hence are not truly parthenogenetic (RIVALS, 1980).

Functional dioecy of this type is confined to the whole subgenus *Ficus*, the majority being monoecious. A primitive representative of the latter, *F. vogelii*, was investigated by NEWTON and LOMO (1979) in Ghana. Its uniform syconia have staminate florets intermingled with pistillate ones. The latter show no clear-cut bimodality but rather a continuous range of style lengths. Oviposition of pollinator *Allotriozoon* concentrates on the shorter-styled ovula. ♂ florets generally fail to mature, unless some wasp progeny is developing in the syconium. One tree produces one to three synchronous syconium generations per year, but blooming times are staggered within populations, allowing wasp (and pollen) intercourse between trees.

Phylogenetic specificity is apparent at the level of Agaonid genera and sections (or lower ranks) of *Ficus*, as taken in account in a new classification of RAMIREZ (1977). Continued inventory of fig faunas confirms that the one wasp/one fig-species system is the rule (WIEBES, 1979). Tropical woods usually harbor various sympatric fig species: JANZEN (1979), in a programmatic essay, asks the question how syconia prevent interference of allospecific Agaonids with the "wrong" pollen. Selective attraction by allomones or mechanical filtering of ostiole may happen, or both. Establishment of a new variety depends on a new wasp, otherwise immediate back-crossing would occur. Figs have many "cuckoos" and cleptoparasites: wasp species which parasistize ovula without pollination (which is done by legitimate visitors); some of them circumvent the ostiole by laying eggs through the syconium wall with their long ovipositor, others are predatory on legitimate immatures, etc. Legitimate Agaonids may be derived from one of these types (WIEBES, 1977).

7. Community Structure and Pollination

a) Distribution of Flower Types

The general character of a community in terms of floral ecology does not necessarily reflect any special condition of the biotope for pollination, but may simply be a result of floral features prevalent in the plant families which participate due to edaphic or climatic preferences. Careful statistics are needed to reveal such camouflages and to recognize significant correlations. OSTLER and HARPER (1978) compared 25 communities of the Wasatch Mts. (USA), DEL MORAL and STANDLEY (1979) those of Western and Eastern Cascade Mts., and PROCTOR (1978) Irish communities. In brief, they concluded that with growing species number of communities, the general diversity of zoophilous types, rates of eutropy, blue and purple colors, and zygomorphism increase, while relative rates of anemophily, generalized zoophily, and yellow flowers decrease. The latter is correlated with the better chances of wind pollination in low diversity communities containing few dominant, gregarious species and prevalent promiscuous visitors (which prefer yellow). Species richness means more intense annidation through competition for a multiple and more choosy pollinator fauna, a type best represented in low elevation open woodlands. Most birdflowers occurred in these and also near the timberline. White corollas prevailed in dark forests and anemophily in the canopy. ANTONOVA (1979) reported on entomophily in Russian steppe oakforests and SWALES (1979) on Arctic and Subarctic vegetation.

Successional age influences a community's pollination structure profoundly, as was shown by PARISH and BAZZAZ (1979). Early herbaceous communities (disturbed habitats) differed from a late one (tall grass prairie) in Illinois in the dominance of autogamy and generalist visitors and in longer mean flowering times/species (±50 d). In the mature community outcrossing euphilic types and specialized visitors prevailed and mean blooming time was shorter (±30 d). An annual community is probably too ephemeral and species composition too fluctuating for the development of a co-adaptive system, while members of mature plant societies have co-evolved long enough to create biotic niches.

The mangrove forest, investigated by TOMLINSON et al. (1979) and PRIMACK and TOMLINSON (1980) in Australasia, consists of large stands containing few species. A pioneer community, it has some traits in common with weeds: frequent inbreeding, abundant fruit set, and early start of flowering in life. *Rhizophora*, most wide-spread and predominant, is basically wind-pollinated. However, other Rhizophoraceae (*Bruguiera*, *Ceriops*) and Combretaceae (*Lumnitzera*, TOMLINSON et al., 1978) underwent an adaptive radiation to bird, moth, or generalist entomophilous flowers.

b) Pollinator Competition

The phenology of the aculeate pollinator fauna in a seasonal Costa Rican forest was followed by HEITHAUS (1979).

> Threehundred and thirty species visited flowers belonging to 53 families (58% of spp. were legumes). Wasps stand out in diversity (138 spp.). Most of the aculeate taxa were uncommon. Social bees made up 50% of counted individuals, but only 12% of the bee spp. Despite species richness temporal staggering of flight periods, a means of partitioning flower resources, is not more intense than in temperate regions; the mean flight periods are even three times as long. Anthophorid bees predominated in dry season, most Halictids and solitary wasps in the wet season (VI-X). Halictids and Megachilids had nonuniform distributions, while Colletids and social Hymenoptera (Apids, Vespids) were aseasonal.

Spatial competition of *Bombus*, *Xylocopa*, and (introduced) *Apis mellifera* was studied by SCHAFFER et al. (1979) in Arizona, where *Agave schottii* was the principal nectar source. Honeybees prevailed in habitats, where the plant was most productive, *Bombus* in intermediate, and carpenter bees in least-quality sites. The same rank order was observed in relation to day times of nectar availability. The superiority of Apids seems a necessity with regard to their social requirements and colony sizes. In places of the Rocky Mts., where two *Bombus* species with different proboscis lengths competed for an *Aconitum* and a *Delphinium*, workers concentrated on the respective host they matched best with mouthparts (INOUYE, 1978). Nectar depletion was the regulator, not aggressive interference. If all foragers of one species at the site were caught, the remaining species soon included the "free" source for exploitation. But alarmed by food shortage, pygmy "house-workers" of the removed species which normally stay in the nest, began to take over foraging.

Africanized honeybees also displaced stingless and other native bees from flower patches in Guiana through nectar depletion, not by chasing them away (ROUBIK, 1978). They were superior in occupying artificial feeders owing to quicker social communication, and resisted aggressive *Trigona* spp. (ROUBIK, 1980). *Bombus* and large solitary bees, when foraging on Composite heads, passively precluded landing of smaller competitors, and the latter, if already on the heads, tended to give way (DECELLES and LAROCA, 1979). A reverse dominance among four Apids was found by KOENIGER and VORWOHL (1979) on baits in Sri Lanka: smallest *Trigona* drove three *Apis* species away, of which, again, respectively smaller species were superior in the "pecking order" (*florea* > *cerana* > *dorsata*). Pollen types of their combs showed little overlap, revealing a smooth partitioning of natural sources. *Apis dorsata* and *A. cerana* seasonally migrate within Sri Lanka, while the other two Apids are stationary.

Single mass-flowering trees (e.g., *Tabebuia*) attract not only many bees, but also birds which prey on the latter. As the bees would tend to stay at the rich source and thus cause geitonogamy or no fertilization at all, GENTRY (1978) argues that the plants indirectly profit, as the birds, by chasing some visitors away, promote fluctuation between conspecific trees. SCHEMSKE (1980a) showed that in ornithophilous *Combretum farinosum* (Costa Rica) roaming flower bird species cause more outcrossing than stationary ones, and that the unusual amounts of nectar this plant produces serve to satiate "tenants", and thus prevent them from chasing the more effective "transients" away.

c) Flower Competition and Seasonality

Various attempts have been made to verify predictions based on the notion that plants escape competition for pollinators by creation or utilization of spatial or temporal niches in a community.

How trees share bats as pollinators in the same season, is illustrated by HOWELL (1977): species avoid undue interspecific pollen transfer (which would result in pollen waste) by opening their flowers at different hours of the night or by depositing on different body parts of bats.

Bog Ericads or rain-forest birdflowers are typical examples of seasonal staggering. The eight Ericaceae of a North American swamp, observed by READER (1977), share a continuous polylectic bee fauna (chiefly bumblebees) by mutually exclusive blooming periods, with peaks spread evenly over the summer. The year-round flowering sequence found in the Costa Rican hermit-pollinated ornithophiles was noted by STILES (1977) not to be disarrayed by climatic irregularities. Sometimes droughts may cause some peak displacement but the rising gaps may be filled by a species whose phenological reaction norm just fits this situation, thus preventing the movement of pollinators out of the community.

The pattern of flowering peaks seems to be governed by competition (trends to minimize overlaps) as well as mutualism: sequential blooming times ensure that a pollinator species remains in the system to the benefit of all involved community members. Both aspects could be demonstrated by WASER (1978) and WASER and REAL (1979) using two subsequently flowering species of *Delphinium* and *Ipomopsis*, visited by the same hummingbirds in alpine meadows of Colorado. Pollen spoilage through interspecific visits during the short overlap period caused noticeable seed set reductions in both plants when compared to nonoverlap periods.

In years of poor blooming in *Delphinium*, many discouraged birds left the site before *Ipomopsis* began to bloom. Consequently the latter also suffered low fecundity despite abundant flowers. THOMSON (1980), working in a similar biotope, confirmed the prediction that a species will, above all, tend to keep its initial overlap short. During dominance of an antecedent competitor, scattered forerunners would be at a disadvantage as pollinators are reluctant to visit rare and novel flowers (unless these mimic the former). A quick rise of flower abundance converts the pollinator and it begins to visit this new source constantly. Once established, fidelity will persevere even if flower density decreases slowly. Whether a flowering curve's rapid rise (its skewness) is constant in a species and hence is truly adaptive, is still uncertain.

When pollinators become scarce near the end of the season, a reverse situation may occur. Pollination and seed set in a late-flowering *Polemonium* was found by ZIMMERMANN (1980) to be pollinator-limited only in the second half of blooming period, whereas seed set was complete in the initial phase, indicating that pollinators were still numerous at that time.

Marked differences of overall flower density curves in disturbed habitats, bog vegetation, and woodland, as well as between woody and herbaceous plants in these associations were ascertained by HEINRICH (1976b). They indicate divergent niche utilization. The phenology of vernal flowers in woodland is dictated by the need to anticipate canopy closure which showed little annual variation in Illinois (SCHEMSKE et al., 1978). These herbs seem to have evolved a "fail safe mechanism" to realize, and end, blooming on time, no matter how their temperature-dependent start may vary. They show great overlap of blooming times and similarity in whitish flower color.

Nest-founding bumblebee queens develop a moderate flower constancy, partitioned among *Dicentra* (MACIOR, 1978a,b), *Podophyllum* (SWANSON and SOHMER, 1976), and other plants, while *Claytonia* (SCHEMSKE, 1977), *Hepatica* (BERNHARDT, 1976), and *Erythronium* spp. (BERNHARDT, 1977) preferently rely on oligolectic *Andrena* spp.

d) Floral Synergism on Community Level

Pure stands of *Hieracium florentinum* were less insect-visited than those of *H. aurantiacum*. However, when flowering amid this favorite, the former's visitation rate was raised significantly, as pollinators did not discriminate between the similar flower heads. Although THOMSON's study concerned apomicts, he uses his results as a model to demonstrate that a less rewarding or less common plant species may benefit from a "magnet" species by flower mimicry. Its gain should outweigh the detriment of heterospecific pollen transfer. A "ring" of nine synchronously flowering convergent ornithophiles in a community in Arizona jointly maintained two species of hummingbirds which foraged on these flowers indiscriminately. Though the latter differed in the place of pollen attachment, about half of sampled stigmata had some heterospecific pollen, one fifth more than conspecific. BROWN and KODRIC-BROWN (1979) are inclined to call this uniformity in birdflowers which lack a predominant model, a case of Müllerian mimicry as the participants profit in a mutualistic way. One member, a nectarless variety of *Lobelia cardinalis* (a species elsewhere nectariferous), is an exception and parasitizes on the local system as a "Batesian mimic". Monarch butterflies in Panama were deceived by (notoriously) "dry" *Epidendrum ibaguense* which mimics (nectariferous) flowers of *Lantana* and *Asclepias* growing nearby (BOYDEN, 1980).

e) Pollination of Sympatric Congeners

How co-occurrent sister species prevent hybridization by post- and pre-pollination barriers has been investigated in various species pairs (or triplets) of mainly North American communities. Intersterility was found to be sufficient in two *Dicentra* spp. which shared nine species of *Bombus* queens as pollinators (MACIOR, 1978b), and also in three milkweeds (KEPHART and HEISER, 1980), with a broader range of visitors. In both cases, flower constancy and slight edaphic differences, leading to spatial separation, kept interspecific transfer low. Two Californian *Cercidium* spp. having similar yellow corollas, both have *Centris* bees as pollinators. However, in UV light the flowers' pattern differ markedly and foragers discriminated between them with ease, remaining constant to one or the other (JONES, 1978). BARBER and ESTES (1978) compared two sympatric *Pyrrhopappus* spp., having almost identical heads, but differing in their flowering peaks, diurnal receptivity hours, and soil types, thus attracting different Apids. In autogamous *Rhinanthus minor*, inaccessibility of the stigma to foraging *Bombus* impedes introgression of *Rh. serotinus*, while the reverse does occur, owing to the latter's protruding style (KWAK, 1978, 1979a,b).

Divergent flower types are another pre-pollination isolation barrier, but secondary pollinators may neglect this. Two *Aquilegia* species studied by MILLER (1978) in the Rockies are respectively ornithophilous and sphingophilous; because *Bombus* which was a prominent co-visitor of both, chiefly collected pollen in the dichogamous flowers, it usually missed the receptive phase. In the largely sympatric *Impatiens pallida* and *I. capensis*, the first is bumblebee, the second bird-adapted. *I. capensis* is mostly frequented by hummingbirds in southern parts (SCHEMSKE, 1978), but only rarely in New Jersey (WALLER, 1980) and in Delaware, where it shares *Bombus vagans* with *I. pallida* as a principal pollinator (RUST, 1977). Difference in color and nectar composition make flower fidelity feasible.

8. Populational Aspects and Compatibility

a) Pollen Flow

From tests on artificial flower patches, WADDINGTON (1979b) corroborated that foraging honeybees tend to proceed in a straightforward direction, a habit which prevents revisitation of flowers just explored. In bumblebees at least, overall course is upwind which allows for easier landing on flowers (WOODELL, 1978). Specific inflorescence heights which differ among various competitive steppe herbs, meet pollinators' inclination to maintain given flight levels and thus promote flower fidelity (WADDINGTON, 1979a). Dwarfism of a certain variety of *Nigella arvensis* (observed by EISIKOWITCH, 1978), seems to be an adaptation to the low flight level adopted by bees during the frequent strong winds in the coastal habitat. A *Melipona* sp. found in the rain-forest of Panama forages primarily in the canopy, but descends to the understory for few days when a mass-flowering *Hybanthus* sp. is in bloom (AUGSPURGER, 1980). On average, a single plant has 226 flowers, but foragers tend to exploit only a few and then move onto neighboring plants. This behavior results in 10% out-crossing. Biotic pollen transfer between widely dispersed rain-forest trees is reviewed by FRANKIE (1976).

WADDINGTON's investigations (1979b) into the distances flown by honeybees between flowers revealed a leptokurtic distribution, i.e., short-

and long-distance flights were relatively better represented than in a normal distribution curve, behavioral tendency observed also in pollinators of various natural populations. Foragers in violets had a mean flight distance of ±50 cm (BEATTIE, 1978). As the plants grow in clumps (produced by ant dispersal of seeds), pollen flux was mainly inside the latter. It was found that in a *Primula* sp. whose dimorphic pollen facilitates direct counting, the mean distance of pollen transferal ranged between 1.11 and 2.5 m (RICHARDS, 1978). LINHART and MENDENHALL (1977), by using dye powder, traced carry-over up to 400 m in a sphingophilous *Lindenia*; flowers they had displaced from moths' usual path along a streamside were ignored. [C^{14}]-labeling of pollen is described by REINKE and BLOOM (1979).

The contribution of pollen transport (apart from seed dispersal) to the gene flow within populations is considered by LEVIN (1978). Overall gene migration is not random; local breeding units ("neighborhoods") are formed. Gene flow may be slow in high density arrays, if flight distances are short. Separate patches inbreed by sib pollination and the resultant gene accumulation leads to divergent environment adjustments. Paucity of conspecifics (decimated by man) is detrimental, as pollinators tend to ignore such plants. Large neighborhood areas with few individuals are natural in far-spaced rain-forest species, e.g., "trap-liners". SCHMITT (1980) estimates that neighborhoods in alpine *Senecio* spp. of Colorado (generalists) comprise ±23 plants and measure 0.75 m Ø, as far as *Bombus* is concerned. Both size and area increased dramatically through co-visiting Lepidoptera (>6000 plants in 200 m^2) who by flying greater distances counteract genetic subdivision of patches. Altitude-dependent heterogenity of pollinators reinforces floral polymorphism in a *Polemonium* species (GALEN and KEVAN, 1980).

PRICE and WASER (1979) detected a significant gradient of mating ability in a *Delphinium* population, when plants at various distances from a donor across a transect of 1 km were hand-pollinated. The result, measured in number of seed/flower and seedling fitness, was best in a (mean) out-crossing distance of 10 m, i.e., apparently between individuals of intermediate genetical status: similar enough, but not identical. Actual pollen carry-over distance by pollinators averaged 1 m.

In anemophiles (POHL, 1930) it is well known that the high number of pollen grains produced per ovule (p/o ratio) is correlated to the mode of transfer to compensate for pollen waste and to maximize seed set. Similarly in zoophiles p/o ratios reflect the extent of out-crossing and the likelihood of cross pollen reaching a stigma. Among the representatives tested by CRUDEN (1977), p/o ratio increased from 4.7 in cleistogamous flowers through various intermediates to 5860 in obligate outbreeders, i.e., 1000-fold. In predominantly self-pollinated weed communities it was low (136), whereas in late succession stages with predictable amphimictic pollination it averaged 7250.

b) Autogamy and Cleistogamy; Apomixis

SOLBRIG and ROLLINS (1977) concluded from field studies of *Leavenworthia* in Florida that inavailability of pollinators accounts for evolutionary shifts to self-fertility and autogamy (at the specific and intraspecific level). The autogamists are vernal colonizers of cedar swamp margins which soon dry out: in order to escape water stress, they reproduce early, at a time before most pollinators emerge. Comparative work on various genera again revealed a clear correlation between autogamy and growth habit.

Among the eight *Anthyllis* taxa tested by COUDERC (1980), herbs were principally autogamous (with precocious anther dehiscence, COUDERC and GORENFLOT, 1978), and shrubby species obligate outbreeders (self-fertile but dichogamous). Annual *Melilotus* spp. are selfers, and in four species with both annual and biennial lines, the former had the higher selfing rate (SANO, 1977). Contrary to earlier views, predominantly autogamous *Rhinanthus* spp. are the derived members of the genus. Although five species in France were all found by CAMPION-BOURGET (1980) to be self-fertile, their capability of producing spontaneous self seed in experimental plots varied between 30% and 100%. For reasons unknown, lateral racemes of vernal and summer forms do not produce seeds. As a weedy intruder in America, *Striga asiatica* is autogamous, in contrast to an allogamous native Nigerian stock (NICKRENT and MUSSELMAN, 1979). Further pertinent work was done on *Mazus* (KIMATA, 1978), *Gilia* (SCHOEN, 1977a), *Centaurium* (UBSDELL, 1979), European orchids (REINHARD, 1977), and *Festuca* (AUQUIER, 1977).

According to LORD (1979a,b, 1980), the proportion of *cleistogamy* in *Lamium amplexicaule* is determined by innate factors and photoperiodism. In cleistogamous buds, anther sizes autonomously increase from lower to upper whorls of inflorescences. The upper flowers are potentially chasmogamous, but long-day regimes only trigger expansion in 50%. Evidently, gibberellinic acid, produced in the full-sized anthers under light control, is responsible. The proportion of cleisto- and chasmogamy in *Impatiens capensis* and *I. pallida* is very flexible and phenotypically conditioned (SCHEMSKE, 1978); increase in light intensity, plant size, and vigor favor chasmogamous flowering, while the formation of cleistogamous buds is less dependent on environment (WALLER, 1980) and predominates in shade, dry stands, and late summer. Formation of cleistogamous flowers costs >100 times less energy than of open ones. They are less exposed to phytophages and seeds mature faster. In races of *Salpiglossis sinuata*, presence or lack of cleistogamous flowers is genetically fixed. Introduced chasmogamous pollen failed to fertilize the latter (CHI WON LEE, 1978).

Cleistogamous flowers of neotropic Malpighiaceae stand out in their extreme simplicity (lacking petals, style, stigma; one indehiscent anther) and an unusual selfing mechanism: pollen tubes grow down inside the filament, then enter the ovary from the bottom (ANDERSON, 1980). CONNOR (1979), in his survey of breeding systems in grasses, lists 70 genera known to produce cleistogamous spikelets, often at ground level. PARGNEY (1978) studied cleistogamy in *Oxalis corniculata*, SELL (1977) in *Ruellia*.

KAUR (1978) concludes from polyembryonic tree seeds that apomixis may be widespread in Malaysian climax rain-forests. Apomixis in grasses REDDY (1977) shows a progression from pseudogamy (i.e., male gamete necessary for endosperm formation) in Pooideae to nucellar apospory, e.g., in Andropogoneae and (PHILIPSON, 1978) in *Cortaderia*. *Citrus* spp. are pseudogamic: ESEN and SOOST (1977).

c) Stigmatic Secretions and Pollen Germination

First direct evidence that seed ferns had pollination droplets: ROTHWELL (1977). A permineralized ovule of *Callospermarion* (Pennsylvanian age) bears a noncellular, micropylar protrusion. The mass, once probably gummose, envelops three pollen-like spores, but not those ascribed to the genus (embryo sac empty and shrunk). The pollination drop of *Ephedra* may be repeatedly secreted during a couple of days from the intact nucellar dome (MOUSSEL, 1980). Secretion is arrested when pollen tubes develop (atop the micropylar tube), and the drop is finally reabsorbed.

ENDRESS (1979, 1980a) described aberrant "noncarpellary pollination" in Monimiaceae, in particular *Tambourissa*: he concludes from ♀ flower structure that pollen does not come in direct contact with the many (lysicarpous) carpels lining the bottom of urn-shaped receptacle, but remains outside of the entrance in a mucilage plug, the "hyperstigma". Receptacular slime also fills the interior, mediating pollen tubes to carpels. Related genera have more open cups with stigmatic pollination.

Although differences between "wet" and "dry" stigmata are not absolute, they are distinct types and their distribution, surveyed by HESLOP-HARRISON and SHIVANNA (1977) over 1000 species, is a taxonomically relevant feature. Binucleate pollen (the sort readily germinating in vitro) relates to the former and trinucleate pollen to the latter type. Their connection with incompatibility systems received further support. WOITTIEZ and WILLEMSE (1979) discuss the array of physical forces causing the adhesion of pollen to stigmata; surface tension of stigmatic fluids is the most important. In the same way, any type of pollenkitt causes pollen to stick to dry papillae. In absence of fluids, electrodynamic forces and gravity come into action. As onion pollen is able to germinate when applied to a stigma prior to onset of secretion, CURRAH and OCKENDON (1978) conclude that the fluid serves merely for pollen capture. In *Phaseolus* (LORD and WEBSTER, 1979), however, where it consists of lipids, amino acids, protein, reducing acids, phenols, and glucose, its importance for germination, tube growth, etc. cannot be denied.

The sequence of ovule fertilization by first pollen tubes and hence seed formation in fabaceous ovaries is not a random one. HOROVITZ et al. (1976) have found that the second ovule (at the peduncular end) was favorite and even-numbered ovules were privileged above odd-numbered.

d) Incompatibility

Review of genetical aspects: Progress 40, p. 276. General: NETTANCOURT (1977, 1979), PANDEY (1979b), HOGENBOOM (1979). Three principal classes of incompatibility (IC) seem to exist: (A) unspecific inaptitude to accept foreign gametes resp. pollen tubes, (B) barriers blocking hybridization among relatives, (C) self-IC to prevent inbreeding. While (A) rests on general "incongruity" of genomes, (B) and (C) rely on S-genes and are "recognitive" in nature. Within seed plants, self-IC seems restricted to angiosperms, a barrier against selfpollination in bisexual flowers, including (ARROYO, 1976) geitonogamy and sib-mating. The interspecific IC in gymnosperms, the sole barrier found in this group, is least understood. After heterospecific pollination, nucellar encyme activity is weak compared to compatible response in *Pinus* ovules (KORMUTAK, 1976). PANDEY presumes (B) is more primitive than (C) which was superimposed on the former at a later stage. Contradicting HOGENBOOM who attributes (B) to incongruity, he considers that both the rejection of similar heterospecific and of self-pollen are genetically and physiologically related reactions, controlled by separate but linked specifities of certain S-alleles. In *Nicotiana*, self-incompatible species are also capable of rejecting pollen from self-fertile sister species, while the latter may function as seed parents with the former (unilateral IC).

Predominance of self-compatibility characterizes the New Zealand flora. Various grasses and few dicots (these with multiple-seeded propagules) are the only known exceptions (PANDEY, 1979a; PRIMACK, 1979). Chiefly selfers persisted after single-seed long-distance arrival, and IC could not establish because of wide-spread polyploidy which, however, does not affect the (digenic) IC system of grasses.

Attempts to attain a better understanding of the intricate pollen/style interactions continue; some fascinating results stress the great importance nature pays to the control of pollination. Electrophysiological signals accompany self-IC reaction in maize (LYSIKOW and DUKHOVNY, 1978) and lilies (SPANJERS, 1978). In the *Lilium* style, a series of short voltage variations with an amplitude of -2 to -6 mV occur 120 min after self-pollination. Further experiments in which IC barriers were anticipated by bud-pollination, show that these must develop in the style like morphological structures. Even intergeneric crosses were possible (rye and weat: THOMAS and ANDERSON, 1978). The immature *Raphanus* and *Cheiranthus* style accepts self (and cross) pollen. But when SHIVANNA et al. (1978) enzymatically removed proteins of a thin stigmatic pellicle already present, both kinds failed to germinate. The conclusion is that two independent systems operate and that these are laid down successively during ontogenesis: (A) a factor allowing germination and entry of tubes (releasing cutinase?) and (B) a (sporophytic) recognition system involving antigens. Evidently, (A) was already at work (as proved by its removal), while (B) had not yet been acquired. In *Petunia* buds, self-pollen tubes grow to full length before the (gametophytic) IC system is established. When ripening, the style undergoes drastic changes in protein pattern (HERRERO and DICKINSON, 1980). Watersoluble, loosely bound proteins of the *Petunia* pollen do not participate in the recognition process: washed pollen also germinates and fertilizes (GILLISSEN and BRANTJES, 1978). The functionality of proteinic coatings and inclusions of "dry" stigmata or pertinent pollen has been further substantiated in sporophytic IC and also in gametophytic systems which operate at the level of (dry) stigma (grasses: J. HESLOP-HARRISON, 1979; *Tradescantia*: HERD and BEADLE, 1980). Treatment with cycloheximide (inhibitor or protein synthesis) made self-pollen compatible in *Brassica* (FERRARI and WALLACE, 1977). Proteinic pellicles overlay the papillar surface in *Helianthus*, grasses, and (PETTITT, 1980) seagrasses, intracuticular chambers contain the matter in *Crocus* and cuticular blisters in *Tradescantia*, here increasing toward the papilla base (where incompatible tubes are halted).

Live and EM observations of pollen/stigma contacts in grasses (J. HESLOP-HARRISON, 1979) and *Crocus* (1977) reveal an extremely rapid sequence of pollen hydratation, germination, and tube penetration.

Less than 5 min elapse between pollen capture and germination in rye. First the grain swells through permeation of water (and probably solubles) from the walls of stigmatic papilla across an intervening meniscus. Upon hydratation, biomembranes of pollen cytoplasma (disintegrated in the dry state) reorganize and recover osmotic capacities within 2 min. Some exudation follows, possibly bringing pollen substances to stigma. The compatible pollen tube may penetrate the stigmatic cuticle within 60 sec after first contact, because its tip rapidly erodes the cuticle, apparently by release of stigma-own esterase activity. The tube then grows down between cuticle and cellulosic wall, causing necrosis of papilla and all subsequently contacted cells. In *Crocus*, where reactions are similar, the tube tip is halted after contact with stigma for 10 min, while the cuticle undergoes erosion. Again, esterase activity occurs and acid hydrolase is produced by the pollen. Enzyme interaction of pollen (esterase, acid phosphatase) and stigmatic pellicle (esterase) was also observed in *Helianthus* by VITHANAGE and KNOX (1977, 1979).

9. Dioecism, Monoecism, Sex Distribution

LLOYD (1979b, 1980) and WILLSON (1979) characterize the various breeding patterns of angiosperms as different parental strategies, in part governed by pollen transmission method. Separation of sexes with all intergrades, sex proportions, and the ability of their regulation arose to optimize fecundity (ROSS, 1980) and outbreeding; dioecy, moreover, often meets differential ecological requirements of sexes including (FREEMAN et al., 1976) habitat preferences. Species with only cosexual flowers produce a fix proportion of gender, risking unnecessary investment: part of population may merely function as pollen donor to complement demands of the rest (HOROWITZ, 1978), a situation likely to occur even in orchids (JANZEN et al., 1980). In contrast, monoecists and dioecists are flexible and can adjust parental output to conditions varying between populations. WILLSON claims that the two factors of DARWIN's sexual selection act as regulators: males (pollen) compete for access to females (styles, eggs) or vice versa; and ♀ units discriminate between ♂ units as sexual partners.

For instance, single-fathered pollen, delivered so as to arrive simultaneously in a single large portion on a stigma, may be most favorable for fertilization, resulting in numerous progeny with this high quality maleness. In situations of low success in pollen donation, female units increase. Long-term advantages, such as fitness through heterozygosity, are denied as selective factors. ROSS and WEIR (1976) simulated genic determination of sex control by computer runs.

Complete *dioecy* (♂ and ♀ in equal proportions) and thus obligate outbreeding predominated in all Californian *Fragaria chiloensis* populations studied by HANCOCK and BRINGHURST (1979). In those of low density, monoecious individuals (capable of geitonogamy) were relatively frequent (38%). Only around 1% of the plants had ⚥ flowers: Homozygotes completely lacking ♂- and ♀-suppressor genes are rare in this octoploid species. Two structurally hermaphroditic *Solanum* taxa of Central America proved the functionally staminate and pistillate morphs in one species: first case of true dioecy in the genus (G. ANDERSON, 1979).

Using *Coussarea* and *Cordia* as examples, BEACH and BAWA (1980) reconstruct occasional origin of dioecy from *heterodistyly* in tubular flowers: shift from long- to short-tongued visitors rendered short anthers and styles unemployed and pollen transfer unidirectional from long anthers to pins; ♀ plants thus evolved from pin, ♂ from thrum-morph. A pertinent mathematical model of gene selection is presented by LLOYD (1979a).

Evolution of dioecy from *gynodioecy* through ♀ sterility of the hermaphrodite morph is evident in *Fuchsia* (ARROYO and RAVEN, 1975) and *Cirsium* (DELANNAY, 1978, 1979). This pathway is probably exceptional ("unstable" gynodioecy) compared to hermaphroditic and monoecious origin, in opinion of ROSS who constructed new digenic model of gynodioecism, and of CHARLESWORTH and CHARLESWORTH (1978) who checked selection theories on both systems with computer calculations. In gynodioecy, the female morph is only maintained in a population, if seed output is more than double that of the bisexual morph. The rareness of androdioecism has genetic reasons: cytoplasmatic factors for ♀ sterility cannot be transmitted (contrary to ♂ sterility which is often inherited this way).

In *gynodioecious Thymus vulgaris*, studied by ASSOUAD et al. (1978) and DOMMÉE et al. (1978), open-pollinated bisexual parents produced exclusively bisexual progeny, ♀ plants gave both morphs though the pistillate ones prevail. Seed set and viability increased from selfed bisexuals over intercrossed bisexuals to intermorph breeding almost twofold. Rates of

♀ plants varied enormously, correlating with habitat: dominance of ♀♀ in closer stands in disturbed grasslands (i.e., diversity of progeny stressed), more hermaphrodites and selfing in rocky sites (adaptation to a stable habitat). Incidence of gynodioecy in one population of *Nemophila menziesii* (a species normally hermaphrodite) lacking the proper pollinator (*Andrena* spp.), is interpreted by GANDERS (1978) as a response to excess autogamy mediated by inapt visitors. Other gynodioecious taxa investigated were *Dianthus* (DEM'YANOVA and POKATAEVA, 1979), *Silene* (BROCKMANN and BOCQUET, 1978), *Mammillaria* (GANDERS and KENNEDY, 1979), and *Echium* (VERESHCHAGINA, 1978).

Sexual segregation may have adaptive significance other than that of outbreeding, such as interferent requirements of male and female efforts, or division of labor regarding fruit and pollen investment. ♀ plants of (gynodioecious) plantains, varying in proportion, are more vigorous than the (unusually self-incompatible!) bisexual morph. Costs are saved, as pollen is not produced and can be invested in seed weight (PRIMACK, 1978; KROHNE et al., 1980). Most cone-bearing, anemochorous gymnosperms are monoecious, while those with fleshy zoochorous seeds are dioecious; this fact is attributed to mode of dispersal by GIVNISH (1980): showy (and costly) seed crops must concentrate on being attractive enough to animals, a capacity best achieved without the concurrent energetical burden of staminate flowers.

A differential mortality of sexes is related to reproduction in *Triplaris* by MELAMPY and HOWE (1977). A monoecious understory tree, *Acer pensylvanicus*, produces chiefly pollen until beyond midlife, turning to femaleness (and fruiting) before death (concomitant with canopy closure). Numbers of ♂-stage trees in a population was six times that of ♀♀ (HIBBS and FISCHER, 1979).

Males and females of four dioecious tree species in a rain-forest community of Costa Rica were distributed at random; individual distances averaged between 4 and 7 m (BAWA and OPLER, 1977). In total, 20% of tree species in this community were dioecious (all entomophilous), 65% hermaphrodites (88% self-incompatible), and 15% monoecious (BAWA, 1979). Counts in deciduous forests of Venezuela were similar (ZAPATA and ARROYO, 1978). Thirteen percent of the indigenous plants of New Zealand are dioecious, including *Rubus* and *Clematis* (GODLEY, 1979).

VAN DER PIJL (1978) discusses how intercourse of visitors is secured even if ♂ and ♀ flowers are dissimilar or if food is confined to one type of flower. Oligoleges are better pollinators of willows than honeybees, where strong fidelity prevents a bee from foraging on the two (dissimilar) sexes (A. MEEUSE, 1978a). Sometimes the nonrewarding sex mimics the rewarding one, provoking "mistake"-pollination (BAKER, 1976): a sterile androecium attracts pollen-seeking bees to pistillate flowers of *Vitis* (BRANTJES, 1978) and *Cupania* (BAWA, 1977). Hawkmoths are deceived by the stigma of nectar- and tubeless ♀ *Jacaratia* blossom which imitates the rim of (more numerous) rewarding, tubate ♂ flowers (BAWA, 1980).

10. Heterostyly

GANDERS (1979c), in a general survey, enumerates 164 genera of 24 families in which heterostyly is known to occur, including the eight genera with reliable tristyly. In his model and in a similar, computerized one by CHARLESWORTH and CHARLESWORTH (1979b), di- and trimorphism developed in a sequence of additive steps. In *disty*, diallelic incompatibility probably preceded structural dimorphism: (1) in a monomorphic, self-fertile, increasingly inbreeding population, pollen of a mutant becomes incompatible to its own and all conspecific stigmata. (2) Another mutation in a hermaphrodite of this population (which is now functionally gynodioecious), creates a style which accepts only the new pollen type. (3) Both new flower types (reciprocally fertile), supersede the original self-compatible morph. (4) Rearranged style and anther positions promote legitimate seed set. (5) All pertinent features become linked in a "super-gene". A fortuitous coadaptation of mutants 1 and 2 remains somewhat mysterious, and structural dimorphism, again found much less efficient than expected, is hardly explicable by success. According to counts of ORNDUFF (1979d, 1980) in English populations of *Primula*, DARWIN's classic example, the two style forms received even less legitimous pollen than with random transferal (results may differ in completely natural biotopes).

Intermorph pollen flow and selfing rates were strongly asymmetrical: pin flowers gave three times more pollen than thrum flowers and more "pin-" than "thrum-pollen" was removed by insects (but only 1%-2% of total yield reached a stigma). Pins received more self- and intramorph pollen than thrums and thrum flowers bore more seed than pin despite similar ovule numbers. Likewise, in *Pulmonaria obscura*, assessed by OLESEN (1979), intramorph (illegitimate) pollen travel outweight intermorph pollinators and transfer from pin to thrum was better than in the opposite direction.

Self-compatibility of heterostyles (in *Amsinckia grandiflora* only detectable in the absence of legitimous pollination: WELLER and ORNDUFF, 1977) is surmised to be secondary. Hence, *Lithospermum cobrense*, sterile with illegitimous pollen, probably gave rise to partially autogamous *L. caroliniense* (GANDERS, 1979b). CHARLESWORTH and CHARLESWORTH (1979a) hypothesize that homostyles could emerge secondarily in a dimorphic group, by rare cross-overs among three sub-units (alleles) within the supergene recombining pin and thrum characters. *Turnera ulmifolia* has a homostylous (long-styled) self-fertile variety (BARRETT, 1978), a weed, but not more so than the main species (including *T. trioniflora*: BENTLEY, 1979), where, remarkably, weediness has not led to a breakdown of distyly and incompatibility.

ROGERS (1979) reports pollen dimorphism in *Linum suffruticosum*, ORNDUFF (1979b) pollen flow and floral characters of little known distylous *Gelseminum* (Logan.), and zygomorphic (!) *Oplonia* (Acanthac., ORNDUFF, 1979c), KEEGAN et al. (1979) those of *Mitchella*. *Erythroxylum coca* is dimorphic. Crossing experiments of GANDERS (1979a) show that the existence of two anther levels has no connection with (previously surmised) tristyly in this and related species.

Presence of two anther whorls and self-compatibility were likely preconditions for the origin of *tristyly*. In the model developed by CHARLESWORTH (1979), the complementary positioning of etages and stigmata passed a distylous stage, mid-stigma coming last. Unlike distyly, the incompatibility system may have succeeded the first structural changes. BARRETT (1977a) found gradations of incompatibility between morphs and etages of a *Pontederia*. While intermorph breeding gave equally high and intramorph crosses equally low seed sets in the three morphs, selfing had divergent results: the short-styled morph was most and the mid-styled least self-incompatible. The two anther etages differed from each other in all morphs as well: those closest to the stigma produced least seeds, the distant ones (in mid-style the upper set) moderate yields upon selfing.

Tristyly in *Eichhornia crassipes* is now certain, as the "missed" short-styled morph has been discovered in native Amazon haunts. A bee (*Ancy-*

loscelis) has been recognized here as the chief pollinator of water hyacinth. Probably due to prevalent vegetative propagation and autogamy, one morph (usually the most self-fertile mid) predominates in all, including native, populations. BARRETT (1977b, 1979) presumes "mid" was the main founder of adventive stands. A "mid plant" bearing alleles for "long", could segregate "longs" by the F_1-generation (except in India, where no seed set occurs), but neither "mid" nor "long" can produce "short", possibly the reason why the latter is absent outside America. Inequality of morph frequencies in *Lythrum salicaria* cannot be attributed to different selfing rates. This follows calculations by HEUCH (1979). The notion that dimorphism in partly trimorphic genera is derived, received further support. In *Oxalis*, section *Ionoxalis* tristylous taxa of S. Mexico gave rise to distylous northern derivatives by loss of mid-styles (WELLER, 1978, 1979, 1980; CHARLESWORTH, 1979). All ten American *Lythrum* species are dimorphic. According to ORDNDUFF (1979a), distyly was attained more than once independently within the genus and within *Nesaea*.

In one section of *Lythrum*, the inner stamen whorl (and short style) of tristylous ancestors disappeared; the remaining six stamens correspond to former mid and long morphs, respectively. Another section has distylous flowers with the two intrafloral stamen levels retained and the mid style lost.

In the *enantiostyly* of *Wachendorfia paniculata*, right- or left-handedness of style and complementary collateral anther positions are genetically fixed and uniform in each individual (in related *Dilatris* and *Cyanella*, they alternate within one cyme). As ORNDUFF (1978) stated, this unique dimorphism resembles distyly in breeding behavior: intermorph pollinations were more fertile than those within morphs and selfing. Self-incompatibility is weak, but in left-styled flowers less so than in right-styled.

Conclusion

A distinctive trait of New Anthecology, especially evident in the foregoing topics, is quantification; no doubt, a successful approach. Statistics have not only substantiated relationships believed for long to exist, but they have – not at least by disproof of other predictions – also revealed new and unexpected patterns of coadaptation in the floral sphere. Nevertheless, in the reviewer's opinion, one principal complaint must be raised. Numerical methods favor the propensity to focus exclusively on dynamics of interactions and to absolutely identify ecology with economy. The target is, as one of these papers frankly declares, the analysis of "... the variety of selection pressures resulting in the adaptive compromises we call organisms". In other words, when adaptiveness is subtracted almost nothing remains. This reasoning disregards essential aspects of a biosystem. Species have inherent and autonomous characters, a fact tacitly admitted if "strategies" are stressed. To reach a less one-sided comprehension of flower relationships, more comparative ecology, based on morphology of larger taxonomic groups, is certainly needed and should be a complementary task of future research in anthecology.

Acknowledgments. The author is grateful to Magister SUSANNE KLENNER (Vienna) for assistance, to JANICE DAWE (Fairbanks), and DIANA MANTELL (Johannisburg) for stylistic suggestions.

References

ABRAHAMSON, W.G., McCREA, K.D.: Rhodora 79, 269-277 (1977). - ACKERMAN, J.D., MESLER, M.R.: Am. J. Botany 66, 820-824 (1979). - ADDICOTT, F.T.: Am. Midland Naturalist 97, 321-332 (1977). - ANDERSON, G.J.: Nature 282, 836-838 (1979). - ANDERSON, W.R.: Biotropica 11, 219-223 (1979); - Science 207, 892-893 (1980). - ANTONOVA, L.A.: Zh. obshch. Biol. 40, 290-294 (1979) (Russian). - ARMBRUSTER, W.S., WEBSTER, G.L.: Biotropica 11, 278-283 (1979). - ARMSTRONG, J.A.: New Zeal. J. Botany 17, 467-508 (1979). - ARROYO, M.T.: Taxon 25, 543-548 (1976). - ARROYO, M.T., RAVEN, P.H.: Evolution 29, 500-511 (1975). - ASSOUAD, M.W., DOMMÉE, B., LUMARET, R., VALDEYRON, G.: Botan. J. Linn. Soc. 77, 29-39 (1978). - AUGSPURGER, C.K.: Evolution 34, 475-488 (1980). - AUQUIER, P.: Bull. Soc. Roy. Botan. Belg. 110, 129-150 (1977).

BAKER, H.G.: Linn. Soc. Symp. Ser. Nr. 2, 161-169 (1976); - Apidologie 8, 349-356 (1977); - Chemical aspects of the pollination biology of woody plants in the tropics, 57-82. In: Tropical trees as living systems, eds. P.B. TOMLINSON, M.H. ZIMMERMANN. Cambridge Univ. Press 1978; - New Zeal. J. Botany 17, 431-440 (1979). - BAKER, H.G., BAKER, I.: Am. J. Botany 66, 591-600 (1979). - BAKER, H.G., OPLER, P.A., BAKER, I.: Botan. Gaz. 139, 322-332 (1978). - BARBER, S.C., ESTES, J.R.: Am. J. Botany 65, 562-566 (1978). - BARRETT, S.C.: New Phytologist 78, 209-220 (1977a); - Biotropica 9, 230-238 (1977b); - Can. J. Botany 56, 1713-1725 (1978); - Evolution 33, 499-510 (1979). - BARROWS, E.M.: Biotropica 9, 133-134 (1977); - ibid. 12, 23-29 (1980). - BAUMANN, E.: Plant Syst. Evol. 130, 43-52 (1978a); - Flora 167, 301-314 (1978b). - BAWA, K.S.: Evolution 31, 52-63 (1977); - New Zeal. J. Botany 17, 521-524 (1979); - Evolution 34, 467-474 (1980). - BAWA, K.S., OPLER, P.A.: Evolution 31, 64-68 (1977). - BAZZAZ, F.A., CARLSON, R.W., HARPER, J.L.: Nature 279, 554-555 (1979). - BEACH, J.H., BAWA, K.S.: Evolution 34, 1138-1143 (1980). - BEATTIE, A.: Plant-animal interactions affecting gene flow in *Viola*, 151-164. In: The Pollination of Flowers by Insects, ed. A.J. RICHARDS. London: Acad. Press 1978. - BELL, C.R., LINDSEY, A.H.: The umbel as a reproductive unit in the Apiaceae, 739-747. In: 2nd Symp. Intern. sur les Ombellifères, Perpignan 1977, appeared 1978. - BENTLEY, B.L.: Biotropica 11, 11-17 (1979). - BERGSTRÖM, G.: Role of volatile chemicals in *Ophrys* pollinator interactions, 207-231. In: Biochemical Aspects of Plant and Animal Coevolution, ed. J.B. HARBORNE. London, New York, San Francisco: Acad. Press 1978. - BERNHARDT, P.: Bull. Torrey Botan. Club 103, 255-258 (1976); - Rhodora 79, 278-282 (1977). - BERNHARDT, P., MONTALVO, E.A.: Brittonia 31, 64-71 (1979). - BOEKE, J.D., ORTIZ-CRESPO, F.I.: Science 201, 167-168 (1978). - BOYDEN, T.C.: Evolution 34, 135-136 (1980). - BRANTJES, N.B.: Vitis 17, 229-233 (1978); - Vakbl. Biol. 59, 4, 47-50 (1979a) (in Dutch); - ibid. 59, 22, 396-403 (1979b) (in Dutch); - De Levende Natuur 82, 9-16 (1980). - BRANTJES, N.B., BOS, J.J.: New Phytologist 84, 139-143 (1980). - BRANTJES, N.B., DE VOS, O.C.: New Phytologist 87, 425-430 (1981). - BRINES, M.L., GOULD, J.L.: Science 206, 571-573 (1979). - BROCKMANN, I., BOCQUET, G.: Ber. Deut. Botan. Ges. 91, 217-230 (1978). - BROWN, J.H., KODRIC-BROWN, A.: Ecology 60, 1022-1035 (1979). - BRUNKE, E.J., HAMMERSCHMIDT, F.J.: Dragoco Report 3, 78-80 (1980). - BUCHMANN,, S.L.: Diss. Abstr. Internat. 39, 12 (1979). - BUCHMANN, S.L., HURLEY, J.P.: J. theor. Biol. 72, 639-657 (1978). - BÜEL, H.: Die Orchidee 29, 106-109 (1978). - BUTTROSE, M.S., GRANT, W.J., LOTT, J.N.: Austral. J. Botany 25, 567-570 (1977).

CAMPION-BOURGET, F.: Rev. Cytol. Biol. végét. Bot. 3, 199-219 (1980). - CARPENTER, F.L.: Oecologia 35, 123-132 (1978). - CARPENTER, F.L., RECHER, H.F.: Am. Naturalist 113, 871-879 (1979). - CHARLESWORTH, D.:

Evolution 33, 486-498 (1979). - CHARLESWORTH, B., CHARLESWORTH, D.: Am. Naturalist 112, 975-997 (1978); - ibid. 114, 499-513 (1979a); - ibid. 114, 467-498 (1979b). - CHI WON LEE, ERICKSON, H.T., JANICK, J.: Am. J. Botany 66, 626-632 (1979). - CLARK, C.: Madroño 26, 22-25 (1979). - CONNOR, H.E.: New Zeal. J. Botany 17, 547-574 (1979). - COUDERC, H.: Bull. Soc. Botan. France 127, 139-149 (1980). - COUDERC, H., GORENFLOT, R.: Bull. Soc. Botan. France 125, 369-378 (1978). - CORBET, S.A.: Bees and the nectar of *Echium vulgare*, 21-30. In: The Pollination of Flowers by Insects. ed. A.J. RICHARDS. London: Acad. Press 1978. - CREPET, W.L.: Bio Science 29, 102-108 (1979). - CROAT, T.B.: Am. J. Botany 67, 888-904 (1980). - CRUDEN, R.W.: Evolution 31, 32-46 (1977). - CRUDEN, R.W., HERMANN-PARKER, S.M.: ibid. 31, 863-866 (1977); - J. Ecol. 67, 155-168 (1979). - CRUDEN, R.W., JENSEN, K.G.: Am. J. Botany 66, 875-879 (1979). - CRUDEN, R.W., KINSMAN, S., STOCKHOUSE, R.E. II, LINHART, Y.B.: Biotropica 8, 204-210 (1976). - CURRAH, L., OCKENDON, D.J.: New Phytologist 81, 419-428 (1978).

DAFNI, A., IVRY, Y.: New Phytologist 83, 181-187 (1979). - DANIEL, M.J.: New Zeal. J. Zool. 6, 357-370 (1979). - DECELLES, P., LAROCA, S.: J. Kansas Entomol. Soc. 52, 483-488 (1979). - DELANNAY, X.: Bull. Soc. Roy. Botan. Belg. 111, 10-18 (1978); - Plant System. Evol. 132, 327-332 (1979). - DEL MORAL, R., STANDLEY, L.A.: Am. J. Botany 66, 26-35 (1979). - DEM'YANOVA, E.G., POKATAEVA, N.V.: Botan. Zh. 62, 1469-1479 (1977) (Russian). - DOMMÉE, B., ASSOUAD, R., VALDEYRON, G.: Botan. J. Linnean Soc. 77, 17-28 (1978). - DUCKER, S.C., PETTITT, J.M., KNOX, R.B.: Australian J. Botany 26, 265-285 (1978).

EHRENFELD, J.G.: Am. Midland Naturalist 101, 87-98 (1979). - EICKWORT, G.C., GINSBERG, H.S.: Ann. Rev. Entomol. 25, 421-446 (1980). - EISIKOWITCH, D.: Insect visiting of two subspecies of *Nigella arvensis* under adverse seaside conditions, 125-132. In: The Pollination of Flowers by Insects, ed. A.J. Richards. London: Acad. Press 1978. - ELVERS, I.: Botan. Notiser 131, 159-160 (1978). - ENDRESS, P.K.: Experientia 35, 454 (1979); - Plant System. Evol. 134, 79-120 (1980a); - Botan. Jahrb. System. 101, 393-433 (1980b). - ERNET, D.: Plant System. Evol. 128, 1-22 (1977). - ESEN, A., SOOST, R.K.: Am. J. Botany 64, 607-614 (1977).

FAEGRI, K.: Trends in research in pollination ecology, 5-12. In: The Pollination of Flowers by Insects, ed. A.J. RICHARDS. London: Acad. Press 1978. - FAEGRI, K., VAN DER PIJL, L.: The Principles of Pollination Ecology, 3rd ed. Oxford, New York: Pergamon Press 1979. - FAHN, A.: Am. J. Botany 66, 977-985 (1979). - FAHN, A., BENOUAICHE, P.: Ann. Botany 44, 85-93 (1979). - FEINSINGER, P., SWARM, L.A.: Biotropica 10, 238-239 (1978). - FEINSINGER, P., LINHART, Y.B., SWARM, L.A., WOLFE, J.A.: Ann. Missouri Botan. Garden 66, 451-471 (1979). - FERRARI, T.E., WALLACE, D.H.: Science 196, 436-438 (1977). - FORD, H.A., PATON, D.C., FORDE, N.: New Zeal. J. Botany 17, 509-519 (1979). - FRANKIE, G.W.: Linnean Soc. Symp. Ser. No. 2, 151-159 (1976). - FREEMAN, D.C., KLIKOFF, L.G., HARPER, K.T.: Science 193, 597-599 (1976).

GANDERS, F.R.: Can. J. Botany 56, 1400-1408 (1978); - Botan. J. Linnean Soc. 78, 11-20 (1979a); - Am. J. Botany 66, 746-748 (1979b); - New Zeal. J. Botany 17, 607-635 (1979c). - GANDERS, F.R., KENNEDY, H.: Madroño 25, 234 (1978/79). - GALEN, C., KEVAN, P.: Am. Midland Naturalist 104, 281-289 (1980). - GENTRY, A.H.: Biotropica 10, 68-69 (1978). - GILLISSEN, L.J., BRANTJES, N.B.: Acta Botan. Neerl. 27, 205-212 (1978). - GIVNISH, T.J.: Evolution 34, 959-972 (1980). - GODLEY, E.J.: New Zeal. J. Botany 17, 441-466 (1979). - GOLDSMITH, T.H.: Science 207, 786-788 (1980). - GOTTSBERGER, G.: Plant System. Evol. Suppl. 1, 211-226 (1977). - GOTTSBERGER, G., SILBERBAUER-GOTTSBERGER, I., EHRENDORFER,

F.: Plant System. Evol. 135, 11-39 (1980). - GOULD, E.: Biotropica 10, 184-193 (1978). - GRACE, J., NELSON, M.: New Phytologist 87, 413-423 (1981). - GRAF, U.: Orchidee 28, 147-149 (1977). - GRANT, V., GRANT, K.A.: Plant System. Evol. 132, 85-90 (1979a); - ibid. 133, 29-37 (1979b). - GRANT, V., CONNELL, W.A.: ibid. 133, 99-102 (1979). - GRANT, V., HURD, P.D.: ibid. 133, 15-28 (1979). - GRANT, V., GRANT, K.A., HURD, P.D. Jr.: Plant System. Evol. 132, 312-320 (1979). - GUILLARMOD, A.J., JUBB, R.A., SKEAD, C.J.: Ann. Missouri Botan. Garden 66, 521-527 (1979). - GUNNING, B.E., HUGHES, J.E.: Austral. J. Plant Physiol. 3, 619-637 (1976).

HAMMER, K.: Flora 167, 41-56 (1978). - HANCOCK, J.F. Jr., BRINGHURST, R.S.: Bull. Torrey Botan. Club 106, 229-231 (1979). - HARBORNE, J.B., SMITH, D.M.: Biochem. System. Ecol. 6, 127-130 (1978a); - ibid. 6, 287-291 (1978b). - HEINRICH, B.: Ecology 57, 874-889 (1976a); - ibid. 57, 890-899 (1976b); - Ecology 60, 245-255 (1979a); - Bumblebee Economics. 246 pp. Cambridge (Mass.), London: Harvard University Press 1979b. - HEITHAUS, E.R.: Ecology 60, 190-202 (1979). - HERD, Y.R., BEADLE, D.J.: Ann. Botany 45, 251-256 (1980). - HERRERO, M., DICKINSON, H.G.: Planta 148, 138-145 (1980). - HESLOP-HARRISON, J.: New Zeal. J. Botany 17, 537-546 (1979). - HESLOP-HARRISON, Y.: Ann. Botany 41, 913-922 (1977). - HESLOP-HARRISON, Y., SHIVANNA, K.R.: ibid. 41, 1233-1258 (1977). - HESSE, M.: Naturkund. Jb. Stadt Linz 23, 9-16 (1977); - Plant System. Evol. 130, 13-42 (1978); - ibid. 131, 277-289 (1979a); - Flora 168, 540-577 (1979b); - Pollen et Spores 21, 399-413 (1979c); - Plant System. Evol. 132, 107-139 (1979d); - ibid. 133, 135-148 (1980a); - ibid. 135, 253-263 (1980b); - Beitr. Elektronenmikrosk. Direktabb. Oberfl. 13, 211-216 (1980c); - Plant System. Evol. 134, 229-267 (1980d). - HEUCH, I.: Ann. Botany 44, 611-616 (1979). - HIBBS, D.E., FISCHER, B.C.: Bull. Torrey Botan. Club 106, 222-227 (1979). - HOGENBOOM, N.G.: Incompatibility and Incongruity in *Lycopersicon*, 435-444. In: The Biology and Taxonomy of the Solanaceae, eds. J.G. HAWKES, R.N. LESTER, A.D. SKELDING. Linnean Soc. London: Acad. Press 1979. - HOLM, E.: The Western Austral. Naturalist 14, 71-74 (1978a); - ibid. 14, 60-62 (1978b); - The Biology of Flowers. 144 pp. Penguin Nature Guides 1979. - HOPPER, S.D., BURBIDGE, A.H.: Austral. J. Botany 26, 335-350 (1978). - HOROVITZ, A.: Am. J. Botany 65, 485-486 (1978). - HOROVITZ, A., MEIRI, L., BEILES, A.: Botan. Gaz. 137, 250-254 (1976). - HOWELL, D.J.: Nature 270, 509-510; 688-689 (1977). - HURD, P.D. Jr., LABERGE, W.E., LINSLEY, E.G.: Smithsonian Contr. in Zoology 310, 1-158 (1980).

INOUYE, D.W.: Ecology 59, 672-678 (1978). - IWARSSON, M.: Botan. Notiser 132, 349-355 (1979). - JANZEN, D.H.: Ann. Rev. Ecol. System. 10, 13-51 (1979); - Evolution 34, 611-612 (1980). - JANZEN, D.H., DE VRIES, P., GLADSTONE, D.E., HIGGINS, M.L., LEWINSOHN, T.M.: Biotropica 12, 72-74 (1980). - JONES, C.E.: Evolution 32, 189-198 (1978).

KAUR, A., HA, C.O., JONG, K., SANDS, V.E., CHAN, H.T., SOEPADMO, E., ASHTON, P.S.: Nature 271, 440-442 (1978). - KAY, Q.O.: The role of preferential and assortative pollination in the maintenance of flower polymorphisms, 175-190. In: The Pollination of Flowers by Insects, ed. A.J. RICHARDS. London: Acad. Press 1978; - Watsonia 12, 339-340 (1979). - KEEGAN, C.R., VOSS, R.H., BAWA, K.S.: Rhodora 81, 567-573 (1979). - KEIGHERY, G.H.: Plant System. Evol. 135, 171-176 (1980). - KENNEDY, H.: Univ. Californ. Publ. Botan. 71, 1-90 (1978). - KEPHART, S.R., HEISER, C.B. Jr.: Evolution 34, 738-746 (1980). - KEVAN, P.G.: Floral coloration, its colorimetric analysis and significance in anthecology, 51-78. In: The Pollination of Flowers by Insects, ed. A.J. RICHARDS. London: Acad. Press 1978; - Am. J. Botany 66, 749-751 (1979). - KIMATA, M.: Plant System. Evol. 129, 243-253 (1978). - KOENIGER, N., VORWOHL, G.: J. Apicult. Res. 18, 95-109 (1979). - KONON, N.T.: Rastit. Resursy 14,

73-77 (1978) (Russian). - KORMUTAK, A.: Biológia (Bratislava) 31, 521-526 (1976). - KROHNE, D.T., BAKER, I., BAKER, H.: Am. Midland Naturalist 103, 269-279 (1980). - KUGLER, H.: Ber. Deut. Botan. Ges. 92, 637-643 (1980a); - Flora 169, 524-529 (1980b). - KWAK, M.M.: Taxon 27, 145-158 (1978); - Variation in *Rhinanthus* (Rhinanthoideae; Scrophulariaceae) in relation to bumblebee pollination, 48-58. In: Proceedings of the Second Symposium on Parasitic Weeds, eds. L.J. MUSSELMAN, A.D. WORSHAM, R.E. EPLEE. Raleigh (N.C.): North Carolina State University 1979a; - Oecologia 41, 1-9 (1979b); - Acta Botan. Neerl. 28, 177-195 (1979c).

LACK, A.: J. Ecol. 66, 287-295 (1978). - LAROCA, S., WINSTON, M.L.: J. Kansas Entomol. Soc. 51, 274-275 (1978). - LEPPIK, E.E.: Phytomorphology 27, 161-168 (1977). - LEROY, J.F.: Le problème de l'origine des angiospermes en rapport avec les pollinisations. Colloque Sci. Intern. Origine des Angiospermes et rapports avec les insectes pollinisateurs, 14 pp., Tananarive 1980. - LEVIN, D.A.: Pollinator behaviour and the breeding structure of plant populations, 133-150. In: The Pollination of Flowers by Insects, ed. A.J. RICHARDS. London: Acad. Press 1978. - LINDAUER, M.: Proc. XV. Intern. Congress of Entomol., 450-460 (1976). - LINHART, Y.B., MENDENHALL, J.A.: Biotropica 9, 143 (1977). - LINSLEY, E.G.: J. Kansas Entomol. Soc. 51, 531-546 (1978). - LLOYD, D.G.: Plant System. Evol. 131, 71-80 (1979a); - New Zeal. J. Botany 17, 595-606 (1979b); - Evolution 34, 123-134 (1980). - LORD, E.M.: Botan. Gaz. 140, 39-50 (1979a); - Ann. Botany 44, 757-766 (1979b); - Am. J. Botany 67, 529-533 (1980). - LORD, E.M., WEBSTER, B.D.: Botan. Gaz. 140, 266-271 (1979). - LUMER, C.: Brittonia 32, 512-517 (1980). - LÜTTGE, U.: Apidologie 8, 305-320 (1977). - LYSIKOV, V.N., DUKHOVNYJ, A.I.: Izvest. Akad. Nauk. Moldav., Biol. Khim. Nauk. 4, 26-30 (1978) (Russian).

MACIOR, L.M.: Am. J. Botany 65, 57-62 (1978a); - Oikos 30, 452-460 (1978b); - Bull. Torrey Botan. Club 105, 268-277 (1978c); - Proc. IV. Intern. Sympos. on Pollination, Md. Agric. Exp. Sta. Spec. Misc. Publ. 1, 441-446 (1978d). - MATHUR, G., MOHAN RAM, H.Y.: Ann. Botany 42, 1473-1476 (1978). - McDADE, L.A., KINSMAN, S.: Evolution 34, 944-958 (1980). - McLELLAN, A.R.: J. Appl. Ecol. 13, 801-811 (1976). - MEEUSE, A.D.: Entomophily in *Salix*: theoretical considerations, 47-50. In: The Pollination of Flowers by Insects, ed. A.J. RICHARDS. London: Acad. Press 1978a. - MEEUSE, A.D.: Proc. Koningl. Nederl. Akad. Wetensch. Amsterdam, Ser. C 81, 300-312, 313-326 (1978b); - Vakbl. Biol. 60, 2, 22-26 (1980). - MEEUSE, B.D.: The physiology of some sapromyiophilous flowers, 97-104. In: The Pollination of Flowers by Insects, ed. A.J. RICHARDS. London: Acad. Press 1978. - MELAMPY, M.N., HAYWORTH, A.M.: Evolution 34, 1144-1154 (1980). - MELAMPY, M.N., HOWE, H.F.: Evolution 31, 867-872 (1977). - MESLER, M.R., ACKERMAN, J.D., LU, K.L.: Am. J. Botany 67, 564-567 (1980). - MICHENER, C.D.: Ann. Missouri Botan. Garden 66, 277-347 (1979). - MICHENER, C.D., WINSTON, M.L., JANDER, R.: Univ. Kansas Sci. Bull. 51, 575-601 (1978). - MILLER, R.B.: Am. J. Botany 65, 406-414 (1978). - MOGFORD, D.J.: J. S. Afr. Botany 46, 121-125 (1980). - MORI, S.A., PRANCE, G.T., BOLTEN, A.B.: Brittonia 30, 113-130 (1978). - MORI, S.A., ORCHARD, J.E., PRANCE, G.T.: Science 209, 400-403 (1980). - MORTON, E.S.: Ann. Missouri Botan. Garden 66, 482-489 (1979). - MOUSSEL, B.: Rev. Cytol. Biol. vég. Botan. 3, 65-89 (1980).

NEEMAN, G., GALIL, J.: New Phytologist 81, 375-380 (1978). - NETTANCOURT, D. De: Incompatibility in Angiosperms. Berlin, Heidelberg, New York: Springer 1977; - Bull. Soc. Botan. France 126, actual. botan., 97-104 (1979). - NEWTON, L.E., LOMO, A.: Botan. J. Linnean Soc. 78, 21-30 (1979). - NICHOLS, R.: Planta 135, 155-159 (1977). - NICKRENT,

D.L., MUSSELMAN, L.J.: Brittonia 31, 253-256 (1979). - NIKLAS, K.J.: Science 211, 275-277 (1981). - NILSSON, L.A.: Botan. Notiser 131, 35-51 (1978a); - ibid. 131, 355-368 (1978b); - ibid. 132, 329-347 (1979a); - ibid. 132, 537-549 (1979b); - ibid. 133, 367-385 (1980).

OLESEN, J.M.: New Phytologist 82, 757-767 (1979). - ORNDUFF, R.: Bull. Torrey Botan. Club 106, 4-8 (1979a); - J. Arnold Arbor. 60, 377-381 (1979b); - ibid. 60, 382-385 (1979c); - Botan. J. Linnean Soc. 78, 1-10 (1979d); - Plant System. Evol. 135, 89-93 (1980). - ORNDUFF, R., DULBERGER, R.: New Phytologist 80, 427-434 (1978). - OSCHE, G.: Biologie unserer Zeit Jg. 9, Nr. 6, 161-170 (1979). - OSTLER, W.K., HARPER, K.T.: Ecology 59, 848-861 (1978).

PANDEY, K.K.: New Zeal. J. Botany 17, 225-226 (1979); - The genus *Nicotiana*: evolution of incompatibility in flowering plants, 421-444. In: The Biology and Taxonomy of the Solanaceae, eds. J.G. HAWKES, N.R. LESTER, A.D. SKELDING. Linnean Soc. London: Acad. Press. 1979. - PARGNEY, J.C.: Can. J. Botany 56, 1262-1269 (1978). - PARRISH, J.A., BAZZAZ, F.A.: Ecology 60, 597-610 (1979). - PAULUS, H.: SB Naturwiss. Verh. Hamburg 2, 51-81 (1978). - PAULUS, H., GACK, C.: Die Orchidee, Sonderheft 1980, Orchideentagung Wuppertal, 55-68. - PETTITT, J.M.: Ann. Botany 45, 257-271 (1980). - PFLUMM, W.: Apidologie 8, 401-412 (1977). - PHILIPSON, M.N.: New Zeal. J. Botany 16, 45-59 (1978). - PIJL, L. VAN DER: Reproductive integration and sexual disharmony in floral functions, 79-88. In: The Pollination of Flowers by Insects, ed. A.J. RICHARDS. London: Acad. Press 1978. - PLAYER, G.: Ecol. Monogr. 49, 73-87 (1979). - PRANCE, G.T.: Biotropica 12, 239 (1980a); - Brittonia 32, 505-507 (1980b). - PRICE, M.V., WASER, N.M.: Nature 277, 294-297 (1979). - PRIMACK, R.B.: New Phytologist 81, 449-458 (1978); - New Zeal. J. Botany 17, 9-13 (1979). - PRIMACK, R.B., TOMLINSON, P.B.: Biotropica 12, 229-231 (1980). - PROCTOR, M.C.: Insect pollination syndromes in an evolutionary and ecosystematic context, 105-116. In: The Pollination of Flowers by Insects, ed. A.J. RICHARDS. London: Acad. Press 1978. - PROCTOR, J., PROCTOR, S.: Nature's use of colour in plants and their flowers, 116 pp. London: Publishing House Peter Lowe 1978. - PYKE, G.H.: Oecologia 36, 281 (1978).

RAMIREZ, B.W.: Ann. Missouri Botan. Garden 64, 296-310 (1977). - RAMIREZ, B.W., GOMEZ, P.L.: Brenesia 14-15, 407-412 (1978). - RAVEN, P.H.: New Zeal. J. Botany 17, 575-593 (1979). - READER, R.J.: Can. J. Botany 55, 2279-2287 (1977). - REDDY, P.S.: Phytomorphology 27, 45-50 (1977). - REINHARD, H.R.: Orchidee 28, 178-182 (1977). - REINKE, D.C., BLOOM, W.L.: Syst. Botany 4, 223-229 (1979). - RICHARDS, A.J. (ed.): The Pollination of Flowers by Insects. Linnean Soc. Symp. Ser. 6, 1-213 (1978). - RICHARDS, A.J., IBRAHIM, H.: Estimation of neighbourhood size in two populations of *Primula veris*, 165-174. In: The Pollination of Flowers by Insects, ed. A.J. RICHARDS. London: Acad. Press 1978. - RIVALS, P.: Bull. Soc. Hist. Nat. Toulouse 115, 335-337 (1980). - ROBERTS, R.B., VALLESPIR, S.R.: Ann. Entomol. Soc. Am. 71, 619-627 (1978). - ROGERS, C.M.: Plant System. Evol. 131, 127-132 (1979). - ROSS, M.D.: Evolution 32, 174-188 (1978); - ibid. 34, 769-778 (1980). - ROSS, M.D., WEIR, B.S.: ibid. 30, 425-441 (1976). - ROUBIK, D.W.: Science 201, 1030-1032 (1978); - Ecology 61, 836-845 (1980). - ROURKE, J.P.: Veld & Flora 66, 54-56 (1980). - ROURKE, J.P., WIENS, D.: Ann. Missouri Botan. Garden 64, 1-17 (1977). - ROTHWELL, G.W.: Science 198, 1251-1252 (1977). - RUST, R.W.: Bull. Torrey Botan. Club 104, 361-367 (1977).

SAID, C.: Bull. Soc. Botan. France 126, 311-324 (1979). - SANO, Y.: Oecol. Plant. 12, 383-394 (1977). - SASTRAPRADJA, S., AMINAH, S.H., LUBIS, I., SASTRAPRADJA, D.: Ann. bogor. 6, 43-55 (1975). - SAZIMA, I.,

SAZIMA, M.: Biotropica 9, 213-215 (1977). - SAZIMA, M.: Revista brasil. Botan. 1, 133-138 (1978). - SAZIMA, M., SAZIMA, I.: Biotropica 10, 100-109 (1978); - Flora 169, 84-88 (1980). - SCHAFFER, W.M., SCHAFFER, M.V.: Ecology 60, 1051-1069 (1979). - SCHAFFER, W.M., JENSEN, D.B., HOBBS, D.E., GUREVITCH, J., TODD, J.R., SCHAFFER, M.V.: Ecology 60, 976-987 (1979). - SCHEMSKE, D.W.: Bull. Torrey Botan. Club 104, 254-263 (1977); - Ecology 59, 596-613 (1978); - Biotropica 12, 169-181 (1980a); - Evolution 34, 489-493 (1980b). - SCHEMSKE, D.W., WILLSON, M.F., MELAMPY, M.N., a.o.: Ecology 59, 351-366 (1978). - SCHICK, B.: Flora 170, 394-432 (1980). - SCHMITT, J.: Evolution 34, 934-943 (1980). - SCHNEPF, E.: Apidologie 8, 295-304 (1977). - SCHNEPF, E., WITZIG, F., SCHILL, R.: Über die Bildung und Feinstruktur des Translators der Pollinarien von *Asclepias curassavica* und *Gomphocarpus fruticosus* (Asclep.). In: Tropische und Subtrop. Pflanzenwelt 25, 7-39. Wiesbaden: Franz Steiner 1979. - SCHOEN, D.J.: Syst. Botany 2, 280-286 (1977a); - Bull. Torrey Botan. Club 104, 234-240 (1977b). - SCOGIN, R.: Botan. Gaz. 140, 29-31 (1979). - SCOGIN, R., YOUNG, D.A., JONES, C.E.: Bull. Torrey Botan. Club 104, 155-159 (1977). - SELL, Y.: Ber. Deut. Botan. Ges. 90, 135-147 (1977). - SHIVANNA, K.R., HESLOP-HARRISON, Y., HESLOP-HARRISON, J.: Acta Botan. Neerl. 27, 107-119 (1978). - SHUEL, R.W.: Can. J. Botany 56, 565-571 (1978). - SHUEL, R.W., TSAO, W.: ibid. 56, 833-842 (1978). - SKVARLA, J.J., RAVEN, P.H., CHISSOE, W.F., SHARP, M.: Pollen Spores 20, 15-143 (1978). - SOHMER, S.H., SHEFTON, D.F.: Brittonia 30, 355-364 (1978). - SOLBRIG, O.T., ROLLINS, R.C.: Evolution 31, 265-281 (1977). - SPANJERS, A.W.: Experientia 34, 36-37 (1978). - START, A.N., MARSHALL, A.G.: Linnean Soc. Symp. Ser. No. 2, 141-150 (1976). - STEAD, A.D., MOORE, K.G.: Ann. Botany 41, 283-292 (1977); - Planta 146, 409-414 (1979). - STEINER, K.E.: Ann. Missouri Botan. Garden 66, 490-502 (1979). - STELLEMAN, P.: The possible role of insect visits in pollination of reputedly anemophilous plants, exemplified by *Plantago lanceolata*, and syrphid flies, 41-46. In: The Pollination of Flowers by Insects, ed. A.J. RICHARDS. London: Acad. Press 1978. - STILES, F.G.: Science 198, 1177-1178 (1977); - Brenesia 15, Suppl., 151-180 (1979). - STIRTON, C.H.: Bothalia 12, 223-224 (1977a); - ibid. 12, 225-227 (1977b); - ibid. 12, 229-230 (1977c). - STRICKLER, K.: Ecology 60, 998-1009 (1979). - SWANSON, S.D., SOHMER, S.H.: Bull. Torrey Botan. Club 103, 223-226 (1976). - SWALES, D.E.: Rhodora 81, 363-407 (1979). - SUSSMAN, R.W., RAVEN, P.H.: Science 200, 731-736 (1978). - SYED, R.A.: Bull. Entomol. Res. 69, 213-224 (1979).

TAYLOR, T.N., LEVIN, D.A.: Grana 15, 91-112 (1975). - TENGÖ, J.: Acta Univers. Upsaliensis, Abstr. Uppsala Diss. Faculty of Science 500, 3-31 (1979). - TENGÖ, J., BERGSTRÖM, G.: Comp. Biochem. Physiol. 57 B, 197-202 (1977). - TESCHNER, W.: Orchidee 27, 261-267 (1976); - Sonderheft "Die Orchidee" (Orchideentagung Wuppertal), 104-115 (1980). - THIEN, L.B.: Biotropica 12, 1-13 (1980). - TOLEDO, V.M.: ibid. 9, 262-267 (1977). - TOLEDO, V.M., HERNANDEZ, H.M.: Ann. Missouri Botan. Garden 66, 503-511 (1979). - THOMAS, J.B., ANDERSON, R.G.: Can. J. Botany 56, 3000-3007 (1978). - THOMSON, J.D.: Am. Midland Naturalist 100, 431-440 (1978); - Ecology 61, 572-579 (1980). - THORP, R.W.: Ann. Missouri Botan. Garden 66, 788-812 (1979). - TORCHIO, P.F., TEPEDINO, V.J.: Evolution 34, 993-1003 (1980). - TOMLINSON, P.B., BUNT, J.S., PRIMACK, R.B., DUKE, N.C.: J. Arnold Arbor. 59, 342-351 (1978). - TOMLINSON, P.B., PRIMACK, R.B., BUNT, J.S.: Biotropica 11, 256-277 (1979).

UBSDELL, R.A.: Watsonia 12, 225-232 (1979). - UHL, N.W., MOORE, H.E. Jr.: Biotropica 9, 170-190 (1977).

VALDEYRON, G., LLOYD, D.G.: Evolution 33, 673-685 (1979). - VALLA, J.J., ANCIBOR, E.: Darwiniana 21, 407-415 (1978). - VERESHCHAGINA, V.A.: Botan. Zh. 63, 183-190 (1978) (Russian). - VITHANAGE, H.M., KNOX, R.B.:

Phytomorphology 27, 168-179 (1977); - Ann. Botany 44, 95-106 (1979). - VOGEL, S.: Evolutionary shifts from reward to deception in pollen flowers, 89-96. In: The Pollination of Flowers by Insects, ed. A.J. RICHARDS. London: Acad. Press 1978; - Florengeschichte im Spiegel blütenökologischer Erkenntnisse, 7-48. In: Vortr. Rhein. Westfäl. Akad. Wiss. N. 291. Westdeutscher Verlag 1980. - VOSS, R., TURNER, M., INOUYE, R., FISHER, M., CORT, R.: Am. Midland Naturalist 103, 262-268 (1980). - VÖTH, W.: Orchidee 31, 159-162 (1980).

WADDINGTON, K.D.: Oecologia 40, 43-50 (1979a); - Am. Midland Naturalist 101, 278-285 (1979b). - WADDINGTON, K.D., HOLDEN, L.R.: Am. Naturalist 114, 179-196 (1979). - WAINWRIGHT, C.M.: Bull. Torrey Botan. Club 105, 24-38 (1978). - WALLER, D.M.: Evolution 34, 747-761 (1980). - WASER, N.M.: Ecology 59, 934-944 (1978). - WASER, N.M., REAL, L.A.: Nature 281, 670-672 (1979). - WEBB, C.J.: Evolution 33, 662-672 (1979). - WEBSTER, G.L., ARMBRUSTER, W.S.: Brittonia 31, 352-357 (1979). - WELLER, S.G.: Syst. Botany 3, 115-126 (1978); - ibid. 4, 57-71 (1979); - Can. J. Botany 58, 1908-1911 (1980). - WELLER, S.G., ORNDUFF, R.: Evolution 31, 47-51 (1977). - WESTERKAMP, C.: Sitzungsber. Ges. Naturf. Freunde Berlin, N. F. 18, 71-79 (1978). - WHIGHAM, D.F., McWETHY, M.: Am. J. Botany 67, 550-555 (1980). - WHITHAM, T.G.: Science 197, 593-596 (1977). - WIEBES, J.T.: The Garden's Bull. Singapore 29, 207-237 (1977); - Ann. Rev. Ecol. Syst. 10, 1-12 (1979). - WIENS, D., ROURKE, J.: Nature 276, 71-73 (1978). - WIENS, D., RENFREE, M., WOOLLER, R.O.: Ann. Missouri Botan. Garden 66, 830-838 (1979). - WILLEMSTEIN, S.C.: Lists of flowers visited by Cetoniidae (Coleoptera) and the Central European Cerambycinae and Lepturinae (Coleoptera-Cerambycidae) based on historical and pollen analytical research. Diss. Rijksherbarium Leiden, May 1978. - WILLIAMS, N.: Selbyana 2, 345-355 (1978). - WILLSON, M.F.: Am. Naturalist 113, 777-790 (1979). - WILLSON, M.F., BERTIN, R.I.: Can. J. Botany 57, 1380-1388 (1979). - WILLSON, M.F., MILLER, L.J., RATHCKE, B.J.: Evolution 33, 52-63 (1979a). - WILLSON, M.F., BERTIN, R.I., PRICE, P.W.: Am. Midland Naturalist 102, 23-35 (1979b). - WOITTIEZ, R.D., WILLEMSE, M.T.: Phytomorphology 29, 57-63 (1979). - WOODELL, S.R.: Directionality in bumblebees in relation to environmental factors, 31-40. In: The Pollination of Flowers by Insects, ed. A.J. RICHARDS. London: Acad. Press 1978.

YOUNG, A.M.: Biotropica 12, 222-223 (1980).

ZAPATA, T.R., ARROYO, M.T.: ibid. 10, 221-230 (1978). - ZIEGLER, H., LÜTTGE, U., STICHLER, W.: Naturwissenschaften 66, 580 (1979). - ZIMMERMANN, M.: Ecology 61, 497-501 (1980).

Professor Dr. STEFAN VOGEL
Botanisches Institut der Universität
Rennweg 14
A-1030 Wien

II. Proteins in Plant Evolution and Systematics

By UWE JENSEN

1. Systematical Significance

There are scarcely any characters in systematics that have generated more controversy during their long period of application than that of the proteins. On the one hand their high a priori significance tempts an experimentator to extend, often inadmissibly, the a posteriori significance of his results. While, on the other hand, exaggerated criticism may arise when results from comparative protein experiments are at odds with the current concepts of relationships. An example of this comes from amino acid sequencing, where initial reflections (BOULTER et al., 1970) have led to a too optimistic view, some a posteriori considerations (CRONQUIST, 1976), however, to an underestimation of its significance in systematics.

For comparative investigations different protein properties have been used. These can range from one taxonomic character per protein molecule (e.g., molecular weight, sedimentation constant, charge, isoelectric point), to several (serological determinants) or even a higher number of characters (e.g. sequence properties).

In *monoproteinic* comparisons the latter properties are preferentially, although not exclusively, used. The advantage of a comparison of homologous proteins from different taxa is that it affords the opportunity of comparing specific properties of defined proteins. In this way, the evolutionary changing of the molecule can be conceived. Unfortunately, to accomplish this, the proteins have to be purified, which often produces difficulties (DERBYSHIRE et al., 1976). This is especially true for leaf material, where polyphenols, tannins, and similar compounds can interfere (TUCKER and FAIRBROTHERS, 1970). In seeds, pigments of the seed coat can reduce the extractable protein quantity, however, the pigment-protein complex can be separated (JENSEN and PENNER, 1980).

Sequence data are monoproteinic data par excellence. Currently cytochrome c, ferredoxin, and plastocyanin have been investigated. The main evolution of these proteins occurred in prokaryotes. This is apparent from the composite evolutionary tree generated by SCHWARTZ and DAYHOFF (1978). Gene duplications were important evolutionary events. SCHULZ (1981) suggested that the multitude of existing proteins can be traced back to approximately 200 polypeptide structures. In higher plants homologous proteins, e.g., ferredoxins, are closely related (SCHWARTZ and DAYHOFF, 1978) and random evolutionary amino acid substitutions have considerable influence (BOULTER et al., 1979). For this reason, plant sequence data have to be interprèted cautiously. Unfortunately, at the present time the limitations cannot be clearly circumvented (BOULTER et al., 1979).

Serological properties are closely, although not always linearly linked to the sequence properties (ATASSI et al., 1970; WALLACE et al., 1973).

WALLACE and BOULTER (1976a,b) and NOBLE et al. (1969) suggest at least four external determinants for the plastocyanin molecule and for *Pisum sativum* vicilin (DAVEY and DUDMAN, 1979). The seed storage proteins aquilegilin and nigellin extracted from *Nigella damascena*, both revealed a minimum of nine determinants (JENSEN and PENNER, 1980). Twelve determinants were detected from the 5-aminolevulinate dehydratase from spinach (LIEDGENS et al., 1980). For comparative purposes small protein molecules with suspected determinant numbers of approximately four are less amenable to qualitative comparisons due to their inherent low variability. However, quantitative determinations of the serological reactivity also fail to reveal convincing results (WALLACE and BOULTER, 1976a,b; MURPHY, 1977). It is the analysis of larger storage proteins which holds out the promise of significant data when monoproteinic (JENSEN and PENNER, 1980) or monoclonal antisera (SCHNEIDER et al., 1981) become more readily available.

Multiproteinic Comparison. The advantage of a comparison of crude protein extracts is based on the simultaneous use of many characters (JENSEN, 1974). For more significant results the comparability has to be improved, i.e., the use of extracts containing the same proteins, extracts yielded from homologous organs, similar ontogenetic stages, similar extraction conditions, etc. Unfortunately in many publications this point has not been regarded sufficiently for the discussion of the data. For various reasons seed storage proteins have been preferred (LEE and FAIRBROTHERS, 1978; JENSEN and PENNER, 1980; WOLFF, 1980). To a minor extent also crude tuber proteins (RICKEMAN and DESBOROUGH, 1978; LESTER, 1979; STEGEMANN, 1979) and crude pollen proteins (CHUPOV, 1978; BRUNNER and FAIRBROTHERS, 1979; PETERSEN and FAIRBROTHERS, 1979) have been used successfully. More information can be yielded from the serological data, if their analysis and interpretation is intensivated (CRISTOFOLINI, 1980).

2. Cytochrome c-Type Proteins

The cytochromes (a, b, c) are among the most widespread proteins in living organisms, being involved in respiration, photosynthesis, chemical synthesis, and other forms of electron transport (TIMKOVICH and DICKERSON, 1973). In particular the mitochondrial respiratory cytochromes c from different groups of organisms have been the subject of comparative investigations. Although there are characteristic differences in the cytochrome c structure among animals, fungi and plants, the corresponding three-dimensional structure has been conserved from their common ancestor (BOULTER and RAMSHAW, 1972). Plant cytochrome c has a molecular weight of ca. 12.700, and contains ca. 111 residues. Principally the amino acid sequences as used in systematics have been determined for 1 prokaryote, 2 protista, 6 fungal and 28 eukaryotic plants. Most of these data were published by the BOULTER group in Durham during the early 1970's.

In 1970, BOULTER et al. were convinced from animal cytochrome c data that also "higher plants amino acid sequence determinations may prove to be very enlightening" and they expressed the hope "that the major branches of a phylogenetic tree of the plant kingdom will be established in the next decade". Indeed, the first results were promising (THOMPSON et al., 1971). However, faced with the impression of the final sequence results, the question arose whether or not the current cytochrome c tree of plants mirrors the phyletic system to a considerable extent as pointed out by BOULTER et al. (1979). CRONQUIST's (1976) monumental criticism has already been discussed by KUBITZKI (1977). Essential points of which have been rebutted by BOULTER et al. (1979). Different methods in data handling (PEACOCK and BOULTER, 1975) have also been the subject of controversy.

Comprehensive sequence data (dendrograms) are given by BOULTER et al. (1972) for 15 higher plant taxa, by RAMSHAW et al. (1974c) for 15 higher plant taxa, lower plants, fungi and animals, by BOULTER (1974a) for 20 higher plant taxa, the 5 monocot species not considered, by BOULTER (1974b) for 25 higher plant taxa, and by BOULTER (1976) for 25 higher plant taxa and 3 algae.

Many of the sequence differences are low and insignificant. However, (1) the relatively high correspondence between *Brassica* and *Cucurbita maxima* cytochrome c (BOULTER, 1976) emphasizes the close relationships suggested, (2) the separation of *Euglena* from *Enteromorpha* and other green algae was confirmed, *Euglena* being more similar with the nonphotosynthetic protozoan *Crithidia*, and (3) a separation of *Spinacia oleracea* and *Fagopyrum esculentum* from the other angiosperms was suggested. The deduced separation and early phylogenetical diverging of the Centrosperms could not be verified by plastocyanin or ferredoxin data. However, the ferredoxin-sequencing results are ambiguous: whereas for *Spinacia* no significant difference from the other three angiosperms was found, *Phytolacca* ferredoxin is well separated (WAKABAYASHI et al., 1978, 1980).

Presently cytochrome c sequence data alone cannot give final decisions about the course of evolution. The knowledge of the chemistry of the molecule, as well as of its changing properties in evolution (e.g., different or convergent selection pressure: WONG et al. (1976), different rate of evolution, generation of allotypes), should be enlarged in future before valuable systematical considerations can be given.

The calculations of the unit evolutionary period have proved to be rather doubtful, and an equal rate of evolutionary changings not always verified (WADA et al., 1975, for ferredoxin). Therefore, the calculations of RAMSHAW et al. (1972), which resulted in an origination of the angiosperms long before the Cretaceous time, are subject to detailed criticism by CRONQUIST (1976).

The soluble cytochrome c-554, involved in the photosynthesis of cyanobacteria and algae, also belongs to the c-type and shows homology with the mitochondrial cytochrome c at each end of the chain (LAYCOCK, 1972). The cytochrome c-554 has been isolated and sequenced (83-89 residues) from cyanobacteria (AMBLER and BARTSCH, 1975; AITKEN, 1976) and eukaryotic algae (LAYCOCK, 1972, 1975; PETTIGREW, 1974). It is premature to draw definitive conclusions from these few sequence comparisons. However, they could be interpreted as indicating a strong (symbiotic? PETTIGREW, 1979) relationship between Cyanobacteria and Rhodophyta, Phaeophyta, Chrysophyta, and a separate origin of Chlorophyta *(Chlorella)* from prokaryotic organisms (AMBLER and BARTSCH, 1975; AITKEN, 1976). Although studies of glucotransferase isozymes (FREDRICK, 1977, 1980) are supportive of this interpretation, ferredoxin data (ANDREW et al., 1976) are inconclusive. Other protein data show *Cyanidium caldarium* to be a transition taxon between the prokaryotes and the red algae (GLAZER et al., 1971; WILLIAMS et al., 1974; FREDRICK, 1976, 1980), which supports their close relationship.

3. Ferredoxin

The ferredoxins are proteins containing nonheme iron and labile sulfur, and are involved in the photosynthetic electron transport. They are single proteins and are present in certain anaerobic bacteria as well as in all photosynthetic organisms (HALL et al., 1973). The "plant-type ferredoxins" found in cyanobacteria and eukaryotic plants have a molecular weight of ca. 11,500, and a residue variability of 93 to 99. Of the four or five cysteins detected for each ferredoxin, three or four are found in one almost invariant section of the molecule which appears to be involved in the stabilizing of the cluster cavity (FUKUYAMA et al., 1980). Altogether about 25 positions are invariant in ferredoxin, at least in regards to the taxa investigated to date. These taxa include those from the cyanobacteria (WADA et al., 1975; TANAKA et al., 1975, 1976; HASE et al., 1978a, 1980), algae (SUGENO and MATSUBARA, 1969; ANDREW et al., 1976; HASE et al., 1978b, 1980) and higher plants (MATSUBARA et al., 1967; BENSON and YASUNOBU, 1969; KERESZTES-NAGY et al., 1969; RAO and MATSUBARA, 1970; HASE et al., 1977a,b; WAKABAYASHI et al., 1978, 1980; DUTTON et al., 1980).

For congeneric species no, or very few differences are observed. This is also true for the *Triticum-Aegilops* group (SHIN et al., 1979). Between species of different genera the numbers of amino acid differences are higher (16 to 23, *Phytolacca* excluded), although no correlation with a close or remote relationship can be observed. Only for higher categories are more relevant data revealed. These are expressed by a ferredoxin sequence dendrogram constructed by WAKABAYASHI et al. (1978). This dendrogram separates three ferredoxin groups: prokaryotes, *Equisetum* species, and angiosperms (including *Scenedesmus*). Within the latter group *Phytolacca* ferredoxins (WAKABAYASHI et al., 1980) are unexpectedly more similar to *Scenedesmus* ferredoxin than to other angiosperm ferredoxins so far investigated.

One of the reasons for the difficult taxonomical interpretation of ferredoxin sequence data is the occurrence of polymorphous ferredoxin molecules in one organism (WAKABAYASHI et al., 1978). Quite frequently two variants or even three (*Phytolacca japonica:* SUSUKI, 1976, cit. WAKABAYASHI et al., 1980) are observed, without any understanding of their possible functional differences (HASE et al., 1975). Considerable amino acid differences between these I- and II-ferredoxins led to the conclusion of WAKABAYASHI et al. (1978) and DUTTON et al. (1980), that these two ferredoxins are derived by gene duplication occurring independently in different phyletic lines. In regard to *Equisetum* the gene duplication must have occurred after the divergence of the horsetails from the progenitors of the angiosperms, but before the separation of the *Equisetum* species. Presuming a rather equal rate of ferredoxin evolution, HASE et al. (1977b) calculate the Permian time for this event.

In other cases an additional microheterogeneity has been detected, e.g., in *Phytolacca esculenta* ferredoxin I by WAKABAYASHI et al. (1980). In these cases the ferredoxin molecule probably is encoded by one gene, and the sequence heterogeneity is due to allelic variants (WAKABAYASHI et al., 1980).

In the face of these ferredoxin variations occurring within one organism, any taxonomic interpretation of comparative sequencing data becomes rather dubious at the present state of knowledge.

Other properties of the ferredoxin molecule have been compared, although less intensively. Small differences in the isoelectric point of S-carbaminomethyl-cysteinyl ferredoxins between *Nicotiana* species are probably dependent on the presence or absence of methionine (KWANYUEN and WILDMAN, 1975). These differences can be of limited use for hybrid analysis. However, their use in elucidating infrageneric differentiation in other genera has yet to be proven. Use of quantitative serological methods has revealed graduated results only among closely related taxa (TEL-OR et al., 1977), however, qualitative determinant comparison methods (OUCHTERLONY technique) have failed (KÜMMEL and KESSLER, 1980) and should be preserved for the differentiation of higher categories.

4. Plastocyanin

Plastocyanin is an enzyme of the photosynthetic electron transfer chain found in green organisms. Its molecular weight is about 10,500. In higher plant plastocyanins, both arginine and tryptophan (SCAWEN and BOULTER, 1974) are absent in those molecules analyzed; copper is fixed in the active centrum.

Concerning comparative physicochemical properties very little is known. Only sequence data have been used for systematic considerations. Completely or partially sequenced plastocyanin molecules are known from 3 cyanobacteria, 1 green alga, 7 pteridophytes, 3 gymnosperms, and 73 angiosperms. Thus the plastocyanin data should give a greater insight into the phylogenetic alterations of the molecule than any other plant protein tested to date.

Only a few plastocyanins have been sequenced completely (HASLETT et al., 1974, 1978; KELLY and AMBLER, 1974; MILNE et al., 1974; RAMSHAW et al., 1974a,b; SCAWEN and BOULTER, 1974; SCAWEN et al., 1974, 1975; AITKEN, 1975). In other cases more or less the 40 N-terminal residues alone were considered. These investigations were chiefly performed by BOULTER and his co-workers at Durham. Comprehended data are partly published by HASLETT et al. (1977), BOULTER et al. (1978, 1979) and GRUND et al. (1981). For these publications topologies for the sequence similarities have been obtained using different compatibility methods (LE QUESNE, 1969; DAYHOFF, 1972; ESTABROOK and LANDRUM, 1975; SNEATH et al., 1975). The high anticipations of the Boulter group for the evaluation of phylogenetic relationships from the sequence data has yet to be realized (CRONQUIST, 1980).

Small sequence differences between very closely related taxa are taxonomically valid: the sequences for two *Brassica oleracea* varieties and two *Heracleum* species are identical (BOULTER et al., 1979). Two *Equisetum* species are well separated, as well as *Pteridium*, *Blechnum*, and *Dryopteris* from *Osmunda* and other plants investigated. Within the Compositae plastocyanin sequence data from 22 members of 8 tribes generally agree with the current classification (BOULTER et al., 1978). Notwithstanding a relatively high sequence heterogeneity, the separate position of the Inuleae + Astereae and the high similarity between the Cynareae and Cichorieae, expressed by the intermixed position in the dendrogram, is obvious. This last statement is another argument against the separation of Cichoriaceae and Asteraceae as distinct families. Many other families have demonstrated distinct plastocyanins, differing from those of other families (BOULTER et al., 1979; CRONQUIST, 1980).

However, there are valid exceptions from these statements. Beside the astonishing high sequence difference between two *Senecio* species, which might be explained by a technical error, there are cases to be considered, where families are readily divorced. *Solanum tuberosum* and *Lycopersicon esculentum* (only cultivated species!) show striking differences from other Solanaceae species including *Solanum crispum* (HASLETT et al., 1978). In addition *Vigna radiata* and *Phaseolus vulgaris* differ markedly from other species of the Fabaceae (BOULTER et al., (1979). Certainly, an especially high separation of the tribe Phaseoleae within the Fabaceae might be assumed as confirmed by storage protein data (DUDMAN and MILLERD, 1975). Nevertheless the plastocyanin dendrogram alone cannot be a correct reflection of the phyletic relationships. Moreover, calculating the similarities between the plastocyanins of all angiosperm taxa and the plastocyanins of five Ranunculaceae taxa, GRUND et al. (1981) noted a higher sequence similarity between some non-Ranunculaceae taxa (three Genisteae species, *Crataegus monogyna*, *Nicotiana tabacum*, *Lavatera arborea*) and Ranunculaceae taxa than that which appeared among the Ranunculaceae taxa themselves. This similarity cannot be explained in terms of systematic relationships.

The overall results suggest a strong expression of neutral, i.e., selectively equivalent, as well as parallel or back mutations in the course of plastocyanin evolution (BOULTER, 1980). Although one mutational event might not always have concerned one isolated amino acid position, the question arises whether the investigation and comparison of particular sequence positions would be more useful. Especially appropriate should be those positions which reveal a rather conservative, versus a highly variable ("covarious", FITCH and MARKOWITZ, 1970) exchange of amino acids without showing heterogeneity. Elimination of the highly variable positions, although raising the significance of the results in the Asteraceae (BOULTER et al., 1978), have no considerable effect on the systematic relevance of the results employing all accessable plastocyanin sequence material. Thus the existence of polymorphous plastocyanins, as reported for a *Malva* sample by HASLETT (unpubl., cit. in HASLETT et al., 1977), has to be considered more severely. The existence of different alleles coding for slightly different plastocyanins as well as post-transcriptional modifications of the molecule cannot be excluded. In consequence, even the influence of a specific selection pressure on different taxa growing in the same habitat might cause convergent exchanges of amino acids in nonrelated taxa. In order to answer these questions, the knowledge of the physicochemical properties of the plastocyanin molecule has to be improved.

Closely linked with the sequence properties are the serological determinants, ca. 4 for each plastocyanin molecule (WALLACE and BOULTER, 1976b). The results of a comparative analysis of 37 angiosperm species using the quantitative microcomplement fixation technique (WALLACE and BOULTER, 1976a,b) were disappointing. The correlation between the serological data and the sequence resemblance was poorer than that observed for other protein systems (WALLACE and BOULTER, 1976b). The serological data themselves, except for a few taxa closely related to the reference family (Urticaceae or Chenopodiaceae), revealed a clustering of low and unvalid similarity values. Remarkably similar trends in the results of these serological investigations as well as of those of fraction I proteins, using the same technique (MURPHY, 1978), raises the supposition that the quantitative microcomplement fixation methods are of inferior value in plant systematics when compared to qualitative immunological methods, such as Ouchterlony or presaturation tests.

5. Fraction-I-Protein ("Rubisco")

Since KAWASHIMA (1969) purified F-I-protein from spinach and tobacco leaves, the physicochemical properties of this protein have been intensively investigated from many prokaryotic and eukaryotic taxa. These comparative investigations show promise of leading to new taxonomic and phylogenetic reflexions. Reviews concerning the evolutionary and taxonomic aspects of F-I-protein have been published by TAKABE and AKAZAWA (1975a), WILDMAN et al. (1975), KUNG (1976), WILDMAN (1979, 1981a), GRAY (1980) and CHEN and WILDMAN (1981).

Fraction-I-protein is identical with ribulose-1,5-bisphosphate (RuBp) carboxylase-oxygenase in that it catalyzes the crucial reactions of both photosynthesis and photorespiration. It is "the most abundant protein in the world" (ELLIS, 1979) and constitutes sometimes more than 50% of the soluble protein in the leaves of C_3 plants. In C_4 plants F-I-proteins (HUBER et al., 1976; KIRCHANSKI and PARK, 1976) and specific RNA (LINK et al., 1978) are found entirely or almost entirely in bundle-sheath cells. In small quantities it has even been found in the endosperm of germinating seeds (OSMOND et al., 1975).

The F-I-protein molecule is composed of large (LS) and small subunits (SS). The most primitive enzyme from *Rhodospirillum rubrum* is composed of only two LS (TABITA and McFADDEN, 1974). These have the capacity to combine with either CO_2 or O_2. Although the ancestors of *R. rubrum* are thought to have originated before O_2 became part of the Earth's atmosphere, their F-I-protein enzymes had already anticipated this event (WILDMAN, 1981b). Other prokaryotes have only six or eight LS (TABITA et al., 1974). Most of the blue-green algae, however, as well as the Thiorhodaceae and some other bacteria, establish a LS_8SS_8 macromolecule.

The ancestral gene coding for the large, catalytic (NISHIMURA and AKAZAWA, 1974) subunit seems to have been established first and expressed in anaerobic bacteria (McFADDEN and TABITA, 1974). The SS, which may have a regulatory function (NISHIMURA et al., 1973; NISHIMURA and AKAZAWA, 1974), evolved later, though probably before the main evolution of the blue-green algae (WILDMAN, 1979).

In eukaryotic plants the molecule is composed of 8 LS, each about 56,000 molecular weight, and 8 SS, each about 13,000, resulting in a molecular weight of about 550,000 for the complete protein (e.g., BAKER et al., 1975, 1977). The chloroplast DNA codes for the primary structure of the LS (CHAN and WILDMAN, 1972; COEN et al., 1977; BOTTOMLEY and WHITFELD, 1979), the translation being on 70S ribosomes (BLAIR and ELLIS, 1973), whereas nuclear DNA contains the genetic information for the SS sequence of amino acids (KAWASHIMA and WILDMAN, 1972; KUNG, 1976; CHEN and SAND, 1979), the translation being on 80S ribosomes outside of the chloroplasts (CRIDDLE et al., 1970). Lack of serological cross-reactivity between the large and small subunits (BOWIEN and MAYER, 1978) emphasizes the dissimilarities.

Several physicochemical properties (electrophoretic mobility, isoelectric point, charge, molecular weight, amino acid sequence, serological determinants) of the whole F-I-protein molecule along with the subunits and tryptic peptides have proven to be useful characters in systematic research. The electrofocusing composition of F-I-protein shows a considerable constancy regardless of the developmental stage of the plant (WILDMAN, 1981b). In some instances, however, the use of different methods might lead to divergences in the series of taxonomic similarities (GATENBY, 1978).

In contrast to the limited evidence of electrophoretical methods (BÖRNER et al., 1976), isoelectric focusing of carboxymethylated F-I-proteins in the presence of 8 M urea has been successfully used for comparative investigations. The monomeric polypeptides of both subunits were well and reproducibly separated. The investigation of many higher plant species has shown the LS to resolve into a cluster of three polypeptides (CHEN et al., 1976b), which are suspected of being modifications introduced by the iodoacetic acid treatment (O'CONNELL and BRADY, 1979) or post-transcriptional products of a single chloroplast DNA due to the basic similarity in their peptide structure (WILDMAN, 1981b). The SS resolve into one, two, three or four polypeptides depending on the plant species. These polypeptides are separate gene products, indicated by sequence heterogeneity (GIBBONS et al., 1975; IWAI et al., 1976; STRØBAEK et al., 1976).

Among conspecific taxa the composition of the LS- and SS-polypeptide structure is generally invariable (CHEN et al., 1975b; UCHIMIYA et al., 1979a,b). However, this should not be construed as complete molecular identity (GRAY, 1977; KUNG et al., 1977; GATENBY, 1978). For the elucidation of infrageneric phylogenies, LS clusters, as well as the more variable SS polypeptides, are mostly adapted (CHEN et al., 1976a; GATENBY and COCKING, 1977, 1978a,b,c; UCHIMIYA and WILDMAN, 1978; UCHIMIYA et al., 1979b; CHEN and WILDMAN, 1981).

The differences within subunits of the F-I-proteins of the genera *Gossypium* and *Nicotiana* could be highly correlated with the genome structure, serological (GRAY, 1977) and morphological characters (CHEN and WILDMAN, 1981). In fact, CHEN et al. (1976a) published a computer-generated evolutionary tree for 63 *Nicotiana* species based on LS and SS data. It now appears that *Nicotiana* radiated prior to the drift of the continents (CHEN et al., 1976a), from an ancient evolutionary center in South America into North America, Africa and – via the Antarctic land bridge – Australia. This is clearly demonstrated by the rather homogeneous F-I-proteins of the 20 Australian species, the recently discovered (MERXMÜLLER and BUTTLER, 1975) African *N. africana* and two relictic Patagonian species, *N. noctiflora* and *N. petunioides* (GRAY, 1977; CHEN and WILDMAN, 1981). Previous to the migration into Australia, amphiploidy should have occurred in South America (CHEN and WILDMAN, 1981). The endemic plant species *Gossypium tomentosum* (Hawaii) and *Nicotiana cordiflora* (Juan Fernandez Islands) are suspected to be introduced and late arrivals (CHEN and WILDMAN, 1981).

In amphiploid hybrid analysis subunit investigations have demonstrated that while only the female parent is responsible for the hybrid LS, both parents contribute to the SS polypeptides. The analysis of a single protein therefore allows the identification of not only the parent species but also the direction of the cross giving rise to an allopolyploid. In *Nicotiana* serological reactions are able to discriminate the identity of the progenitors (GRAY, 1978).

Important results that have been presented for different plant taxa include: The scheme of U (1935) concerning the amphidiploid origin of *Brassica oleracea*, *B. nigra* and *B. campestris* has been confirmed (GATENBY and COCKING, 1978c) and the identity of the male and female parents has been indicated (UCHIMIYA and WILDMAN, 1978). The multiple SS composition of *L. esculentum* and other *Lycopersicon* species might have an allopolyploid origin (GATENBY and COCKING, 1978a), however, much of this genome duplication has probably been eliminated in subsequent evolution (UCHIMIYA et al., 1979b). The European potato could have originated by crossing of introduced (♀?) Chilean potato (ssp. *tuberosum*) with the (♂?) Andean potato (ssp. *andigena*) (GATENBY and COCKING, 1979b). The patterns of inheritance have also been demonstrated in intraspecific hybrids of *Gossypium* (CHEN and MEYER, 1979). The B-genome of the tetraploid emmer must have been contributed by a female parent (CHEN et al., 1975a), but not from a contemporary form of *Aegilops speltoides* (VEDEL et al., 1978).

Concerning taxa with only one SS polypeptide (e.g., Poaceae), rather than employing isoelectric focusing, the less expensive immunoelectrophoresis (REICHENBÄCHER et al., 1977) can be used to discriminate species or genera. These can serve as marker techniques for the chloroplast genomes provided that the influence of the SS on the electrophoretic mobility is negligible. Tryptic peptide analyses of LS- and SS-subunits have been performed (KAWASHIMA et al., 1974: *Nicotiana*; KAWASHIMA and TANABE, 1975: Solanaceae), but apparently are of questionable significance.

The use of amino acid sequencing for comparative purposes is in its infancy. Only the 120 amino acid residues of the spinach SS have been completely sequenced (MARTIN, 1979). For six species the N-terminal SS-peptide sequences have been published (GIBBONS et al., 1975; HASLETT et al., 1976; POULSEN et al., 1976; STRØBAEK et al., 1976). LS peptides have been sequenced for spinach by STRINGER and HARTMAN (1978) and for barley by POULSEN et al. (1979). The value of F-I-protein sequence data as regards systematic considerations, cannot as yet be fully judged. It is also of note (KAWASHIMA and WILDMAN, 1971) that its systematic value is different for SS (high variability) versus LS (low variability).

In serological experiments comparison of LS produce quite different similarity values when contrasted with those of the SS, due to the higher evolutionary rate of the latter (BROWN et al., 1976). Therefore the serological identity, for example, of *Nicotiana* species (KAWASHIMA et al., 1971) must be realized by rather similar SS of closely related species. However, the partial serological identity, usually observed between more distantly related eukaryotic taxa (e.g., between *Nicotiana* and *Beta:* GRAY and WILDMAN, 1976; *Nicotiana* and *Spinacia:* KAWASHIMA and WILDMAN, 1971; *Spinacia* and *Chlorella:* SUGIYAMA et al., 1969; *Euglena* and *Lactuca*, *Cucurbita*, *Tetragena:* RABINOWITZ et al., 1975) or even between eukaryotic and prokaryotic taxa (e.g., *Euglena* and *Anabaena:* LORD et al., 1975; *Spinacia* and *Chromatium:* AKAZAWA et al., 1972; TAKABE and AKAZAWA, 1975b) must be realized by LS determinants identical to the F-I-proteins compared. The missing serological cross-reaction between *Rhodospirillum rubrum* and, for example, spinach F-I-protein (AKAZAWA et al., 1969) is connected with the rather primitive enzyme constitution of this species (2 LS only). MURPHY (1978) compared complete F-I-protein molecules from more than 50 species of gymnosperms and angiosperms serologically. Unfortunately the taxa could not be discriminated satisfactorily using the quantitative micro-complement fixation technique. Apparently the poor result is not worth the considerable technical expense.

The idea of the phylogenetical clock has been recovered by UCHIMIYA et al. (1979b) and CHEN and WILDMAN (1981), who compared the fossil records with the presumed rate of evolution of the LS and SS polypeptides. Judging from the differentiation found in the genera *Gossypium* and *Nicotiana*, these taxa might be as old as the Lemnaceae (ca. 50 million years, from fossil records), whereas *Oenothera* (13 identical structured species) should be phylogenetically younger. LS and SS polypeptides show a difference in their evolutionary rate. Using Australian *Nicotiana* species with identical LS polpypetides, CHEN et al. (1976a) calculated an average rate of 0.27 mutations/100 amino acids/10^8 years, which would represent the smallest mutation rate for a protein ever recorded. The protein differences between infrageneric taxa are mostly contributed by the SS polypeptides (GRAY and KEKWICK, 1974; KAWASHIMA and TANABE, 1975; KAWASHIMA et al., 1976). Their evolution was less conservative than the evolution of the LS polypeptides (KWOK and WILDMAN, 1974). Again from the Australian *Nicotiana* species CHEN et al. (1976a) estimated a rate of 3.04 mutations/100 amino acids/10^8 years for the SS, which is comparable with the rate calculated for cytochrome c. These calculations are correct, however, only when one assumes an equal rate of F-I-polypeptides evolution. This, indeed, has still to be proved.

6. Isozymes

Isozymes (incl. allozymes) are used to an increasing extent in the elucidation of the phylogenetic relationships of clones, cultivars, varieties, populations, hybrids, or other infraspecific or infrageneric taxa. Since small gene-dependent discrepancies in isozyme properties are expressed by a more or less different electrophoretic behavior, electrophoresis (in starch- or polyacrylamide gels) in combination with a staining reaction based on the catalytic activities of the isozymes has been the preferable technique. Isoelectric focusing has been used only sparingly (NAKAI and TSUNEWAKI, 1971; GARCIA and TSUNEWAKI, 1977; KATO and TOKUMASU, 1979). Immunological techniques are powerful tools for the investigation of the evolutionary behavior of an entire enzyme. From investigations of glutamate dehydrogenase (PAHLICH et al., 1980) a degree of evolutionary stability can be suggested that corresponds to F-I-protein. Comparison of amino acid sequences of isozymes is at the very beginning (MAZZA and WELINDER, 1980). A critical recent review was published by HURKA (1980).

a) Isozyme Banding Comparison

The use of zymograms for solving systematic problems is in principal as good or as problematic as the use of electropherograms of crude protein extracts. The similarity values between the banding patterns of different taxa were used as taxonomic criteria. Most of the isozymes exhibit only one or few bands of one or two loci (e.g., MDH, ADH, GDH, LAP, GOT), which frequently are identical in different taxa. For this reason, taxonomically they are less useful than esterases and peroxidases (PAINE et al., 1973), which in consequence of broad substrate specifities exhibit both inter- and intraspecific polymorphism (MITRA et al., 1970). Consequently for taxonomic contributions, esterase and peroxidase enzyme systems have predominated. Isozymes have been extracted from various plant tissues, i.e., from seeds, cotyledons, seedlings, roots, leaves, anthers, and pollen. These investigations focus mostly on the angiosperms, algae and bryophytes have seldom been studied.

b) Allozyme Comparison

The very complexity of isozyme patterns compared can obscure a great deal of the potential information content of zymograms (GOTTLIEB, 1977b; TORRES et al., 1978a). Therefore during the last years the crucial emphasis has been displaced from only statistical comparisons of complex isozyme patterns to precise genetic analyses of allozyme bands, thus permitting insight into single gene/enzyme systems. The systematic aspects (GOTTLIEB, 1977b) and the theoretical suppositions (JOHNSON, 1977) have been discussed.

The usefulness of such studies has been convincingly demonstrated, for example, for *Chenopodium* by CRAWFORD and WILSON (1977, 1979). Although these species are phenotypically plastic and taxonomically difficult, the gene loci comparison using electrophoretical separation of the allozymes revealed genetically distinct taxa and demonstrated the absence of interspecific hybridization. A remarkable contribution to the origin of cultivated citrus species has been made by TORRES et al. (1978a), who used the three isozyme systems glutamate oxalacetate transaminase, phosphoglucose isomerase, and phosphoglucose mutase as genetic markers. Their data, similar to the amylase isozyme data of ESEN and SCORA (1977) confirmed the possible hybrid origin of *Citrus aurantium* (sour orange) from *C. grandis* (pummelo) × *C. reticulata* (mandarins) as well as *C. paradisi* (grapefruits) from *C. grandis* × *C. sinensis* (sweet orange). Also conspecific populations and culticars (TORRES et al., 1978b for avocado) have been investigated by "indirect" genetic analysis (RICK and FOBES, 1975; CRAWFORD, 1979). The evaluation of the potential of genetic resources within populations

(GOTTLIEB, 1975, 1977b), especially of crop plants (BROWN et al., 1978) is of general utility. From such formal genetic analyses the number of enzyme subunits can be postulated and their allelic state or two-loci origin elucidated (WEEDEN and GOTTLIEB, 1979). Since gene duplication (GOTTLIEB, 1977a) can be considered a unique event, all congeneric species which possess the duplicated gene should form a monophyletic assemblage (GOTTLIEB, 1977b; GOTTLIEB and WEEDEN, 1979 for *Clarkia*). For ADH an ancestral gene must have been duplicated at least prior to the divergence of monocot and dicot plants (JAASKA and JAASKA, 1980). Assuming the rate of gene substitution to be the same for each enzyme locus, the expected number of amino acids per protein differences among species, and hence the time of phylogenetic divergence, can be roughly calculated from electrophoresis data (NEI, 1971, 1974; LEVY and LEVIN, 1975 for *Oenothera*).

c) Genetic Polymorphism

The isozyme pattern variability within populations can be used to explain the amount of genetic polymorphism. Respective to the isozymes tested monomorphic or quasi-monomorphic populations (MORAN and MARSHALL, 1978), highly polymorphic populations in which no two individuals appear to be genetically similar (IMAM and ALLARD, 1965; JAIN and MARSHALL, 1967), and all situations between both extremes are known. Populations with relatively broad edaphic tolerances are electrophoretically and genetically more variable than those of edaphically restricted species (BABBEL and SELANDER, 1974; for *Hymenopappus* and *Lupinus* species). Within *Hordeum spontaneum*, the patterns of variability are in part predictable ecologically and, therefore, presumably adaptive (NEVO et al., 1979). Natural selection, operating in different ways in different environments, seems to be the prominent factor responsible for the observed patterns of genic variability (CLEGG and ALLARD, 1972), although the role of environmental heterogeneity in genetic polymorphism is not as yet completely understood (HEDRICK et al., 1976). The impact of breeding systems on genotypic distribution within a population and hence on the complexity of isozyme patterns has been frequently discussed (e.g., by ALLARD, 1975; RICK and FOBES, 1975; JAIN, 1976; GOTTLIEB, 1977b; PHILLIPS and BROWN, 1977), and has been reviewed in detail by HURKA (1980).

d) Additivity of Bands in Hybrids

In general, electrophoretic banding patterns are additive in hybrids. In consequence of this in allopolyploids (although not observed in autopolyploids of, for example, *Ricinus communis*: TIMKO et al., 1980) an enzyme multiplicity can be expected (MITRA and BHATIA, 1971). This might explain the adaptability of hybrids to an extended range of environments (CUBADDA et al., 1975; GOTTLIEB, 1976). Exceptions from simple additivity, however, were observed (e.g., YADAVA et al., 1979), and these exceptions may be caused by novel heteromers in hybrids (HURKA, 1980), silent alleles, or even technical difficulties. For the triploid endosperm a maternal gene dosage effect can influence the additivity of the banding patterns (KIM et al., 1979).

e) Systematically Relevant Results

From the multitude of results reported only those concerning new aspects for the origin of the hexaploid *Triticum aestivum* are mentioned. From acid phosphatase and esterase isozyme data JAASKA (1976, 1980) suggested that both *T. boeoticum* and *T. urartu* might have contributed the A-genome to the tetraploid wheats. A special NADP-dependent aromatic alcohol dehydrogenase, which is under the genetic control of the B-genome, is present in all taxa of emmer and dinkel wheats. This dehydrogenase is characteristic of only one diploid species, namely *Aegilops speltoides* s.l. (JAASKA, 1978a). Therefore, this species should be genetically most closely related (but not identical: VEDEL et al., 1978) to the ancestral diploid which has contributed its B-genome (by the female parent: CHEN et al., 1975) to the origin of the tetraploid emmer wheats. The progenitor of the AB-genome, characterized by a special esterase zymogram (NAKAI, 1979), should be transferred to *Triticum carthlicum* (BUSHUK and KERBER, 1978). A special α-amylase zymogram also distinguishes the varieties *typica*, *meyeri* and *strangulata* of *Aegilops squarrosa* (NAKAI, 1979) as being the expected male parent D-donors for the origin of the hexaploid wheats (CHEN et al., 1975a; VEDEL et al., 1978). From the present geographical distribution for the varieties mentioned, the area from the Georgian SSR to northern Iran could be the birthplace of hexaploid wheats (NAKAI, 1979).

f) Limitations of Taxonomic Interpretation

It cannot be predicted a priori which enzyme system is most useful for a special taxonomic study (HURKA, 1980; SCHIAVO et al., 1980; YAMAMOTO and PLITMANN, 1980). In many investigations peroxidase systems are used successfully. In different species and amphidiploids of *Brassica*, however, the peroxidase isozyme patterns did not reveal considerable variation (YADAVA et al., 1979). Lactate dehydrogenase often is monomorph (BOSBACH, 1978, for *Capsella bursa-pastoris*), in *Picea glauca*, however, this enzyme varies considerably (TSAY and TAYLOR, 1978). In *Aegilops speltoides* a large variation in esterase patterns is reported, but no variations in *A. caudata* (NAKAI, 1973).

The isozyme patterns of mixed plant material from populations, varieties, or species represent the sum of more or less all individual enzymes existing in the range of the entire taxon. In principal the elucidation of the enzyme pattern of single seeds (TE NIJENHUIS, 1971), needles (PRUS-GLOWACKI and SZWEYKOWSKI, 1977) or other tissues is advisable, but care should be taken for the appearance of specific patterns for each organ. In most cases, however, nothing is known about the degree of homogeneity of the samples used. Furthermore, the gene expression may be different for different ontogenetical stages (THOMAS and NEUCERE, 1974), due to differential activation in development. Therefore care should be taken to assure phenological similarity of the samples (KELLEY and ADAMS, 1977).

Environmental conditions can also influence the isozyme banding patterns (De JONG, 1973; HAYWARD and McADAM, 1977), particularly in sensitive systems as peroxidases (ALMGÅRD and CLAPHAM, 1975) and esterases (WOLFF, 1980). This is not contradictory to the common conception that isozyme patterns are less modifiable than many other nonproteinic plant features.

The limitations caused by the influence of different techniques are remarkable. Electrophoresis tends to underrepresent genetic diversity (AVISE, 1975; AYALA, 1975; McLEOD et al., 1979). Different gels (GOTTLIEB, 1971; CLARKSON et al., 1974) and different electrophoretic approaches should be used to overcome the risk of a contradictory evidence regarding systematic relationships (PAYNE and FAIRBROTHERS, 1973; SHECHTER and De WET, 1975). In this way more emphasis should be given to electrophoretic differences rather than to evidence of similarity (GOTTLIEB, 1977b).

7. Pollen and Spore Proteins

Studies of pollen proteins can elucidate similarities or dissimilarities of hay fever proteins in respect to their allergenic reactivity (LEE and DICKINSON, 1979). Also in those studies geared to taxonomic aspects, taxa of wind-pollinated plants were generally used because of the facility in obtaining sufficient material. Especially fresh pollen is recommended to avoid conformational changes of the proteins (JOHNSON and FAIRBROTHERS, 1975). Although proteins originating from the sporophytic tapetum might be co-extracted (HOWLETT et al., 1973; KNOX et al., 1975), the prominent proteins are derived from the haploid plant material.

Using crude pollen extracts a complex banding pattern will occur. However, isoelectric focusing has proven to be efficient for the unequivocal differentiation of *Typha* species and hybrids (KRATTINGER et al., 1979). By serological methods infraspecific taxa are difficult to distinguish from each other (VILLAMIL and FAIRBROTHERS, 1974). For species within genera, families or orders, however, the differentiation capability for pollen proteins (as well as for fern spores: PETERSEN and FAIRBROTHERS, 1971) is better, revealing results comparable with those from crude seed proteins (BRUNNER and FAIRBROTHERS, 1979: Betulaceae; CHUPOV, 1978; PETERSEN and FAIRBROTHERS, 1979: Amentiferae). Concerning interfamiliar comparisons, the results of CHUPOV (1978) and PETERSEN and FAIRBROTHERS (1979) are in favor of close connections between Juglandales and Fagales, while Juglandaceae and Anacardiaceae should be widely separated.

8. Other Nonstorage Proteins

Other proteins, when studied more intensively, might be of significance for systematics. This would include the phloem exudate proteins (SABNIS and HART, 1976; GIETL and ZIEGLER, 1980), the leghemoglobins in the Fabaceae (HURRELL et al., 1977), virus inhibitor proteins (GRASSO and SHEPHERD, 1978) and ribosomal proteins of algae (GÖTZ and ARNOLD, 1980a, b). For individual proteins complete or partial sequences have been determined (MOSSÉ, 1973; LEHTOVAARA et al., 1980: *Pisum sativum*-leghemoglobins). LYNN et al. (1980) demonstrated homology of the N-terminal sequences of latex proteases from *Asclepias* and *Carica*.

An interesting aspect is the use of enzyme activity in comparative studies: FLÜCK and JAFFE (1974) for cholinesterases in Characeae and angiosperms; GROSS et al. (1975) for hydroxycinnamate co-enzyme A ligase in higher plants; GUTIERREZ et al. (1976) for PEP carboxykinase in Poaceae subfam. Panicoideae; IBRAHIM (1977) for coniferyl alcohol glucosyltransferase in higher plants; JOEL et al. (1978) for laccase in Anacardiaceae. Such data are equivalent with rough screening data of the substrates themselves. They are systematically less important since high quantitative differences (enzyme activity) might mask nonidentified qualitative differences (molecular structure). However, concerning well-defined proteins within particular taxonomic groups useful contributions can be expected. Examples of this are the investigations done on shikimic acid-activated dehydroquinate hydrolase, which has been found to occur throughout the families Juncaceae, Poaceae, and Cyperaceae (BOUDET et al., 1977), and the examination of α-galactosidase activity in legumes (HANKINS et al., 1980).

For the elucidation of the phylogenetic position of the Cyanobacteria, Rhodophyta, and Cryptophyta, the comparative investigation of the phycobiliproteins has already promoted strong arguments. By comparing physical and serological properties as well as partial protein sequences it has been shown that (1) the α- and β-subunits arose probably by gene duplication early in the evolution of the biliproteins (GLAZER et al., 1976) and before the origin of eukaryotes, (2) these pigments in Cyanobacteria and Rhodophyta are strikingly similar, e.g., in C-phycocyanin sequences (WILLIAMS et al., 1974), thus confirming a close evolutionary relationship. In contrast to this the biliproteins in the Cryptomonads show striking differences (McCOLL et al., 1976).

9. Crude Tuber Proteins (Potato)

Zone electrophoresis has been shown to be the preferred method, if closely related species or conspecific taxa are to be compared (RICKEMAN and DESBOROUGH, 1978; STEGEMANN, 1979). It has also been determined that an electrophoretic pattern is unique for any given clone or variety (CZUPRYN and TOCZKO, 1974). With improved electrophoretic techniques such a high degree of reproducibility can be obtained that it formed the basis for the standardization of the *Index of European potato varieties*, which combined genetic and electrophoretic information for 530 cultivars (STEGEMANN and LOESCHCKE, 1976, 1977). At generic *(Solanum)* or higher level (Solanaceae) serological methods are systematically more significant (LESTER, 1979).

10. Seed Storage Proteins

a) Purified Proteins or Defined Fractions

α) Fabaceae. In *Pisum sativum* about 80% of the total seed protein is composed of two storage proteins, legumin and vicilin (CASEY, 1979), which are serologically different (MILLERD et al., 1971; SCHOLZ et al., 1974) and therefore probably nonhomologous. Their properties are already well known (DERBYSHIRE et al., 1976) with heterogeneity being observed for vicilin (MANTEUFFEL and SCHOLZ, 1975; DAVEY and DUDMAN, 1979) as well as for legumin (UTSUMI and MORI, 1980). Number and properties of the subunits have been revealed by SDS gel electrophoresis and isoelectric focusing (HIGGINS and SPENCER, 1977; CASEY, 1979; CROY et al., 1979; PRZYBYLSKA et al., 1979). Once founded on such detailed knowledge concerning protein properties, the comparison of proteins of different taxa, as well as the demonstration of their homology and the evaluation of their grade of similarity, achieves greater significance in systematics.

Within species some molecular characteristics seem to be identical or nearly identical, e.g., molecular weight, charge, main subunit structure, and serological determinants (SCHLESIER and SCHOLZ, 1975; CROY et al., 1979).

Other properties, especially those associated with the subunits, differ. An example of this is shown in the results produced by urea and SDS PAGE electrophoresis of the 2-mercaptoethanol reduced legumin polypeptides for various *Pisum* lines (PRZYBYLSKA et al., 1979). Also the carbohydrate content of these glycoproteins varies considerably (DAVEY and DUDMAN, 1979; GATEHOUSE et al., 1980; HURKMAN and BEEVERS, 1980). The carbohydrate probably does not participate in the serological determinants, as suggested at least for one *Pisum sativum* vicilin (DAVEY and DUDMAN, 1979).

Between different species some important characters of proteins are either identical or so strikingly similar that they demonstrate the homology of these proteins. *Pisum sativum* and *Vicia faba* legumins have a similar disulfide bridged subunit construction with basic subunits of molecular weight of about 20,000 and acidic subunits of about 40,000 (KITAMURA et al., 1976; CASEY, 1979; CROY et al., 1979). For the latter N-terminal leucine and threonine have been demonstrated (WRIGHT and BOULTER, 1974; CASEY, 1979). For these legumins as well as for the probably homologous proteins arachin, from *Arachis hypogaea* (TOMBS and LOWE; 1967), and glycinin, from *Glycine max* (CATSIMPOOLAS, 1969), a congruous quaternary protein structure has been proposed. The homology is confirmed by a remarkable similarity in the tryptic peptide maps (CROY et al., 1979) and identical (DUDMAN and MILLERD, 1975) or almost identical serological reactions (CROY et al., 1979). However, different molecular weights of the native molecules and the subunits (CROY et al., 1979) express the influence of equivocal evolutionary changes; the amount of such differences obviously must not correlate with the phylogenetic relationships.

The molecular structural data (molecular weight, charge, subunit structure) can be applied optimally either within conspecific or congeneric taxa. Simple electrophoretic banding patterns of cultivars, crossings, and segregants can be interpreted in terms of distinct gene expressions (BLIXT et al., 1980). For less closely related taxa, properties of the tertiary structure become more valid. In the complete absence of comparative sequence data (which correspond with the tertiary structure) the serological data merit attention.

The serological data of DUDMAN and MILLERD (1975) are in good harmony with the current conception of relationships, which show the following decreasing similarities: *Vicia faba* (reference) – *Pisum sativum* (legumin and vicilin unchanged serologically) – *Lens culinaris* and two *Lathyrus* species (legumin unchanged, the probably faster evolved vicilin highly similar) – *Cicer arietinum*, *Abrus precatorius* and taxa of many tribes of the Fabaceae (difficulties in yielding any serological correspondence with *Vicia faba* legumin and vicilin) – Genisteae and Phaseoleae (with serologically completely different legumin and vicilin proteins). Also fingerprint patterns of the globulin fractions, predominated by legumin and vicilin, emphasize the isolated position of the Phaseoleae within the Fabaceae, which is confirmed by the plastocyanin sequences (BOULTER et al., 1979), DNA hybridization studies (BOLTON et al., 1965) and the occurrence of specific Phaseoleae proteins (BARKER et al., 1976; HALL et al., 1977; BROWN et al., 1981).

Although a missing serological cross-reaction indicates altered molecule structures, nevertheless similar proteins with homologous sequences may exist. This has been demonstrated for the storage proteins of the Vicieae and *Glycine max*, which share similar properties (GILROY et al., 1979), as well as for storage proteins ("conglutin") of *Lupinus* (tribe Genisteae), which DANIELSSON (1949) has suggested to be vicilin- and legumin-like.

β) Poaceae. In several taxa of the Poaceae, globulins also constitute the bulk of storage proteins (LUTHE and PETERSON, 1977; KHAVKIN et al., 1978) and have been used for taxonomic comparisons (KHAVKIN et al., 1979: *Zea mays* the reference antiserum). Their chemical properties and probably physiological functions are much like those of the dicots (DERBYSHIRE et al., 1976).

The major storage proteins of most cereal grains, however, are alcohol-soluble prolamins and the alkali-soluble glutelins. In the past the prolamins generally are reported to be almost exclusively located in seeds of the Poaceae, however, evidence now has been presented (DARMENCY et al., 1980) that suggest they might have a much wider distribution and thus occur also in diverse dicot species. They frequently represent about half the content of the seed protein (MOSSÉ, 1966). Although prolamins reveal considerable infravarietal biochemical polymorphism (KIM et al., 1978; HAGEN and RUBENSTEIN, 1980; VITALE et al., 1980) they appear to be useful for the investigation of cogeneric species relationships and the identification of polyploids and varieties (WRIGLEY and SHEPHERD, 1973, 1974; JOHNSON, 1975; DHALIWAL, 1977; DOLL and BROWN, 1979; KIM and MOSSÉ, 1979; KIM et al., 1979). Two-dimensional gel techniques (SHEWRY et al., 1977) as well as SDS PAGE are useful for the study of the protein subunits (SHEWRY et al., 1978). Serological techniques have been proved useful for the comparison of nongliadins of the prolamin fraction (KONAREV, 1981). Rarely have glutelins been investigated (e.g., JULIANO and BOULTER, 1976; BURNOUF and BOURIQUET, 1980). For the present, the protein fractions investigated often seem either to be complex or contaminated, which thus limits, considerably, their value for systematic work.

γ) Other Angiosperms. Since the biological role of seeds is much the same throughout the flowering plants, it would appear likely that specific storage proteins have evolved and that equivalent homologously related proteins might exist in a wide range of plants (DERBYSHIRE et al., 1976). This is especially true for two types, the 11S (or 12S) and 7S proteins, which are readily observed within Fabaceae taxa, and which have been reported for other families (DERBYSHIRE et al., 1976; REICHELT et al., 1980). The 11S protein legumin of the Vicieae has been recently demonstrated to be probably homologous with a prominent seed globulin of the Cucurbitaceae (HARA et al., 1978; GILROY et al., 1979). Two major storage proteins were also reported for the Ranunculaceae (JENSEN and PENNER, 1980) and Scrophulariaceae (BÜTTNER and JENSEN, 1981), for one of which a similar serological determinant structure could be demonstrated. DERBYSHIRE et al. (1976) discussed the suspected homologies of both 11S and 7S types of proteins. This cannot be proven from electrophoresis, molecular weight, subunit structure, and other physical data alone. The considerable variability of the 11S and 7S proteins and the polymorphism already within one plant are shown to be the main areas of uncertainty.

Because of the deficiency of complete sequence data, serological cross-reactivity can demonstrate the suspected protein homology (METZGER et al., 1968). Theoretically, serological determinant convergences, i.e., the occurrence of congruous determinants on different proteins, can never be excluded. This is especially apparent in those cases where the same subunits are involved in the quaternary structure of different proteins (DERBYSHIRE et al., 1976). Until recently, no evidence of this possibility has been presented. Using serological cross-reactions the *Digitalis purpurea* storage globulin tubiflorin has been demonstrated for many Tubiflorae and related taxa (JENSEN and BÜTTNER, 1981). Moreover, the *Nigella damascena* and *Aquilegia vulgaris* major storage proteins ("H-globulin" = nigellin; "N-globulin" = aquilegilin) occur frequently in dicot and monocot taxa (BÜTTNER and JENSEN, 1981) and also account for positive cross-reactions between distantly related taxa (JENSEN and PENNER, 1980). These reactions have previously been called "asystematic" or "antisystematic" reactions (FROHNE et al., 1961). According to our present knowledge, it is most likely that many angiosperm families are characterized by only two nonhomologous major storage proteins, and that within the angiosperms only a few common major storage proteins occur, having developed a more or less distinct grade of polymorphism during their molecular evolution. The clarification of their distribution promises new, useful data for the angiosperm phylogeny. For this purpose intensive studies of the physicochemical properties are presupposed. However, the use of unclearly defined fractions (e.g., "albumins" and "globulins") and rough separation methods should be avoided.

b) Crude Seed Extracts

The use of crude seed extracts has a long history in comparative protein research. In consequence of this its methodology has been continually updated and improved. Concerning *serological* investigations mainly qualitative methods for evaluations of the antigenic reactivity have been used (Ouchterlony immunodiffusion, immunelectrophoresis, presaturation tests).

Rocket immunelectrophoresis (LAURELL, 1972) was recommended as an especially sensitive method for systematical serological research (LEE, 1977; FAIRBROTHERS, 1980). The quantitative Boyden-procedure has been resumed by LEE and FAIRBROTHERS (1978) and a Polish group (PRUS-GLOWACKI, 1975; ANIOL, 1976a,b,c,d; NOWACKI and JAWORKSI, 1978). The "hyperreaction" occurring in the experiments of ANIOL (1976c) was already observed by HAMMOND (1955) and JENSEN (1968) and explained by a different turbidity value for different protein precipitates in a complex mixture.

Usually the serological reactivity of the seed storage proteins varies little or not at all within a species (JENSEN, 1968; CRISTOFOLINI and CHIAPELLA, 1977). However, using high titered antisera, congeneric species can be separated (PRUS-GLOWACKI, 1975: *Lupinus*; RUDNICKA, 1975: *Trifolium*; ANIOL, 1976a,b,c,d: *Aegilops*, *Secale*, *Elymus*, *Agropyron*; NOWACKI and JAWORSKI, 1978: *Lupinus*; EL TINAY et al., 1979: *Acacia*; SIMON, 1979a, b: *Prosopis*, *Medicago*; KLOZOVA et al., 1981: *Allium*). Likewise valuable results are reported when comparing species within a tribe (CRISTOFOLINI and CHIAPELLA, 1977: Genisteae), related tribes (SMITH and LESTER, 1980: *Zea* and related taxa), a family (CHUPOV and CUTJAVINA, 1978: Liliaceae; JOHN, 1978: Primulaceae) or a strictly conceived order (BRUNNER and FAIRBROTHERS, 1978: Cornales; KOLBE, 1978: Capparales; LEE and FAIRBROTHERS, 1978: Gentianales). If rather distantly related taxa are compared, using undefined crude protein extracts, low reactivities and a high risk for heterogeneous materials may be responsible for results of less significance (FROHNE and JOHN, 1978; KOLBE and JOHN, 1979, 1980; JOHN and KOLBE, 1980).

The following includes the more significant recent results. The Amaryllidaceae are serologically significantly separated from the Liliaceae s.l. Within the Liliaceae the Lilioideae are isolated. The group *Agave-Yucca* differs significantly from the *Dracaena* group and can be connected with the Liliaceae (CHUPOV and CUTJAVINA, 1978). The Cucurbitaceae, Datiscaceae and Begoniaceae are distinctly separated from the Flacourtiaceae, Violaceae, Passifloraceae and Turneraceae. The position of the Loasaceae within the Violales remains doubtful (KOLBE and JOHN, 1979). The incorporation of the Tropaeolaceae into the Capparales was not supported by KOLBE (1978). Concerning the Brassicaceae the position of *Sisymbrium*, *Isatis*, and *Bunias* within the tribe Sisymbreae is confirmed; *Alliaria* however is different. The Brassiceae are close to the Sisymbreae and the separation of siliqua- and silicula-bearing genera cannot be supported. *Alyssum* is closer to *Arabis* (Arabideae) than to *Lunaria* (Alysseae). *Barbaraea* is separated from *Arabis* in the Arabeae (KOLBE, 1973, 1978). The Primulaceae, Myrsinaceae, and Theophrastaceae (Primulales) are serologically similar. Connections with Ericaceae, not Plumbaginaceae were detected (FROHNE and JOHN, 1978). The seed proteins of the iridoid-producing families Hydrangeaceae, Escalloniaceae, and Roridulaceae (iridoids uncertain) are similar with those of the Cornales and support the separation from the Saxifragales (GRUND and JENSEN, 1981). The genus *Aralidium* should be excluded from the Cornaceae but included into the Cornales (FAIRBROTHERS, 1980).

The use of seed protein profiles obtained by *electrophoresis* in plant systematics has been greatly expanded in the last decade (LADIZINSKY and HYMOWITZ, 1979: short review). The complex electrophoresis banding patterns are, however, not well adapted to genetic studies, even when special bands can be identified (ORF and HYMOWITZ, 1976). For comparison purposes they must be examined for the number of similar bands in relation to the overall number of bands, thus giving an index of similarity (SOKAL and SNEATH; 1963; VAUGHAN and DENFORD, 1968). Such data are highly adapted for numerical taxonomic and variance analyses (ROBINSON and MAGARRITY, 1975; HILL, 1977).

The risk of erroneous homologization can be reduced by comparing closely related taxa and by using different or combined separation techniques (two-dimensional electrophoresis or additional iso-electrofocusing). As many as 84 different bands within a genus *(Bulnesia)* have been identified (COMAS et al., 1979). The banding pattern is highly stable and affected only slightly by environmental conditions or seasonal fluctuations (CRAWFORD and JULIAN, 1976). Generally water-soluble proteins were extracted, which eventually were roughly separated into the albumin and globulin fractions, a process which precedes electrophoresis. Occasionally the discriminating capacity of the albumins is higher (PHELAN and VAUGHAN, 1976; PRZYBYLSKA et al., 1977). In addition isoenzyme banding patterns are included in the calculations of similarities (PHELAN and VAUGHAN, 1976: *Brassica;* MISSET, 1977: Fabaceae). For the identification of wheat cultivars a rough nomenclature for the protein bands has been developed (AUTRAN and BOURDET, 1975; BUSHUK and ZILLMAN, 1978).

In contrast to serological data the use of electrophoretic data is restricted to conspecific or congeneric populations, varieties, cultivars, hybrids, or polyploids. Between distantly related taxa the patterns generally tend to be too diverse and thus incapable of comparison. MISSET (1977), however, revealed an acceptable dendrogram for Fabaceae species.

Intraspecific variation may be small or lacking (CRAWFORD and JULIAN, 1976; LADIZINSKY and HAMEL, 1980, for *Cajanus* and *Atylosia* species). In other cases the comparison of different banding patterns helps in the understanding of intraspecific variability and, connected with morphologic and geographic characteristics, in the understanding of the evolutionary differentiation (SHECHTER, 1975: *Sorghum bicolor*-cultivars; SHECHTER and De WET, 1975; CRAWFORD, 1976: *Chenopodium fremontii*-populations; GASQUEZ and COMPOINT, 1976: *Echinochloa crus galli*-populations). For single genotypes the proteins of single seeds should be compared (PRZYBYLSKA et al., 1977).

New arguments for the origin of hybrids and allopolyploids can be revealed. BOOTH and RICHARD (1978) differentiated diploids, tetraploids, and hexaploids from the *Hordeum murinum* aggregate electrophoretically, and indicated the putative parents of the polyploids. Usually a simple addition of the two parental patterns was found for both storage and enzyme proteins in the hybrid or amphiploid (WAINES and JOHNSON, 1975; HOUTS and HILLEBRAND, 1976: *Galeopsis tetrahit*). It is also possible to simulate a hybrid or amphiploid protein pattern by an in vitro mixture of the parental proteins (JOHNSON, 1975; EDMONDS and GLIDEWELL, 1977; BUSHUK and KERBER, 1978). Extra protein bands (DESBOROUGH and PELOQUIN, 1966; WAINES and JOHNSON, 1971) have been interpreted as an expression of parental silent genes.

Proteins of congeneric species are frequently compared electrophoretically (e.g., GRAY, 1975: *Tripsacum;* TALI, 1975: *Vicia;* CRAWFORD and JULIAN, 1976: *Chenopodium;* LUBIS et al., 1977: *Vigna;* SAHAI and RANA, 1977: *Phaseolus* and *Vigna;* YANG et al., 1977: *Larrea;* HEKMAT-SHOAR et al., 1978: *Suaeda;* LADIZINSKY, 1979a,b: *Lens, Trigonella*). The similarity index of such taxa is still sufficient high to enable such comparisons.

11. Conclusion

In spite of a high a priori significance for protein characters, they cannot yet fully withstand a posteriori testings. Therefore, all statements such as "protein comparisons enable exact declarations about the natural relationship" (REICHENBÄCHER et al., 1976) are rather uncritical. Nevertheless, contributions of protein characters can no longer be omitted from modern systematical work. Increasing knowledge of the physicochemical properties and new and more sensitive methods in protein chemistry are raising the significance of comparative protein research, and thus helping to express evolutionary relationships.

References

AITKEN, A.: Biochem. J. 149, 675-683 (1975); - Nature 263, 793-796 (1976). - AKAZAWA, T., SATO, K., SUGIYAMA, T.: Arch. Biochem. Biophys. 132, 255-261 (1969). - AKAZAWA, T., KONDO, H., SHIMAZUE, T., NISHIMURA, M., SUGIYAMA, T.: Biochemistry 11, 1298-1303 (1972). - ALLARD, R.W.: Genetics 79, 115-126 (1975). - ALMGÅRD, G., CLAPHAM, D.: Swed. J. Agric. Res. 5, 61-67 (1975). - AMBLER, R.P., BARTSCH, R.G.: Nature 253, 285-288 (1975). - ANDREW, P.W., ROGERS, L.J., BOULTER, D., HASLETT, B.G.: Eur. J. Biochem. 69, 243-248 (1976). - ANIOL, A.: Genet. Polon. 17, 113-121 (1976a); - ibid. 17, 123-131 (1976b); - ibid 17, 343-351 (1976c); ibid. 17, 333-342 (1976d). - ATASSI, M.Z., TARLOWSKI, D.P., PAULL, J.H.: Biochim. Biophys. Acta 221, 623-635 (1970). - AUTRAN, J.-C., BOURDET, A.: Ann. Amélior. Plantes 25, 277-301 (1975). - AVISE, J.C.: Syst. Zool. 23, 465-481 (1975). - AYALA, F.J.: Evol. Biol. 8, 1-78 (1975).

BABBEL, G.R., SELANDER, R.K.: Evolution 28, 619-630 (1974). - BAKER, T.S., EISENBERG, D., EISERLING, F.A., WEISSMAN, L.: J. Mol. Biol. 91, 391-399 (1975). - BAKER, T.S., EISENBERG, D., EISERLING, F.: Science 196, 293-295 (1977). - BARKER, R.D.J., DERBYSHIRE, E., YARWOOD, A., BOULTER, D.: Phytochemistry 15, 751-757 (1976). - BENSON, A.M., YASUNOBU, K.T.: J. Biol. Chem. 244, 955-963 (1969). - BLAIR, G.E., ELLIS, R.J.: Biochim. Biophys. Acta 319, 223-234 (1973). - BLIXT, S., PRZYBYLSKA, J., ZIMNIAK-PRZYBYLSKA, Z.: Gen. Pol. 21, 153-160 (1980). - BOLTON, E.T., BRITTEN, R.J., COWIE, D.B., ROBERTS, R.B., SZAFRANSKI, P., WARING, M.J.: Carnegie Inst. Year Book 64, 313 (1965). - BOOTH, T.A., RICHARDS, A.J.: Bot. J. Linn. Soc. 76, 115-125 (1978). - BÖRNER, T., JAHN, G., HAGEMANN, R.: Biochem. Physiol. Pflanzen 169, 179-181 (1976). - BOSBACH, K.: Ph. D. Thesis, Univ. Münster 1978. - BOTTOMLEY, W., WHITFELD, P.: Eur. J. Biochem. 93, 31-39 (1979). - BOUDET, A.M., BOUDET, A., BOUYSSOU, H.: Phytochemistry 16, 919-922 (1977). - BOULTER, D.: Syst. Zool. 22, 549-553 (1974a); - The use of amino acid sequence data in the classification of higher plants, 211-216. In: Chemistry in Botanical Classification (Nobel Symp. 25), eds. G. BENDZ, J. SANTESSON. New York-London: Academic Press 1974b; - Coll. Int. C.N.R.S. 371-377 (1976); - The evaluation of present results and future possibilities of the use of amino acid sequence data in phylogenetic studies with specific reference to plant proteins, 235-240. In: Chemosystematics: Principles and Practice, eds. F.A. BISBY, J.G. VAUGHAN, C.A. WRIGHT. London-New York: Academic Press 1980. - BOULTER, D., RAMSHAW, J.A.M.: Phytochemistry 11, 553-561 (1972). - BOULTER, D., LAYCOCK, M.V., RAMSHAW, J.A.M., THOMPSON, E.W.: Taxon 19, 561-564 (1970). - BOULTER, D., RAMSHAW, J.A.M., THOMPSON, E.W., RICHARDSON, M., BROWN, R.H.: Proc. R. Soc. London Ser. B 181, 441-455 (1972). - BOULTER, D., GLEAVES, J.T.,

HASLETT, B.G., PEACOCK, D., JENSEN, U.: Phytochemistry 17, 1585-1589 (1978). - BOULTER, D., PEACOCK, D., GUISE, A., GLEAVES, J.T., ESTABROOK, G.: Phytochemistry 18, 603-608 (1979). - BOWIEN, B., MAYER, F.: Eur. J. Biochem. 88, 97-107 (1978). - BROWN, A.H.D.: Theor. Appl. Genet. 52, 145-157 (1978). - BROWN, A.H.D., NEVO, E., ZOHARY, D., DAGAN, O.: Genetica 49, 97-108 (1978). - BROWN, J.W.S., MA, Y., BLISS, F.A., HALL, T.C.: Theor. Appl. Genet. 59, 83-88 (1981). - BROWN, R.H., ARMITAGE, T.L., MERRETT, M.J.: Plant Physiol. 58, 773-776 (1976). - BRUNNER, F., FAIRBROTHERS, D.E.: Serol. Mus. 53, 2-5 (1978); - Bull. Torrey Bot. Club 106, 97-103 (1979). - BURNOUF, T., BOURIQUET, R.: Theor. Appl. Genet. 58, 107-111 (1980). - BUSHUK, W., KERBER, E.R.: Can. J. Plant Sci. 58, 1019-1024 (1978). - BUSHUK, W., ZILLMAN, R.R.: Can. J. Plant Sci. 58, 505-515 (1978). - BÜTTNER, C., JENSEN, U.: Biochem. Syst. Ecol. 9 (1981).

CASEY, R.: Biochem. J. 177, 509-520 (1979). - CATSIMPOOLAS, N.: FEBS Lett. 4, 259-261 (1969). - CHAN, P.-H., WILDMAN, S.G.: Biochim. Biophys. Acta 277, 677-680 (1972). - CHEN, K., MEYER, V.G.: J. Hered. 70, 431-433 (1979). - CHEN, K., SAND, S.A.: Science 204, 179-180 (1979). - CHEN, K., WILDMAN, S.G.: Pl. Syst. Evol. 138, 89-113 (1981). - CHEN, K., GRAY, J.C., WILDMAN, S.G.: Science 190, 1304-1306 (1975a). - CHEN, K., KUNG, S.D., GRAY, J.C., WILDMAN, S.G.: Biochem. Genet. 13, 771-778 (1975b). - CHEN, K., JOHAL, S., WILDMAN, S.G.: Role of chloroplast and nuclear DNA genes during evolution of Fraction-I-protein, 3-11. In: Genetics and Biogenesis of Chloroplasts and Mitochondria, eds. T. BÜCHER, W. NEUPERT, W. SEBALD, S. WERNER. Amsterdam: North Holland/Elsevier 1976a. - CHEN, K., KUNG, S.D., GRAY, J.C., WILDMAN, S.G.: Plant Sci. Lett. 7, 429-434 (1976b). - CHUPOV, V.S.: Bot. Zh. (Leningrad) 63, 1579-1585 (1978). - CHUPOV, V.S., CUTJAVINA, N.G.: Bot. Zh. (Leningrad) 63, 473-493 (1978). - CLARKSON, R.B., HUANG, F.H., CECH, F.C., GINGERICH, L.A.: Biochem. Syst. Ecol. 2, 59-66 (1974). - CLEGG, M.T., ALLARD, R.W.: Proc. Nat. Acad. Sci. 69, 1820-1824 (1972). - COEN, D.M., BEDBROOK, J.R., BOGORAD, L., RICH, A.: Proc. Nat. Acad. Sci. 74, 5487-5491 (1977). - COMAS, C.I., HUNZIKER, J.H., CRISCI, J.V.: Biochem. Syst. Ecol. 7, 303-308 (1979). - CRAWFORD, D.J.: Biochem. Syst. Ecol. 4, 169-172 (1976); - Bull. Torrey Bot. Club 106, 257-261 (1979). - CRAWFORD, D.J., JULIAN, E.A.: Am. J. Bot. 63, 302-308 (1976). - CRAWFORD, D.J., WILSON, H.D.: Syst. Bot. 2, 180-190 (1977); - Am. J. Bot. 66, 237-244 (1979). - CRIDDLE, R.S., DAU, B., KLEINKOPF, G.E., HUFFAKER, R.C.: Biochem. Biophys. Res. Commun. 41, 621-627 (1970). - CRISTOFOLINI, G., CHIAPELLA, L.F.: Taxon 26, 43-56 (1977). - CRONQUIST, A.: Brittonia 28, 1-27 (1976); - Chemistry in plant taxonomy: an assessment of where we stand, 1-27. In: Chemosystematics: Principles and Practice, eds. F.A. BISBY, J.G. VAUGHAN, C.A. WRIGHT. London-New York: Academic Press 1980. - CROY, R.R.D., DERBYSHIRE, E., KRISHNA, T.G., BOULTER, D.: New Phytol. 83, 29-35 (1979). - CUBADDA, R., BOZZINI, A., QUATTRUCCI, E.: Theor. Appl. Genet. 45, 290-293 (1975). - CZYPRYN, M., TOCZKO, K.: Acta Soc. Bot. Polon. 43, 491-498 (1974).

DANIELSSON, C.E.: Biochem. J. 44, 387-400 (1949). - DARMENCY, H., GASQUEZ, J., MOSSÉ, J.: C. R. Acad. Sci. Paris 290, 435-438 (1980). - DAVEY, R.A., DUDMAN, W.F.: Aust. J. Plant Physiol. 6, 435-447 (1979). - DAYHOFF, M.O.: Atlas of protein sequence and structure, 5. ed. Silver Spring, Maryland: National Biomedical Res. Found. 1972. - De JONG, D.W.: Am. J. Bot. 60, 846-852 (1973). - DERBYSHIRE, E., WRIGHT, D.J., BOULTER, D.: Phytochemistry 15, 3-24 (1976). - DESBOROUGH, S., PELOQUIN, S.J.: Phytochemistry 5, 727-733 (1966). - DHALIWAL, H.S.: Theor. Appl. Genet. 51, 71-79 (1977). - DOLL, H., BROWN, A.H.D.: Can. J. Genet. Cytol. 21, 391-404 (1979). - DUDMAN, W.F., MILLERD, A.: Biochem. Syst. Ecol. 3, 25-33 (1975). - DUTTON, J.E., ROGERS, L.J., HASLETT, B.G., TAKRURI, I.T.H., GLEAVES, J.T., BOULTER, D.: J. Exp. Bot. 31, 379-391 (1980).

EDMONDS, J.M., GLIDEWELL, S.M.: Plant Syst. Evol. 127, 277-291 (1977). - ELLIS, R.J.: Trends Biochem. Sci. 241-244 (1979). - EL TINAY, A.H., KARAMALLA, K.A., EL AMIN, H.M., SHIGIDI, M.T.A., ISHAG, K.E.A.: J. Exp. Bot. 30, 607-615 (1979). - ESEN, A., SCORA, R.W.: Am. J. Bot. 64, 305-309 (1977). - ESTABROOK, G.F., LANDRUM, L.: Taxon 24, 609-613 (1975).

FAIRBROTHERS, D.E.: Taxon 29, 412-416 (1980). - FITCH, W.M., MARKOWITZ, E.: Biochem. Genet. 4, 579-593 (1970). - FLÜCK, R.A., JAFFE, M.J.: Phytochemistry 13, 2475-2480 (1974). - FREDRICK, J.F.: Plant Cell Physiol. 17, 317-322 (1976); - Phytochemistry 16, 1571-1573 (1977); - Phytochemistry 19, 539-542 (1980). - FROHNE, D., JOHN, J.: Biochem. Syst. Ecol. 6, 315-322 (1978). - FROHNE, D., MORITZ, O., JENSEN, U.: Flora 150, 332-352 (1961). - FUKUYAMA, K., HASE, T., MATSUMOTO, S., TSUKIHARA, T., KATSUBE, Y., TANAKA, N., KAKUDO, M., WADA, K., MATSUBARA, H.: Nature 286, 522-523 (1980).

GARCIA, A., TSUNEWAKI, K.: Japan. J. Genet. 52, 379-386 (1977). - GASQUEZ, J., COMPOINT, J.P.: Ann. Amélior. Plantes 26, 345-355 (1976). - GATEHOUSE, J.A., CROY, R.R.D., BOULTER, D.: Biochem. J. 185, 497-503 (1980). - GATENBY, A.A.: Biochim. Biophys. Acta 534, 169-172 (1978). - GATENBY, A.A., COCKING, E.C.: Plant Sci. Lett. 10, 97-101 (1977); - ibid. 13, 171-176 (1978a); - ibid. 12, 177-181 (1978b); - ibid 12, 299-303 (1978c). - GIBBONS, G.C., STRØBAEK, S., HASLETT, B., BOULTER, D.: Experientia 31, 1040-1041 (1975). - GIETL, C., ZIEGLER, H.: Biochem. Physiol. Pflanz. 175, 50-57 (1980). - GILROY, J., WRIGHT, D.J., BOULTER, D.: Phytochemistry 18, 315-316 (1979). - GLAZER, A.N., COHEN-BAZIRE, G., STANIER, R.Y.: Proc. Natl. Acad. Sci. USA 68, 3005-3008 (1971). - GLAZER, A.N., APELL, G.S., HIXSON, C.S., BRYANT, D.A., RIMON, S., BROWN, D.M.: Proc. Natl. Acad. Sci. USA 73, 428-431 (1976). - GOTTLIEB, L.D.: BioScience 21, 939-944 (1971); - Evolution 29, 213-225 (1975); - Biochemical Consequences of Speciation in Plants, 123-140. In: Molecular Evolution, ed. F.J. AYALA. 1976; - Genetics 86, 289-307 (1977a); - Ann. Miss. Bot. Gard. 64, 161-180 (1977b). - GOTTLIEB, L.D., WEEDEN, N.F.: Evolution 33, 1024-1039 (1979). - GÖTZ, H., ARNOLD, C.-G.: Biochem. Physiol. Pflanz. 175, 1-8 (1980a); - Planta 149, 19-26 (1980b). - GRASSO, S., SHEPHERD, R.J.: Phytopathology 68, 199-205 (1978). - GRAY, J.C.: Plant Syst. Evol. 128, 53-69 (1977); - Plant Syst. Evol. 129, 177-183 (1978); - Fraction I protein and plant phylogeny, 167-193. In: Chemosystematics: Principles and Practice, eds. F.A. BISBY, J.G. VAUGHAN, C.A. WRIGHT. London-New York: Academic Press 1980. - GRAY, J.C., KEKWICK, R.G.O.: Eur. J. Biochem. 44, 481-489 (1974). - GRAY, J.C., WILDMAN, S.G.: Plant Sci. Lett. 6, 91-96 (1976). - GRAY, J.R.: Biochem. Syst. Ecol. 2, 211-213 (1975). - GROSS, G.G., MANSELL, R.L., ZENK, M.H.: Biochem. Physiol. Pflanz. 168, 41-51 (1957). - GRUND, C., JENSEN, U.: Plant Syst. Evol. 137, 1-22 (1981). - GRUND, C., GILROY, J., GLEAVES, J.T., JENSEN, U., BOULTER, D.: Phytochemistry 20, 1559-1565 (1981). - GUTIERREZ, M., EDWARDS, G.E., BROWN, W.V.: Biochem. Syst. Ecol. 4, 47-49 (1976).

HAGEN, G., RUBENSTEIN, I.: Plant Sci. Lett. 19, 217-223 (1980). - HALL, D.O., CAMMACK, R., RAO, K.K.: Pure Appl. Chem. 34, 553-577 (1973). - HALL, T.C., McLEESTER, R.C., BLISS, F.A.: Plant Physiol. 59, 1122-1124 (1977). - HAMMOND, D.: Serol. Mus. 14, 1-3 (1955). - HANKINS, C.N., KINDINGER, J.I., SHANNON, L.M.: Plant Physiol. 66, 375-378 (1980). - HARA, I., OHMIYA, M., MATSUBARA, H.: Plant Cell Physiol. 19, 237-243 (1978). - HASE, T., WADA, K., MATSUBARA, H.: J. Biochem. 78, 605-610 (1975); - ibid. 82, 267-276 (1977a); - ibid. 82, 277-286 (1977b). - HASE, T., WAKABAYASHI, S., WADA, K., MATSUBARA, H.: J. Biochem. 83, 761-770 (1978a). - HASE, T., WAKABAYASHI, S., WADA, K., MATSUBARA, H., JÜTTNER, F., RAO, K.K., FRY, I., HALL, D.O.: FEBS Lett. 96, 41-44 (1978b). - HASE, T., MATSUBARA, H., BEN-AMOTZ, A., RAO, K.K., HALL, D.O.: Phytochemistry 19, 2065-2070 (1980). - HASLETT, B.G., BAILEY,

C.J., RAMSHAW, J.A.M., SCAWEN, M.D., BOULTER, D.: Biochem. Soc. Transact. 2, 1329-1331 (1974). - HASLETT, B.G., YARWOOD, A., EVANS, I.M., BOULTER, D.: Biochim. Biophys. Acta 420, 122-132 (1976). - HASLETT, B.G., GLEAVES, T., BOULTER, D.: Phytochemistry 16, 363-365 (1977). - HASLETT, B.G., EVANS, I.M., BOULTER, D.: Phytochemistry 17, 735-739 (1978). - HAYWARD, M.D., McADAM, N.J.: Z. Pflanzenzüchtung 79, 59-68 (1977). - HEDRICK, P.W., GINEVAN, M.E., EWING, E.P.: Annu. Rev. Ecol. Syst. 7, 1-32 (1976). - HEKMAT-SHOAR, H., BILLARD, J.-P., BOUCAUD, J.: C. R. Acad. Sci. Paris 286, 53-56 (1978). - HIGGINS, T.J.V., SPENCER, D.: Plant Physiol. 60, 655-661 (1977). - HILL, R.J.: Bull. Torrey Bot. Club 104, 93-101 (1977). - HOUTS, K.P., HILLEBRAND, G.R.: Am. J. Bot. 63, 156-165 (1976). - HOWLETT, B.J., KNOX, R.B., HESLOP-HARRISON, J.: J. Cell Sci. 13, 603-619 (1973). - HUBER, S.C., HALL, T.C., EDWARDS, G.E.: Plant Physiol. 57, 730-733 (1976). - HURKA, H.: Enzymes as a taxonomic tool: a botanist's view, 103-121. In: Chemosystematics: Principles and Practice, eds. F.A. BISBY, J.G. VAUGHAN, C.A. WRIGHT. London-New York: Academic Press 1980. - HURKMAN, W.J., BEEVERS, L.: Planta 150, 82-88 (1980). - HURRELL, J.G.R., THULBORN, K.R., BROUGHTON, W.J., DILWORTH, M.J., LEACH, S.J.: FEBS Lett. 84, 244-246 (1977).

IBRAHIM, R.K.: Z. Pflanzenphysiol. 85, 253-262 (1977). - IMAM, A.G., ALLARD, R.W.: Genetics 51, 49-62 (1965). - IWAI, S., TANABE, Y., KAWASHIMA, N.: Biochem. Biophys. Res. Commun. 73, 993-996 (1976).

JAASKA, V.: Eesti NSV Teaduste Akad. Toimet. 25, 132-145 (1976); - Theor. Appl. Genet. 53, 209-217 (1978); - Theor. Appl. Genet. 56, 273-284 (1980). - JAASKA, V., JAASKA, V.: Biochem. Physiol. Pflanz. 175, 570-577 (1980). - JAIN, S.K.: Annu. Rev. Ecol. Syst. 7, 469-495 (1976). - JAIN, S.K., MARSHALL, D.R.: Am. Natur. 101, 19-33 (1967). - JENSEN, U.: Bot. Jahrb. 88, 204-268 (1968). - JENSEN, U., BÜTTNER, C.: Taxon 30, 404-419 (1981). - JENSEN, U., PENNER, R.: Biochem. Syst. Ecol. 8, 161-170 (1980). - JOEL, D.M., MARBACH, I., MAYER, A.M.: Phytochemistry 17, 796-797 (1978). - JOHN, J.: Biochem. Syst. Ecol. 6, 323-327 (1978). - JOHN, J., KOLBE, K.-P.: Biochem. Syst. Ecol. 8, 241-248 (1980). - JOHNSON, B.L.: Can. J. Genet. Cytol. 17, 21-39 (1975). - JOHNSON, G.B.: Annu. Rev. Ecol. Syst. 8, 309-328 (1977). - JOHNSON, R.G., FAIRBROTHERS, D.E.: Biochem. Syst. Ecol. 3, 205-208 (1975). - JULIANO, B.O., BOULTER, D.: Phytochemistry 15, 1601-1606 (1976).

KATO, M., TOKUMASU, S.: Euphytica 28, 339-349 (1979). - KAWASHIMA, N.: Plant Cell Physiol. 10, 31-40 (1969). - KAWASHIMA, N., TANABE, Y.: Biochem. Syst. Ecol. 2, 193-199 (1975). - KAWASHIMA, N., WILDMAN, S.G.: Biochim. Biophys. Acta 229, 749-760 (1971); - Biochim. Biophys. Acta 262, 42-49 (1972). - KAWASHIMA, N., KWOK, S.-Y., WILDMAN, S.G.: Biochim. Biophys. Acta 236, 578-586 (1971). - KAWASHIMA, N., TANABE, Y., IWAI, S.: Biochim. Biophys. Acta 371, 417-431 (1974); - Biochim. Biophys. Acta 427, 70-77 (1976). - KELLEY, W.A., ADAMS, R.P.: Am. J. Bot. 64, 1092-1096 (1977). - KELLY, J., AMBLER, R.P.: Biochem. J. 143, 681-690 (1974). - KERESZTES-NAGY, S., PERINI, F., MARGOLIASH, E.: J. Biol. Chem. 244, 981-995 (1969). - KHAVKIN, E.E., MISHARIN, S.I., MARKOV, Y.Y., PESHKOVA, A.A.: Planta 143, 11-20 (1978). - KHAVKIN, E.E., MISHARIN, S.I., MOZGOVA, E.A.: Z. Pflanzenphysiol. 92, 177-181 (1979). - KIM, S.I., MOSSÉ, J.: Can. J. Genet. Cytol. 21, 309-318 (1979). - KIM, S.I., CHARBONNIER, L., MOSSÉ, J.: Biochim. Biophys. Acta 537, 22-30 (1978). - KIM, S.I., SAUR, L., MOSSÉ, J.: Theor. Appl. Genet. 54, 49-54 (1979). - KIRCHANSKI, S.J., PARK, R.B.: Plant Physiol. 58, 345-349 (1976). - KITAMURA, K., TAKAGI, T., SHIBASAKI, K.: Agric. Biol. Chem. 40, 1837-1844 (1976). - KLOZOVA, E., TURKOVA, V., PITTEROVA, K., HADACOVA, V.: Biol. Plant (Praha) 23, 9-15 (1981). - KNOX, R.B., HESLOP-HARRISON, J., HESLOP-HARRISON, Y.: Pollenwall proteins: localization and characterization of gametophytic and sporophytic fractions, 177-187.

In: Biol. J. Linn. Soc. 7, Suppl. 1: The Biology of the Male Gamete, eds. J.G. DUCKETT, P.A. RACEY. New York: Academic Press 1975. - KOLBE, K.-P.: Ph. D. Thesis, University of Kiel 1973; - Bot. Jahrb. Syst. 99, 468-489 (1978). - KOLBE, K.-P., JOHN, J.: Bot. Jahrb. Syst. 101, 3-15 (1979); - Biochem. Syst. Ecol. 8, 249-256 (1980). - KONAREV, A.V.: Theor. Appl. Genet. 59, 117-121 (1981). - KRATTINGER, K., RAST, D., KARESCH, H.: Biochem. Syst. Ecol. 7, 125-128 (1979). - KUBITZKI, K.: Systematics and evolution of seed plants, 192-238. In: Progress in Botany, Vol. 39, eds. H. ELLENBERG, K. ESSER, H. MERXMÜLLER, E. SCHNEPF, H. ZIEGLER. Berlin-Heidelberg-New York: Springer 1977. - KÜMMEL, H., KESSLER, E.: Arch. Microbiol. 126, 15-19 (1980). - KUNG, S.-D.: Science 191, 429-434 (1976). - KUNG, S.-D., LEE, C.-I., WOOD, D.D., MOSCARELLO, M.A.: Plant Physiol. 60, 89-94 (1977). - KWANYUEN, P., WILDMAN, S.G.: Biochem. Biophys. Acta 405, 167-174 (1975). - KWOK, S.Y., WILDMAN, S.G.: J. Mol. Evol. 3, 103-108 (1974).

LADIZINSKY, G.: Bot. Gaz. 140, 449-451 (1979a); - Plant Syst. Evol. 133, 87-94 (1979b). - LADIZINSKY, G., HAMEL, A.: Euphytica 29, 313-317 (1980). - LADIZINSKY, G., HYMOWITZ, T.: Theor. Appl. Genet. 54, 145-151 (1979). - LAURELL, C.-B.: Scand. J. Clin. Lab. Invest. 29, Suppl. 124, 21-37 (1972). - LAYCOCK, M.V.: Can. J. Biochem. 50, 1311-1325 (1972); - Biochem. J. 149, 271-279 (1975). - LEE, Y.S.: Syst. Bot. 2, 169-179 (1977). - LEE, Y.S., DICKINSON, D.B.: Am. J. Bot. 66, 245-252 (1979). - LEE, Y.S., FAIRBROTHERS, D.E.: Taxon 27, 159-185 (1978). - LEHTOVAARA, P., LAPPALAINEN, A., ELLFOLK, N.: Biochim. Biophys. Acta 623, 98-106 (1980). - LE QUESNE, W.J.: Syst. Zool. 18, 201-205 (1969). - LESTER, R.N.: The use of protein characters in the taxonomy of *Solanum* and other Solanaceae, 285-304. In: The Biology and Taxonomy of the Solanaceae, eds. J.G. HAWKES, R.N. LESTER, A.D. SKELDING. Linn. Soc. Symp. Ser. 7, 1979. - LEVY, M., LEVIN, D.A.: Genetics 79, 493-512 (1975). - LIEDGENS, W., GRÜTZMANN, R., SCHNEIDER, H.A.W.: Naturforscher 35c, 958-962 (1980). - LINK, G., COEN, D.M., BOGORAD, L.: Cell 15, 725-731 (1978). - LORD, J.M., CODD, G.A., STEWART, W.D.P.: Plant Sci. Lett. 4, 377-383 (1975). - LUBIS, I., LUBIS, S.H.A., SASTRAPRADJA, D., SASTRAPRADJA, S.: Ann. Bogor. 6, 125-132 (1977). - LUTHE, D.S., PETERSON, D.M.: Plant Physiol. 59, 836-841 (1977). - LYNN, K.R., YAGUCHI, M., ROY, C.: Biochim. Biophys. Acta 624, 579-580 (1980).

MANTEUFFEL, R., SCHOLZ, G.: Biochem. Physiol. Pflanz. 168, 277-285 (1975). - MARTIN, P.G.: Aust. J. Plant Physiol. 6, 401-408 (1979). - MATSUBARA, H., SASAKI, R.M., CHAIN, R.K.: Biochemistry 57, 439-445 (1967). - MAZZA, G., WELINDER, K.G.: Eur. J. Biochem. 108, 481-489 (1980). - McCOLL, R., BERNS, D.S., GIBBONS, O.: Arch. Biochem. Biophys. 177, 265-275 (1976). - McFADDEN, B.A., TABITA, F.R.: Biosystems 6, 93-112 (1974). - McLEOD, M.J., ESHBAUGH, W.H., GUTTMAN, S.I.: Bull. Torrey Bot. Club 106, 326-333 (1979). - MERXMÜLLER, H., BUTTLER, K.P.: Mitt. Bot. Staatssamml. München 12, 91-104 (1975). - METZGER, H., SHAPIRO, M.B., MOSIMANN, J.E., VINTON, J.E.: Nature 219, 1166-1168 (1968). - MILLERD, A., SIMON, M., STERN, H.: Plant Physiol. 48, 419-425 (1971). - MILNE, P.R., WELLS, J.R.E., AMBLER, R.P.: Biochem. J. 143, 691-701 (1974). - MISSET, M.-T.: Saussurea 8, 1-18 (1977). - MITRA, R., BHATIA, C.R.: Genet. Res. Cambr. 18, 57-69 (1971). - MITRA, R., JAGANNATH, D.R., BHATIA, C.R.: Phytochemistry 9, 1843-1850 (1970). - MORAN, G.F., MARSHALL, D.R.: Aust. J. Biol. Sci. 31, 283-291 (1978). - MOSSÉ, J.: Alcohol-soluble proteins of cereal grains, 1663-1669. In: Symp. Proc.: Improving the nutritional Value of Cereal Grains by the use of Gene Mutations and Selective Breeding (50th Annual Meeting of the Fed. Am. Soc. Exp. Biol., Atlantic City, N.J.) 1966; - Physiol. Vég. 11, 361-384 (1973). - MURPHY, T.M.: Phytochemistry 17, 439-443 (1978).

NAKAI, Y.: Seiken Ziho 24, 45-73 (1973); - Japan. J. Genet. 54, 175-189 (1979). - NAKAI, Y., TSUNEWAKI, K.: Japan. J. Genet. 46, 321-336

(1971). - NEI, M.: Am. Natur. 105, 385-398 (1971); - The theory and estimation of genetic distances, 45-54. In: Genetic Structure of Populations, ed. N.E. MORTON. Honolulu: Univ. Hawaii Press 1974. - NEVO, E., ZOHARY, D., BROWN, A.H.D., HABER, M.: Evolution 33, 815-833 (1979). - NISHIMURA, M., AKAZAWA, T.: Biochemistry 13, 2277-2281 (1974). - NISHIMURA, M., TAKABE, T., SUGIYAMA, T., AKAZAWA, T.: J. Biochem. 74, 945-954 (1973). - NOWACKI, E., JAWORSKI, A.: Genet. Polon. 19, 153-163 (1978).

O'CONNELL, P.B.H., BRADY, C.J.: Proc. Aust. Biochem. Soc. 12, 25 (1979). - ORF, J.H., HYMOWITZ, T.: Genet. Newslett. 3, 27-28 (1976). - OSMOND, C.B., AKAZAWA, T., BEEVERS, H.: Plant Physiol. 55, 226-230 (1975).

PAHLICH, E., OTT, W., SCHAD, B.: J. Exp. Bot. 31, 419-423 (1980). - PAYNE, R.C., FAIRBROTHERS, D.E.: Am. J. Bot. 60, 182-189 (1973). - PAYNE, R.C., OLIVEIRA, A.R., FAIRBROTHERS, D.E.: Biochem. Syst. 1, 59-61 (1973). - PEACOCK, D., BOULTER, D.: J. Mol. Biol. 95, 513-527 (1975). - PETTIGREW, G.W.: Biochem. J. 139, 449-459 (1974); - Structural features of Protozoan cytochromes. In: Biochemistry and Physiology of Protozoa, Vol. 1, eds. M. LEVANDOWSKY, S.H. HUTNER. New York-London: Academic Press 1979. - PETERSEN, F.P., FAIRBROTHERS, D.E.: Syst. Bot. 4, 230-241 (1979). - PETERSEN, R.L., FAIRBROTHERS, D.E.: Am. Midl. Nat. 85, 437-457 (1971). - PHELAN, J.R., VAUGHAN, J.G.: Biochem. Syst. Ecol. 4, 173-178 (1976). - PHILLIPS, M.A., BROWN, A.H.D.: Aust. J. Biol. Sci. 30, 337-344 (1977). - POULSEN, C., STRØBAEK, S., HASLETT, B.G.: Studies on the primary structure of the small subunit of ribulose-1,5-diphosphate carboxylase, 17-24. In: Genetics and Biogenesis of Chloroplasts and Mitochondria, eds. T. BÜCHER, W. NEUPERT, W. SEBALD, S. WERNER. Amsterdam: Elsevier 1976. - POULSEN, C., MARTIN, B., SVENDSEN, I.: Carlsberg Res. Commun. 44, 191-199 (1979). - PRUS-LOWACKI, W.: Genet. Polon. 16, 37-46 (1975). - PRUS-LOWACKI, W., SZWEYKOWSKI, J.: Bull. Soc. Amis Sci. Lett. Poznan, Sér. D. 17, 15-27 (1977). - PRZYBYLSKA, J., BLIXT, S., HURICH, J., ZIMNIAK-PRZYBYLSKA, Z.: Genet. Polon. 18, 27-38 (1977). - PRZYBYLSKA, J., HURICH, J., ZIMNIAK-PRZYBYLSKA, Z.: Genet. Polon. 20, 517-528 (1979).

RABINOWITZ, H., REISFELD, A., SAGHER, D., EDELMAN, M.: Plant Physiol. 56, 345-350 (1975). - RAMSHAW, J.A.M., RICHARDSON, D.L., MEATYARD, B.T., BROWN, R.H., RICHARDSON, M., THOMPSON, E.W., BOULTER, D.: New Phytol. 71, 773-779 (1972). - RAMSHAW, J.A.M., SCAWEN, M.D., BAILEY, C.J., BOULTER, D.: Biochem. J. 139, 583-592 (1974a). - RAMSHAW, J.A.M., SCAWEN, M.D., BOULTER, D.: Biochem. J. 141, 835-843 (1974b). - RAMSHAW, J.A.M., PEACOCK, D., MEATYARD, B.T., BOULTER, D.: Phytochemistry 13, 2783-2789 (1974c). - RAO, K.K., MATSUBARA, H.: Biochem. Biophys. Res. Commun. 38, 500-506 (1970). - REICHELT, R., SCHWENKE, K.-D., KÖNIG, T., PÄHTZ, W., WANGERMANN, G.: Biochem. Physiol. Pflanz. 175, 653-663 (1980). - REICHENBÄCHER, D., RICHTER, J., SPAAR, D.: Arch. Züchtungsforsch. 6, 177-184 (1976); - Biochem. Physiol. Pflanz. 171, 299-306 (1977). - RICK, C.M., FOBES, J.F.: Bull. Torrey Bot. Club 102, 376-384 (1975). - RICKEMAN, V.S., DESBOROUGH, S.L.: Theor. Appl. Genet. 52, 217-220 (1978). - ROBINSON, P.J., MEGARRITY, R.G.: Aust. J. Agric. Res. 26, 467-479 (1975). - RUDNICKA, A.: Genet. Polon. 16, 261-269 (1975).

SABNIS, D.D., HART, J.W.: Planta 130, 211-218 (1976). - SAHAI, S., RANA, R.S.: New Phytol. 79, 527-534 (1977). - SCAWEN, M.D., BOULTER, D.: Biochem. J. 143, 257-264 (1974). - SCAWEN, M.D., RAMSHAW, J.A.M., BROWN, R.H., BOULTER, D.: Eur. J. Biochem. 44, 299-303 (1974). - SCAWEN, M.D., RAMSHAW, J.A.M., BOULTER, D.: Biochem. J. 147, 343-349 (1975). - SCHIAVO, F.L., MELA, L., RONCHI, V.N., TERZI, M.: Plant Sci. Lett. 18, 45-55 (1980). - SCHLESIER, B., SCHOLZ, G.: Kulturpflanze 23, 157-166 (1975). - SCHNEIDER, H.A.W., LIEDGENS, W.: Z. Naturforsch. 36c, 44-50

(1981). - SCHOLZ, G., RICHTER, J., MANTEUFFEL, R.: Biochem. Physiol. Pflanz. 166, 163-172 (1974). - SCHULZ, G.E.: Angew. Chem. 93, 143-151 (1981). - SCHWARTZ, R.M., DAYHOFF, M.O.: Science 199, 395-403 (1978). - SHECHTER, Y.: Bull. Torrey Bot. Club 102, 334-339 (1975). - SHECHTER, Y., De WET, J.M.J.: Am. J. Bot. 62, 254-261 (1975). - SHEWRY, P.R., PRATT, H.M., CHARLTON, M.J., MIFLIN, B.J.: J. Exp. Bot. 28, 597-606 (1977). - SHEWRY, P.R., PRATT, H.M., MIFLIN, B.J.: J. Sci. Fd. Agric. 29, 587-596 (1978). - SHIN, M., YOKOYAMA, Z., ABE, A., FUKUSAWA, H.: J. Biochem. 85, 1075-1081 (1979). - SIMON, J.-P.: Aliso 9, 483-497 (1979a); - Bot. Gaz. 140, 452-460 (1979b). - SMITH, J.S.C., LESTER, R.N.: Econ. Bot. 34, 201-218 (1980). - SNEATH, P.H.A., SACKIN, M.J., AMBLER, R.P.: Syst. Zool. 24, 311-332 (1975). - SOKAL, R.R., SNEATH, P.H.: Principles of Numerical Taxonomy. 359 pp. San Francisco: Freeman 1963. - STEGEMANN, H.: SDS-PAGE, merits and limits (Review), 306-328. In: Electrokinetic Separation Methods, eds. RIGHETTI, VAN OSS, VANDERHOFF. Amsterdam: Elsevier 1979. - STEGEMANN, H., LOESCHCKE, V.: Mitt. Biol. Bundesanst. Land-Forstwirtsch. Berlin-Dahlem 168, 1-214 (1976); - Potato Res. 20, 101-110 (1977). - STRINGER, C.D., HARTMAN, F.C.: Biochem. Biophys. Res. Commun. 80, 1043-1048 (1978). - STRØBAEK, S., GIBBONS, G.C., HASLETT, B., BOULTER, D., WILDMAN, S.G.: Carlsberg Res. Commun. 41, 335-343 (1976). - SUGENO, K., MATSUBARA, H.: J. Biol. Chem. 244, 2979-2989 (1969). - SUGIYAMA, T., MATSUMOTO, C., AKAZAWA, T., MIYACHI, S.: Arch. Biochem. Biophys. 129, 597-602 (1969).

TABITA, F.R., McFADDEN, B.A.: J. Biol. Chem. 249, 3459-3464 (1974). - TABITA, F.R., STEVENS, S.E., QUIJANO, R.: Biochem. Biophys. Res. Commun. 61, 45-52 (1974). - TAKABE, T., AKAZAWA, T.: Plant Cell Physiol. 16, 1049-1060 (1975a); - Biochemistry 14, 46-50 (1975b). - TALI, V.: Eesti NSV Teaduste Akad. Toim. 24, 35-41 (1975). - TANAKA, M., HANIU, M., ZEITLIN, S., YASUNOBU, K.T., EVANS, M.C.W., RAO, K.K., HALL, D.O.: Biochem. Biophys. Res. Commun. 64, 399-407 (1975). - TANAKA, M., HANIU, M., YASUNOBU, K.T., RAO, K.K., HALL, D.O.: Biochem. Biophys. Res. Commun. 69, 759-765 (1976). - TEL-OR, E., CAMMACK, R., RAO, K.K., ROGERS, L.J., STEWART, W.D.P., HALL, D.O.: Biochim. Biophys. Acta 490, 120-131 (1977). - Te NIJENHUIS, B.: Euphytica 20, 498-507 (1971). - THOMAS, D.L., NEUCERE, N.J.: Am. J. Bot. 61, 457-463 (1974). - THOMPSON, E.W., NOTTON, B.A., RICHARDSON, M., BOULTER, D.: Biochem. J. 124, 787-791 (1971). - TIMKO, M.P., VASCONCELOS, A.C., FAIRBROTHERS, D.E.: Biochem. Gen. 18, 171-183 (1980). - TIMKOVICH, R., DICKERSON, R.E.: J. Mol. Biol. 79, 39-56 (1973). - TOMBS, M.P., LOWE, M.: Biochem. J. 105, 181-187 (1967). - TORRES, A.M., SOOST, R.K., DIEDENHOFEN, U.: Am. J. Bot. 65, 869-881 (1978a). - TORRES, A.M., DIEDENHOFEN, U., BERGH, B.O., KNIGHT, R.J.: Am. J. Bot. 65, 134-139 (1978b). - TSAY, R.C., TAYLOR, I.E.P.: Can. J. Bot. 56, 80-90 (1978). - TUCKER, A.O., FAIRBROTHERS, D.E.: Phytochemistry 9, 1399-1403 (1970).

UCHIMIYA, H., WILDMAN, S.G.: J. Heredity 69, 299-303 (1978). - UCHIMIYA, H., CHEN, K., WILDMAN, S.G.: Plant Sci. Lett. 17, 63-66 (1979a); - Biochem. Genet. 17, 333-341 (1979b). - UTSUMI, S., MORI, T.: Biochim. Biophys. Acta 621, 179-189 (1980).

VAUGHAN, J.G., DENFORD, K.E.: J. Exp. Bot. 19, 724-732 (1968). - VEDEL, F., QUETIER, F., DOSBA, F., DOUSSINAULT, G.: Plant Sci. Lett. 13, 97-102 (1978). - VILLAMIL, C.B., FAIRBROTHERS, D.E.: Biochem. Syst. Ecol. 2, 15-20 (1974). - VITALE, A., SOAVE, C., GALANTE, E.: Plant Sci. Lett. 18, 57-64 (1980).

WADA, K., HASE, T., MATSUBARA, H.: J. Biochem. 78, 637-639 (1975). - WAINES, J.G., JOHNSON, B.L.: Wheat Inf. Serv. 32, 22-24 (1971); - Can. J. Genet. Cytol. 17, 1-8 (1975). - WAKABAYASHI, S., HASE, T., WADA, K., MATSUBARA, H., SUZUKI, K., TAKAICHI, S.: J. Biochem. 83, 1305-1319

(1978). - WAKABAYASHI, S., HASE, T., WADA, K., MATSUBARA, H., SUZUKI, K.: J. Biochem. 87, 227-236 (1980). - WALLACE, D.G., BOULTER, D.: Phytochemistry 15, 137-141 (1976a); - Immunochemistry 13, 831-835 (1976b). - WALLACE, D.G., KING, M.-C., WILSON, A.C.: Syst. Zool. 22, 1-13 (1973). - WEEDEN, N.F., GOTTLIEB, L.D.: Biochem. Genet. 17, 287-296 (1979). - WILDMAN, S.G.: Arch. Biochem. Biophys. 196, 598-610 (1979); - Molecular aspects of wheat evolution: Rubisco composition, 61-74. In: Wheat Science – Today and Tomorrow, eds. L.T. EVANS, W.J., PEACOCK. Cambridge: University Press 1981a; Manuskript 1981b. - WILDMAN, S.G., CHEN, K., GRAY, J.C., KUNG, S.D., KWANYUEN, P., SAKANO, K.: Evolution of ferredoxin and fraction I protein in the genus *Nicotiana*, 309-329. In: Genetics and Biogenesis of Mitochondria and Chloroplasts, eds. P.S. PERLMAN, C.W. BIRKY, T.J. BYER. Columbus: Ohio State Univ. Press 1975. - WILLIAMS, V.P., FREIDENREICH, P., GLAZER, A.N.: Biochem. Biophys. Res. Commun. 59, 462-466 (1974). - WOLFF, G.: Theor. Appl. Genet. 57, 225-232 (1980). - WONG, A.K.C., LIU, T.S., WANG, C.C.: J. Mol. Biol. 102, 287-295 (1976). - WRIGHT, D.J., BOULTER, D.: Biochem. J. 141, 413-418 (1974). - WRIGLEY, C.W., SHEPHERD, K.W.: Ann. N.Y. Acad. Sci. 209, 154-162 (1973); - Aust. J. Agric. Res. 14, 796-804 (1974).

YADAVA, J.S., CHOWDHURY, J.B., KAKAR, S.N., NAINAWATEE, H.S.: Theor. Appl. Genet. 54, 89-91 (1979). - YAMAMOTO, K., PLITMANN, U.: Japan. J. Genet. 55, 151-164 (1980). - YANG, T.W., HUNZIKER, J.H., POGGIO, L., NARANJO, C.A.: Plant Syst. Evol. 126, 331-346 (1977).

Professor Dr. UWE JENSEN
Lehrstuhl für Pflanzenökologie
und Systematik
Universität Bayreuth
D 8580 Bayreuth

Subject Index

Encyclopedia of Plant Physiology

New Series · Editors: A. Pirson, M. H. Zimmermann

Volume 7

Physiology of Movements

Editors: W. Haupt, M. E. Feinleib
With contributions by numerous experts
1979. 185 figures, 19 tables. XVII, 731 pages
ISBN 3-540-08776-1

Plant movements cover a wide field of plant physiology. This volume emphasizes those fields where substantial progress in understanding has been made, or where major new aspects are evoring. It is devoted particularly to the mechanisms of perception, transduction and responce common to completely diverse types of movement. On the other hand, these steps of the reaction chain may be different even within a given type of movement for different examples. This volume clarifies and analyzes the signal chains that lead to movements in plants.

Volume 8

Secondary Plant Products

Editors: E. A. Bell, B. V. Charlwood
With contributions by numerous experts
1980. 176 figures, 44 tables and numerous schemes and formulas. XVI, 674 pages
ISBN 3-540-09461-X

The first comprehensive exposition of this important and timely field, illuminates recent research results on a variety of secondary plant by-products. Using numerous illustrations and tables, Drs. Bell and Charlwood discuss the biochemical and physiological phenomena involved in the synthesis and accumulation of compounds such as alkaloids, isoprenoids, plant phenolics, non-protein amino acids, amines, cyanogenic glycosides, glucosinolates, and betalains.
Secondary Plant Products is sure to become a standard reference to all botanists, biochemists, pharmacologists, and pharmaceutical chemists.

Volume 9

Hormonal Regulation of Development I

Molecular Aspects of Plant Hormones
Editor: J. MacMillan
With contributions by numerous experts
1980. 126 figures. XVII, 681 pages
ISBN 3-540-10161-6

This initial volume of the three-part **Hormonal Regulation of Development** concentrates on the molecular and subcellular aspects of the main classes of native plant hormones. It contains a survey of the different chemical groups of plant hormones, their structures, homologues, occurence, purification and identification. In a break with tradition, the main groups of hormones are treated together within each topic, and their characteristics compared and contrasted. The role and importance of bioassays is evaluated. Progress in our knowledge of biosynthesis and metabolism of plant hormones, as well as present data and theories on the molecular mechanisms of hormone action and effects on tissues are described. The necessary chemical and biochemical background for the succeeding two volumes is provided.
Topics are reviewed authoritatively and in a comprehensive yet readable manner. The text includes numerous illustrations and tables, as well as an extensive bibliography and subject index.

In Preparation

Volume 10

Hormonal Regulation of Development II

From the Cell to the Whole Plant
Editor: T. K. Scott
With contributions by numerous experts
ISBN 3-540-10196-9

The contributions to Volume 10 will trace the roles of hormones at advancing levels of cellular organization in a plant. The cellular processes of increase and change, the interrelations of cells in tissues, of tissues in organs, and of organs in the whole plant will be considered in turn. During this progressive treatment of levels of organization, the relevant basic properties of hormones will be introduced and illustrated.

Volume 11

Hormonal Regulation of Development III

Role of Environmental Factors
Editors: R. P. Pharis, D. M. Reid
With contributions by numerous experts
ISBN 3-540-10197-7

This volume will cover the interrelationships between hormones and factors in the environment of the tissues and organs, and factors in the environment surrounding the whole plants. The role of hormones as the endogenous messengers between the sensors of environmental fluctuations and the eventual changes in growth and development will be considered. Photomorphogenesis will be treated in a separate volume later in the series.

Encyclopedia of Plant Physiology

New Series · Editors: A. Pirson, M. H. Zimmermann

Physiological Plant Ecology

Editors of Volume12A–D: O. L. Lange, P. S. Nobel, C. B. Osmond, H. Ziegler

Volume 12 A:

Responses to the Physical Environment

1981. 109 figures, 40 tables. Approx. 670 pages
ISBN 3-540-10763-0

Contents:
Fundamentals of Radiation and Temperature Relations. – Photosynthetically Active Radiation. – Responses to Different Quantum Flux Densities. – Non-Photosynthetic Responses to Light Quality. – Responses to Photoperiod. – Plant Response to Solar Ultraviolet Radiation. – Responses to Ionizing Radiation. – The Aquatic Environment. – Responses to Light in Aquatic Plants. – Functional Responses of Macrophytes to Temperature. – Responses of Microorganisms to Temperature. – Responses to Extreme Temperatures. Cellular and Sub-Cellular Bases. – Ecological Significance of Resistance to Low Temperature. – Ecological Significance of Resistance to High Temperature. – Wind as an Ecological Factor. – Fire as an Ecological Factor. – The Soil Environment. – Author Index. – Taxonomic Index. – Subject Index.

Volume 12 B:

Water Relations and Carbon Assimilation

ISBN 3-540-10906-4
In preparation

Contents:
Water in the Soil-Plant-Atmosphere Continuum. – Water in Tissues and Cells. – Water Uptake by Roots. – Water Uptake by Organs Other Than Roots. – Transport and Storage of Water. – Resistances of Plant Surfaces to Water Loss: Transport Properties of Cutin, Suberin and Associated Lipids. – Stomatal Responses, Water Loss and CO_2-Assimilation Rates of Plants in Contrasting Environments. – Mathematical Models of Plant Water Loss and Plant Water Relations. – Physiological Responses to Moderate Water Stress. – Desiccation Tolerance. – Frost Drought and its Ecological Significance. – Environmental Aspects of the Germination of Seeds. – Environmental Aspects of the Germination of Spores. – Physiological Responses to Flooding. – Functional Significance of Different Pathways of CO_2 Fixation in Photosynthesis. – Modelling of Photosynthetic Response to Environmental Conditions. – Water Use and Optimization of Carbon Assimilation. – Plant Life Forms and their Carbon, Water and Nutrient Relations.

Volume 12 C:

Responses to the Chemical and Biological Environment

ISBN 3-540-10907-2
In preparation

Contents:
The Ionic Environment and Plant Ionic Relations. – Osmoregulation. – Halotolerant Eukaryotes. – Halotolerant Prokaryotes. – Physiology and Ecology of Nitrogen Nutrition. – Influence of Limestone, Silicates and Soil pH on Plants. – Toxicity and Tolerance in the Responses of Plants to Metals. – Ecophysiology of N_2-fixing Systems. – Ecophysiology of Mycorrhizal Symbioses. – Ecophysiology of Lichen Symbioses. – Interactions between Plants and Animals in Marine Systems. – Ecophysiology of Carnivorous Plants. – Host-Parasite Interactions in Higher Plants. – Virus-Ecology – 'Struggle' of the Genes. – Ecophysiology of Zoophilic Pollination. – Physiological Ecology of Fruits and Their Seeds. – Physiological and Ecological Implications of Herbivory. – Interactions between Plants.

Volume 12 D:

Ecosystem Processes: Mineral Cycling, Productivity and Pollution

ISBN 3-540-10908-0
In preparation

Contents:
Nutrient Allocation in Plants Communities: Mineral Cycling in Terrestrial Ecosystems. – Nutrient Cycling in Freshwater Ecosystems. – Nutrient Cycling in Marine Ecosystems. – Modelling of Growth and Production. – Productivity of Agricultural Systems. – Productivity of Grassland and Tundra. – Productivity of Desert and Mediterranean Climate Plants. – Productivity of Temperate Deciduous and Evergreen Forests. – Productivity of Tropical Forest and Tropical Woodland. – Phytoplankton Productivity in Aquatic Ecosystems. – Effects of Biocides and Artificially Introduced Growth Regulators: Physiological Basis. – Effects of Biocides and Artificially Introduced Growth Regulators: Ecological Implications. – Eutrophication Processes and Pollution of Fresh Water Ecosystems Including Waste Heat. – Ecophysiological Effects of Atmospheric Pollutants. – Ecophysiological Effects of Changing Atmospheric CO_2 Concentration. – Man's Influence on Ecosystem Structure, Operation, and Ecophysiological Processes.

MIX
Papier aus verantwortungsvollen Quellen
Paper from responsible sources
FSC® C105338

If you have any concerns about our products,
you can contact us on
ProductSafety@springernature.com

In case Publisher is established outside the EU,
the EU authorized representative is:
Springer Nature Customer Service Center GmbH
Europaplatz 3, 69115 Heidelberg, Germany

Printed by Libri Plureos GmbH
in Hamburg, Germany